PLENUM PRESS HANDBOOKS OF HIGH-TEMPERATURE MATERIALS

No. 2 – PROPERTIES INDEX

PLENUM PRESS HANDBOOKS OF
HIGH-TEMPERATURE MATERIALS

No. 1 — MATERIALS INDEX
Peter T. B. Shaffer

No. 2 — PROPERTIES INDEX
G. V. Samsonov

PLENUM PRESS HANDBOOKS OF
HIGH-TEMPERATURE MATERIALS

No. 2

PROPERTIES INDEX

by

G. V. Samsonov

with a foreword by

Henry H. Hausner

Authorized translation from the Russian

Ꝑ

SPRINGER SCIENCE+BUSINESS MEDIA, LLC
1964

Library of Congress Catalog Card Number : 64-17207

This compendium was published in Russian under the title TUGOPLAVKIE SOEDINENIYA — SPRAVOCHNIK PO SVOISTVAM I PRIMENENIYU (Refractory Compounds — Handbook of Properties and Applications) by Metallurgizdat, the State Scientific-Technical Press for Ferrous and Nonferrous Metallurgy, in Moscow in 1963. For the English edition the author has added new reference material published in 1962 and 1963 and has revised the tabulated data accordingly.

Г. В. САМСОНОВ

ТУГОПЛАВКИЕ СОЕДИНЕНИЯ
*Справочник по свойствам
и применению*

© *Springer Science+Business Media New York 1964*
Originally published by Consultants Bureau Enterprises, Inc. New York in 1964
Softcover reprint of the hardcover 1st edition 1964

ISBN 978-1-4899-5093-2   ISBN 978-1-4899-5091-8 (eBook)
DOI 10.1007/978-1-4899-5091-8

FOREWORD

Developmental work on refractory compounds is going on all over the world, wherever application of high-temperature materials is considered of importance in technology. The immense amount of data on refractory compounds already available can be found in books, scientific and engineering journals, and in reports, but it is extremely difficult to find one's way through this jungle of information. It was thus highly desirable that an attempt be made to collect data on refractory compounds in the form of tables, as a guide to this literature.

It is fortunate that Professor G. V. Samsonov, the leading expert on refractory compounds in the Soviet Union, undertook the compilation of these data in tabular form by properties. His immense knowledge is based on his own outstanding research work and his great familiarity with the international literature. He is one of the very few experts in the world who are qualified to survey the wide, complex, and complicated field of refractory compounds. Professor Samsonov's attempts at proposing principles in the classification of refractory compounds* have been appreciated by everyone working with such materials.

In the spring of 1963, Professor Samsonov published his data book, "Refractory Compounds—Reference Book on Their Properties and Uses," in Moscow. The undersigned, who received a copy of Professor Samsonov's book, recommended translation of it into English because he was not aware of any other book on this subject as valuable as this one.

Professor Samsonov, who completed his manuscript some time during 1962, not only permitted translation of the book, but also agreed to bring its content up to date, i.e., to June 1963. The present English version of the book therefore contains considerably more recent data than the original Russian edition. Two hundred and twenty-nine references were added to the previous 1100-odd international references, which included 380 from the USSR. Approximately half of all the references are from the literature since 1956. The data are organized according to the properties of more than 600 different materials, such as borides, carbides, nitrides, silicides, etc. (oxides being excluded).

Professor Samsonov is one of the leading metallurgists in the Soviet Union. He is active in the Institute of Powder Metallurgy and Special Alloys of the Ukrainian Academy of Sciences in Kiev, where other eminent metallurgists such as I. M. Fedorchenko and I. N. Frantsevich also contribute their efforts.

New York, N. Y.   Henry Hausner
December 15, 1963

*G. V. Samsonov, "Principles in the Classification of Refractory Compounds," in "Soviet Powder Metallurgy" (a translation of "Poroshkovaya Metallurgiya") No. 2, pp. 73-76, 1962, Consultants Bureau, New York.

PREFACE TO THE AMERICAN EDITION

The rapid and continuing expansion of research into the properties of refractory compounds has rendered necessary the introduction of a number of additions and corrections, despite the fact that the reference book appeared in Russian in the USSR in 1963.

During this brief interval, a number of new refractory phases were discovered and their crystal structure and many of their physical properties studied. There has been a substantial increase of information on the homogeneity ranges of refractory compounds of various composition, their heat capacities, melting points, and heats of formation from the elements. Considerably more data have become available on thermal conductivity and thermal expansion, while the additions concerning vapor pressure, rate of evaporation, and the parameters of diffusion of nonmetals into metals with the formation of refractory phases were so extensive that the corresponding tables had to be almost completely reconstructed. The tabulated data on electrical and magnetic properties have been greatly increased, and new fields of application of refractory compounds have been noted.

The present edition of the reference book also includes a number of new diagrams of the systems formed by transition metals and nonmetals.

The list of references has been correspondingly increased by more than 200 references, mainly to take into account new papers published in 1962 and 1963, and also to include a certain number of previously omitted sources.

All this has made the reference book more complete and in a number of cases more reliable.

The author hopes that the reference book will be useful to American readers who are interested in refractory compounds and that it will contribute to the further development of this field of the modern science of materials.

G. V. S.

PREFACE TO THE RUSSIAN EDITION

The Program of the Communist Party of the Soviet Union, adopted by the historical Twenty-Second Congress, points out that the most important overall national problem is that of speeding up scientific and technical progress to the maximum possible extent.

For technical progress, it is essential to develop and put into service new materials with improved special characteristics and properties— corrosion-resistant, heat-resistant, semiconductor, light, extremely strong and hard materials—making it possible to mechanize and automate technical processes, create fundamentally new constructions, and resolve the most complex technical problems, such as new methods of converting thermal, nuclear, solar, and chemical energy into electrical energy and the control of thermonuclear reactions and processes in plasma.

The development of modern engineering involves increases in all the parameters of technological processes, such as temperatures, stresses, velocities, and also the need to satisfy the requirements of new fields such as rocket engineering and modern electrical engineering.

In the solution of the problems set, an ever-increasing part is played by refractory, hard, corrosion-resistant, and heat-resistant compounds and alloys; the efforts of numerous scientific, technical, and design organizations are directed toward the creation and application of these materials. The fruitful work of these organizations depends primarily and indispensably on a knowledge of the properties of refractory compounds and the possibilities of using them in the different branches of industry. But the information on the properties of refractory compounds, which is available in an extensive literature, has not been systematically classified; this constitutes a considerable obstacle to wide circles of scientific and engineering workers in acquiring a knowledge of these compounds, and to progress in corresponding research and development work.

Individual partial collections of information on the properties and applications of refractory compounds are to be found in the following treatises: P. Schwarzkopf and R. Kieffer, "Refractory Hard Metals" (1957); G. V. Samsonov and Ya. S. Umanskii, "Hard Compounds of Refractory Metals" (1957); G. V. Samsonov, "Silicides and Their Use in Engineering" (1959); I. Campbell, "High-Temperature Technology" (1956); G. V. Samsonov, L. Ya. Markovskii, A. F. Zhigach, and M. G. Valyashko, "Boron, Its Compounds and Alloys" (1960); G. V. Samsonov and K. I. Portnoi, "Alloys Based on Refractory Compounds" (1961); R. Kieffer and F. Benesovsky, "Hartstoffe," (1963);* and also in a number of special reference books. The use of the above-mentioned treatises for day-to-day work is difficult, however—particularly since some of them are intended for the

*Springer, Vienna.

fairly narrow circle of readers specializing in the field of refractory compounds.

In this reference book, the author, on the basis of the work of the laboratory [†] under his direction and published data, has attempted to generalize information on the properties and application of refractory compounds.

The book contains data on the physical, technical, mechanical, chemical, and refractory properties of refractory compounds which are currently most widely used in technical developments and which offer the most promise of further application in solving the problems of modern engineering.

In this first and rather complex attempt, it is naturally difficult to avoid some methodological shortcomings and omissions, and the author will be grateful if these are pointed out to him.

The author is grateful to many Soviet investigators and also certain foreign authors—R. Kieffer, F. Benesovsky (Austria), B. Aronsson, S. Rundquist (Sweden), I. Batsek (Czechoslovakia), F. Eisenkolb (German Democratic Republic), V. Rutovsky, A. Stolar (Polish People's Republic)— who kindly placed at his disposal numerous data and references to work on the properties of refractory compounds.

The author is also grateful to colleagues in the Department of the Metallurgy of Rare Metals and Refractory Compounds of the Academy of Sciences of the Ukrainian SSR who were very helpful in the preparation of this reference book, particularly Yu. B. Paderno, A. S. Bolgaru, L. L. Vereikinaya, G. N. Dubrovskaya, V. V. Fesenko, and Prof. Dr. B. M. Tsarev, who looked through the section on thermionic emission properties, and to the reviewers Prof. Dr. Chem. Sc. B. F. Ormont, Prof. Dr. Tech. Sc. A. N. Krestovnikov, and Cand. Tech. Sc. M. Yu. Bal'shin for valuable comments and advice, most of which was accepted and is reflected in the final form of the book.

[†]Laboratory of the Metallurgy of Rare Metals and Refractory Compounds of the Institute of Cermets and Special Alloys of the Academy of Sciences of the Ukrainian SSR.

X

CONTENTS

INTRODUCTION

It is difficult to define the term "refractory compound," since any sub-division into refractory and nonrefractory compounds is arbitrary and presupposes the fixing of some melting point boundary above which chemical compounds are considered to be refractory. Such a boundary has been repeatedly established and has gradually been shifted to regions of higher and higher temperatures, from 1000°C in the second half of the nineteenth century to 2000°C in the first half of the twentieth century, and currently it is very often assumed to be 3000°C.

However, the expression "refractory compound" is currently gradually losing its original meaning and is becoming deeper and more fundamental, encompassing a whole complex of properties, including high hardness, brittleness, and heat of formation, as well as specific electrical and magnetic properties, as determined by the electronic structure of the corresponding compounds and the position of their components in the periodic system of the elements.

From this point of view, which nowadays is becoming more firmly established, a refractory compound need not always be one of very high melting point;* it may denote symbolically a substance possessing a combination of other properties, e.g., high hardness, low vapor tension and rate of evaporation, resistance to the action of chemically corrosive media, etc. The principal tenor of the concept denoted traditionally by the term "refractory compound" is becoming increasingly the character of the chemical bond between the components of the compounds, which is mainly metallic or covalent with a small proportion of ionic bond. Such types of bond occur as a rule in compounds of metals (mainly transition metals or metals similar to them, according to a number of criteria) with nonmetals of the type of boron, carbon, silicon, nitrogen, sulfur, phosphorus, etc., not having excessively high ionization potentials, which would result in the formation of an ionic bond, and also in compounds of both nonmetals and certain metals with each other.

Although the physical and chemical properties of refractory compounds have not been sufficiently studied, it is nevertheless possible to suggest principles of their scientific classification as bases for the subsequent extension of research and approach to solution of the problem of producing refractory compounds having predetermined properties.

The accompanying periodic table shows the arrangement of the components of refractory compounds in the periodic system of the elements. These components comprise elements of the odd subgroups of groups II-VII, group VIII, Lanthanides, actinides, and also light nonmetals of

*It may be pointed out that the Russian equivalent of "refractory" is literally "difficultly fusible."

1

2

Group

Period	Row	I	II	III	IV	V	VI	VII	O	VIII
I	1	Hydrogen **H** 1 1.0080						Hydrogen (H)	Helium **He** 2 4.002	
II	2	Lithium **Li** 3 6.940	Beryllium **Be** 4 9.013	Boron **B** 5 10.82	Carbon **C** 6 12.010	Nitrogen **N** 7 14.008	Oxygen **O** 8 16	Fluorine **F** 9 19.00	Neon **Ne** 10 20.183	
III	3	Sodium **Na** 11 22.997	Magnesium **Mg** 12 24.32	Aluminium **Al** 13 26.97	Silicon **Si** 14 28.09	Phosphorus **P** 15 30.974	Sulfur **S** 16 32.066	Chlorine **Cl** 17 35.457	Argon **Ar** 18 39.944	
IV	4	Potassium **K** 19 39.100	Calcium **Ca** 20 40.08	Scandium **Sc** 21 44.96	Titanium **Ti** 22 47.90	Vanadium **V** 23 50.95	Chromium **Cr** 24 52.01	Manganese **Mn** 25 54.93		Iron **Fe** 26 55.85 — Cobalt **Co** 27 58.94 — Nickel **Ni** 28 58.69
V	5	Copper **Cu** 29 63.542	Zinc **Zn** 30 65.377	Gallium **Ga** 31 69.72	Germanium **Ge** 32 72.60	Arsenic **As** 33 74.91	Selenium **Se** 34 78.96	Bromine **Br** 35 79.916	Krypton **Kr** 36 83.80	
VI	6	Rubidium **Rb** 37 85.48	Strontium **Sr** 38 87.63	Yttrium **Y** 39 88.92	Zirconium **Zr** 40 92.12	Niobium **Nb** 41 92.91	Molybdenum **Mo** 42 95.95	Technetium **Tc** 43 [98.91]		Ruthenium **Ru** 44 101.7 — Rhodium **Rh** 45 102.91 — Palladium **Pd** 46 106.7
VII	7	Silver **Ag** 47 107.880	Cadmium **Cd** 48 112.41	Indium **In** 49 114.76	Tin **Sn** 50 118.70	Antimony **Sb** 51 121.76	Tellurium **Te** 52 127.61	Iodine **I** 53 126.91	Xenon **Xe** 54 131.3	
VIII	8	Cesium **Cs** 55 132.91	Barium **Ba** 56 137.36	Lanthanum **La** 57 138.92	Hafnium **Hf** 72 178.60	Tantalum **Ta** 73 180.88	Tungsten **W** 74 183.92	Rhenium **Re** 75 186.31		Osmium **Os** 76 190.2 — Iridium **Ir** 77 193.1 — Platinum **Pt** 78 195.23
VIII	9	Gold **Au** 79 197.2	Mercury **Hg** 80 200.61	Thallium **Tl** 81 204.39	Lead **Pb** 82 207.21	Bismuth **Bi** 83 209.00	Polonium **Po** 84 [210]	Astatine **At** 85 [210]	Radon **Rn** 86 222	
VII	10	Francium **Fr** 87 [223]	Radium **Ra** 88 226.05	Actinium **** Ac** 89 227						$\leftarrow EO_4$

Hydrogen compound → Oxide →	**S a l t l i k e**			**G a s e o u s**				
Hydrogen compound	—	—	EH_3	EH_4	EH_3	EH_2	EH	
Oxide	E_2O	EO	E_2O_3	EO_2	E_2O_5	EO_3	E_2O_7	

• L a n t h a n i d e s •

| 58 Cerium **Ce** 140.13 | 59 Praseodymium **Pr** 140.92 | 60 Neodymium **Nd** 144.27 | 61 Promethium **Pm** [145] | 62 Samarium **Sm** 150.43 | 63 Europium **Eu** 152.0 | 64 Gadolinium **Gd** 156.9 | 65 Terbium **Tb** 159.2 | 66 Dysprosium **Dy** 162.46 | 67 Holmium **Ho** 164.94 | 68 Erbium **Er** 167.2 | 69 Thulium **Tu** 169.4 | 70 Ytterbium **Yb** 173.04 | 71 Lutecium **Lu** 174.99 |

** A c t i n i d e s

| 90 Thorium **Th** 232.12 | 91 Protactinium **Pa** 231 | 92 Uranium **U** 238.07 | 93 Neptunium **Np** [237] | 94 Plutonium **Pu** [244] | 95 Americium **Am** [243] | 96 Curium **Cm** [243] | 97 Berkelium **Bk** [247] | 98 Californium **Cf** [251] | 99 Einsteinium **Es** [254] | 100 Fermium **Fm** [253] | 101 Mendelevium **Mv** [256] | 102 Nobelium **No** [254] | 103 Lawrencium **Lw** [257] |

Metallic components of refractory compounds   Nonmetallic components of refractory compounds

periods II and III (B, C, N, O, Si, P, S) and aluminum. These components combine with one another to form the following three fundamental classes of refractory compounds:

1. compounds of metals with nonmetals; these compounds include borides, carbides, nitrides, oxides, silicides, phosphides, sulfides;
2. compounds of nonmetals with each other; in particular, these compounds may include carbides, nitrides, sulfides, phosphides of boron and silicon, and also alloys of boron and silicon;
3. compounds of metals with each other, known as intermetallic compounds.

Compounds of the first of the above-mentioned classes are conveniently called "metal-like" refractory compounds, in view of their external and, particularly, internal resemblance to metals and intermetallic compounds. The chemical bond in the lattices of these compounds, in addition to the s and p electrons of the metallic and nonmetallic components, respectively, is also formed by the electrons of the deeper, incomplete d and f levels of the transition metals, to which belong almost all metallic components of the metal-like refractory compounds. Isolated atoms of metals of the odd subgroup of group II, the alkaline earth metals, do not have any electrons in the d and f shells, but in compounds with nonmetals, energy states corresponding to these shells may occur.

Thus, a necessary condition for the formation of metal-like refractory compounds is the participation in the bonds of incomplete d and f electron levels or the possibility of their formation in the compounds; in other words, this condition is reduced in the majority of cases to the requirement that the metallic components belong to the transition elements. As a qualitative criterion of the degree of participation in the bond and the determination of the distribution of electron concentration in the lattice, it is possible to use the quantity $1/Nn$, where n is the number of electrons in the incomplete level and N is the principal quantum number of this level, as proposed by the author in 1953 (the proposal considered the electrons to be a degenerate gas in the Coulomb field of atomic nuclei or atomic cores). Another criterion is the ability of atoms of the nonmetals to give off valence electrons, which may characterize the ionization potentials of these atoms.

The electron density between the atom cores in the crystal lattice and the character of its distribution depend on the number n of electrons in the incomplete electron level, the principal quantum number N of this level and the ionization potentials φ_i of the atoms of nonmetals. An increase in the criterion $1/Nn$, i.e., a scattering or acceptor capacity of the atoms of a transition metal, produces a displacement of the relative electron concentration toward the metallic atom (for φ_i = const), while an increase in φ_i for constant acceptor ability of the core of the metal atom produces a displacement toward the atom of the nonmetal with a corresponding increase in the proportion of ionic bond. Thus, a variation in the values of $1/Nn$ and φ_i produces diversity but (in binary compounds) the number of combinations of the above-mentioned criteria is not infinitely large. This

3

in its turn determines the peculiar "continuously" discrete character of the variation of the type of bond and correspondingly of the physico-chemical properties of the metal-like refractory compounds, which underlines the dialectic nature of the unity of continuity and discreteness of the interatomic reaction in crystals and in particular in these compounds.

Consequently, metal-like refractory compounds are heterodesmic in the character of their chemical bond, the proportion of one type of bond or other being determined, as stated above, by the criteria and features of the crystal structure of these phases.

It may be assumed that in borides, where the boron atoms are isolated from each other (Me_2B), the valence electrons of the boron remain predominantly on the free d levels of the metal atom if this atom has a sufficiently high acceptor capacity; typical metallic phases, similar to intermetallic compounds, are then formed. In the formation of pairs, chains, networks, and skeleton structural elements from boron atoms in the phases Me_3B_4, MeB_2, MeB_4, MeB_6, etc., a considerable proportion of the p electrons of the boron is expended in the formation of covalent B—B bonds and a smaller proportion is transferred to the general electron aggregate, ensuring a metallic bond in the lattice. The proportion of metallic bond in borides, therefore, decreases with increase in the B/Me ratio.

In silicides, formed with the participation of atoms of silicon, which has a very low ionization potential, these peculiarities are intensified, and while practically all silicides having isolated silicon atoms possess metallic properties, the higher silicide phases of metals not having very high acceptor characteristics (iron, manganese, rhenium, chromium) are semiconductors.

In carbides, on the contrary, the proportion of metallic bond increases as the result of the higher ionization potential of carbon, and the carbides of metals having high $1/Nn$ values possess typical metallic properties (TiC, ZrC, HfC, VC).

In nitrides, the proportion of ionic bond increases correspondingly, especially in nitrides of metals having a low acceptor capacity (molybdenum, tungsten, rhenium), while the nitrides of niobium, tantalum, and chromium show a combination of metallic bond and ionic bond, with some preponderance of the latter. A decrease in nitrogen content of nitride phases, within their homogeneity ranges, results in a strengthening of the Me—Me bonds, a weakening of the bonds of the metal atom cores with the nitrogen, and hence the possibility of the appearance of fairly wide breaks in the energy states in the lattice; this determines the semiconductor properties of nitrides with lattices deficient in nitrogen.

It must be pointed out that, due to the lower ionization potential of the oxygen atom compared with nitrogen, the proportion of ionic bond in metal oxides having high acceptor characteristics (titanium, zirconium, hafnium, vanadium) is rather less than in the corresponding nitrides, and this must be particularly pronounced for the lower oxides.

Passing now to the phosphides, the decisive significance for the character of the chemical bond and crystal structure is provided by the low ionization potential of phosphorus, which is much lower than that of carbon

4

and nitrogen, but higher than that of boron and silicon, as well as the greater size of the phosphorus atom, which must result in a greater expansion of the crystal lattices and in a corresponding weakening of the strength of their bonds. Compared with nitrides, phosphides have a smaller proportion of ionic bond, which is diminished still further by reducing the phosphorus content in phosphide phases, i.e., by increasing the Me/P ratio.

Bearing in mind that the difference in the ionization potentials of phosphorus and carbon is to some extent compensated by the geometric factor of the expansion of the lattices of phosphides, due to the larger radius of the phosphorus atom, it is to be expected that the distribution of electron density in phosphides ought to be on the whole similar to the distribution in lattices of the carbides of the corresponding metals; this is confirmed by the results of x-ray diffraction analysis and the determination of some physical properties of phosphides. The lower bond strength in phosphides, together with an electron density distribution similar to that of carbides, is expressed in the tendency of phosphides to dissociate and in their relatively low melting points and hardness.

Similar considerations also apply to the refractory metal-like lanthanides and actinides with deeply situated incomplete electron levels.

The second class of refractory compounds is formed by compounds of metals with each other or the so-called nonmetallic refractory compounds. All these compounds, like the metal-like compounds, are characterized by a heterodesmic character of the bond, but with predominance of the covalent bond, and they have semiconductor properties as well as high electrical resistance at room temperature; as a rule, these compounds have a structure with layer, chain, or skeleton structural groups or patterns, and either melt with decomposition or decompose before reaching the melting point.

The accompanying table shows a number of currently known compounds of this class.

Element	Ionization potential, eV	Si	B	S	P	C	N
Si	8.14	Si	Si_xB	Si_xS	SiP	SiC	Si_3N_4
B	8.28	B_xSi	B	B_xS	BP	B_xC	BN
S	10.42	S_xSi	B_xS	S	S_xP	–	–
P	10.43	SiP	BP	S_xP	P	–	–
C	11.24	SiC	B_xC	–	–	C (diamond)	–
N	14.51	Si_3N_4	BN	–	–	–	–

In crystals of the elements situated along the diagonal of the table the width of the energy gaps increases in the direction indicated by the arrows, while in the nonmetallic compounds formed by these elements it may be assumed that there will be an increase in the proportion of ionic bond with increase in the difference in ionization potentials of the components (from Si_xB to Si_3N_4, from Si_xN to BN, etc.).

Finally, three elements of the periodic system occupy an intermediate position with regard to the ability to form refractory metal-like and non-metallic compounds. These elements, beryllium, magnesium, and aluminum, are capable of forming fairly refractory semiconductor compounds with nonmetals (beryllium, magnesium, aluminum borides, aluminum nitride, magnesium silicides), and they may also enter into the composition of intermetallic compounds of the beryllide, aluminide, etc., type.

The third class of refractory compounds comprises compounds of metals with each other—intermetallic compounds. A special branch of chemistry which has arisen in the last few years, metal chemistry, is devoted to the study of these compounds.

The classification of refractory compounds in the above-mentioned three classes, as proposed by the author, is based on concepts of the periodic regularity of the change in character of the chemical bond with variation in the acceptor capacity of the atoms of the transition metals and the ionization potentials of the atoms of nonmetals.

On the basis of this classification, it is possible to explain a number of the properties of refractory compounds and also the directions of their variations; consequently, it becomes possible to estimate approximately the as yet uninvestigated properties of the refractory compounds themselves and their mutual alloys.

The author considered it best not to include in the reference book the properties of certain little-studied compounds rarely used in practice. Thus, in the presentation of the information on carbides, borides, nitrides, and other classes of metal-like compounds, no data are given on the refractory compounds of metals of the platinum group; for the sulfides, data are given only for the class of sulfides of the rare-earth metals and actinides, in most of which the properties of refractory compounds in the wide sense are most clearly expressed, the proportion of ionic bond, in particular, being small. It was, however, found expedient to consider also the properties of oxysulfides of the rare-earth metals and actinides, which are very similar to the properties of sulfides and are obtained simply by replacement of two atoms of sulfur in a sesquisulfide by two atoms of oxygen. This is one of the few exceptions where the tables of the reference book give the properties of ternary and not binary compounds.

The book gives no data on the properties of oxides, since there is a special literature on this subject, including the book by S. G. Tresvyatskii and A. M. Cherepanov,* the new treatise by E. Rishkewitsh,† the book by W. Espe,‡ and others.

On the other hand, the author considered it useful to include data on the properties of such a refractory element as boron, as well as on the new technical form of graphite, pyrographite, a material of importance in high-temperature engineering. Data on ordinary graphites are not given, since they are presented with sufficient completeness in special literature.

*S. G. Tresvyatskii and A. M. Cherepanov, Highly Refractory Oxide Materials and Components, Metallurgizdat, 1957.
†E. Rishkewitsh, Treatise of Refractory Oxides, Academic Press, New York, 1960.
‡W. Espe, Werkstoffkunde der Hochvakuumtechnik, VEB Deutscher Verlag der Wissenschaften, Berlin, Vol. I. Metalle und metallschleifende Werkstoffe, 1959; Vol. II. Silikatwerkstoffe, 1960.

Chapters I-V of the reference book give information of a general character on refractory compounds, data on their crystal structure, specific gravity, thermochemical, thermal, electrical and magnetic, optical, mechanical, chemical, and refractory properties.

In Chapters VI and VII, the author has mainly attempted to provide some idea of the resistance of refractory compounds to the action of different chemical reagents and molten media, and also to oxidation.

Chapter VIII gives concise tables of information on current and prospective fields of application of refractory compounds in different branches of industry.

In each chapter, the following sequence in order of classes of compounds has been adopted: metal-like borides, carbides, nitrides, silicides, phosphides, sulfides, and nonmetallic compounds.

The reference book gives the most reliable data and indicates the literature sources in which duplicate values have been obtained by different investigators; they may be of interest to scientific workers conducting research in fields of study of refractory compounds.

The authenticity of the data given in the tables has been determined mainly by the reliability of the methods used in obtaining the numerical values of the property and the purity and condition of the specimens, and also from statistical indications. The number of reliable data may not therefore always include the most recent; in a number of cases, preference may have been given to results obtained some decades ago and not to recent, sometimes incidental values, found for specimens of indefinite phase composition. To diminish the well-known subjective character of such an estimate in the determination of the reliability of certain data, use was made of the previously ascertained regularity of their variations, associated for example with the atomic number of the elements, electronegativity values, and acceptor capacity of the atom cores, and also of the actual experience acquired by the author and the group of workers in the laboratory under his direction.

The data obtained in the Laboratory of the Metallurgy of Rare Metals and Refractory Compounds of the Institute of Cermets and Special Alloys of the Academy of Sciences of the Ukrainian SSR or by other investigators of specimens prepared in the laboratory, and not previously published, are indicated by an asterisk in the "source" column with a reference in the remarks column to the name of the investigator.

The tables give fairly complete lists of literature sources for the same property for each substance, apart from the foregoing considerations, to indicate the statistical reliability of the data and the degree of investigation involved; this enables definite problems to be put to laboratories and individual research workers engaged in the study of refractory compounds and their synthesis. Thus, for example, the tables show that very little study has been given to some electrical and especially optical and mechanical properties of almost all the refractory compounds; a knowledge of these properties is absolutely essential for the solution of a number of practical problems.

A number of phases whose properties are given in the tables, although their actual existence and individuality have not been established with certainty, are indicated by a question mark; all other details and characteristics are given in the remarks and notes.

The Appendix gives a collection of currently known phase diagrams of binary systems in which refractory compounds are formed; these diagrams will enable the reader to compare the properties of the phases with their position in the corresponding systems.

The following remarks must be made on the material of the individual tables.

The table "Electronic Structure of Isolated Atoms," unlike similar tables as usually published, includes some refinements and additions in accordance with work done in recent years; in particular the electronic configuration of the terbium atom has been refined. The table also gives the electronic structures proposed for the elements with atomic numbers from 98 to 103.

The table "Ionization Potentials of the Atoms" is taken from the well-known tables of Kaye and Laby [347], completed with a number of new data, in particular the ionization potentials of osmium, iridium, erbium, thorium, and uranium.

The table "Ratio of the Radii of Some Atoms of Nonmetals and Metals" is given to facilitate comparison by the reader of the corresponding ratios, with the condition of formation of interstitial phases $r_{Me} : r_x$, found by Hägg.

For convenience in using the book, the table "Composition of Refractory Compounds" has also been included in an attempt to provide a list of all the currently known refractory compounds of the classes considered in the book. For most of the compounds listed, data regarding their properties are given.

The table "Homogeneity Range" gives the most reliable data on the homogeneity ranges of metal-like refractory compounds. It should be noted that these data are as yet not very numerous; Soviet and other investigators have only recently commenced intensive efforts to obtain exhaustive data in this area.

The table "Crystal Structure" gives the fundamental parameters of the crystal structure of refractory compounds. It not only gives data for crystal phases having a well-investigated structure, but also information of an incomplete character; for example, for some phases, limited information is given on the type of structure or space group, etc. In a number of cases, in addition to the ordinary phases, the table gives what are known as Nowotny phases, i.e., silicide phases, the structure of which is stabilized only in the presence of carbon, nitrogen, or boron. Such phases, having a D_{8_8} type of structure [space group $(D_{6h}^3 - C6/mcm)$], include, for example, Me_5Si_3 silicides (of titanium, zirconium, hafnium, vanadium, and others).

The table "Density" gives values determined by the pycnometer method and values calculated from x-ray data. In a number of cases, the density has been calculated from the values of the lattice parameters, and the corresponding literature references relate to values of the constants of the crystal lattices of the phases. All the information is given for phases of limit (maximum) composition relative to the nonmetal content.

8

Chapter I concludes with data on the temperature stability ranges of refractory compounds, obtained principally from the phase diagrams of the corresponding binary systems.

Chapter II contains information concerning the fundamental thermal and thermodynamic properties of refractory compounds. The heats of formation have been taken from original work and partly from reviews and also from well-known reference works.* In cases where the information has been taken from other reference books, the references are made to the latter and not to the original work used for the compilation of such books.

In the table "Entropy of Compounds," the values of the absolute entropy of compounds are given, as well as the entropy of formation of refractory compounds, calculated on the basis of the values for the entropy of elements. The data relative to some borides, marked by an asterisk, have been calculated by means of Istmen's well-known semi-empirical formula. The tables "Free Energy of Formation of Refractory Compounds" (ΔF according to Helmholtz) and "Heat Capacity" were compiled on the basis of the most reliable data and compared with the last reference data of E. Kubaschewski and O. Evans [928]; the values of the heat capacity of refractory compounds have been converted to 20°C for convenience of the reader.

The very limited data in the table "Heats of Dissociation" are given on the basis of sources found to be the most reliable; they are, however, provisional, since there is not sufficient clarity concerning the concept of the mechanism of sublimation and dissociation, composition of the vapor, etc. The value of the energy of sublimation of boron, the energy being converted for the formation of monatomic vapor, has been assumed to be equal to 141 kcal at 25°C; other values relating to sublimation with the formation of a set of molecules $B_x + B_y$, etc., and amounting to 90 and 101 kcal, have been omitted.

In the table "Melting Point," the temperatures are given for convenience in degrees Celsius and Kelvin, and cases of decomposition on melting are indicated in the remarks column.

The boiling points given in the tables are principally calculated values, obtained by using the most reliable equations for the temperature-dependence of the vapor pressure; however, due to inadequately clear concepts of the composition of the vapor, these data, like the data on heats of sublimation and dissociation, are quite tentative.

The vapor pressure constants and rates of evaporation given in the table are not, strictly speaking, physical, and they depend substantially on the experimental conditions, degree of roughness of the surface of the specimens, their porosity, and so forth; they have, however, considerable practical value in determining the possibility of utilizing refractory compounds in a vacuum at high temperatures and their corresponding service life. Most of these data have been obtained by the use of the Langmuir

*O. Kubaschewski and E. L. Evans, Metallurgical Thermochemistry, New York, Pergamon, 1958. Thermal Constants of Inorganic Substances, edited by E. V. Britske and A. F. Kapustinskii, Izd. Akad. Nauk SSSR, 1949.

method, since work on the determination of the parameters of evaporation by the effusion method, which is more reliable and clearer in physical interpretation, was commenced only recently in the USSR and abroad. A number of corresponding data are included in the book.

The thermal conductivity values have been obtained principally by the steady-state method; most of the data have been converted by the authors of the different publications, including the author of the present reference work, to nonporous condition of the specimens, using the fairly reliable extrapolation to zero porosity according to Kingery's formula $\varkappa_{P=0} = \varkappa_P (1-P)$, where P is the porosity in fractions of unity [239]. However, despite the absence in a number of publications of any reference to the condition of the specimens and their porosity, the author has considered it possible to reproduce the thermal conductivity data given in these publications as tentative values, since information on the thermal conductivity of refractory compounds is extremely scanty.

The coefficients of thermal expansion in most investigations have been determined with the use of the dilatometer, and only in individual cases [891, 920], and in work on uranium silicides and other work, by the x-ray structural analysis method. Data obtained in the laboratory using a quartz dilatometer are marked by an asterisk; the satisfactory reliability of such data is also noted by Nowotny [92].

The values for the energy of the crystal lattice were calculated on the basis of the formula of E. S. Sarkisov [822], originally used by him for calculating the energies of crystal lattices of ionic compounds. The essence of Sarkisov's formula, however, derived on the assumption of the collectivization of electrons between the cores of atoms of the two components, makes it more suitable for calculation of the energy of the lattices of metal-like compounds with a high proportion of metallic bond; this has been shown* by G. V. Samsonov [457], O. I. Shulishova [823], and G. V. Samsonov and O. I. Shulishova [880].

The values of the atomization† energy have been taken entirely from the calculations made by B. F. Ormont according to the formula given on p. 133.

The characteristic temperatures were partly obtained from data of x-ray structural studies, carried out mainly by Ya. S. Umanskii and partly by calculation from Lindeman's well-known formula.

The averaged values of the root-mean-square amplitude of the elastic vibrations of structural complexes of the crystal lattices of refractory compounds were calculated from the Debye—Waller relationship [372], by means of the characteristic temperatures and the masses of the vibrating complexes:

$$U_s^2 = \frac{3h^2 T}{4\pi^2 mk\theta^2}\left[\Phi(x) + \frac{x}{4}\right]$$

where U_s^2 is the square of the root mean square of the vibration amplitude of the structural complex; θ is the characteristic temperature; m is the

*The work was conducted in the Laboratory of the Metallurgy of Rare Metals and Refractory Compounds of the Institute of Cermets and Special Alloys of the Academy of Sciences of the Ukrainian SSR.

†See p. 133 for an explanation of this term—Ed.

10

mass of the vibrating element, normally assumed to be equal to the mass of a molecule of the compound; $\Phi(x)$ is the Debye function [here $x = \theta(T)$]; T is the absolute temperature; k and h are known constants.

The table "Parameters of the Diffusion of Nonmetals into Metals with the formation of Refractory Compounds" mainly gives data on the parameters of reaction diffusion, i.e., diffusion occurring with the formation of the corresponding compounds; in a number of cases, however, data have been utilized on the ordinary heterodiffusion of metalloids into transition metals. The majority of the data have been obtained by the classical method, and the minority by the use of radioactive isotopes.

Chapter III of the book gives data on the electrical and magnetic properties of refractory compounds. It should be noted that electrical properties are sensitive to structure and depend substantially on the method of measurement and also on the purity of the specimens. Therefore, a large proportion of the data presented, for example those on electrical conductivity, bear a provisional and tentative character; in addition, the degree of reliability is largely determined by the statistical numbers of determinations in the different investigations. It is nevertheless possible to assume that in the course of time the values of the electrical resistance of a number of compounds, at least those having metallic conductivity, will vary immaterially (will decrease somewhat); this of course cannot be said of the nonmetallic compounds which are semiconductors.

The situation regarding the determination of data on the superconductivity of refractory compounds is particularly unfavorable, since they are difficult to obtain without ferromagnetic impurities. Evidently, it will have to be assumed that compounds whose transition point is less than 2-3°K are not superconductors.

The data regarding the thermoelectric properties will also have to be looked upon as tentative, since these properties depend materially on the method of measurement and also on the state and purity of the specimens. The thermionic emission properties are characterized by the more accurate data obtained in the USSR by investigators, principally of the school of Prof. B. M. Tsarev.

The very limited data for the width of the forbidden bands of semiconductor refractory compounds, principally of certain silicides and nonmetallic compounds, do not permit one to regard the information given as complete.

The magnetic properties of refractory compounds are given from original work and compared with the data of the well-known reference book by Foëx [368].

Chapter IV, dealing with optical properties, contains information on the color of some refractory compounds, emission coefficients, and absorption spectra in the infrared region. The color of the compounds naturally depends on their composition, especially their degree of dispersion, etc. The data given in the corresponding table provide a qualitative idea of the color of a number of refractory compounds in the disperse state, and in the author's opinion may be useful.

The emission coefficients of refractory compounds were determined principally by the method of comparing the brightness temperature and true temperature on heating the powders according to the model of an absolutely black body, as described in [216,933].

The data on the absorption spectra in the infrared region are quite limited; nevertheless, the author has considered it necessary to give them to underline the importance of this method of the physical investigation of refractory compounds, which gives important practical results, despite the fact that insufficient attention is given to it.

Chapter V gives the mechanical properties of refractory compounds. These data, the least reliable and to some extent representing "semi-qualitative characteristics," point to the fact that the main problem of organizations associated with the investigation of refractory compounds ought to be the development of methods of determining the mechanical properties and making the corresponding measurements. The data on tensile strength, compressive strength, and shear strength are very unreliable; somewhat more reliable are the data on the modulus of elasticity, particularly owing to the work carried out in recent years under the guidance of I. N. Frantsevich, Academician of the Academy of Sciences of the Ukrainian SSR.

Hardness on the mineralogical scale has been given because it is of interest to crystallographers and mineralogists.

The author also considered it appropriate to include in the book data on the Rockwell and Vickers hardness of refractory compounds, despite the unreliability of these figures. These data are of interest since in engineering the hardness of welded-on and diffused coatings of refractory compounds is usually determined in terms of Rockwell and Vickers hardness values.

The microhardness values may be regarded as sufficiently reliable; almost all of them were obtained with the Russian instrument PMT-3, and only a few with the Knoop tester.

The data on the chemical (Ch. VI) and refractory (Ch. VII) properties of refractory compounds have mainly a qualitative and descriptive character. They are, however, quite important for the use of refractory compounds in a number of branches of modern engineering and are given with maximum permissible completeness, which may be of material assistance to designers and technologists.

The tables "Resistance of Compact Refractory Compounds to the Action of Chemical Reagents" gives mainly data on interaction in the solid phase in mixtures and also on contact reactions (the latter cases are noted specially in the remarks to the table).

In Chapter VIII, "Examples of the Applications of Refractory Compounds," in cases relating to the actual utilization the fundamental properties of the corresponding alloys are given.

These data, however, do not claim to be at all exhaustive and cannot take the place of special treatises, in which problems of application are dealt with at considerable length. This chapter is merely intended to acquaint the reader with the most important examples of the use of refractory compounds in various fields of engineering and scientific research.

In conclusion, it should be pointed out that the investigation of the properties of refractory compounds is not yet being pursued sufficiently systematically and completely. This reference book is only a first attempt to classify the available data and to indicate those properties of refractory compounds which are as yet little developed.

Chapter I

GENERAL INFORMATION, STOICHIOMETRY, AND CRYSTAL-CHEMICAL PROPERTIES

ELECTRONIC STRUCTURE OF ISOLATED ATOMS

Atomic number	Element	K 1 — s 0	L 2 — s 0	L 2 — p 1	M 3 — s 0	M 3 — p 1	M 3 — d 2	N 4 — s 0	N 4 — p 1	N 4 — d 2	N 4 — f 3	O 5 — s 0	O 5 — p 1	O 5 — d 2	O 5 — f 3	P 6 — s 0	P 6 — p 1	P 6 — d 2	P 6 — f 3	Q 7 — s 0	Q 7 — p 1	Q 7 — d 2	Q 7 — f 3
1	H	1																					
2	He	2																					
3	Li	2	1																				
4	Be	2	2																				
5	B	2	2	1																			
6	C	2	2	2																			
7	N	2	2	3																			
8	O	2	2	4																			
9	F	2	2	5																			
10	Ne	2	2	6																			
11	Na	2	2	6	1																		
12	Mg	2	2	6	2																		
13	Al	2	2	6	2	1																	
14	Si	2	2	6	2	2																	
15	P	2	2	6	2	3																	
16	S	2	2	6	2	4																	
17	Cl	2	3	6	2	5																	
18	Ar	2	2	6	2	6																	
19	K	2	2	6	2	6	—	1															
20	Ca	2	2	6	2	6	—	2															
21	Sc	2	2	6	2	6	1	2															

Atomic number	Element	K 1 — s 0	L 2 — s 0	L 2 — p 1	M 3 — s 0	M 3 — p 1	M 3 — d 2	N 4 — s 0	N 4 — p 1	N 4 — d 2	N 4 — f 3	O 5 — s 0	O 5 — p 1
22	Ti	2	2	6	2	6	2	2					
23	V	2	2	6	2	6	3	2					
24	Cr	2	2	6	2	6	5	1					
25	Mn	2	2	6	2	6	5	2					
26	Fe	2	2	6	2	6	6	2					
27	Co	2	2	6	2	6	7	2					
28	Ni	2	2	6	2	6	8	2					
29	Cu	2	2	6	2	6	10	1					
30	Zn	2	2	6	2	6	10	2					
31	Ga	2	2	6	2	6	10	2	1				
32	Ge	2	2	6	2	6	10	2	2				
33	As	2	2	6	2	6	10	2	3				
34	Se	2	2	6	2	6	10	2	4				
35	Br	2	2	6	2	6	10	2	5				
36	Kr	2	2	6	2	6	10	2	6				
37	Rb	2	2	6	2	6	10	2	6	—	—	1	
38	Sr	2	2	6	2	6	10	2	6	—	—	2	
39	Y	2	2	6	2	6	10	2	6	1	—	2	
40	Zr	2	2	6	2	6	10	2	6	2	—	2	
41	Nb	2	2	6	2	6	10	2	6	4	—	1	
42	Mo	2	2	6	2	6	10	2	6	5	—	1	
43	Tc	2	2	6	1	6	10	2	6	6	—	1	
44	Ru	2	2	6	2	6	10	2	6	7	—	1	
45	Rh	2	2	6	2	6	10	2	6	8	—	1	
46	Pd	2	2	6	2	6	10	2	6	10	—		
47	Ag	2	2	6	2	6	10	2	6	10	—	1	
48	Cd	2	2	6	2	6	10	2	6	10	—	2	
49	In	2	2	6	2	6	10	2	6	10	—	2	1
50	Sn	2	2	6	2	6	10	2	6	20	—	2	2
51	Sb	2	2	6	2	6	10	2	6	10	—	2	3

ELECTRONIC STRUCTURE OF ISOLATED ATOMS (continued)

Shell / Subshell key — K(1), L(2), M(3), N(4), O(5), P(6), Q(7)

Atomic number	Element	1s	2s	2p	3s	3p	3d	4s	4p	4d	4f	5s	5p	5d	5f	6s	6p	6d	6f	7s	7p	7d	7f
52	Te	2	2	6	2	6	10	2	6	10	—	2	4										
53	I	2	2	6	2	6	10	2	6	10	—	2	5										
54	Xe	2	2	6	2	6	10	2	6	10	—	2	6										
55	Cs	2	2	6	2	6	10	2	6	10	—	2	6		—	1							
56	Ba	2	2	6	2	6	10	2	6	10	—	2	6		—	2							
57	La	2	2	6	2	6	10	2	6	10	—	2	6	1	—	2							
58	Ce	2	2	6	2	6	10	2	6	10	2	2	6		—	2							
59	Pr	2	2	6	2	6	10	2	6	10	3	2	6		—	2							
60	Nd	2	2	6	2	6	10	2	6	10	4	2	6		—	2							
61	Pm	2	2	6	2	6	10	2	6	10	5	2	6		—	2							
62	Sm	2	2	6	2	6	10	2	6	10	6	2	6		—	2							
63	Eu	2	2	6	2	6	10	2	6	10	7	2	6		—	2							
64	Gd	2	2	6	2	6	10	2	6	10	7	2	6	1	—	2							
65	Tb	2	2	6	2	6	10	2	6	10	8	2	6	1	—	2							
66	Dy	2	2	6	2	6	10	2	6	10	10	2	6		—	2							
67	Ho	2	2	6	2	6	10	2	6	10	11	2	6		—	2							
68	Er	2	2	6	2	6	10	2	6	10	12	2	6		—	2							
69	Tu	2	2	6	2	6	10	2	6	10	13	2	6		—	2							
70	Yb	2	2	6	2	6	10	2	6	10	14	2	6		—	2							
71	Lu	2	2	6	2	6	10	2	6	10	14	2	6	1		2							
72	Hf	2	2	6	2	6	10	2	6	10	14	2	6	2		2							
73	Ta	2	2	6	2	6	10	2	6	10	14	2	6	3		2							
74	W	2	2	6	2	6	10	2	6	10	14	2	6	4		2							
75	Re	2	2	6	2	6	10	2	6	10	14	2	6	5		2							
76	Os	2	2	6	2	6	10	2	6	10	14	2	6	6		2							
77	Ir	2	2	6	2	6	10	2	6	10	14	2	6	7		2							
78	Pt	2	2	6	2	6	10	2	6	10	14	2	6	9	—	1							
79	Au	2	2	6	2	6	10	2	6	10	14	2	6	10	—	1							

ELECTRONIC STRUCTURE OF ISOLATED ATOMS (continued)

Shell / Subshell designations:
K (1): s 0 — L (2): s 0, p 1 — M (3): s 0, p 1, d 2 — N (4): s 0, p 1, d 2, f 3 — O (5): s 0, p 1, d 2, f 3 — P (6): s 0, p 1, d 2, f 3 — Q (7): s 0, p 1, d 2, f 3

Atomic number	Element	K(1) s	L(2) s	L(2) p	M(3) s	M(3) p	M(3) d	N(4) s	N(4) p	N(4) d	N(4) f	O(5) s	O(5) p	O(5) d	O(5) f	P(6) s	P(6) p	P(6) d	P(6) f	Q(7) s	Q(7) p	Q(7) d	Q(7) f
80	Hg	2	2	6	2	6	10	2	6	10	14	2	6	10	—	2							
81	Tl	2	2	6	2	6	10	2	6	10	14	2	6	10	—	2	1						
82	Pb	2	2	6	2	6	10	2	6	10	14	2	6	10	—	2	2						
83	Bi	2	2	6	2	6	10	2	6	10	14	2	6	10	—	2	3						
84	Po	2	2	6	2	6	10	2	6	10	14	2	6	10	—	2	4						
85	At	2	2	6	2	6	10	2	6	10	14	2	6	10	—	2	5						
86	Rn	2	2	6	2	6	10	2	6	10	14	2	6	10	—	2	6						
87	Fr	2	2	6	2	6	10	2	6	10	14	2	6	10	—	2	6	—	—	1			
88	Ra	2	2	6	2	6	10	2	6	10	14	2	6	10	—	2	6	—	—	2			
89	Ac	2	2	6	2	6	10	2	6	10	14	2	6	10	—	2	6	1	—	2			
90	Th	2	2	6	2	6	10	2	6	10	14	2	6	10	—	2	6	2	—	2			
91	Pa	2	2	6	2	6	10	2	6	10	14	2	6	10	2	2	6	1	—	2			
92	U	2	2	6	2	6	10	2	6	10	14	2	6	10	3	2	6	1	—	2			
93	Np	2	2	6	2	6	10	2	6	10	14	2	6	10	4	2	6	1	—	2			
94	Pu	2	2	6	2	6	10	2	6	10	14	2	6	10	5	2	6	1	—	2			
95	Am	2	2	6	2	6	10	2	6	10	14	2	6	10	6	2	6	1	—	2			
96	Cm	2	2	6	2	6	10	2	6	10	14	2	6	10	7	2	6	1	—	2			
97	Bk	2	2	6	2	6	10	2	6	10	14	2	6	10	9	2	6	—	—	2			
98	Cf	2	2	6	2	6	10	2	6	10	14	2	6	10	10	2	6	—	—	2			
99	Es	2	2	6	2	6	10	2	6	10	14	2	6	10	11	2	6	—	—	2			
100	Fm	2	2	6	2	6	10	2	6	10	14	2	6	10	12	2	6	—	—	2			
101	Mv	2	2	6	2	6	10	2	6	10	14	2	6	10	13	2	6	—	—	2			
102	No	2	2	6	2	6	10	2	6	10	14	2	6	10	14	2	6	—	—	2			
103	Lw	2	2	6	2	6	10	2	6	10	14	2	6	10	14	2	6	1	—	2			

IONIZATION POTENTIALS OF THE ATOMS

Atom	Electrons of the outer shell							
	I	II	III	IV	V	VI	VII	VIII
	Work of detachment of electrons, eV							
H	13.54	—	—	—	—	—	—	—
He	24.54	54.16	—	—	—	—	—	—
Li	5.37	75.3	121.9	—	—	—	—	—
Be	9.30	18.12	153.1	216.6	—	—	—	—
B	8.28	24.99	37.70	258.0	338.5	—	—	—
C	11.24	24.28	47.55	64.1	390.1	487.4	—	—
N	14.51	29.41	47.36	77.0	97.3	549	663	—
O	13.57	34.75	54.8	77.5	113.3	137.3	735	867
F	17.46	34.71	62.3	87.3	114.8	156.5	184.2	949
Ne	21.47	40.67	63.2	97.1	127.0	159.1	206.6	237.9
Na	5.09	46.65	71.3	99.0	139.1	173.9	210.5	263.6
Mg	7.63	15.10	79.4	109.4	142.2	188.5	227.9	269.0
Al	5.94	18.85	28.35	119.6	154.9	192.7	245.1	289.2
Si	8.14	16.29	33.35	44.84	167.4	207.9	250.5	309.1
P	10.43	19.75	30.08	51.1	64.6	222.8	268.3	315.7
S	10.42	23.25	34.89	47.32	72.2	87.5	285.7	336.2
Cl	13.01	23.85	39.67	53.5	68.0	96.5	113.8	356.1
Ar	15.68	27.64	40.94	59.7	75.7	92.1	124.1	143.2
K	4.32	31.45	46.00	61.7	83.3	101.4	119.7	155.0
Ca	6.25	11.87	51.1	68.1	86.1	110.5	130.6	150.7
Sc	6.7	12.8	26.19	74.5	93.9	114.2	141.3	163.4
Ti	6.81	13.6	28.39	45.40	101.7	123.5	145.9	175.7
V	6.74	15.13	30.31	48.35	68.7	132.8	156.9	181.3
Cr	6.7	16.41	32.12	50.9	72.4	96.0	167.6	193.9
Mn	7.41	14.5	33.97	53.4	75.8	100.7	127.4	206.3
Fe	7.83	15.9	31.69	55.9	79.0	104.9	133.1	162.8
Co	7.8	17.47	33.77	53.2	82.2	108.9	138.2	169.6
Ni	7.6	18.88	35.92	56.0	79.1	112.9	143.1	175.7
Cu	7.67	20.33	37.93	58.9	82.7	109.3	148.0	181.5
Zn	9.37	18.04	40.00	61.6	86.3	113.7	148.8	187.5
Ga	5.97	20.39	30.66	64.3	89.8	118.3	149.2	182.7
Ge	8.10	15.95	33.68	45.51	93.3	122.6	154.7	189.1
As	10.05	18.88	28.30	49.25	62.6	127.0	159.9	195.7
Se	9.75	21.57	32.11	43.03	67.1	81.9	165.3	201.9
Br	11.82	21.47	35.60	47.77	60.1	87.2	103.5	208.2
Kr	13.94	24.28	35.71	52.1	65.9	79.6	109.6	127.3
Rb	4.19	27.14	39.32	52.5	71.1	86.4	101.5	134.3
Sr	5.68	10.86	42.98	56.9	71.7	92.7	109.4	125.8
Y	6.6	12.3	20.46	61.5	77.1	93.5	116.7	134.8
Zr	6.92	13.97	22.64	34.83	82.6	99.9	117.9	143.2
Nb	—	13.48	24.7	37.7	51.9	106.3	125.3	144.7

IONIZATION POTENTIALS OF THE ATOMS (continued)

Atom	\multicolumn{8}{c}{Electrons of the outer shell}
	I	II	III	IV	V	VI	VII	VIII
	\multicolumn{8}{c}{Work of detachment of electrons, eV}
Mo	7.2	15.17	27.00	40.53	55.6	71.7	132.7	153.2
Ru	7.5	16.37	28.62	46.52	62.9	80.6	99.6	119.3
Rh	7.7	18.07	31.03	45.63	66.7	85.2	105.0	125.8
Pd	8.30	19.85	33.36	48.77	65.6	89.9	110.5	132.2
Ag	7.58	21.50	35.79	51.8	69.6	88.7	116.2	138.7
Cd	8.94	16.80	38.00	55.0	73.4	93.5	114.7	145.5
In	5.76	18.76	27.85	57.8	77.4	98.2	120.5	143.7
Sn	7.54	14.56	30.45	40.72	80.9	103.1	126.1	150.5
Sb	8.35	17.01	25.22	44.02	55.4	107.3	132.0	157.2
Te	8.89	19.33	28.39	37.73	59.5	71.9	137.0	164.1
I	10 43	19.11	31.40	41.70	52.1	76.8	90.2	170.0
Xe	12.08	21.18	31.33	45.46	56.9	68.3	96.0	110.4
Cs	3.86	23.37	33.97	45.55	61.5	74.1	86.4	117.1
Ba	5.21	9.96	36.75	48.80	61.8	79.5	93.1	106.4
La	5.59	11.38	19.1	52.2	65.7	80.0	99.5	114.1
Ce	6.54	14.8	—	33.3	69.7	84.6	100.2	121.5
Pr	5.76	—	--	—	—	89.3	105.5	122.4
Nd	6.31	—	—	—	—	—	111.0	128.5
Sm	6.55	11.4	—	—	—	—	—	—
Eu	5.64	11.4	—	—	—	—	—	—
Gd	6.65	—	—	—	—	—	—	—
Tb	6.74	—	—	—	—	—	—	---
Dy	6.82	—	—	—	—	—	—	—
Yb	6.25	12.11	—	—	—	—	—	—
Hf	—	14.8	—	—	—	—	—	—
Ta	—	—	22.27	33.08	—	—	—	—
W	8.1	—	24.08	35.36	—	—	—	—
Re	7.8	13.17	25.96	37.71	50.6	64.5	79.0	—
Os	8.7	—	—	—	—	—	—	—
Ir	9.2	—	—	—	—	—	—	—
Pt	8.8	17.37	28.55	41.13	54.8	75.3	91.9	109.3
Au	9.20	18.84	30.46	43.52	57.8	73.1	96.4	114.4
Hg	10.41	18.55	32.43	45.98	60.8	76.9	93.7	119.7
Tl	6.08	20.29	29.63	48.50	63.9	80.5	98.2	116.5
Pb	7.37	14.91	31.97	42.46	67.1	84.3	102.6	121.8
Bi	7.25	16.72	25.41	45.46	57.0	88.1	107.0	127.0
Rn	10.69	20.02	29.78	43.78	55.1	66.8	96.7	111.2
Fr	4.0	21.5	—	—	—	—	—	—
Ra	5.21	10.19	34.26	46.41	58.5	76.0	89.3	102.8
Th	—	—	29.4	—	—	—	—	—
U	4.5—5.0	—	—	—	—	—	—	—

RATIO OF THE RADII OF SOME ATOMS
OF NONMETALS AND METALS

Metal (Me)	Radius of metal atom R_{Me}, A	Metalloid (X)					
		N	C	B	S	P	Si
		Radius of metalloid atom R_X, A					
		0.70	0.77	0.80	1.04	1.10	1.17
		R_X/R_{Me}					
Be	1.12	0.62	0.69	0.71	0.93	0.98	1.04
Mg	1.60	0.44	0.48	0.50	0.65	0.69	0.73
Ca	1.97	0.36	0.39	0.41	0.53	0.56	0.59
Sr	2.15	0.32	0.36	0.37	0.48	0.51	0.54
Ba	2.22	0.32	0.35	0.36	0.47	0.50	0.53
Sc	1.62	0.43	0.48	0.49	0.64	0.68	0.72
Y	1.80	0.39	0.43	0.44	0.58	0.61	0.65
La	1.87	0.37	0.41	0.43	0.56	0.59	0.62
Ce	1.82	0.38	0.42	0.44	0.57	0.60	0.64
Pr	1.82	0.38	0.43	0.44	0.57	0.60	0.64
Nd	1.82	0.38	0.42	0.44	0.57	0.50	0.64
Pm	1.81	0.39	0.42	0.44	0.57	0.61	0.65
Sm	1.85	0.38	0.42	0.63	0.56	0.59	0.63
Eu	2.08	0.34	0.37	0.38	0.50	0.53	0.56
Gd	1.80	0.39	0.43	0.44	0.58	0.61	0.65
Tb	1.77	0.40	0.44	0.45	0.59	0.62	0.66
Dy	1.77	0.40	0.44	0.45	0.59	0.62	0.66
Ho	1.76	0.40	0.44	0.45	0.59	0.62	0.66
Er	1.75	0.40	0.44	0.46	0.59	0.63	0.67
Tu	1.74	0.40	0.44	0.46	0.60	0.63	0.67
Yb	1.93	0.36	0.40	0.41	0.54	0.57	0.61
Lu	1.74	0.40	0.44	0.46	0.60	0.63	0.67
Ti	1.47	0.48	0.52	0.66	0.71	0.75	0.80
Zr	1.60	0.44	0.48	0.50	0.65	0.69	0.73
Hf	1.58	0.44	0.49	0.51	0.66	0.70	0.74
V	1.34	0.52	0.57	0.60	0.78	0.82	0.87
Nb	1.46	0.48	0.53	0.55	0.71	0.75	0.80
Ta	1.46	0.48	0.53	0.55	0.71	0.75	0.80
Cr	1.36	0.51	0.57	0.59	0.70	0.81	0.86
Mo	1.39	0.50	0.55	0.58	0.75	0.79	0.84
W	1.39	0.50	0.55	0.58	0.75	0.79	0.84
Mn	1.31	0.53	0.59	0.61	0.79	0.84	0.89
Tc	1.36	0.51	0.57	0.59	0.76	0.81	0.86
Re	1.37	0.51	0.56	0.58	0.76	0.80	0.85
Fe	1.26	0.56	0.61	0.63	0.82	0.87	0.93

RATIO OF THE RADII OF SOME ATOMS
OF NONMETALS AND METALS (continued)

Metal (Me)	Radius of metal atom R_{Me}, A	Metalloid (X)					
		N	C	B	S	P	Si
		Radius of metalloid atom R_X, A					
		0.70	0.77	0.80	1.04	1.10	1.17
		R_X / R_{Me}					
Ru	1.34	0.52	0.57	0.60	0.78	0.82	0.87
Os	1.35	0.52	0.57	0.59	0.77	0.81	0.87
Co	1.25	0.56	0.62	0.64	0.83	0.88	0.94
Rh	1.34	0.52	0.57	0.60	0.78	0.82	0.87
Ir	1.36	0.51	0.57	0.59	0.76	0.81	0.86
Ni	1.24	0.56	0.62	0.64	0.84	0.89	0.94
Pd	1.37	0.51	0.56	0.58	0.76	0.80	0.85
Pt	1.38	0.51	0.56	0.58	0.75	0.80	0.85
Th	1.80	0.39	0.43	0.44	0.58	0.61	0.65
U	1.52	0.46	0.51	0.53	0.68	0.72	0.77

Remarks. 1. For the metals, the radii are those calculated by Pauling for the coordination number 12. 2. For Pm and Tc, the radii have been estimated. 3. For the metalloids, the radii are the covalent radii as calculated by Pauling.

COMPOSITION OF REFRACTORY COMPOUNDS

Phase	Molecular weight	Metalloid content, %	
		atomic	weight
Be_5B	55.89	16.67	19.36
Be_2B	28.84	33.33	37.5
BeB_2	30.65	66.67	70.61
BeB_4	52.23	80.0	82.76
BeB_6	73.94	85.71	87.81
BeB_9	106.39	90.0	91.53
BeB_{12}	138.85	92.28	93.51
MgB_2	45.96	66.67	47.09
MgB_6	89.24	85.71	72.74
MgB_{12}	154.16	92.31	84.23

Phase	Molecular weight	Metalloid content, %	
		atomic	weight
CaB_6	105.00	85.71	61.83
SrB_6	152.55	85.71	42.56
BaB_6	202.28	85.71	32.09
AlB_2	48.62	66.67	44.51
AlB_4 [1*] (?)	70.26	80.00	61.60
AlB_{10}	135.18	90.91	80.05
AlB_{12}	156.82	92.31	82.80
ScB_2	66.74	66.67	32.42
ScB_4	88.38	80.00	48.9
ScB_6	110.02	85.71	59.00
ScB_{12}	179.94	92.31	74.2
YB_2	110.56	66.67	19.57
YB_3	121.38	75.0	26.74
YB_4	132.20	80.0	32.81
YB_6	153.84	85.71	42.19
YB_{12}	218.76	92.31	59.25
YB_{70}	846.32	98.4	89.00
LaB_3	171.38	75.0	18.94
LaB_4	182.20	80.0	23.75
LaB_6	203.84	85.71	31.35
CeB_4	183.41	80.0	23.59
CeB_6	205.05	85.71	31.66
PrB_3	173.38	75.00	18.72
PrB_4	184.20	80.00	23.49
PrB_6	205.89	85.71	31.54
NdB_4	187.55	80.0	23.1
NdB_6	209.19	85.71	31.03
SmB_4	193.63	80.0	28.79
SmB_6	215.27	85.71	30.16
EuB_6	216.92	85.71	29.92
GdB_3	189.72	75.0	17.11
GdB_4	200.54	80.0	21.58
GdB_6	222.18	85.71	29.22
TbB_4	202.21	80.0	21.40
TbB_6	223.85	85.71	29.00
DyB_4	205.79	80.0	21.03
DyB_6	227.43	85.71	28.55
DyB_{12}	292.30	92.30	44.60
HoB_4	208.22	80.0	20.79
HoB_6	229.86	85.71	28.24
HoB_{12}	294.78	92.30	44.40
ErB_4	210.55	80.0	20.56
ErB_6	232.19	85.71	27.96
ErB_{12}	297.04	92.30	43.6

[1*] See [629].

COMPOSITION OF REFRACTORY COMPOUNDS (continued)

Phase	Molecular weight	Metalloid content, %	
		atomic	weight
TuB_4	212.22	80.0	20.39
TuB_6	233.86	85.71	27.76
TuB_{12}	299.24	92.30	43.40
YbB_3	205.50	75.0	15.80
YbB_4	216.32	80.0	20.00
YbB_6	237.96	85.71	27.28
LuB_2	196.63	66.67	11.00
LuB_4	218.27	80.0	19.03
LuB_6	239.91	85.71	27.06
LuB_{12}	304.83	92.30	42.60
ThB_2 (?)	253.69	66.67	8.53
ThB_4	275.33	80.0	15.72
ThB_6	296.97	85.71	21.86
UB	248.89	50.0	4.35
UB_2	259.71	66.67	8.37
UB_4	281.35	80.00	15.38
UB_{12}	367.91	92.31	35.29
PuB	252.82	50.0	4.28
PuB_2	263.64	66.67	8.21
PuB_4	285.28	80.0	15.17
PuB_6	306.42	85.71	21.18
Ti_2B	106.62	33.33	10.15
TiB	58.72	50.0	18.43
TiB_2	69.54	66.67	31.12
Ti_2B_5(?)	149.90	71.43	36.09
TiB_{12}(?)	177.74	92.31	73.05
ZrB	102.04	50.0	10.60
ZrB_2	112.86	66.67	19.25
ZrB_{12}	221.06	92.31	58.74
HfB	189.32	50.0	5.71
HfB_2	200.14	66.67	10.81
V_3B_2	174.49	40.0	12.40
VB	61.77	50.0	17.52
V_3B_4	196.13	57.14	22.07
VB_2	72.55	66.67	29.81
Nb_2B	196.64	33.33	5.50
Nb_3B_2	300.37	40.0	7.20
NbB	103.74	50.0	10.43
Nb_3B_4	322.01	57.14	13.44
NbB_2	114.55	66.67	18.89
Ta_2B	372.72	33.33	2.90
Ta_3B_2	564.49	40.0	3.83
TaB	191.77	50.0	5.64
Ta_3B_4	586.13	57.14	7.39
TaB_2	202.59	66.67	10.68
Cr_4B	218.86	20.0	4.94

COMPOSITION OF REFRACTORY COMPOUNDS (continued)

Phase	Molecular weight	Metalloid content, %	
		atomic	weight
Cr_2B	114.84	33.33	9.42
Cr_5B_3	292.51	37.50	11.10
CrB	62.83	50.0	17.24
Cr_3B_4	199.31	57.14	21.72
CrB_2	73.65	66.67	29.38
$Cr_2B_5(?)$	158.12	71.46	34.22
CrB_6	116.93	85.21	55.52
Mo_2B	202.72	33.33	5.34
Mo_3B_2	309.49	40.0	6.99
MoB	106.77	50.0	10.13
MoB_2	117.59	66.67	18.40
Mo_2B_5	246.00	71.43	21.99
$MoB_4(?)$	139.23	80.0	31.08
W_2B	378.54	33.33	2.86
WB	194.62	50.0	5.56
W_2B_5	421.82	71.43	12.83
$WB_4(?)$	227.12	80.0	19.06
Mn_4B	230.58	20.0	4.69
Mn_2B	120.70	33.33	8.96
MnB	65.76	50.0	16.45
Mn_3B_4	208.10	57.14	20.80
MnB_2	76.58	66.67	28.26
Re_3B	569.48	25.0	1.90
Re_7B_3	1336.00	30.0	2.43
Re_2B_5	426.54	71.43	12.68
ReB_3	218.68	66.67	14.84
Fe_2B	122.52	33.33	8.83
FeB	66.67	50.0	16.23
Co_3B	187.64	25.0	5.77
Co_2B	128.70	33.33	8.4
CoB	69.76	50.0	15.51
CoB_2	80.58	66.67	26.86
Ni_3B	186.95	25.0	5.79
Ni_2B	128.24	33.33	8.44
Ni_3B_2	197.71	40.0	10.95
Ni_4B_3	267.30	42.86	12.14
NiB	69.53	50.00	15.56
$NiB_2 (?)$	80.35	66.67	26.93
Be_2C	30.04	33.33	39.98
BeC_2	33.04	66.67	72.71
Mg_2C_3	84.67	60.0	42.56
MgC_2	48.34	66.67	49.69
CaC_2	64.10	66.67	37.48

COMPOSITION OF REFRACTORY COMPOUNDS (continued)

Phase	Molecular weight	Metalloid content, %	
		atomic	weight
SrC_2	111.65	66.67	21.52
BaC_2	161.38	66.67	14.88
Al_4C_3	143.95	42.86	25.03
ScC	56.97	50.0	21.08
Y_3C	278.76	25.0	4.30
YC	100.92	50.0	11.89
Y_2C_3	213.84	60.0	16.84
YC_2	112.92	66.67	21.25
La_2C_3	313.84	60.00	11.47
LaC_2	162.94	66.67	14.74
Ce_2C_3	316.29	60.0	11.39
CeC_2	164.15	66.67	14.63
Pr_2C_3	317.84	60.00	11.36
PrC_2	164.94	66.67	14.56
Nd_2C_3	324.54	60.00	11.10
NdC_2	168.29	66.67	14.27
Sm_3C	463.29	25.00	2.59
Sm_2C_3	336.86	60.00	10.68
SmC_2	174.45	66.67	13.77
Gd_2C_3	349.8	60.00	10.29
GdC_2	180.9	66.67	13.27
Tb_2C_3	354.43	60.00	10.16
TbC_2	182.32	66.67	13.10
Dy_3C	499.38	25.00	2.41
Dy_2C_3	360.92	60.00	9.97
DyC_2	186.46	66.67	12.87
Ho_3C	506.82	25.00	2.38
Ho_2C_3	365.91	60.00	9.83
HoC_2	188.94	66.67	12.71
Er_3C	513.61	25.00	2.34
Er_2C_3	370.43	60.0	9.74
ErC_2	191.22	66.67	12.55
Tu_3C	520.21	25.00	2.31
Tu_2C_3	374.83	60.0	9.68
TuC_2	193.42	66.67	12.42
YbC_2	197.04	66.67	12.18
Lu_3C	536.97	25.00	2.25
Lu_2C_3	386.01	60.00	9.34
LuC_2	199.01	66.67	12.06
ThC	244.06	50.0	4.92
ThC_2	256.07	66.67	9.38
UC	250.08	50.0	4.80
U_2C_3	512.17	60.0	7.04

Phase	Molecular weight	Metalloid content, %	
		atomic	weight
UC_2	262.09	66.67	9.16
Pu_2C	500.01	33.33	2.40
PuC	254.01	50.0	4.72
Pu_2C_3	524.03	60.0	6.88
TiC	59.91	50.0	20.05
ZrC	103.23	50.0	11.64
HfC	190.51	50.0	6.31
V_2C	113.91	33.33	10.54
VC	62.96	50.0	19.08
Nb_2C	197.83	33.33	6.07
NbC	104.92	50.0	11.45
Ta_2C	373.91	33.33	3.21
TaC	192.96	50.0	6.22
$Cr_{23}C_6$	1268.30	20.69	5.68
Cr_7C_3	400.10	30.0	9.01
Cr_3C_2	180.05	40.0	13.34
Mo_2C	203.91	33.33	5.89
MoC	107.96	50.0	11.13
W_2C	397.73	33.33	3.16
WC	195.87	50.00	6.13
$Mn_{23}C_6$	1335.69	20.69	5.39
Mn_3C	176.83	25.0	6.79
Mn_7C_3	298.72	30.00	8.57
Mn_5C_2	420.61	28.57	8.04
Fe_4C	235.41	20.00	5.10
Fe_3C	179.56	25.0	6.69
Fe_5C_2	303.25	28.57	7.91
Fe_2C	123.71	33.33	9.71
Co_3C	188.83	25.0	6.86
Co_2C	129.89	33.33	9.25
Ni_3C	188.14	25.0	6.38
Be_3N_2	55.06	40.0	50.89
Ca_3N_2	148.26	40.0	18.89
Mg_3N_2	100.98	40.0	27.75
Sr_3N_2	290.91	40.0	9.63
Ba_3N_2	440.10	40.0	6.37
AlN	40.99	50.0	34.18
ScN	58.97	50.0	23.76
YN	102.93	50.0	13.61
LaN	152.93	50.0	9.16
CeN	154.14	50.0	9.09
PrN	154.93	50.0	9.04
NdN	158.28	50.0	8.85
SmN	164.44	50.0	8.52

COMPOSITION OF REFRACTORY COMPOUNDS (continued)

Phase	Molecular weight	Metalloid content, %	
		atomic	weight
EuN	166.01	50.0	8,44
GdN	171.27	50.0	8,18
TbN	172.94	50.0	8.10
DyN	176.52	50.0	7.94
HoN	178.95	50.0	7,83
ErN	181.28	50.0	7,73
TuN	182.95	50.0	7,66
YbN	187.05	50.0	7.49
LuN	189.00	50.0	7.41
ThN	246.05	50.0	5.69
Th_3N_4	752.18	57.14	7.42
Th_2N_3	506.12	60.0	8.30
UN	252.07	50.0	5.55
U_2N_3	518.17	60.0	8.11
UN_2	266.07	66.67	10.52
NpN	251.00	50.0	5.58
PuN	256.00	50.0	5.47
PuN_2	270.02	66.67	10.37
Ti_2N	109.80	33.33	12,81
TiN	61.90	50.0	22.63
ZrN	105.22	50.0	13,31
HfN	192.51	50.0	7.28
V_3N	166.85	25.0	8,39
VN	64.95	50.0	21.56
Nb_2N	199.82	33.33	7.01
$NbN_{0.75}$	103.41	42.85	10.13
$NbN_{0.98}$	106.62	49.50	12.89
NbN	106.91	50.0	13.10
Ta_2N	375.90	33.33	3.73
TaN	194.95	50.0	7,19
Cr_2N	118,02	33.33	11.86
CrN	66,01	50.0	21.21
Mo_3N	301.85	25.0	4.64
Mo_2N	205.90	33.33	6.80
MoN	109.95	50.0	12.73
W_2N	318.72	33.33	4.39
WN	197.86	50.0	7.08
Mn_4N	233.76	20.0	6.00
Mn_5N_2	302.80	28.57	9.25
Mn_2N	123.88	33.33	11.30
Mn_3N_2	192.82	40.0	14.52
Mn_6N_5	399.76	45,5	17,50
Re_2N	368.62	33.33	3.62
Fe_4N	237.40	20.0	5.90

COMPOSITION OF REFRACTORY COMPOUNDS (continued)

Phase	Molecular weight	Metalloid content, %	
		atomic	weight
Fe_3N	181.55	25.0	7.71
Fe_2N	125.70	33.33	11.14
Co_3N	190.82	25.0	7.34
Co_2N	131.88	33.33	10.62
Ni_3N	190.14	25.0	7.37
Mg_2Si	76.73	33.33	36.61
Ca_2Si	108.25	33.33	25.95
$CaSi$	68.17	50.00	41.20
$CaSi_2$	96.26	66.67	58.36
Sr_2Si	203.35	33.33	13.79
$SrSi$	115.72	50.0	24.28
$SrSi_2$	143.81	66.67	39.07
$BaSi$	165.45	50.0	16.98
$BaSi_2$	193.54	66.67	29.03
$BaSi_3$	221.63	75.0	38.02
Sc_5Si_3	309.68	37.5	27.2
Sc_3Si_5	275.60	62.5	51.0
$ScSi_2$	101.14	66.67	55.6
Y_5Si_3	528.87	37.52	15.93
YSi	117.01	50.0	24.01
Y_3Si_5	407.06	62.5	34.5
YSi_2	145.10	66.67	38.72
$LaSi$	167.02	50.0	16.82
$LaSi_2$	195.10	66.67	28.79
$CeSi$	168.22	50.00	16.70
$CeSi_2$	196.31	66.67	28.62
$PrSi_2$	197.10	66.67	28.50
$NdSi_2$	200.45	66.67	28.02
$SmSi_2$	206.53	66.67	27.19
$EuSi_2$	208.18	66.67	26.99
$GdSi_2$	213.44	66.67	26.32
$DySi_2$	218.69	66.67	25.69
$YbSi_2$	229.22	66.67	24.60
Th_3Si_2	770.33	40.0	7.47
$ThSi$	260.14	50.0	10.80
$ThSi_2$	288.23	66.67	19.49
U_3Si	724.24	25.0	3.78
U_3Si_2	752.33	40.0	7.29
USi	266.16	50.0	10.55
USi_2	294.25	66.67	10.09
USi_3	322.34	75.0	26.14
$NpSi_2$	293.18	66.67	19.16
$PuSi_2$	298.18	66.67	18.84
Ti_5Si_3	323.77	37.5	26.03
$TiSi$	75.99	50.0	36.97
$TiSi_2$	104.08	66.67	53.98
$Zr_4Si(?)$	392.97	20.00	7.15
Zr_2Si	210.53	33.33	13.34
Zr_5Si_3	540.37	37.5	15.59

COMPOSITION OF REFRACTORY COMPOUNDS (continued)

Phase	Molecular weight	Metalloid content, %	
		atomic	weight
Zr_3Si_2	329.84	40.0	16.76
$ZrSi$	119.31	50.00	23.54
$ZrSi_2$	147.40	66.67	38.11
Hf_2Si	385.09	33.33	7.29
Hf_5Si_3	976.77	37.5	8.63
$HfSi$	206.59	50.0	13.59
$HfSi_2$	234.68	66.67	23.93
V_3Si	180.94	25.0	15.53
V_5Si_3	339.02	37.5	24.86
VSi_2	107.13	66.67	52.44
Nb_4Si	399.73	20.0	7.03
Nb_5Si_3	548.82	37.5	15.36
$NbSi_2$	149.09	66.67	37.68
$Ta_{4.5}Si$	342.37	18.18	3.35
Ta_2Si	389.99	33.33	7.20
Ta_5Si_3	989.02	37.5	8.52
$TaSi_2$	237.13	66.67	23.69
Cr_3Si	184.12	25.0	15.26
Cr_5Si_3	344.32	37.5	24.47
$CrSi$	80.10	50.0	35.07
$CrSi_2$	108.19	66.67	51.93
Mo_3Si	315.94	25.0	8.89
Mo_5Si_3	564.02	37.5	14.94
$MoSi_2$	152.16	66.67	36.93
W_3Si	579.67	25.0	4.85
W_5Si_3	1003.57	37.5	8.39
WSi_2	240.04	66.67	23.40
Mn_3Si	192.88	25.0	14.55
Mn_5Si_3	358.92	37.5	23.46
$MnSi$	83.03	50.0	33.85
Mn_3Si_5	305.24	62.5	46.01
$MnSi_2$	110.08	66.67	50.58
Re_2Si	586.75	25.0	4.79
Re_5Si_3	1015.37	37.5	8.30
$ReSi$	214.31	50.0	13.10
$ReSi_2$	242.40	66.67	23.17
Fe_3Si	195.64	25.0	14.36
Fe_5Si_3	363.52	37.5	23.18
$FeSi$	83.94	50.0	33.46
$FeSi_2$	112.03	66.67	50.15
Co_3Si	204.91	25.0	13.71
Co_2Si	145.97	33.33	19.24
$CoSi$	87.03	50.0	32.28
$CoSi_2$	115.12	66.67	48.80

COMPOSITION OF REFRACTORY COMPOUNDS (continued)

Phase	Molecular weight	Metalloid content, %	
		atomic	weight
$CoSi_2$	143.21	75.0	58.84
Ni_3Si	204.22	25.0	13.76
Ni_5Si_2	349.73	28.6	16.07
Ni_2Si	145.51	33.33	19.31
Ni_3Si_2	232.31	40.0	24.19
$NiSi$	86.80	50.0	32.37
$NiSi_2$	114.87	66.67	48.91
Be_3P_2	88.99	40.0	69.62
Mg_3P_2	134.91	40.0	45.92
Ca_3P_2	182.19	40.0	34.00
Sr_3P_2	324.84	40.0	19.07
Ba_3P_2	474.03	40.0	13.07
BaP_2	199.31	66.67	31.08
AlP	57.96	50.0	53.45
ScP	76.08	50.0	41.65
YP	119.90	50.0	25.85
LaP	169.90	50.0	18.24
CeP	171.11	50.0	18.11
PrP	171.90	50.0	18.02
NdP	175.25	50.0	17.68
SmP	181.33	50.0	17.08
$ThP_{0.75}$	255.28	42.86	9.10
Th_3P_4	820.05	57.14	15.11
UP	269.05	50.0	11.52
U_3P_4	838.11	57.14	14.78
Np_3P_4	834.90	57.14	14.84
PuP	272.98	50.0	11.27
Ti_3P	174.68	25.0	17.74
TiP	78.88	50.0	39.28
ZrP	122.20	50.0	25.35
ZrP_2 (?)	153.18	66.67	40.44
HfP	209.58	50.0	14.79
V_3P	183.83	25.0	16.85
VP	81.93	50.0	37.81
NbP	123.89	50.0	25.01
TaP	211.93	50.0	14.62
Cr_3P	187.01	25.0	16.57
Cr_2P	135.0	33.33	22.95
CrP	82.99	50.0	37.33
CrP_2	113.96	66.67	54.36
Mo_3P	318.83	25.0	9.72
$MoP_{0.75}$	119.15	43.0	19.41
MoP	126.93	50.0	24.41
MoP_2 (?)	157.90	66.67	39.23
WP	214.90	50.0	14.42
WP_2 (?)	245.81	66.67	25.20
Mn_3P	195.80	25.0	15.82

COMPOSITION OF REFRACTORY COMPOUNDS (continued)

Phase	Molecular weight	Metalloid content, %	
		atomic	weight
Mn_2P	140.86	33.33	21.99
Mn_3P_2	226.77	40.00	27.32
MnP	85.92	50.00	36.06
Re_2P	403.60	33.33	7.68
ReP	217.29	50.00	14.26
ReP_2	248.26	66.67	24.95
ReP_3 (?)	279.23	75.00	33.28
Fe_3P	198.53	25.00	15.61
Fe_2P	142.68	33.33	21.71
FeP	86.83	50.00	35.68
FeP_2 (?)	117.80	66.67	52.59
Co_2P	148.86	33.33	20.81
CoP	89.92	50.0	34.45
CoP_3 (?)	151.86	75.0	61.19
Ni_3P	207.05	25.0	14.96
$Ni_{12}P_5$	859.16	29.41	18.03
Ni_2P	148.36	33.33	20.88
NiP_3 (?)	151.62	75.0	61.29
Sc_2S_3	186.12	60.0	51.69
YS	120.99	50.0	26.51
Y_5S_7	669.06	58.33	33.55
Y_2S_3	274.04	60.0	35.11
YS_2	153.05	66.67	41.90
Y_2O_2S *	241.91	20.0	13.25
LaS	170.99	50.0	18.75
La_3S_4	545.04	57.14	23.55
La_2S_3	374.04	60.0	25.72
LaS_2	203.05	66.67	31.58
La_2O_2S	341.91	20.0	9.38
CeS	172.20	50.0	18.62
Ce_3S_4	548.65	57.14	23.38
Ce_2S_3	376.46	60.00	25.55
CeS_2	204.26	66.67	31.40
Ce_2O_2S	344.33	20.00	9.31
PrS	172.99	50.0	18.54
Pr_3S_4	551.02	57.14	23.28
Pr_2S_3	378.05	60.0	25.40
Pr_2O_2S	345.91	20.0	9.27
NdS	176.34	50.0	18.18
Nd_3S_4	561.07	57.14	22.86
Nd_2S_3	384.74	60.0	25.0
Nd_2O_2S	352.61	20.0	9.09

* In oxysulfides, the S content is given.

COMPOSITION OF REFRACTORY COMPOUNDS (continued)

Phase	Molecular weight	Metalloid content, %	
		atomic	weight
SmS	182.50	50.0	17 57
Sm₃S₄	579.55	57.14	22.13
Sm₂S₃	397.06	60.0	24.23
Sm₂O₂S	364.93	20.0	8.79
EuS	184.07	50.0	17.42
Eu₃S₄	584.28	57.14	21.95
Eu₂S₃.₈₁	426.17	65.58	28.65
Fu₂O₂S	368.07	20.0	8.71
GdS	189.33	50.0	16.94
Gd₂S₃	410.72	60.0	23.42
GdS₂	221.39	66.67	28.97
Gd₂O₂S	378.59	20.0	8.47
Tb₂O₂S	381.93	20.0	8.40
Dy₅S₇	1037.01	58.33	21.65
Dy₂S₃	421.22	60.00	22.84
DyS₂	226.64	66.67	28.30
Dy₇O₂S	389.09	20.0	8.24
Ho₂O₂S	393.95	20.0	8.14
ErS	199.34	50.0	16.09
Er₅S₇	1060.81	58.33	21.16
Er₂S₃	430.74	60.0	22.33
Er₂O₂S	398.61	20.0	8.05
Tu₂O₂S	401.95	20.0	7.98
YbS	205.10	50.0	15.63
Yb₄S₄	647.36	57.14	19.80
Yb₂S₃	442.28	60.0	21.75
Yb₂O₂S	410.15	20.0	7.82
Lu₂O₂S	414.05	20.0	7.74
Ac₂S₃	550.20	60.0	17.49
ThS	264.12	50.0	12.14
Th₂S₃	560.30	60.00	17.17
Th₄S₇	1152.66	63.64	19.47
ThS₂	296.18	66.67	21.65
ThOS	296.12	33.33	10.83
PaOS	295.07	33.33	10.87
US	270.14	50.0	11.87
U₂S₃	572.34	60.0	16.81
U₃S₅	874.54	62.5	18.33
US₂	302.20	66.67	21.22
UOS	286.14	33.33	11.21
Np₂S₃	570.20	60.0	16.87
NpOS	285.07	33.33	11.25

Phase	Molecular weight	Metalloid content, %	
		atomic	weight
PuS	274.07	50.0	11.70
Pu_2S_3	580.20	60.0	16.58
Pu_2O_2S	548.07	20.0	5.85
Am_2S_3	582.20	60.0	16.52
$B_{13}C_2$ *	164.68	13.34	14.60
B_4C *	55.29	20.0	21.72
SiC *	40.10	50.0	29.95
BN **	24.82	50.0	56.44
Si_3N_4 **	140.30	57.14	39.94
$B_{12}Si$ ***	157.93	7.69	17.78
B_6Si ***	93.01	14.29	30.20
B_4Si ***	71.37	20.0	39.36
B_3Si ***	60.55	25.0	46.39
BP ****	41.80	50.0	74.10
$B_{13}P_2$ ****	202.62	13.34	30.55

* Carbon content. ** Nitrogen content.
*** Silicon content. **** Phosphorus content.

HOMOGENEITY RANGE

Phase	Metalloid content, %		Ref.	Year	Remarks
	atomic	weight			
LaB_6	85.8—88	57.8—62.3	[1030]	1961	
TiB_2	66.67—75	31.1—40.4	[31]	1949	
ZrB_2	~65.6—~66.3	~18.2—~19.0	[1168]	1962	
VB_2	~66—~68	~29.2—~31.1	[222]	1959	
NbB	Narrow	Narrow	[28]	1951	
NbB_2	~64—~74	~17.2—~24.9	[223]	1959	
TaB_2	~63—~72	~9.2—~13.3	[223]	1959	
Mo_2B	None	None	[48]	1947	
Mo_3B_2	"	"	[49]	1952	
α-MoB	48—51	9.7—10.0	[49]	1952	[48]
β-MoB	51—52	10.0—10.7	[49]	1952	[48]
MoB_2	70—71.4	19.5—20.8	[49]	1952	[48]
Mo_2B_5	70—71.4	19.5—20.8	[49]	1952	[48]
W_2B	Very narrow	Very narrow	[48]	1947	
WB	48—51	5.2—5.8	[48]	1947	
W_2B_5	68—71.4	11.1—12.8	[48]	1947	
Ni_3B	Very narrow	Very narrow	[227]	1958	
TiC	18—50	5.2—20.0	[31]	1949	
ZrC	21—50	3.3—11.6	[63]	1953	At 1600° 35-50% (atom.) C [982]
HfC	36—50	3.7—6.3	[64]	1954	$HfC_{0.56—1.0}$; [982]
V_2C	27—33.3	8.0—10.5	[65]	1954	

Phase	Metalloid content, %		Ref.	Year	Remarks
	atomic	weight			
VC	38.7—44.2	12.9—15.8	[1058]	1961	δ-phase: [1269, 1335]
VC	44.2—47.8	15.8—17.83	[1058]	1961	ε-phase: [65, 1269, 1335]
Nb_2C	26.5—33.3	4.5—6.1	[627]	1959	$NbC_{0.36-0.50}$; [66]
Nb_2C	28.1—33.66	4.8—6.1	[695]	1959	$NbC_{0.39-0.51}$ at 1600°C; [1270, 1083, 1335]
NbC	41.1—47.6	8.3—10.5	[627]	1959	$NbC_{0.70-0.91}$; [66, 1085]
NbC	41.8—50	8.5—11.45	[695]	1959	$NbC_{0.72-1.0}$ at 1600°C; [1270]
Ta_2C	22—33	1.8—3.2	[666]	1954	
TaC	36—50	3.6—6.2	[666]	1954	
Mo_2C	31.2—33.3	5.4—5.9	[1]	1957	
W_2C	29.8—33.3	2.7—3.2	[1]	1957	
Ti_2N	Narrow	Narrow	[1205]	1962	
TiN	37.5—50	14.9—22.6	[1205]	1962	[1]
ZrN	46—50	11.5—13.3	[592]	1956	
V_3N	25—33	8.4—11.9	[90]	1949	
VN	41—50	16.0—21.6	[90]	1949	
Nb_2N	28.5—33.5	5.7—7.1	[91]	1954	
$NbN_{0.75}$	42.9—44.0	10.2—10.6	[955]	1960	
$NbN_{0.98}$	46.8—49.5	11.55—12.85	[955]	1960	
NbN	50.0—50.6	13.1—13.3	[955]	1960	
Ta_2N	28.5—31	3.0—3.4	[92]	1954	$TaN_{0.4-0.45}$
TaN	44.5—47.3	5.8—6.5	[835]	1959	$TaN_{0.8-0.9}$, [92]
Cr_2N	32—33.3	11.3—11.8	[93]	1934	
Mo_2N	32—33	6.4—6.7	[95]	1954	
Mn_4N	18—20	5.8—6.1	[626]	1957	< 400°; [1207]
Mn_2N	28.3—34.7	9.14—11.9	[1207]	1963	400°; [626]
Mn_2N	28.1—32.9	9.1—11.1	[1207]	1963	500°
Mn_2N	— —30.9	— —10.3	[1207]	1963	600°
Mn_3N_2	38.2—41.0	13.1—15.1	[1207]	1963	400°; [626]
Mn_3N_2	38.5—40.8	13.8—14.9	[1207]	1963	500°
Mn_3N_2	— —40.5	— —14.8	[1207]	1963	600°
Re_2N	30—33.3	10—11.2	[600]	1951	
CosN	26—26.7	7.7—8	[697]	1959	
U_3Si	25.6—26.1	3.9—4.0	[641]	1958	Silicon content higher than corresponds to U_3Si
V_3Si	range ~1%	—	[1249]	1963	At 800°
α-Nb_5Si_3	36.6—40.4	15—17	[1069]	1961	[1142]
$NbSi_2$	65.1—68.7	31.9—40	[1069]	1961	[1142]
Cr_3Si	18—31	10.6—19.5	[1329]	1961	At 1000°C
Cr_5Si_3	36—41	23.3—27.2	[1329]	1961	Same
CrSi	~49—~51	~34.1—~36.0	[1329]	1961	"
$CrSi_2$	~66—~70	~51.2—~55.6	[1329]	1961	"
$MoSi_2$	65.8—66.7	36.0—36.9	[675]	1960	
$FeSi_2$	68.3—72.0	53.5—56.5	[1232]	1962	1080°, α-lebauite
TiP	48.0—48.5	37.4—37.9	[130]	1954	
α-ZrP	Very narrow	Very narrow	[130]	1954	
β-ZrP	48.0—48.5	23.9—24.2	[130]	1954	
HfP	Very narrow	Very narrow	[1239]	1962	
Yb_3S_4	57.08—59.35	19.77—21.29	[1012]	1961	
B_4C	17.6—29.5	19.2—31.7	[773]	1960	Carbon content

CRYSTAL STRUCTURE

Phase	Unit cell	Space group[1*]	Structure type	a, Å	b, Å	c, Å	α	c/a	Ref.	Year	Remarks
Be₄B	Tetrag.	D_{4h}^7 — P4/nmm	—	3.365	—	7.050	—	2.093	[563]	1960	[863, 865, 866, 1201]
Be₂B	Cubic	O_h^5 — Fm3m	CaF₂	4.661	—	—	—	—	[540]	1955	[863, 865, 866]
BeB₂	Hexag.	D_{6h}^1 — C6/mmm	AlB₁₂	9.79	—	9.55	—	0.98	[865]	1961	[864, 866]
BeB₆	Tetrag.	D_{4h}^{12} — C4/nnm	β-B	10.16	—	14.28	—	1.41	[865]	1961	[864, 866]
BeB₁₂	Hexag.	—	AlB₂	5.08	—	8.80	—	1.73	[927]	1960	[1183]
MgB₂	Cubic	D_{6h}^1 — C6/mmm	CaB₆	3.083	—	3.520	—	1.142	[541]	1955	[1, 3, 12, 15, 476, 846; 1115]
CaB₆	"	O_h^1 — Pm3m	CaB₆	4.144	—	—	—	—	[891]	1961	[4, 11, 15, 846, 1115]
SrB₆	"	O_h^1 — Pm3m	CaB₆	4.195	—	—	—	—	[891]	1961	[1115]
BaB₆	"	O_h^1 — Pm3m	CaB₆	4.268	—	—	—	—	[891]	1961	[58, 1152, 1153, 1156]
AlB₂	Hexag.	D_{6h}^1 — C6/mmm	AlB₂	3.009	—	3.262	—	1.08	[574]	1956	Pseudotetrag.[1156,1321]
AlB₁₀	Rhombic	—	—	8.881	9.100	5.690	—	—	[577]	1948	[1156,1321,1328]
AlB₁₂	Monocl.	—	—	8.505	10.98	7.378	143°39′	—	[576]	1939	[576,1154,1155,1156]
α-AlB₁₂	Tetrag.	D_4^4 — $P4_12_1$	—	10.161	—	14.283	—	1.41	[577]	1948	Pseudotetrag. [576,716,1154,1155,1156]
β-AlB₁₂	Rhombic	D_{2h}^{28} — Imma	—	12.34	12.631	10.161	—	—	[577]	1948	
γ-AlB₁₂	"	—	—	16.6	17.5	10.2	—	—	[1156]	1960	[1321]
ScB₂	Hexag.	D_{6h}^1 — C6/mmm	AlB₂	3.146	—	3.517	—	1.118	[6]	1958	T. S. Verkhoglyadova
ScB₄	Tetrag.	—	—	7.70	—	3.64	—	0.473	[1109]	1963	
ScB₁₂	Cubic	O_h^1 — Pm3m	—	7.422	—	—	—	—	[1110]	1963	
YB₂	Hexag.	—	—	3.298	—	3.843	—	1.165	[8]	1959	[8]
YB₃ (?)	Tetrag.	—	—	3.78	—	3.55	—	0.94	[8]	1956	
YB₄	"	D_{4h}^5 — P4/mbm	UB₄	7.111	—	4.107	—	0.565	[1110]	1959	[8, 9, 846]

1* O. T. Khorpyakov, Yu. B. Paderno, V. P. Dzeganovs'kii. Standard X-Ray Patterns for Hard and Refractory Compounds [in Ukrainian], Vid. Akad. Nauk URSR, Kiev, 1961.

CRYSTAL STRUCTURE (continued)

Phase	Unit cell	Space group[1]*	Structure type	a, Å	b, Å	c, Å	α	c/a	Ref.	Year	Remarks
YB_6	Cubic	$O_h^1 - Pm3m$	CaB_6	4.101	—	—	—	—	[891]	1961	[10—13, 62, 846, 1110]
YB_{12}	"	—	—	7.500	—	—	—	—	[1110]	1959	[1283]
YB_{70}	Tetrag.	—	—	11.75	—	12.62	—	1.07	[1110]	1959	
$LaB_4(?)$	"	—	—	3.82	—	3.96	—	1.04	[9]	1956	
LaB_4	"	$D_{4h}^5 - P4/mbm$	UB_4	7.3240	—	4.1811	—	0.57	[1030]	1961	[846, 14]
LaB_6	Cubic	$O_h^1 - Pm3m$	CaB_6	4.1561	—	—	—	—	[1030]	1961	[3, 4, 9, 12, 13, 15, 846, 891]
CeB_4	Tetrag.	$D_{4h}^5 - P4/mbm$	UB_4	7.205	—	4.090	—	0.558	[17]	1950	[846]
CeB_6	Cubic	$O_h^1 - Pm3m$	CaB_6	4.138	—	—	—	—	[891]	1961	[3, 4, 11, 12, 15, 476, 846]
PrB_3	Pseudocubic	—	—	3.81	—	—	—	—	[9]	1956	[846]
PrB_4	Tetrag.	$D_{4h}^5 - P4/mbm$	UB_4	7.20	—	4.11	—	0.571	[9]	1956	
PrB_6	Cubic	$O_h^1 - Pm3m$	CaB_6	4.131	—	—	—	—	[891]	1961	[4, 11, 13, 476, 846]
NdB_4	Tetrag.	$D_{4h}^5 - P4/mbm$	UB_4	7.219	—	4.1020	—	0.568	[477]	1959	[4, 11, 13, 15, 846, 891]
NdB_6	Cubic	$O_h^1 - Pm3m$	CaB_6	4.1260	—	—	—	—	[477]	1959	
SmB_4	Tetrag.	$D_{4h}^5 - P4/mbm$	UB_4	7.174	—	4.0696	—	0.557	[477]	1959	[9, 846]
SmB_6	Cubic	$O_h^1 - Pm\,3\,m$	CaB_6	4.1333	—	—	—	—	[477]	1959	[4, 10, 13, 346, 891]
EuB_6	"	$O_h^1 - Pm\,3\,m$	CaB_6	4.182	—	—	—	—	[891]	1961	[13, 14, 19, 846]
GdB_3	Tetrag.	—	—	3.79	—	3.63	—	0.958	[9]	1956	[7, 9, 20, 846]
GdB_4	"	$D_{4h}^5 - P4/mbm$	UB_4	7.144	—	4.0479	—	0.567	[477]	1959	[7, 9, 11, 13, 15, 846, 891]
GdB_6	Cubic	$O_h^1 - Pm\,3\,m$	CaB_6	4.1078	—	—	—	—	[477]	1959	

CRYSTAL STRUCTURE (continued)

Phase	Unit cell	Space group¹*	Structure type	a, Å	b, Å	c, Å	α	c/a	Ref.	Year	Remarks
TbB₄	Tetrag.	D_{4h}^5 — P4/mbm	UB₄	7.118	—	4.0286	—	0.566	[477]	1959	[21]
TbB₆	Cubic	O_h^1 — Pm3m	CaB₆	4.1020	—	—	—	—	[477]	1959	[13, 22, 891]
DyB₄	Tetrag.	D_{4h}^5 — P4/mbm	UB₄	7.101	—	4.0174	—	0.573	[477]	1959	[23, 846]
DyB₆	Cubic	O_h^1 — Pm3m	CaB₆	4.0976	—	—	—	—	[477]	1959	[23, 846]
DyB₁₂	"	O_h^5—Fm3m	UB₁₂	7.501	—	—	—	—	[1283]	1961	[1318]
HoB₄	Tetrag.	D_{4h}^5 — P4/mbm	UB₄	7.086	—	4.0079	—	0.566	[477]	1959	[20, 23]
HoB₆	Cubic	O_h^1 — Pm3m	CaB₆	4.096	—	—	—	—	[477]	1959	[13, 23]
HoB₁₂	"	O_h^5—Fm3m	UB₁₂	7.492	—	—	—	—	[1293]	1961	[1318]
ErB₄	Tetrag.	D_{4h}^5 — P4/mbm	UB₄	7.071	—	3.9972	—	0.565	[477]	1959	[16, 21, 846]
ErB₆	Cubic	O_h^1 — Pm3m	CaB₆	4.101	—	—	—	—	[13]	1959	[4, 11, 13, 476, 846]
ErB₁₂	"	O_h^5—Fm3m	UB₁₂	7.484	—	—	—	—	[1283]	1961	[1318]
TuB₄	Tetrag.	D_{4h}^5 — P4/mbm	UB₄	7.06	—	3.99	—	0.57	[24]	1961	
TuB₆	Cubic	O_h^1 — Pm3m	CaB₆	4.110	—	—	—	—	[24]	1961	
TuB₁₂	"	O_h^5—Fm3m	UB₁₂	7.476	—	—	—	—	[1283]	1961	[1318]
YbBₛ	Tetrag.	—	—	3.77	—	3.56	—	0.94	[9]	1956	[846]
YbB₄	"	D_{4h}^5 — P4/mbm	UB₄	7.01	—	4.00	—	0.57	[9]	1956	
YbB₆	Cubic	O_h^1 — Pm3m	CaB₆	4.1468	—	—	—	—	[477]	1959	[11, 12, 13, 15, 9, 846, 891]
LuB₂	Hexag.	D_{6h}^1—C6/mmm	AlB₂	3.246	—	3.704	—	1,141	[1209]	1963	
LuB₄	Tetrag.	D_{4h}^5 — P4/mbm	UB₄	6.983	—	3.930	—	0.562	[20]	1958	[16]
LuBc	Cubic	O_h^1 — Pm3m	CaB₆	4.11	—	—	—	—	[16]	1958	
LuB₁₂	"	O_h^5—Fm3m	UB₁₂	7.464	—	—	—	—	[1283]	1961	[1318]
ThB₂ (?)	Cubic fc	—	—	5.58	—	—	—	—	[389]	1953	
ThB₄	Tetrag.	D_{4h}^5 — P4/mbm	UB₄	7.256	—	4.113	—	0.567	[17]	1950	[479, 1097]

37

CRYSTAL STRUCTURE (continued)

Phase	Unit cell	Space group[1a]	Structure type	a, Å	b, Å	c, Å	α	c/a	Ref.	Year	Remarks
ThB_6	Cubic	$O_h^1 - Pm3m$	CaB_6	4.109	—	—	—	—	[891]	1961	[9, 15, 25, 27, 846, 1097]
UB	"	$O_h^5 - Fm3m$	NaCl	4.88	—	—	—	—	[832]	1959	
UB_2	Hexag.	$D_{6h}^1 - C6/mmm$	AlB_2	3.136	—	3.988	—	1.272	[28]	1951	[1097]
UB_4	Tetrag.	$D_{4h}^5 - P4/mbm$	UB_4	7.080	—	3.978	—	0.562	[15]	1954	[17, 29, 1097]
UB_{12}	Cubic	$O_h^5 - Fm3m$	UB_{12}	7.472	—	—	—	—	[995]	1961	[15,29,1283,1318]
PuB	"	$O_h^5 - Fm3m$	NaCl	4.92	—	—	—	—	[719]	1960	
PuB_2	Hexag.	$D_{6h}^1 - C6/mmm$	AlB_2	3.18	—	3.90	—	1.23	[719]	1960	
PuB_4	Tetrag.	$D_{4h}^5 - P4/mbm$	UB_4	7.10	—	4.014	—	0.57	[719]	1960	
PuB_6	Cubic	$O_h^1 - Pm3m$	CaB_6	4.115—4.140	—	—	—	—	[719]	1960	
TiB	Rhombic	$D_{2h}^{16} - Pbnm$	FeB	6.12	3.06	4.56	—	—	[720]	1954	[935, 936]
TiB_2	Hexag.	$D_{6h}^1 - C6/mmm$	AlB_2	3.026	—	3.213	—	1.062	[31]	1949	[935, 936]
Ti_2B_5	"	$D_{6h}^4 - C6/mmc$	W_2B_5	2.98	—	13.98	—	4.70	[30]	1952	[935, 936]
ZrB	Cubic	$O_h^5 - Fm3m$	NaCl	4.65	—	—	—	—	[32]	1953	[935, 936]
ZrB_2	Hexag.	$D_{6h}^1 - C6/mmm$	AlB_2	3.162	—	3.523	—	1.114	[34]	1958	[36, 565]
ZrB_{12}	Cubic	$O_h^5 - Fm3m$	UB_{12}	7.408	—	—	—	—	[35]	1952	[1283,1318]
HfB	"	$O_h^5 - Fm3m$	NaCl	4.62	—	—	—	—	[36]	1953	[935, 936]
HfB_2	Hexag.	$D_{6h}^1 - C6/mmm$	AlB_2	3.141	—	3.470	—	1.105	[36]	1953	
V_3B_2	Tetrag.	$D_{4h}^5 - P4/mbm$	U_3Si_2	5.746	—	3.032	—	0.528	[37]	1958	
VB	Rhombic	$D_{2h}^{17} - Cmcm$	TaB	3.07	8.15	2.97	—	—	[222]	1959	[38, 535]
V_3B_4	"	$D_{2h}^{25} - Immm$	Ta_3B_4	3.030	13.18	2.986	—	—	[39]	1956	
VB_2	Hexag.	$D_{6h}^1 - P6/mmm$	AlB_2	3.001	—	3.061	—	1.020	[222]	1959	[40, 721]

Phase	Unit cell	Space group¹*	Structure type	a, Å	b, Å	c, Å	α	c/a	Ref.	Year	Remarks
Nb₂B	Tetrag.	D_{4h}^{18} — I4/mcm	CuAl₂	6.185	—	—	—	—	[41]	1950	
Nb₃B₂	"	D_{4h}^{5} — P4/mbm	U₃Si₂	—	—	3.281	—	0.530	[37]	1958	[28, 41]
NbB	Rhombic	D_{2h}^{17} — Cmcm	TaB	3.297	8.72	3.166	—	—	[223]	1959	[935]
Nb₃B₄	"	D_{2h}^{25} — Immm	Ta₃B₄	3.312	14.11	3.143	—	—	[11]	1950	[935]
NbB₂	Hexag.	D_{6h}^{1} — C6/mmm	AlB₂	3.089	—	3.303	—	1.06	[28]	1951	[42]
Ta₂B	Tetrag.	D_{4h}^{18} — I4/mcm	CuAl₂	5.778	—	4.864	—	0.841	[28]	1951	
Ta₃B₂	"	D_{4h}^{5} — P4/mbm	U₃Si₂	6.184	—	3.187	—	0.523	[37]	1958	
TaB	Rhombic	D_{2h}^{17} — Cmcm	TaB	3.276	8.669	3.157	—	—	[42]	1949	[42]
Ta₃B₄	"	D_{2h}^{25} — Immm	Ta₃B₄	3.29	14.0	3.13	—	—	[42]	1949	
TaB₂	Hexag.	D_{6h}^{1} — C6/mmm	AlB₂	3.078	—	3.265	—	1.050	[28]	1951	
Cr₄B	Rhombic	D_{2h}^{24} — Fddd	Mn₄B	4.26	7.38	14.71	—	—	[43]	1953	[41, 935]
Cr₂B	Tetrag.	D_{4h}^{18} — I4/mcm	CuAl₂	5.180	—	4.316	—	0.832	[43]	1953	(Cr₃B₂)
Cr₅B₃	"	D_{4h}^{18} — I4/mcm	Cr₅B₃	5.46	—	10.64	—	1.945	[43]	1953	[45, 69, 722]
CrB	Rhombic	D_{2h}^{17} — Cmcm	TaB	2.969	7.858	3.002	—	—	[44]	1949	
Cr₃B₄	"	D_{2h}^{25} — Immm	Ta₃B₄	2.984	13.02	2.953	—	—	[41]	1950	[40]
CrB₂	Hexag.	D_{6h}^{1} — C6/mmm	AlB₂	2.970	—	3.074	—	1.035	[46]	1958	
CrB₆ (?)	Tetrag.	—	—	5.468	—	7.152	—	1.313	[956]	1959	[48]
Mo₂B	"	D_{4h}^{18} — I4/amd	CuAl₂	5.543	—	4.735	—	0.854	[49]	1952	Transition temperature 2000°
α-MoB	"	D_{4h}^{19} — I4/amd	MoB	3.110	—	16.97	—	5.45	[49]	1952	
β-MoB	Rhombic	D_{2h}^{17} — Cmcm	TaB	3.16	8.61	3.08	—	—	[49]	1952	

CRYSTAL STRUCTURE (continued)

Phase	Unit cell	Space group[1*]	Structure type	a, Å	b, Å	c, Å	α	c/a	Ref.	Year	Remarks
MoB_2	Hexag.	D_{6h}^1 — C6/mmm	AlB_2	3.05	—	3.113	—	1.01	[49]	1952	[50]
Mo_2B_5	Rhombohedr.	D_{3d}^5 — R3m	Mo_2B_5	3.011	—	20.93	—	6.95	[49]	1952	[48]
MoB_4	Tetrag.	D_{4h}^5 — P4/mbm	UB_4	—	—	—	—	—	[1007]	1961	
W_2B	"	D_{4h}^{18} — I4/mcm	$CuAl_2$	5.564	—	4.740	—	0.852	[48]	1947	[41]
α-WB	"	D_{4h}^{19} — I4/amd	MoB	3.115	—	16.93	—	5.42	[48]	1947	[41]
β-WB	Rhombic	D_{2h}^{17} — Cmcm	TaB	3.19	8.46	3.07	—	—	[51]	1952	Transition temperature 1850—1960°
W_2B_5	Hexag.	D_{6h}^4 — C6/mmc	W_2B_5	2.982	—	13.87	—	4.65	[48]	1947	[41]
WB_4	Tetrag.	D_{4h}^5 — P4/mbm	UB_4	6.34	—	4.50	—	0.71	[1007]	1961	
Mn_4B	Rhombic	D_{2h}^{24} — Fddd	Mn_4B	14.53	7.293	4.209	—	—	[841]	1959	[52]
Mn_2B	Tetrag.	D_{4h}^{18} — I4/mcm	$CuAl_2$	5.148	—	4.208	—	0.82	[841]	1959	[52]
MnB	Rhombic	D_{2h}^{16} — Pbnm	FeB	4.145	5.560	2.977	—	—	[841]	1959	[52]
Mn_3B_4	"	D_{2h}^{25} — Immm	Ta_3B_4	3.302	12.86	2.960	—	—	[52]	1950	
MnB_2	Hexag.	D_{6h}^1 — C6/mmm	AlB_2	3.009	—	3.039	—	1.01	[938]	1960	
ReB	Rhombic	D_{2h}^{17} — Cmcm	ReB	5.890	9.313	7.258	—	—	[849]	1960	[723]
Re_7B_3	Hexag.	C_{6v}^4 — C6mc	Cr_7C_3	7.504	—	4.772	—	0.651	[723]	1960	
ReB_2	"	D_{6h}^4 — C6/mmc	ReB_2	2.900	—	7.478	—	2.578	[1061]	1962	[723]
Re_2B_5	"	D_{6h}^4 — C6/mmc	W_2B_5	2.97	—	13.8	—	4.65	[53]	1958	
Fe_2B	Tetrag.	D_{4h}^{18} — I4/mcm	$CuAl_2$	5.109	—	4.249	—	0.842	[11]	1960	
FeB	Rhombic	D_{2h}^{16} — Pnma	FeB	4.061	5.506	2.952	—	—	[87]	1954	[52]

CRYSTAL STRUCTURE (continued)

Phase	Unit cell	Space group¹*	Structure type	a, Å	b, Å	c, Å	α	c/a	Ref.	Year	Remarks
Co_3B	Rhombic	D_{2h}^{16} — $Pbnm$	Fe_3C	4.411	5.235	6.635	—	—	[54]	1958	[55, 56]
Co_2B	Tetrag.	D_{4h}^{18} — $I4/mcm$	$CuAl_2$	5.016	—	4.220	—	0.841	[59]	1933	
CoB	Rhombic	D_{2h}^{16} — $Pbnm$	FeB	3.956	5.253	3.043	—	—	[59]	1933	[55, 561]
Ni_3B	"	D_{2h}^{16} — $Pbnm$	Fe_3C	5.211	6.619	4.389	—	—	[54]	1958	[551]
Ni_2B	Tetrag.	D_{4h}^{18} — $I4/mcm$	$CuAl_2$	4.993	—	4.249	—	0.851	[41]	1950	
Ni_4B_3	Rhombic	D_{2h}^{16} — $Pbnm$	$o-Ni_4B_3$	11.953	2.981	6.569	—	—	[551]	1959	
Ni_4B_3	Monocl.	C_{2h}^{6} — $C2/c$	$m-Ni_4B_3$	6.430	4.882	7.818	103°18'	—	[551]	1959	
NiB	Rhombic	D_{2h}^{16} — $Pbnm$	FeB	2.925	7.396	2.966	—	—	[60]	1952	
Be_2C	"	O_{h}^{5} — $Fm3m$	CaF_2	4.342	—	—	—	—	[527]	1956	[517, 526]
Mg_2C_3	Hexag.	—	—	7.45	—	10.61	—	1.424	[514]	1948	
MgC_2	Tetrag.	D_{4h}^{17} — $P4/mmm$	ThC_2	5.55	—	5.03	—	0.906	[514]	1942	[513, 514, 842]
CaC_2—I	"	D_{4h}^{17} — $I4/mmm$	CaC_2	5.48	—	6.37	—	1.16	[58]	1954	
CaC_2—II	Tricl.	CI	—	8.42	11.84	3.94	$\alpha=93.4°$ $\beta=92.5°$ $\gamma=89.9°$	—	[1246]	1962	[842]
CaC_2—III CaC_2—IV	Monocl. Cubic	$P2_1/c$ T_h^6—$Pa3$	FeS_2[pyrite]	8.36 5.889	4.20	11.25	$\beta=96.3°$	—	[1263] [1246]	1961 1962	>450±2°C [842] [842,1013]
CaC_2	Tetrag.	D_{4h}^{17} — $I4/mmm$	CaC_2	3.88	—	6.37	—	1.64	[58]	1954	[514, 513, 842]
SrC_2	"	D_{4h}^{17} — $I4/mmm$	CaC_2	4.11	—	6.68	—	1.63	[58]	1954	[513, 514]
BaC_2	"	D_{4h}^{17} — $I4/mmm$	CaC_2	4.40	—	7.06	—	1.60	[58]	1954	[58, 513, 514]
Al_4C_3	Rhombohedr.	D_{3d}^{5} — $R3m$	Al_4C_3	8.53	—	—	28°17'	—	[512]	1934	

CRYSTAL STRUCTURE (continued)

Phase	Unit cell	Space group[1*]	Structure type	a, Å	b, Å	c, Å	α	c/a	Ref.	Year	Remarks
ScC	Cubic	O_h^5 — Fm3m	NaCl	4.51	—	—	—	—	[875]	1961	[78, 1078]
YC₂	Tetrag.	D_{4h}^{17} — I4/mmm	CaC₂	3.80	—	6.57	—	—	[838]	1931	[837, 843]
La₂C₃	Cubic	T_d^6 — Ī43d	Pu₂C₃	8.8185	—	—	—	—	[837]	1958	[844]
α-LaC₂	Tetrag.	D_{4h}^{17} — I4/mmm	CaC₂	3.934	—	6.572	—	1.67	[570]	1959	Up to 1750°; [513, 570, 843, 1005]
β-LaC₂	Cubic	O_h^5 — Fm3m	FeS₂	6.0	—	—	—	—	[1005]	1960	> 1750°
Ce₂C₃	"	T_d^6 — Ī43d	Pu₂C₃	8.455	—	—	—	—	[518]	1955	[837]
CeC₂	Tetrag.	D_{4h}^{17} — I4/mmm	CaC₂	3.878	—	6.488	—	1.673	[837]	1958	[518, 838, 843]
Pr₂C₃	Cubic	T_d^6 — Ī43d	Pu₂C₃	8.6072	—	—	—	—	[837]	1958	[513, 838, 843]
PrC₂	Tetrag.	D_{4h}^{17} — I4/mmm	CaC₂	3.85	—	6.38	—	1.66	[58]	1954	
Nd₂C₃	Cubic	T_d^6 — Ī43d	Pu₂C₃	8.5478	—	—	—	—	[837]	1958	[58, 513, 837, 838, 843]
NdC₂	Tetrag.	D_{4h}^{17} — I4/mmm	CaC₂	3.823	—	6.405	—	1.675	[837]	1958	
Sm₃C	Cubic	O_h^1 — Pm3m	Fe₄N	5.172	—	—	—	—	[837]	1958	[58, 838, 843]
Sm₂C₃	"	T_d^6 — Ī43d	Pu₂C₃	8.4257	—	—	—	—	[837]	1958	
SmC₂	Tetrag.	D_{4h}^{17} — I4/mmm	CaC₂	3.770	—	6.331	—	1.679	[837]	1958	
Gd₃C	Cubic	O_h^1 — Pm3m	Fe₄N	5.126	—	—	—	—	[837]	1958	[843]
Gd₂C₃	"	T_d^6 — Ī43d.	Pu₂C₃	8.3407	—	—	—	—	[837]	1958	
GdC₂	Tetrag.	D_{4h}^{17} — I4/mmm	CaC₂	3.718	—	6.275	—	1.688	[837]	1958	

CRYSTAL STRUCTURE (continued)

Phase	Unit cell	Space group[1a]	Structure type	a, Å	b, Å	c, Å	α	c/a	Ref.	Year	Remarks
Tb₃C	Cubic	O_h^1 — $Pm3m$	Fe₄N	5.107	—	—	—	—	[837]	1958	
Tb₂C₃	,,	T_d^6 — $\bar{I}43d$	Pu₂C₃	8.2617	—	—	—	—	[837]	1958	
TbC₂	Tetrag.	D_{4h}^{17} — $I4/mmm$	CaC₂	3.690	—	6.217	—	1.685	[837]	1958	[843]
Dy₃C	Cubic	O_h^1 — $Pm3m$	Fe₄N	5.079	—	—	—	—	[837]	1958	
Dy₂C₃	,,	T_d^6 — $\bar{I}43d$	Pu₂C₃	8.198	—	—	—	—	[837]	1958	In homogeneity range, low C
DyC₂	Tetrag.	D_{4h}^{17} — $I4/mmm$	CaC₂	3.699	—	6.176	—	1.683	[837]	1958	[843]
Ho₃C	Cubic	O_h^1 — $Pm3m$	Fe₄N	5.061	—	—	—	—	[837]	1958	
Ho₂C₃	,,	T_d^6 — $\bar{I}43d$	Pu₂C₃	8.176	—	—	—	—	[837]	1958	
HoC₂	Tetrag.	D_{4h}^{17} — $I4/mmm$	CaC₂	3.643	—	6.139	—	1.685	[837]	1958	[843]
Er₃C	Cubic	O_h^1 — $Pm3m$	Fe₄N	5.034	—	—	—	—	[837]	1958	
ErC₂	Tetrag.	D_{4h}^{17} — $I4/mmm$	CaC₂	3.620	—	6.094	—	1.683	[837]	1958	[843]
Tu₃C	Cubic	O_h^1 — $Pm3m$	Fe₄N	5.016	—	—	—	—	[837]	1958	
TuC₂	Tegrag.	D_{4h}^{17} — $I4/mmm$	CaC₂	3.600	—	6.047	—	1.689	[837]	1958	
Yb₃C	Cubic	O_h^1 — $Pm3m$	Fe₄N	4.993	—	—	—	—	[837]	1958	[843]
YbC₂	Tetrag.	D_{4h}^{17} — $I4/mmm$	CaC₂	3.637	—	6.109	—	1.680	[837]	1958	
Lu₃C	Cubic	O_h^1 — $Pm3m$	Fe₄N	4.965	—	—	—	—	[837]	1958	
LuC₂	Tetrag.	D_{4h}^{17} — $I4/mmm$	CaC₂	3.563	...	5.964	—	1.674	[837]	1958	
ThC	Cubic	O_h^5 — $Fm3m$	NaCl	5.338	...	—	—	—	[567]	1958	[80, 531, 771, 1278]

CRYSTAL STRUCTURE (continued)

Phase	Unit cell	Space group[1*]	Structure type	a, Å	b, Å	c, Å	α	c/a	Ref.	Year	Remarks
ThC_2	Monocl.	—	—	6.53	4.24	6.56	104°	—	[81]	1951	[513, 1278], pseudotetrag. $a = 5.86$; $c = 5.29$
UC	Cubic	$O_h^5 - Fm3m$	NaCl	4.951	—	—	—	—	[83]	1952	[564, 1014, 1120, 1172, 1173]
U_2C_3	"	$T_d^6 - I43d$	Pu_2C_3	8.088	—	—	—	—	[83]	1952	[1120, 1124]
$\alpha\text{-}UC_2$	Tetrag.	$D_{4h}^{17} - I4/mmm$	CaC_2	3.524	—	5.999	—	1.702	[82]	1948	At < 1820° [84, 727, 1120, 1124, 1125]
$\beta\text{-}UC_2$	Cubic	$O_h^5 - Fm3m$	CaF_2	5.45	—	—	—	—	[1005]	1960	At > 1820° [1120, 1124, 1125]
Pu_2C (?)	"	$O_h^5 - Fm3m$	NaCl	4.920	—	—	—	—	[85]	1949	[86]
PuC	"	$O_h^5 - Fm3m$	NaCl	4.91	—	—	—	—	[617]	1949	[85, 86, 1098]
Pu_2C_3	"	$T_d^6 - I43d$	Pu_2C_3	8.145	—	—	—	—	[85]	1949	[86, 1098]
TiC	"	$O_h^5 - Fm3m$	NaCl	4.324	—	—	—	—	[564]	1960	19.02%, 0.19%, [31]
ZrC	"	$O_h^5 - Fm3m$	NaCl	4.688	—	—	—	—	[982]	1960	[63, 564]
HfC	"	$O_h^5 - Fm3m$	NaCl	4.635	—	—	—	—	[770]	1959	[36, 64, 521, 523, 524, 564, 982]
V_2C	Pseudohexag. (rhombic)	$D_{3d}^3 - \bar{C}3m$	Mo_2C	2.906	—	4.597	—	1.578	[65]	1954	[728, 729, 1251, 1269]
VC	Cubic	$O_h^5 - Fm3m$	NaCl	4.182	—	—	—	—	[65]	1954	[564, 1058, 1269]
Nb_2C	Pseudohexag. (rhombic)	$D_{3d}^3 - \bar{C}3m$	Mo_2C	3.128	—	4.974	—	1.590	[627]	1959	[66, 313, 1083, 1251]

CRYSTAL STRUCTURE (continued)

Phase	Unit cell	Space group[1]*	Structure type	a, Å	b, Å	c, Å	α	c/a	Ref.	Year	Remarks
NbC	Cubic	$O_h^5 - Fm\bar{3}m$	NaCl	4.469	—	—	—	—	[627]	1959	[66,313,522,525, 564,1032]
Ta₂C	Pseudohexag (rhombic)	$D_{3d}^3 - C\bar{3}m$	Mo₂C	3.104	—	4.941	—	1.591	[67]	1955	[1251]
TaC	Cubic	$O_h^5 - Fm\bar{3}m$	NaCl	4.456	—	—	—	—	[564]	1960	[67]
Cr₂₃C₆	*	$O_h^5 - Fm\bar{3}m$	Cr₂₃C₆	10.638	—	—	—	—	[68]	1933	
Cr₇C₃	Hexag.	$C_{6v}^4 - P6_3mc$	Cr₇C₃	13.98	—	4.523	—	0.324	[69]	1935	[730]
Cr₃C₂	Rhombic	$D_{2h}^{16} - Pbnm$	Cr₃C₂	2.891	5.62	11.96	—	—	[70]	1933	[71]*
Mo₂C	Pseudohexag (rhombic)	$D_{3d}^3 - C\bar{3}m$	Mo₂C	3.002	—	4.729	—	1.575	[910]	1960	
γ-MoC	Hexag.	$D_{3h}^1 - C\bar{6}m2$	MoC	2.898	—	2.809	—	0.969	[71]	1952	[571, 572]
W₂C	Pseudohexag (rhombic)	$D_{3d}^3 - C\bar{3}m$	Mo₂C	2.98	—	4.71	—	1.58	[72]	1951	[1251,1252]
WC	Hexag.	$D_{3h}^1 - C\bar{6}m2$	MoC	2.900	—	2.831	—	0.971	[72]	1951	[1126,1127]
Mn₂₃C₆	Cubic	$O_h^5 - Fm\bar{3}m$	Cr₂₃C₆	10.61	—	—	—	—	[731]	1954	[73]
Mn₃C	Rhombic	$D_{2h}^{16} - Pbnm$	Fe₃C	4.530	5.080	6.772	—	—	[84]	1948	[731]
Mn₅C₂	Monocl.	$C2/\bar{c}$	Mn₅C₂	5.086	4.578	11.66	97.75°	—	[731]	1954	[1280]
Mn₇C₃	Hexag.	$C_{6v}^4 - C6mc$	Cr₇C₃	13.90	—	4.54	—	0.33	[731]	1954	[74]
Fe₄C	Cubic	—	—	3.75	—	—	—	—	[624]	1957	[772]
Fe₃C	Rhombic	$D_{2h}^{16} - Pbnm$	Fe₃C	4.5235	5.0890	6.7353	—	—	[76]	1948	[733,1119]
Fe₂C	Hexag.	—	—	2.757	—	4.346	—	—	[882]	1961	[75, 732]
Fe₅C₂	Monocl.	$C2/\bar{c}$	Mn₅C₂	11.56	4.573	5.058	97.44ª	—	[1280]	1962	
Co₃C	Rhombic	$D_{2h}^{16} - Pbnm$	Fe₃C	4.53	5.09	6.74	—	—	[79]	1938	
Co₂C	"	—	—	2.904	—	4.368	—	—	[569]	1951	[58, 519, 520]

* According to [1251], Mo₂C has rhombic lattice structure group $D_{2h}^{14} - Pbcn$ with parameters: a = 4.72₄; b = 6.00₄; c = 5.19₉ Å.

Phase	Unit cell	Space group[1*]	Structure type	a, Å	b, Å	c, Å	α	c/a	Ref.	Year	Remarks
Ni_3C	Hexag.	—	—	2.646	—	4.320	—	1.633	[735]	1951	[58, 77, 78, 734]
Be_3N_2	Cubic	$T_h^7 - Ia3$	Mn_2O_3	8.13	—	—	—	—	[58]	1954	[531]
Mg_3N_2	"	$T_h^7 - Ia3$	Mn_2O_3	9.95	—	—	—	—	[58]	1954	—
α-Ca_3N_2	"	$T_h^7 - Ia3$	Mn_2O_3	11.40	—	—	—	—	[58]	1954	780—1195°, [467]
β-Ca_3N_2	Hexag.	—	—	3.57	—	4.13	—	1.16	[736]	1954	Up to 780°, [467]
AlN	"	$C_{6v}^4 - C6mc$	ZnS	3.104	—	4.965	—	1.600	[58]	1954	[1187,1188,1189]
ScN	Cubic	$O_h^5 - Fm3m$	NaCl	4.44	—	—	—	—	[58]	1954	[1094]
YN	"	$O_h^5 - Fm3m$	NaCl	4.877	—	—	—	—	[607]	1957	—
LaN	"	$O_h^5 - Fm3m$	NaCl	5.30	—	—	—	—	[605]	1956	[58, 389, 1099]
CeN	"	$O_h^5 - Fm3m$	NaCl	5.011	—	—	—	—	[58]	1954	[1099]
PrN	"	$O_h^5 - Fm3m$	NaCl	5.155	—	—	—	—	[58]	1954	[1099]
NdN	"	$O_h^5 - Fm3m$	NaCl	5.15	—	—	—	—	[605]	1956	[58, 1099]
SmN	"	$O_h^5 - Fm3m$	NaCl	5.046	—	—	—	—	[605]	1956	[606, 623]
EuN	"	$O_h^5 - Fm3m$	NaCl	5.01	—	—	—	—	[605]	1956	[606]
GdN	"	$O_h^5 - Fm3m$	NaCl	4.999	—	—	—	—	[605]	1956	[58, 608]
TbN	"	$O_h^5 - Fm3m$	NaCl	4.933	—	—	—	—	[605]	1956	—
DyN	"	$O_h^5 - Fm3m$	NaCl	4.905	—	—	—	—	[605]	1956	—
HoN	"	$O_h^5 - Fm3m$	NaCl	4.874	—	—	—	—	[605]	1956	—
ErN	"	$O_h^5 - Fm3m$	NaCl	4.839	—	—	—	—	[605]	1956	—

CRYSTAL STRUCTURE (continued)

Phase	Unit cell	Space group[1*]	Structure type	a, Å	b, Å	c, Å	α	c/a	Ref.	Year	Remarks
TuN	Cubic	$O_h^5 - Fm3m$	NaCl	4.809	—	—	—	—	[605]	1956	
YbN	"	$O_h^5 - Fm3m$	NaCl	4.786	—	—	—	—	[605]	1956	
LuN	"	$O_h^5 - Fm3m$	NaCl	4.766	—	—	—	—	[605]	1956	
ThN	"	$O_h^5 - Fm3m$	NaCl	5.21	—	—	—	—	[97]	1952	
Th₃N₄	"	—	—	4.55	—	—	—	—	[737]	1952	
Th₂N₃	Hexag.	$D_{3h}^3 - C\bar{3}m$	La₂O₃	3.87	—	6.16	—	1.59	[97]	1952	[82,1139,1206]
UN	Cubic	$O_h^5 - Fm3m$	NaCl	4.890	—	—	—	—	[581]	1958	[1139]
α-U₂N₃	"	$T_h^7 - Ia3$	Mn₂O₃	10.678	—	—	—	—	[82]	1948	[1139]
β-U₂N₃	Hexag.	$D_{3h}^3 - C\bar{3}m$	La₂O₃	3.69	—	5.83	—	1.58	[622]	1956	
UN₂	Cubic	$O_h^5 - Fm3m$	CaF₂	5.31	—	—	—	—	[82]	1948	
NpN	"	$O_h^5 - Fm3m$	NaCl	4.897	—	—	—	—	[85]	1949	
PuN	"	$O_h^5 - Fm3m$	NaCl	4.905	—	—	—	—	[85]	1949	[615, 617]
PuN₂	"	$O_h^5 - Fm3m$	CaF₂	—	—	—	—	—	[615]	1950	
Ti₂N	Tetrag.	$D_{4h}^{14} - P4_2/mnm$	TiO₂	4.9452	—	3.0342	—	0.614	[1205]	1962	[87]
TiN	Cubic	$O_h^5 - Fm3m$	NaCl	4.22	—	—	—	—	[531]	1952	[31]
ZrN	"	$O_h^5 - Fm3m$	NaCl	4.567	—	—	—	—	[88]	1950	[531]
HfN	"	$O_h^5 - Fm3m$	NaCl	4.52	—	—	—	—	[36]	1953	[89]
V₃N	Hexag.	$D_6^6 - C6_32$	εFe₃N	2.835	—	4.541	—	1.60	[90]	1949	
VN	Cubic	$O_h^5 - Fm3m$	NaCl	4.126	—	—	—	—	[90]	1949	
Nb₂N	Hexag.	$C_{6v}^4 - C6/mC$	ZnS	3.054	—	5.005	—	1.64	[955]	1960	[739,1084]
NbN₀.₇₉	Tetrag.	—	—	4.395	—	4.338	—	0.99	[955]	1960	[1084]

CRYSTAL STRUCTURE (continued)

Phase	Unit cell	Space group[1*]	Structure type	a, Å	b, Å	c, Å	α	c/a	Ref.	Year	Remarks
$NbN_{0.98}$	Hexag.	—	—	2.968	—	5.535	—	1.865	[91]	1954	Below 1230°
$NbN_{0.94}$	Cubic	$O_h^5 - Fm3m$	NaCl	4.388	—	—	—	—	[955]	1960	Above 1230°
NbN	Hexag	$D_{3h}^1 - C6m2$	MoC	2.956	—	11.274	—	3.81	[91]	1954	[454,739,741,1084]
Ta_2N	Pseudohexag. (rhombic)	$D_{3d}^3 - C\bar{3}m$	Mo_2C	3.048	—	4.918	—	1.614	[92]	1954	[1028,1251]
TaN	Hexag.	$D_{6h}^1 - C6/mmm$	CoSn	5.185	—	2.908	—	0.561	[92]	1954	[1028]
Cr_2N	"	$D_{3h}^1 - C\bar{6}m2$	NiAs	4.818	—	4.490	—	0.931	[697]	1959	[93]
CrN	Cubic	$O_h^5 - Fm3m$	NaCl	4.148	—	—	—	—	[93]	1934	[697, 950]
Mo_3N	Tetrag.	—	—	4.188	—	4.024	—	0.962	[94]	1930	
Mo_2N	Cubic	$O_h^5 - Fm3m$	NaCl	4.168	—	—	—	—	[94]	1930	
MoN	Hexag.	$D_{6h}^4 - C6/mmc$	MoC	5.737	—	5.619	—	0.980	[95]	1954	
W_2N	Cubic	$O_h^5 - Fm3m$	NaCl	4.118	—	—	—	—	[593]	1959	[95, 1045]
WN	Hexag.	$D_{3h}^1 - C6m2$	MoC	2.899	—	2.832	—	0.977	[95]	1954	[589, 851,1045]
$W_{2.56}N_4$	"	D_{6h}^4	—	2.87	—	11.0	—	3.84	[1208]	1963	Superlattice
Mn_4N	Cubic	$O_h^1 - Pm3m$	Fe_4N	3.857	—	—	—	—	[744]	1955	[743, 1207]
Mn_2N (Mn_5N_2)	Hexag.	$D_d^3 - C\bar{3}m$	CdJ_2	2.834	—	4.541	—	1.62	[697]	1959	[96,1207]
Mn_3N_2	Tetrag.	$D_{4h}^{17} - I4/mmm$	γ-Mn	4.220	—	4.140	—	0.98	[743]	1951	
Mn_6N_5	"	$D_{4h}^{17} - I4/mmm$	γ-Mn	—	—	—	—	—	[1207]	1963	
$ReN_{0.43}$	Cubic	$O_h^1 - Pm3m$	Fe_4N	3.92	—	—	—	—	[600]	1951	[603]
Fe_4N	"	$O_h^1 - Pm3m$	Fe_4N	3.795	—	—	—	—	[744]	1955	[603]
FeN	Hexag.	$D_6^6 - C6s2$	FeN	2.69	—	4.36	—	1.60	[57]	1950	[602, 745]
Fe_2N	Rhombic	—	Fe_2N	5.525	4.827	4.422	—	—	[745]	1955	[602, 746]
Co_3N	Hexag.	$D_6^6 - C6s2$	Fe_3N	2.658—2.666	—	4.351—4.359	—	1.637—1.635	[58]	1954	$Co_3N_{1.05-1.11}$

48

CRYSTAL STRUCTURE (continued)

Phase	Unit cell	Space group[1*]	Structure type	a, Å	b, Å	c, Å	α	c/a	Ref.	Year	Remarks
Co_2N	Rhombic	D_{2h}^{12} — Pmzn	—	2.853	4.606	4.344	—	—	[519]	1951	[604]
Ni_3N	Hexag.	D_6^6 — C6$_3$2	Fe_3N	2.660	—	4.304	—	1.618	[747]	1958	[557, 748]
Mg_2Si	Cubic	O_h^5 — Fm3m	CaF_2	6.338	—	—	—	—	[566]	1926	[58]
Ca_2Si	Rhombic	D_{2h}^{16} — Pbnm	$PbCl_2$	9.002	2.667	4.799	—	—	[98]	1955	[1019]
$CaSi$	,,	D_{2h}^{17} — Cmcm	TaB	3.91	4.59	10.79	—	—	[768]	1950	[1019]
$CaSi_2$	Rhombohedr.	D_{3d}^5 — R3̄m	$CaSi_2$	10.4	—	—	21°30'	—	[58]	1954	Corrugated layers of Si atoms [1272]
$SrSi$	Rhombic	D_{2h}^{17} —Cmcm	CaSi	4.83	11.33	4.04	—	—	[1271]	1962	
$BaSi_2$	Hexag.	D_{6h} — C6/mmm	AlB_2	4.39	—	4.83	—	1.10	[939]	1959	
Sc_5Si_3	,,	D_{6h}^3 —C6/mcm	Mn_5Si_3	7.861	—	5.812	—	0.739	[1122]	1962	[846]
Y_5Si_3	,,	D_{6h}^3 — C6/mcm	Mn_5Si_3	8.403	—	6.303	—	—	[850]	1960	[1019]
YSi	Rhombic	D_{2h}^{17} — Cmcm	CaSi	4.25	10.526	3.826	—	—	[538]	1959	
α-YSi_2	,,	$D_{2h}^{9,8}$ — Imma	α-YSi$_2$	4.04	3.95	13.33	—	—	[749]	1959	Below 450°,[545]
β-YSi_2	Tetrag.	D_{4h}^{19} — I4/amd	α-ThSi$_2$	4.04	—	13.42	—	3.32	[749]	1959	Above 450°
β-$LaSi_2$	,,	D_{4h}^{19} — I4/amd	α-ThSi$_2$	4.31	—	13.80	—	3.20	[846]	1960	[99, 100, 545]
$CeSi$	Rhombic	D_{2h}^{16} — Pbnm	FeB	—	—	—	—	—	[167]	1959	[1019]
β-$CeSi_2$	Tetrag.	D_{4h}^{19} — I4/amd	α-ThSi$_2$	4.27	—	13.88	—	3.25	[846]	1960	[99, 101, 545]
α-$PrSi_2$	Rhombic	D_{2h}^{28} — Imma	α-YSi$_2$	4.23	4.20	13.68	—	—	[749]	1959	Below —120° 545]
β-$PrSi_2$	Tetrag.	D_{4h}^{19} — I4/amd	α-ThSi$_2$	4.20	—	13.76	—	3.21	[846]	1960	Above— 120°, [99, 545]
α-$NdSi_2$	Rhombic	D_{2h}^{28} — Imma	α-YSi$_2$	4.18	4.15	13.56	—	—	[846]	1960	[545]

49

CRYSTAL STRUCTURE (continued)

Phase	Unit cell	Space group[a]	Structure type	a, Å	b, Å	c, Å	α	c/a	Ref.	Year	Remarks
β-NdSi₂	Tetrag.	D_{4h}^{19} — I4₁/amd	α-ThSi₂	4.103	—	13.53	—	3.30	[99]	1952	
α-SmSi₂	Rhombic	D_{2h}^{28} — Imma	α-YSi₂	4.105	4.035	13.46	—	—	[846]	1960	Below 380°,[545,749]
β-SmSi₂	Tetrag.	D_{4h}^{19} — I4₁/amd	α-ThSi₂	4.041	—	13.33	—	3.30	[99]	1952	Above 380° [545]
β-EuSi₂	"	D_{4h}^{19} — I4₁/amd	α-ThSi₂	4.29	—	13.66	—	3.18	[846]	1960	
α-GdSi₂	Rhombic	D_{2h}^{28} — Imma	α-YSi₂	4.09	4.01	13.44	—	—	[846]	1960	Below 400°,[545,749]
β-GdSi₂	Tetrag.	D_{4h}^{19} — I4₁/amd	α-ThSi₂	4.10	—	13.61	—	3.32	[749]	1959	Above 400°
TbSi₂-n	Hexag.	D_{6h}^{1} — P6/mmm	AlB₂	3.847	—	4.146	—	1.078	[1116]	1963	
α-DySi₂	Rhombic	D_{2h}^{28} — Imma	α-YSi₂	4.04	3.95	13.33	—	—	[749]	1959	Below 540°,[545,846]
β-DySi₂	Tetrag.	D_{4h}^{19} — I4₁/amd	α-ThSi₂	4.03	—	13.38	—	3.32	[749]	1959	Above 540°
DySi₂-n	Hexag.	D_{6h}^{1} — P6/mmm	AlB₂	3.83	—	4.11	—	1.073	[1116]	1963	
HoSi₂-n	"	D_{6h}^{1} — P6/mmm	AlB₂	3.816	—	4.107	—	1.076	[1116]	1963	
ErSi₂-n	"	D_{6h}^{1} — P6/mmm	AlB₂	3.799	—	4.089	—	1.076	[1116]	1963	
TuSi₂-n	"	D_{6h}^{1} — P6/mmm	AlB₂	3.773	—	4.070	—	1.079	[1116]	1963	
YbSi₂-n	"	D_{6h}^{1} — P6/mmm	AlB₂	3.771	—	4.098	—	1.087	[1116]	1963	
LuSi₂-n	"	D_{6h}^{1} — P6/mmm	AlB₂	3.745	—	4.050	—	1.081	[1116]	1963	
Th₃Si₂	Tetrag	D_{4h}^{5} — P4/mbm	U₃Si₂	7.835	—	4.154	—	0.530	[102]	1956	[1019]
ThSi	Rhombic	D_{2h}^{16} — Cbnm	FeB	5.89	7.88	4.15	—	—	[102]	1956	[102, 549]
β-ThSi₂	Hexag.	D_{6h}^{1} — P6/mmm	AlB₂	4.136	—	4.126	—	0.99	[1025]	1961	[421]
α-ThSi₂	Tetrag.	D_{4h}^{19} — I4₁/amd	ThSi₂	4.135	—	14.375	—	3.48	[1025]	1961	[1097]
U₃Si	"	D_{4h}^{18} — I4/mcm	U₃Si	6.029	—	8.696	—	1.442	[104]	1952	[1097]
U₃Si₂	"	D_{4h}^{5} — P4/mbm	U₃Si₂	7.330	—	3.903	—	0.532	[104]	1952	[1097]
USi	Rhombic	D_{2h}^{16} — Pbnm	FeB	5.66	7.66	3.91	—	—	[104]	1952	[1097]
α-USi₂	Tetrag.	D_{4h}^{19} — I4₁/amd	ThSi₂	3.97	—	13.71	—	3.45	[101]	1949	[1097]

CRYSTAL STRUCTURE (continued)

Phase	Unit cell	Space group[1a]	Structure type	a, Å	b, Å	c, Å	α	c/a	Ref.	Year	Remarks
β-USi$_2$	Hexag.	D_{6h}^{1} — C6/mmm	AlB$_2$	3.85	—	4.06	—	1.05	[101]	1949	[549, 1097]
USi$_3$	Cubic	O_h^{1} — Pm3m	Cu$_3$Au	4.04	—	—	—	—	[104]	1952	[1097]
NpSi$_2$	Tetrag.	D_{4h}^{19} — I4/amd	ThSi$_2$	3.96	—	13.67	—	3.45	[101]	1949	
β-PuSi$_2$	Hexag.	D_{6h}^{1} — C6/mmm	AlB$_2$	3.884	—	4.082	—	1.051	[105]	1955	
α-PuSi$_2$	Rhombic	D_{4h}^{19} — I4/amd	ThSi$_2$	3.967	—	13.72	—	3.46	[935]	1960	
Ti$_5$Si$_3$	Hexag.	D_{6h}^{3} — C6/mcm	Mn$_5$Si$_3$	7.465	—	5.162	—	0.692	[106]	1951	Nowotny phase
TiSi	Rhombic	C_{2v}^{1} — Pm2m	TiSi	4.97	3.62	6.48	—	—	[1029]	1961	[107, 1019]
TiSi$_2$	"	D_{2h}^{24} — Fddd	TiSi$_2$	8.236	4.773	8.523	—	—	[101]	1949	[421, 826]
Zr$_2$Si	Tetrag.	D_{4h}^{18} — I4/mcm	CuAl$_2$	6.568	—	5.36	—	0.82	[751]	1954	[640, 750]
Zr$_5$Si$_3$	Hexag.	D_{6h}^{3} — C6/mcm	Mn$_5$Si$_3$	7.87	—	5.54	—	0.704	[750]	1953	Nowotny phase
Zr$_3$Si$_2$	Tetrag.	D_{4h}^{5} — P4/mbm	U$_3$Si$_2$	7.081	—	3.701	—	0.525	[826]	1955	[827]
ZrSi	Rhombic	D_{2h}^{16} — Pbnm	FeB	6.69	3.77	5.29	—	—	[108]	1954	[752, 1019]
ZrSi$_2$	"	D_{2h}^{17} — Cmcm	ZrSi$_2$	3.724	14.76	3.67	—	—	[108]	1954	[109, 110, 111, 753]
Hf$_2$Si	Tetrag.	D_{4h}^{18} — I4/mcm	CuAl$_2$	6.48	—	5.21	—	0.805	[418]	1958	Nowotny phase
Hf$_5$Si$_3$	Hexag.	D_{6h}^{3} — C6/mcm	Mn$_5$Si$_3$	7.89	—	5.55	—	0.704	[418]	1958	
HfSi	Rhombic	D_{2h}^{16} — Pbnm	MnP	6.855	3.753	5.191	—	—	[418]	1958	[111, 1019]
HfSi$_2$	"	D_{2h}^{17} — Cmcm	ZrSi$_2$	3.677	14.550	3.649	—	—	[935]	1960	[111, 1092]
V$_3$Si	Cubic	O_h^{3} — Pm3n	β-W	4.712	—	—	—	—	[113]	1939	[680]
V$_5$Si$_3$	Tetrag	D_{4h}^{18} — I4/mcm	W$_5$Si$_3$	9.43	—	5.75	—	0.504	[114]	1956	[115, 680]
V$_5$Si$_3$	Hexag.	D_{6h}^{3} — C6/mcm	Mn$_5$Si$_3$	7.121	—	4.832	—	0.679	[754]	1954	Nowotny phase, [680]
VSi$_2$	"	D_{6}^{4} — C6$_2$2	CrSi$_2$	4.562	—	6.359	—	1.394	[116]	1941	[680]

51

CRYSTAL STRUCTURE (continued)

Phase	Unit cell	Space group[1*]	Structure type	a, Å	b, Å	c, Å	α	c/a	Ref.	Year	Remarks
Nb_4Si	Hexag.	D_6^6 — $C6_32$	ε-Fe_3N	3.59	—	4.26	—	1.24	[117]	1959	[242, 680,1121]
α-Nb_5Si_3	Tetrag.	D_{2h}^{11} — $I\bar{4}2m$	Nb_5Si_3	6.56	—	11.88	—	1.815	[118]	1955	[680, 1069,1142]
β-Nb_5Si_3	"	D_{4h}^{18} — $I4/mcm$	W_5Si_3	10.00	—	5.07	—	0.51	[115]	1955	[680]
Nb_5Si_3	Hexag.	D_{6h}^3 — $C6/mcm$	Mn_5Si_3	7.505	—	5.227	—	0.699	[829]	1957	Nowotny phase [680]
$NbSi_2$	"	D_6^4 — $C6_22$	$CrSi_2$	4.795	—	6.589	—	1.374	[116]	1941	[680, 1069, 1142]
$Ta_{4.5}Si$	"	D_{6h}^4 — $C6/mmc$	—	6.105	—	4.918	—	1.61	[750]	1953	[106]
Ta_2Si	Tetrag.	D_{4h}^{18} — $I4/mcm$	$CuAl_2$	6.157	—	5.039	—	0.81	[750]	1953	
Ta_5Si_3	"	D_{4h}^{18} — $I4/mcm$	W_5Si_3	9.88	—	5.06	—	0.51	[935]	1960	[119, 830, 755]
Ta_5Si_3	"	D_{2d}^{11} — $I\bar{4}2m$	Nb_5Si_3	6.50	—	11.84	—	1.82	[118]	1955	
Ta_5Si_3	Hexag.	D_{6h}^3 — $C6/mcm$	Mn_5Si_3	7.459	—	5.215	—	0.699	[750]	1953	Nowotny phase
$TaSi_2$	"	D_6^4 — $C6_22$	$CrSi_2$	4.782	—	6.565	—	1.373	[116]	1941	
Cr_3Si	Cubic	O_h^3 — $Pm3n$	β-W	4.555	—	—	—	—	[120]	1933	
Cr_5Si_3	Tetrag.	D_{4h}^{18} — $I4/mcm$	W_5Si_3	9.170	—	4.636	—	0.506	[119]	1956	[755, 1265]
Cr_5Si_3	Hexag.	D_{6h}^3 — $C6/mcm$	Mn_5Si_3	6.98	—	4.725	—	0.68	[755]	1955	Nowotny phase
$CrSi$	Cubic	T^4 — $P2_13$	$FeSi$	4.629	—	—	—	—	[120]	1933	[756, 1019]
$CrSi_2$	Hexag.	D_6^4 — $C6_22$	$CrSi_2$	4.431	—	6.634	—	1.476	[756]	1956	[1141]
Mo_3Si	Cubic	O_h^3 — $Pm3n$	β-W	4.890	—	—	—	—	[121]	1950	
Mo_5Si_3	Tetrag.	D_{4h}^{18} — $I4/mcm$	W_5Si_3	9.642	—	4.905	—	0.509	[119]	1956	[124, 130, 755, 1265]

CRYSTAL STRUCTURE (continued)

Phase	Unit cell	Space group[1*]	Structure type	a, Å	b, Å	c, Å	α	c/a	Ref.	Year	Remarks
Mo₅Si₃	Hexag.	D_{6h}^3 — C6/mcm	Mn₅Si₃	7.256	—	4.982	—	0.689	[829]	1957	Nowotny phase
MoSi₂	Tetrag.	D_{4h}^{17} — I4/mmm	MoSi₂	3.203	—	7.887	—	2,463	[116]	1941	
W₃Si	Cubic	O_h^3 — Pm3n	β-W	4.910	—	—	—	—	[123]	1959	
W₅Si₃	Tetrag.	D_{4h}^{18} — I4/mcm	W₅Si₃	9.605	—	4.964	—	0.52	[119]	1956	[124, 755]
W₅Si₃	Hexag.	D_{6h}^3 — C6/mcm	Mn₅Si₃	7.16	—	4.83	—	0.68	[829]	1957	Nowotny phase
WSi₂	Tetrag.	D_{4h}^{17} — I4/mmm	MoSi₂	3.218	—	7.896	—	2.454	[757]	1927	
Mn₃Si	Cubic	O_h — Im3m	α-Fe	2.65	—	—	—	—	[120]	1933	
Mn₅Si₃	Hexag.	D_{6h}^3 — C6/mcm	Mn₅Si₃	6.898	—	4.802	—	0.696	[125]	1933	[1019]
MnSi	Cubic	T^4 — P2₁3	FeSi	4.557	—	—	—	—	[125]	1933	
Re₅Si₃	Tetrag.	D_{4h}^{18} — I4/mcm	W₅Si₃	9.53	—	4.81	—	0.50	[127]	1959	[126, 1019]
ReSi	Cubic	T^4 — P2₁3	FeSi	4.774	—	—	—	—	[127]	1959	[128]
ReSi₂	Tetrag.	D_{4h}^{17} — I4/mmm	MoSi₂	3.131	—	7.676	—	2.451	[127]	1959	
FeSi	Cubic	O_h^5 — Fm3m	BiF₃	5.65	—	—	—	—	[758]	1953	
Fe₅Si₃	Hexag.	D_{6h}^3 — C6/mcm	Mn₅Si₃	6.7551	—	4.7163	—	0.699	[828]	1943	
FeSi	Cubic	T^4 — P2₁3	FeSi	4.489	—	—	—	—	[58]	1954	[935, 1019]
FeSi₂	Tetrag.	D_{4h}^1 — P4/mmm	FeSi₂	2.679	—	5.120	—	1.911	[758]	1953	[935,1068,1071, 1141,1143,1232]
Co₂Si	Rhombic	D_{2h}^{16} — Pbnm	Co₂Si	7.095	4.908	3.730	—	—	[58]	1954	[1019]
CoSi	Cubic	T^4 — P2₁3	FeSi	4.447	—	—	—	—	[756]	1956	
CoSi₂	"	O_h^5 — Fm3m	CaF₂	5.365	—	—	—	—	[426]	1960	

Phase	Unit cell	Space group[1]*	Structure type	a, Å	b, Å	c, Å	α	c/a	Ref.	Year	Remarks
Ni$_3$Si	Cubic	O_h^1 — Pm3m	Cu$_3$Au	3.507	—	—	—	—	[72]	1951	
Ni$_5$Si$_2$	Rhombohedr.	—	—	6.665	—	12.274	—	—	[935]	1960	[1264]
Ni$_2$Si	Hexag.	D_{6h}^4 — C6/mmc	NiAs	3.797	—	4.898	—	1.290	[759]	1952	Above 1214°
Ni$_2$Si	Rhombic	D_{2h}^{16} — Pbnm	PbCl$_2$	7.03	4.99	3.72	—	—	[759]	1952	Below 1214°
Ni$_3$Si$_2$	”	—	—	12.24	10.81	6.93	—	—	[935]	1960	[1264]
NiSi	”	—	—	5.62	5.18	3.34	—	—	[427]	1951	[1019]
NiSi$_2$	Cubic	O_h^5 — Fm3m	CaF$_2$	5.406	—	—	—	—	[759]	1952	
AlP	”	T_d^2 — F43m	ZnS	5.42	—	—	—	—	[58]	1954	
ScP	”	O_h^5 — Fm3m	NaCl	5.312	—	—	—	—	[1128]	1963	
YP	”	O_h^5 — Fm3m	NaCl	5.661	—	—	—	—	[1128]	1963	[1129]
LaP	”	O_h^5 — Fm3m	NaCl	6.013	—	—	—	—	[58]	1954	
CeP	”	O_h^5 — Fm3m	NaCl	5.897	—	—	—	—	[58]	1954	
PrP	”	O_h^5 — Fm3m	NaCl	5.860	—	—	—	—	[58]	1954	
NdP	”	O_h^5 — Fm3m	NaCl	5.826	—	—	—	—	[58]	1954	
SmP	”	O_h^5 — Fm3m	NaCl	5.760	—	—	—	—	[623]	1956	
ThP$_{0.7}$	”	O_h^5 — Fm3m	NaCl	5.818	—	—	—	—	[142]	1938	
Th$_3$P$_4$	”	T_d^6 — I43d	Th$_3$P$_4$	8.60	—	—	—	—	[143]	1938	
UP	”	O_h^5 — Fm3m	NaCl	5.60	—	—	—	—	[58]	1954	
U$_3$P$_4$	”	T_d^6 — I43d	Th$_3$P$_4$	8.22	—	—	—	—	[58]	1954	
PuP	”	O_h^5 — Fm3m	NaCl	5.649	—	—	—	—	[144]	1957	
Ti$_3$P	Tetrag.	S_4^2 — I4	Fe$_3$P	10.00	—	5.017	—	0.502	[129]	1954	
TiP	Hexag.	D_{3h}^1 — C6m2	γ-MoC	3.487	—	11.65	—	3.34	[130]	1954	
α-ZrP	Cubic	—	—	5.27	—	—	—	—	[130]	1954	

CRYSTAL STRUCTURE (continued)

Phase	Unit cell	Space group[1*]	Structure type	a, Å	b, Å	c, Å	α	c/a	Ref.	Year	Remarks
β-ZrP	Hexag.	$D_{3h}^1 - C\bar{6}m2$	γ-MoC	3.677	—	12.52	—	3.40	[130]	1954	
HfP	"	$D_{3h}^1 - C\bar{6}m2$	γ-MoC	3.65	—	12.37	—	3.390	[1239]	1962	
V₃P	Tetrag.	$S_4^2 - I\bar{4}$	Fe₃P	—	—	—	—	—	[131]	1948	
VP	Hexag.	$D_{6h}^4 - C6/mmc$	NiAs	3.18	—	6.22	—	1.96	[130]	1954	[493]
α-NbP	Tetrag.		α-NbP	3.32	—	5.69	—	1.71	[130]	1954	
β-NbP	"	$D_{4h}^{16} - P4/ncm$	β-NbP	3.325	—	11.38	—	3.42	[130]	1954	
α-TaP	"		α-NbP	3.32	—	5.69	—	1.71	[130]	1954	
β-TaP	"	$D_{4h}^{16} - P4/ncm$	β-NbP	3.33	—	11.39	—	3.42	[130]	1954	
Cr₃P	Rhombohedr.	$S_4^2 - I\bar{4}$	Fe₃P	9.144	—	4.567	—	0.50	[130]	1954	
Cr₂P	Rhombic	$D_3^2 - P321$		—	—	—	—	—	[132]	1948	
CrP	Rhombic	$D_{2h}^{16} - Pbnm$	MnP	6.108	5.362	3.112	—	—	[130]	1954	[1277]
Mo₃P	Tetrag.	$S_4^2 - I\bar{4}$	Fe₃P	9.729	—	4.923	—	0.51	[130]	1954	[1277]
MoP	Hexag.	$D_{3h}^1 - C\bar{6}m2$	WC	3.23	—	3.20	—	0.99	[130]	1954	
MoP₂	Rhombic	$Cmc2_1$	MoP₂	3.145	11.184	4.984	—	—	[1277]	1963	[1277]
WP	"	$D_{2h}^{16} - Pbnm$	MnP	6.219	5.717	3.238	—	—	[130]	1954	
WP₂	"	$Cmc2_1$	MoP₂	3.166	11.161	4.973	—	—	[1277]	1963	
Mn₃P	Tetrag.	$S_4^2 - I\bar{4}$	Fe₃P	9.00	—	4.57	—	0.52	[133]	1957	
Mn₂P	Hexag.	$D_{3h}^3 - H\bar{6}m2$	Fe₂P	6.074	—	3.454	—	0.569	[760]	1959	
Mn₃P₂	Cubic		Mn₂O₃	—	—	—	—	—	[134]	1950	
MnP	Rhombic	$D_{2h}^{16} - Pbnm$	MnP	5.905	5.249	3.161	—	—	[58]	1954	
Re₂P	"	$C23$		5.540	2.939	10.040	—	—	[1017]	1961	
Fe₃P	Tetrag.	$S_4^2 - I\bar{4}$	Fe₃P	9.090	—	4.446	—	0.489	[135]	1928	
Fe₂P	Hexag.	$D_{3h}^3 - H\bar{6}m2$	Fe₂P	5.865	—	3.456	—	0.59	[136]	1959	[135, 760]

55

CRYSTAL STRUCTURE (continued)

Phase	Unit cell	Space group¹*	Structure type	a, Å	b, Å	c, Å	α	c/a	Ref.	Year	Remarks
FeP	Rhombic	D_{2h}^{16} — Pbnm	MnP	5.785	5.177	3.089	—	—	[137]	1930	[1054]
FeP₂	"	D_{2h}^{12} — Pnnm	FeS₂	4.975	5.657	2.725	—	—	[58]	1954	
Co₂P	"	D_{2h}^{16} — Pbnm	PbCl₂	6.608	5.644	3.512	—	—	[136]	1959	[58, 760]
CoP	"	D_{2h}^{16} — Pbnm	MnP	5.588	5.066	3.274	—	—	[58]	1954	[1054]
CoP₃	Cubic	T_h^5 — Im3	CoAs₃	7.706	—	—	—	—	[138]	1959	
Ni₃P	Tetrag.	S_4^2 — I4̄	FeP	8.646	—	4.387	—	0.507	[139]	1955	
Ni₁₂P₅	"	C_{4h}^5 — I4/m	—	8.646	—	5.070	—	0.586	[138]	1959	
Ni₂P	Hexag.	D_{3h}^2 — H6̄m2	Fe₂P	5.864	—	3.385	—	0.577	[138]	1959	
NiP₂	Monocl.	C_{2h}^6 — C2/c	—	6.366	5.615	6.071	126°13'	—	[1016]	1961	
NiP₃	Cubic	T_h^5 — Im3	CoAs₃	7.819	—	—	—	—	[138]	1959	
Sc₂S₃	Tetrag.	—	—	10,37	—	31,11	—	~3	[1226]	1961	
YS	Cubic	O_h^5 — Fm3m	NaCl	5.466	—	—	—	—	[145]	1956	
Y₅S₇	Monocl.	—	—	12.67	3.81	11,45	74°	—	[146]	1956	
δ-Y₂S₃	"	—	—	10.17	4.02	17.47	81.17°	—	[502]	1959	
YS₂	Tetrag.	—	YS₂	7.71	—	7.89	—	1.02	[146]	1956	
Y₂O₂S	Hexag.	D_{3d}^3 — C3̄m	La₂O₃	3.78	—	6.56	—	1.73	[147]	1955	
LaS	Cubic	O_h^5 — Fm3m	NaCl	5.840	—	—	—	—	[150]	1957	[148, 149]
γ-La₂S₃	"	T_d^6 — I4̄3d	Th₃P₄	8.723	—	—	—	—	[151]	1949	
La₃S₄	"	T_d^6 — I4̄3d	Th₃P₄	8.748	—	—	—	—	[159]	1956	
LaS₂	"	—	LaS₂	8.20	—	—	—	—	[1225]	1956	
La₂O₂S	Hexag.	D_{3d}^3 — C3̄m	La₂O₃	3.927	—	6.894	—	1.76	[152]	1949	[153, 154]
CeS	Cubic	O_h^3 — Fm3m	NaCl	5.763	—	—	—	—	[145]	1956	[148, 155, 156]

CRYSTAL STRUCTURE (continued)

Phase	Unit cell	Space group¹*	Structure type	a, Å	b, Å	c, Å	α	c/a	Ref.	Year	Remarks
CeS_4	Cubic	$T_d^6 - I\bar{4}3d$	Th_3P_4	8.606	—	—	—	—	[156]	1950	[151]
$\gamma\text{-}Ce_2S_3$	"	$T_d^6 - I\bar{4}3d$	Th_3P_4	8.618	—	—	—	—	[156]	1950	[151]
CeS_2	"	—	LaS_2	8.12	—	—	—	—	[157]	1956	[1225]
Ce_2O_2S	Hexag.	$D_{3d}^3 - C\bar{3}m$	La_2O_3	4.008	—	6.833	—	1.71	[158]	1951	[117]
PrS	Cubic	$O_h^5 - Fm3m$	$NaCl$	5.747	—	—	—	—	[149]	1956	[148]
PrS_4	"	$T_d^6 - I\bar{4}3d$	Th_3P_4	8.611	—	—	—	—	[159]	1956	[158]
$\gamma\text{-}Pr_2S_3$	"	$T_d^6 - I\bar{4}3d$	Th_3P_4	8.611	—	—	—	—	[159]	1956	
PrS_2	"	—	LaS_2	8.08	—	—	—	—	[1225]	1956	[153]
Pr_2O_2S	Hexag.	$D_{3d}^3 - C\bar{3}m$	La_2O_3	3.974	—	6.825	—	1.72	[154]	1958	[148]
NdS	Cubic	$O_h^5 - Fm3m$	$NaCl$	5.690	—	—	—	—	[149]	1956	
Nd_3S_4	"	$T_d^6 - I\bar{4}3d$	Th_3P_4	8.541	—	—	—	—	[159]	1956	
$\gamma\text{-}Nd_2S_3$	"	$T_d^6 - I\bar{4}3d$	Th_3P_4	8.699	—	—	—	—	[159]	1956	
NdS_2	"	—	LaS_2	8.04	—	—	—	—	[1225]	1956	[153]
Nd_2O_2S	Hexag.	$D_{3d}^3 - C\bar{3}m$	La_2O_3	3.946	—	6.790	—	1.72	[154]	1958	[153]
SmS	Cubic	$O_h^5 - Fm3m$	$NaCl$	5.863	—	—	—	—	[149]	1956	[623]
Sm_3S_4	"	$T_d^6 - I\bar{4}3d$	Th_3P_4	8.563	—	—	—	—	[159]	1956	
$\gamma\text{-}Sm_2S_3$	"	$T_d^6 - I\bar{4}3d$	Th_3P_4	8.465	—	—	—	—	[159]	1956	
Sm_2O_2S	Hexag.	$D_{3d}^3 - C\bar{3}m$	La_2O_3	3.893	—	6.717	—	1.72	[154]	1958	[153]
$SmS_{1.94}$	Tetrag.	—	YS_2	7.96	—	7.96	—	1.00	[1227]	1959	
SmS_2	Cubic	—	LaS_2	7.96	—	—	—	—	[1225]	1956	
EuS	"	$O_h^5 - Fm3m$	$NaCl$	5.970	—	—	—	—	[369]	1959	
Eu_3S_4	"	$T_d^6 - I\bar{4}3d$	Th_3P_4	8.537	—	—	—	—	[369]	1959	
$Eu_2S_{3.81}$	Tetrag.	—	—	7.86	—	8.03	—	1.02	[369]	1959	
Eu_2O_2S	Hexag.	$D_{3d}^3 - C\bar{3}m$	La_2O_3	3.872	—	6.686	—	1.72	[154]	1958	

CRYSTAL STRUCTURE (continued)

Phase	Unit cell	Space group¹*	Structure type	a, Å	b, Å	c, Å	α	c/a	Ref.	Year	Remarks
ŮdS	Cubic	$O_h^5 - Fm3m$	NaCl	?	—	—	—	—	[160]	1957	
Gd₂S₃	"	$T_d^6 - \bar{I}43d$	Th₃P₄	8.387	—	—	—	—	[160]	1957	
GdS₂	Tetrag.	—		7.850	—	7.96	—	1.01	[160]	1957	
Gd₂O₂S	Hexag.	$D_{3d}^3 - C\bar{3}m$	La₂O₃	3.850	—	6.668	—	1.73	[160]	1957	
Tb₂O₂S	"	$D_{3d}^3 - C\bar{3}m$	La₂O₃	3.825	—	6.626	—	1.73	[154]	1958	[160]
Dy₅S₇	Monocl.	—		12.84	3.81	11.61	74°	—	[160]	1957	
γ-Dy₂S₃	Cubic	$T_d^6 - \bar{I}43d$	Th₃P₄	8.292	—	—	—	—	[160]	1957	
δ-Dy₂S₃	Monocl.	—		10.170	4.02	17.57	—	—	[160]	1957	
DyS₂	Tetrag.	—		7.690	—	7.85	—	1.02	[160]	1957	
Dy₂O₂S	Hexag.	$D_{3d}^3 - C\bar{3}m$	La₂O₃	3.8029	—	6.603	—	1.74	[154]	1958	
Ho₂O₂S	"	$D_{3d}^3 - C\bar{3}m$	La₂O₃	3.782	—	6.580	—	1.74	[154]	1958	
Er₅S₇	Monocl.	—		12.63	3.77	11.47	74°	—	[160]	1957	
δ-Er₂S₃	"	—		10.07	4.00	17.33	—	—	[160]	1957	
ErS	Cubic	$O_h^5 - Fm3m$	NaCl	5.624	—	—	—	—	[502]	1959	
Er₂O₂S	Hexag	$D_{3d}^3 - C\bar{3}m$	La₂O₃	3.7601	—	6.552	—	1.74	[154]	1958	[160]
Tu₂O₂S	"	$D_{3d}^3 - C\bar{3}m$	La₂O₃	3.747	—	6.538	—	1.75	[154]	1958	
YbS	Cubic	$O_h^5 - Fm3m$	NaCl	5.673	—	—	—	—	[1012]	1961	For YbS₁.₁₃
Yb₃S₄	Rhombic	—		12.81	12.97	3.84	—	—	[1012]	1961	
Yb₂S₃	Hexag.	—		6.784	—	18.29	—	2.702	[1012]	1961	[161]
Yb₂O₂S	"	$D_{3d}^3 - C\bar{3}m$	La₂O₃	3.723	—	6.503	—	1.75	[154]	1958	
Lu₂O₂S	"	$D_{3d}^3 - C\bar{3}m$	La₂O₃	3.709	—	6.486	—	1.74	[154]	1958	
Ac₂S₃	Cubic	$T_d^6 - \bar{I}43d$	Th₃P₄	8.99	—	—	—	—	[151]	1949	

CRYSTAL STRUCTURE (continued)

Phase	Unit cell	Space group[1]*	Structure type	a, Å	b, Å	c, Å	α	c/a	Ref.	Year	Remarks
ThS	Cubic	O_h^5 — Fm3m	NaCl	5.682	—	—	—	—	[155]	1949	[1048]
Th₂S₃	Rhombic	D_{2h}^{16} — Pbnm	Sb₂S₃	10.99	10.85	3.96	—	—	[155]	1949	[1048]
Th₄S₇	Hexag.	C_{6h}^2 — C63/m	Th₇₋ₓS₁₂	11.041	—	3.983	—	0.36	[162]	1949	[1048]
ThS₂	Rhombic	D_{2h}^{16} — Pbnm	PbCl₂	4.268	7.264	8.617	—	—	[155]	1949	
ThOS	Tetrag.	D_{4h}^7 — P4/nmm	—	3.963	—	6.747	—	1.70	[155]	1949	
PaOS	"	D_{4h}^7 — P4/nmm	—	3.832	—	6.704	—	1.75	[163]	1950	
US	Cubic	O_h^5 — Fm3m	NaCl	5.484	—	—	—	—	[155]	1949	
U₂S₃	Rhombic	D_{2h}^{16} — Pbnm	Sb₂S₃	10.34	10.58	3.86	—	—	[164]	1955	[155]
U₃S₅	"	—	—	7.41	8.06	11.70	—	—	[164]	1955	
α-US₂	Tetrag.	—	—	10.26	—	6.30	—	0.61	[165]	1953	
β-USi₂	Rhombic	—	—	4.12	7.11	7.46	—	—	[165]	1953	
γ-USi₂	Hexag.	—	—	7.238	—	4.059	—	0.56	[166]	1955	
UOS	Tetrag.	D_{4h}^7 — P4/nmm	—	3.843	—	6.694	—	1.74	[155]	1949	
Np₂S₃	Rhombic	D_{2h}^{16} — Pbnm	Sb₂S₃	10.32	10.62	3.86	—	—	[155]	1949	
NpOS	Tetrag.	D_{4h}^7 — P4/nmm	—	3.824	—	6.654	—	1.74	[155]	1949	
PuS	Cubic	O_h^5 — Fm3m	NaCl	5.536	—	—	—	—	[155]	1949	
Pu₂S₃	"	T_d^6 — I43d	Th₃P₄	8.454	—	—	—	—	[151]	1949	
Pu₂O₂S	Hexag.	T_{3d}^3 — C3m	Ce₂O₃	3.926	—	6.769	—	1.72	[152]	1949	
Am₂S₃	Cubic	T_d^6 — I43d	Th₃P₄	8.445	—	—	—	—	[151]	1949	
B₄C	Rhombohedr.	D_{3d}^5 — R3m	B₄C	5.598	—	12.12	—	2.165	[168]	1954	[1033] (B₁₂C₃)

CRYSTAL STRUCTURE (continued)

Phase	Unit cell	Spacegroup[1*]	Structure type	a, Å	b, Å	c, Å	α	c/a	Ref.	Year	Remarks
B_{6.5}C	Rhombohedr.	D_{3d}^5–R$\bar{3}$m	B_4C	5.630	–	12.19	–	2.16	[168]	1954	[1003]($B_{13}C_2$)
α-SiC_I	"	C_{3v}^5–R3m	–	12.73	–	–	13°55'	–	[1193]	1948	[58, 1191]
α-SiC_II	Hexag.	C_{6v}^4–C6mc	–	3.080	–	15.098	–	4.91	[1193]	1948	[58, 1191]
α-SiC_III	"	C_{6v}^4–C6mc	–	3.080	–	10.081	–	3.52	[1193]	1948	[58, 1191]
α-SiC_IV	Rhombohedr.	C_{3v}^5–R$\bar{3}$m	–	17.718	–	–	9°58'	–	[58]	1954	[1191]
α-SiC_V	"	C_{3v}^5–R$\bar{3}$m	–	42.84	–	–	4°07'	–	[58]	1954	[1191]. Num. symb[23 33 33]_3
α-SiC_VI	"	C_{3v}^5–R$\bar{3}$m	–	27.759	–	–	6°21.5'	–	[58]	1954	[1191]
α-6iC_VII	"	C_{3v}^5–R$\bar{3}$m	–	22.735	–	–	7°46'	–	[1194]	1947	[1191]
α-6iC_VIII	Hexag.	C_{6v}^4–C6mc	–	3.079	–	20.147	–	6.55	[1192]	1952	[1191]
α-SiC_IX	Rhombohedr.	C_{3v}^5–R$\bar{3}$m	–	73.053	–	–	2°25'	–	[1194]	1947	[1191]
α-SiC_X	Hexag.	C_{3v}^1–C3m	–	3.079	–	25.184	–	8.19	[1195]	1951	[1191]
α-SiC_XI	Rhombohedr.	C_{3v}^5–R$\bar{3}$m	–	42.84	–	–	4°07'	–	[1192]	1952	[1191]. Num. symb[22 22 22 23]_3
α-SiC_XII	"	C_{3v}^5–R$\bar{3}$m	–	62.984	–	–	2°48'	–	[1192]	1952	[1191]
α-6iC_XIII	"	C_{3v}^5–R$\bar{3}$m	–	70.537	–	–	2°30'	–	[1192]	1952	[1191]
α-SiC_XIV	"	C_{3v}^5–R$\bar{3}$m	–	3.079	–	98.11	–	31.9	[1196]	1955	[1191]; parameters in hexag.aspect
α-SiC_XV	Hexag.	C_{3v}^1–C3m	–	3.079	–	40.01	–	13.1	[1196]	1955	[1191]

CRYSTAL STRUCTURE (continued)

Phase	Unit cell	Spacegroup[1*]	Structure type	a, Å	b, Å	c, Å	α	c/a	Ref.	Year	Remarks
α-SiC XVI	Rhombohedr.	C_{3v}^5–R$\bar{3}$m	–	118.363	–	–	1°30'	–	[1197]	1954	[1191]
α-SiC XVII	"	C_{3v}^5–R$\bar{3}$m	–	329.873	–	–	0°32'	–	[1197]	1954	[1191]
α-SiC XVIII	Hexag.	C_{3v}^1–C3m	–	3.079	–	3.073	–	0.999	[1322]	1953	[1191]
α-SiC XIX	"	C_{3v}^1–C3m	–	3.079	–	67.99$_6$	–	22	[1198]	1958	[1191]
α-SiC XX	"	C_{6v}^4–C6mc	–	3.076	–	5.048	–	1.64	[1199]	1959	[1191]
β-SiC	Cubic fc	T_d^2–F$\bar{4}$3m	ZnS[2*]	4.358	–	–	–	–	[58]	1954	[1191]
α-BN	Hexag.	D_{3h}^1–P6m2	Graphite	2.504	–	6.674	–	2.665	[763]	1958	[169, 762, 1260]
β-BN	Cubic	T_d^2–F$\bar{4}$3m	ZnS[3*]	3.615	–	–	–	–	[170]	1957	[171]
γ-BN	Hexag.	C_{6v}^4–C6mc	ZnS[4*]	2.55	–	4.20	–	1.645	[1282]	1963	–
α-Si₃N₄	"	C_{3v}^4–H3c	–	7.76	–	5.64	–	0.725	[548]	1959	[172, 625, 763, 854]
β-Si₃N₄	"	D_3^2–P63/m	–	7.59	–	2.92	–	0.385	[548]	1959	[765, 766]
B₃Si	Tetrag.	–	–	2.829	–	4.765	–	1.63	[173]	1955	[1052]
B₄Si	Hexag.	D_3^7–R32	B₄C	6.330	–	12.736	–	2.012	[911]	1961	[764, 833, 912, 1033]
B₆Si	Rhombic	–	–	14.392	18.267	9.88	–	–	[392]	1959	[174, 275]
B₁₂Si	"	–	AlB₁₂	–	–	–	–	–	[175]	1958	–
BP	Cubic	T_d^2–F$\bar{4}$3m	ZnS	4.538	–	–	–	–	[767]	1958	[763, 1319, 1320]
B₁₃P₂	Hexag.	D_3^7–R32	B₄C	5.984	–	11.850	–	1.980	[911]	1961	[1033, 1319]
Pyro-graphite	"	–	–	260	–	200	–	0.75	[1323]	1962	–

2* Sphalerite.
3* Zinc blende.
4* Wurtzite.

61

Phase	Pycnometer	X-Ray	Ref.	Year	Remarks
Be$_5$B	2.06—2.14	—	[563]	1960	
Be$_2$B	2.15—2.22	1.91	[540]	1955	
BeB$_2$	2.35	—	[540]	1955	[864, 865, 866]
BeB$_6$	2.33	—	[540]	1955	[864, 865, 866, 927]
BeB$_{12}$	2.36	2.42	[927]	1960	
MgB$_2$	2.48—2.67	2.63	[542]	1955	[541]
MgB$_6$	2.45—2.47	—	[542]	1955	
MgB$_{12}$	2.44	—	[542]	1955	
CaB$_6$	2.49	2.44	[3]	1956	[4, 11, 12, 217, 475, 476]
SrB$_6$	3.39	3.42	[11]	1950	[4, 15, 217]
BaB$_6$	4.26	4.25	[3]	1956	[4, 11, 12, 217, 476, 15]
AlB$_2$	3.17	3.15	[697]	1959	
AlB$_{10}$	2.72	2.537	[577]	1958	[1154]
AlB$_{12}$	2.79	—	[578]	1956	[575,577]
α-AlB$_{12}$	—	2.557	[577]	1948	
β-AlB$_{12}$	—	2.557	[577]	1948	
γ-AlB$_{12}$	—	2.56	[1156]	1960	
ScB$_2$	3.65	3.67	[694]	1960	[6]
YB$_2$	—	2.91	[8]	1956	
YB$_3$	—	3.97	[8]	1956	
YB$_4$	—	4.36	[8]	1956	
YB$_6$	3.64	3.67	[15]	1954	[10, 13, 15, 218]
YB$_{12}$	—	3.444	[1283]	1961	[1318]
LaB$_3$(?)	—	4.92	[9]	1956	
LaB$_4$	—	5.44	[14]	1958	
LaB$_6$	4.76	4.72	[9]	1956	[3, 4, 9, 13, 12, 15, 218]
CeB$_4$	—	5.74	[17]	1950	
CeB$_6$	4.69	4.80	[3]	1956	[4, 11, 12, 218, 476]
PrB$_3$(?)	—	5.20	[9]	1956	
PrB$_4$	—	5.74	[9]	1956	
PrB$_6$	4.53	4.84	[13]	1959	[4, 9, 11, 476]
NdB$_4$	—	5.83	[477]	1959	
NdB$_6$	4.86	4.94	[13]	1959	[4, 11, 12, 13, 476, 477]
SmB$_4$	—	6.14	[477]	1959	[9]
SmB$_6$	—	5.08	[18]	1959	[9, 13, 477, 848]
EuB$_6$	—	4.95	[19]	1958	[13, 14]
GdB$_3$(?)	—	6.03	[9]	1956	
GdB$_4$	—	6.47	[9]	1956	
GdB$_6$	5.0	5.30	[13]	1959	[9, 11, 13, 15, 218, 477, 846]
TbB$_4$	—	6.50	[21]	1959	[477]
TbB$_6$	—	5.36	[21]	1959	[13, 477]
DyB$_4$	—	6.74	[477]	1959	
DyB$_6$	—	5.49	[477]	1959	
DyB$_{12}$	—	4.600	[1283]	1961	[1318]
HoB$_4$	—	6.79	[477]	1959	[20]

Phase	Pycnometer	X-Ray	Ref.	Year	Remarks
HoB$_6$	—	5.52	[477]	1959	[21]
HoB$_{12}$	—	4.655	[1283]	1961	[1318]
ErB$_4$	—	6.99	[477]	1959	[21]
ErB$_6$	5.58	5.58	[13]	1959	[4, 11, 12, 218, 476]
ErB$_{12}$	—	4.706	[1283]	1961	[1318]
TuB$_4$	—	7.09	[24]	1961	
TuB$_6$	5.55	5.59	[24]	1961	
TuB$_{12}$	—	4.756	[1283]	1961	[1318]
YbB$_3$ (?)	—	6.74	[9]	1956	
YbB$_4$	—	7.31	[9]	1956	
YbB$_6$	4.37	5.57	[11]	1950	[9, 12, 13, 15, 218, 477]
LuB$_2$	—	9.76	[1209]	1963	
LuB$_4$	—	7.52	[20]	1958	
LuB$_6$	—	5.74	[16]	1958	
LuB$_{12}$	—	4.868	[1283]	1961	[1318]
ThB$_4$	7.5	8.45	[17]	1950	[478]
ThB$_6$	6.4	7.10	[9]	1956	[11, 12, 15, 217, 218, 478]
UB	—	14.20	[832]	1959	
UB$_2$	—	12.69	[28]	1952	
UB$_4$	9.32	9.37	[15]	1954	[17, 29, 217]
UB$_{12}$	5.65	5.87	[15]	1954	[29, 480, 1283]
PuB	—	14.10	[719]	1960	
PuB$_2$	—	12.81	[719]	1960	
PuB$_4$	—	9.36	[719]	1960	
PuB$_6$	—	7.31	[719]	1960	
TiB	5.09	5.26	[87]	1954	[30]
TiB$_2$	4.50	4.52	[178]	1959	[31]
ZrB	5.7	6.7	[32]	1953	
ZrB$_2$	6.17	6.09	[33]	1949	[178, 273, 36]
ZrB$_{12}$	3.70	3.63	[35]	1952	[1283]
HfB	—	11.6	[36]	1953	
HfB$_2$	10.5	11.2	[36]	1953	
V$_3$B$_2$	—	5.83	[222]	1959	
VB	—	5.44	[38]	1952	
V$_3$B$_4$	—	5.46	[39]	1956	
VB$_2$	5.28	5.10	[217]	1929	[33, 273]
Nb$_3$B$_2$	—	8.00	[222]	1959	
NbB	—	7.60	[28]	1951	
Nb$_3$B$_4$	—	7.32	[11]	1950	
NbB$_2$	6.97	7.00	[178]	1959	[33]
Ta$_2$B	—	15.16	[28]	1951	
Ta$_3$B$_2$	—	15.0	[222]	1959	
TaB	14.0	14.29	[42]	1949	
Ta$_3$B$_4$	13.50	13.60	[42]	1949	
TaB$_2$	12.38	12.62	[178]	1959	[42, 273]
Cr$_4$B	—	6.24	[43]	1953	
Cr$_2$B	6.11	6.57	[46]	1958	[43]
Cr$_5$B$_3$	6.10	6.12	[46]	1958	[43]

DENSITY, g/cm^3 (continued)

Phase	Pycnometer	X-Ray	Ref.	Year	Remarks
CrB	6.05	6.11	[46]	1958	[45]
Cr$_3$B$_4$	—	5.76	[46]	1958	
CrB$_2$	5.22	5.60	[46]	1958	
CrB$_6$	—	3.60	[956]	1961	
Mo$_2$B	9.10	9.31	[49]	1952	
MoB	8.2—8.3	8.77	[49]	1952	
MoB$_2$	—	7.78	[50]	1951	
Mo$_2$B$_5$	7.01	7.48	[49]	1952	
MoB$_4$	4.8	4.96	[1007]	1961	
W$_2$B	16.0	10.72	[48]	1947	
α-WB	15.3	16.0	[48]	1947	
W$_2$B$_5$	17.0	13.1	[48]	1947	
WB$_4$	8.3	8.40	[1007]	1961	
Mn$_4$B	6.60	6.87	[696]	1959	
Mn$_2$B	7.20	7.19	[697]	1959	[841]
MnB	6.45	6.37	[697]	1959	[841]
Mn$_3$B$_4$	6.12	5.99	[440]	1960	[697]
MnB$_2$	—	5.37	[938]	1960	
Re$_3$B	—	19.66	[849]	1960	
Re$_2$B$_5$	—	13.56	[53]	1958	
ReB$_3$	—	11.66	[723]	1960	
Fe$_2$B	—	7.32	[11]	1960	
FeB	7.15	6.71	[697]	1959	
Co$_3$B	—	8.80	[56]	1959	[54, 55]
Co$_2$B	7.9—8.33	8.05	[497]	1959	
CoB	7.25	7.32	[497]	1959	
Ni$_3$B	8.17	8.19	[836]	1960	[227]
Ni$_2$B	7.9	8.03	[836]	1960	[697]
Ni$_3$B$_2$(?)	7.5	—	[836]	1960	
Ni$_4$B$_3$	—	7.56	[551]	1959	
NiB	6.5	7.13	[836]	1960	[60]
Be$_2$C	2.26	2.44	[526]	1952	[346, 527]
Mg$_2$C$_3$	—	2.21	[346]	1956	
MgC$_2$	—	2.07	[346]	1956	
CaC$_2$-I	—	2.23	[1247]	1959	[513, 1246]
CaC$_2$-II	—	2.17	[1246]	1962	[513]
CaC$_2$-III	2.15	2.17	[1248]	1961	[513, 1263]
SrC$_2$	3.19	3.26	[513]	1930	
BaC$_2$	3.75	3.90	[513]	1930	
Al$_4$C$_3$	2.95	2.99	[467]	1952	[346]
ScC	3.60	4.12	[875]	1961	
Y$_3$C	4.73	5.4	[1112]	1961	
YC	3.50	—	[879]	1961	
Y$_2$C$_3$	3.66	—	[879]	1961	
YC$_2$	4.13	4.58	[467]	1952	[570, 837, 838, 879]
La$_2$C$_3$	—	6.079	[570]	1959	[837]
α-LaC$_2$	5.02	5.35	[513]	1930	[570, 837]
β-LaC$_2$	—	5.0	[1112]	1961	
Ce$_2$C$_3$	—	6.969	[570]	1959	[837]

Phase	Pycnometer	X-Ray	Ref.	Year	Remarks
CeC_2	5.23	5.56	[513]	1930	[570, 837, 838]
Pr_2C_3	—	6.621	[570]	1959	[837]
PrC_2	5.10	5.75	[346]	1956	[513, 570, 837, 838]
Nd_2C_3	—	6.902	[570]	1959	[837]
NdC_2	5.15	6.08	[346]	1956	[513, 570, 837, 838]
Sm_3C	—	5.36	[837]	1958	
Sm_2C_3	—	7.477	[570]	1959	[837]
SmC_2	5.86	6.50	[346]	1956	[570, 837, 838]
Gd_3C	—	5.95	[837]	1958	
Gd_2C_3	—	8.024	[570]	1959	[837]
GdC_2	5.45	6.94	[513]	1930	[570, 837]
Tb_3C	—	6.09	[837]	1958	
Tb_2C_3	—	8.335	[570]	1959	[837]
TbC_2	—	7.176	[570]	1959	[837]
Dy_3C	—	6.33	[837]	1958	
DyC_2	—	7.45	[570]	1959	[837]
Ho_3C	—	6.49	[837]	1958	
Ho_2C_3	—	8.892	[570]	1959	[837]
HoC_2	—	7.701	[570]	1959	[837]
Er_3C	—	6.68	[837]	1958	
ErC_2	—	7.954	[570]	1959	[837]
Tu_3C	—	6.83	[837]	1958	
TuC_2	—	8.175	[570]	1959	[837]
Yb_3C	—	7.08	[837]	1958	
YbC_2	—	8.097	[570]	1959	[837]
Lu_3C	—	7.28	[837]	1958	
LuC_2	—	8.728	[570]	1959	[837]
ThC	—	10.64	[531]	1952	[80]
ThC_2	9.6	10.61	[346]	1956	[513, 1278]
UC	12.97	13.63	[856]	1960	[82, 1014]
U_2C_3	12.7	12.88	[83]	1951	
UC_2	11.28	11.79	[346]	1956	[82, 84]
PuC	—	13.99	[617]	1949	
Pu_2C_3	—	12.7	[86]	1952	
TiC	4.93	4.92	[31]	1949	[178, 346, 552]
ZrC	6.73	6.66	[178]	1959	[63]
HfC	11.8—12.6	12.67	[64]	1954	
V_2C	—	5.75	[65]	1954	[1269]
VC	5.36	5.48	[346]	1956	[65, 89, 178, 552,1269]
Nb_2C	7.86	7.85	[66]	1954	
NbC	7.56	7.82	[66]	1954	[178, 552]
Ta_2C	14.8	14.9	[66]	1954	[346]
TaC	14.3	14.4	[66]	1954	[178, 467]
$Cr_{23}C_6$	6.97	6.99	[467]	1952	[76]
Cr_7C_3	6.92	6.92	[76]	1948	

Phase	Pycnometer	X-Ray	Ref.	Year	Remarks
Cr_3C_2	6.68	6.74	[567]	1952	[70, 76, 552]
Mo_2C	8.9	9.167	[910]	1960	[346, 482, 483, 1037]
MoC	8.4	8.88	[76]	1948	[483, 552]
W_2C	17.2	17.34	[72]	1951	
WC	15.5—15.7	15.77	[72]	1951	[178, 552]
$Mn_{23}C_6$	—	7.53	[73]	1944	
Mn_3C	6.89	7.53	[1]	1957	
Mn_5C_2	—	7.36	[731]	1954	
Mn_7C_3	—	7.35	[346]	1956	[74]
Fe_3C	7.67	7.69	[1]	1957	[882]
Fe_2C	—	7.16	[882]	1961	[75]
Co_3C	—	8.07	[346]	1956	[79]
Co_2C	—	7.67	[519]	1951	
Ni_3C	7.96	7.55	[77]	1931	[346]
Be_3N_2	2.72	2.71	[697]	1959	
Mg_3N_2	2.71—2.74	2.72	[57]	1950	
Ca_3N_2	2.63	2.56	[697]	1959	
AlN	3.05	2.84	[467]	1952	
ScN	4.2	4.21	[1112]	1961	[697]
YN	5.60	5.89	[607]	1957	
LaN	—	6.90	[1081]	1959	[1112]
CeN	—	8.09	[1081]	1959	[1112]
PrN	—	7.49	[1081]	1959	[1112]
NdN	—	7.70	[1081]	1959	[1112]
SmN	—	8.495	[1112]	1961	
EuN	—	8.78	[605]	1956	[1112]
GdN	—	9.10	[1081]	1959	[1112]
TbN	—	9.57	[605]	1956	[1112]
DyN	—	9.93	[605]	1956	[1112]
HoN	—	10.19	[605]	1956	[1112]
ErN	—	10.45	[605]	1956	[1112]
TuN	—	10.78	[605]	1956	[1112]
YbN	—	11.33	[605]	1956	[1112]
LuN	—	11.52	[605]	1956	[1112]
ThN	—	11.50	[605]	1956	
Th_2N_3	—	10.51	[97]	1952	
UN	—	14.32	[82]	1948	[1139]
U_2N_3	—	11.24	[82]	1948	[1139]
UN_2	—	11.73	[697]	1959	
NpN	—	14.2	[697]	1959	
PuN	—	14.23	[617]	1949	
Ti_2N	4.86	4.91	[1205]	1962	[87]
TiN	5.43	5.44	[178]	1959	[31, 1205]
ZrN	7.09	7.35	[178]	1959	[343, 484]
HfN	—	13.84	[89]	1925	
V_3N	5.967	5.987	[90]	1949	

DENSITY, g/cm^3 (continued)

Phase	Pycnometer	X-Ray	Ref.	Year	Remarks
VN	6.040	6.102	[90]	1949	
Nb$_2$N	8.33	8.31	[91]	1954	
NbN	8.40	8.41	[91]	1954	
Ta$_2$N	—	15.81	[92]	1954	
TaN	15.46	15.86	[485]	1954	[178]
Cr$_2$N	—	6.51	[93]	1934	
CrN	5.8—6.1	6.18	[467]	1952	[93]
Mo$_2$N	8.04	—	[94]	1930	
MoN	8.60	9.18	[95]	1954	
W$_2$N	12.2	—	[95]	1954	
WN	12.08—12.12	15.94	[95]	1954	
Mn$_5$N$_2$	—	6.21	[697]	1959	
Mn$_4$N	—	6.76	[744]	1955	
Mn$_2$N	6.2—6.6	6.51	[467]	1952	[697]
Mn$_3$N$_2$	—	6.6	[697]	1959	
Fe$_4$N	6.57	7.21	[467]	1952	
Fe$_2$N	6.35	7.08	[467]	1952	
Co$_3$N	7.1	—	[467]	1952	
Co$_2$N	6.4—6.5	7.66	[697]	1959	[467]
Ni$_3$N	7.66	7.91	[557]	1943	
Mg$_2$Si	—	2.0	[566]	1926	
Ca$_2$Si	—	2.12	[98]	1955	
CaSi	—	3.21	[117]	1959	
CaSi$_2$	—	2.41	[58]	1954	
SrSi	3.39	3.47	[1271]	1962	
BaSi$_2$	—	3.98	[939]	1959	
Sc$_5$Si$_3$	—	1.98	[1122]	1962	
YSi	4.33	4.53	[538]	1959	
Y$_5$Si$_3$	—	4.56	[566]	1926	
α-YSi$_2$	4.5	4.52	[846]	1960	
β-LaSi$_2$	5.05	5.10	[846]	1960	
β-CeSi$_2$	5.31	5.45	[99]	1952	
β-PrSi$_2$	5.46	5.64	[99]	1952	
β-NdSi$_2$	4.7	5.84	[846]	1960	
α-SmSi$_2$	5.14	6.13	[846]	1960	
β-EuSi$_2$	—	5.50	[846]	1960	
α-GdSi$_2$	6.4	6.43	[846]	1960	
α-DySi$_2$	5.2	6.8	[846]	1960	
Th$_3$Si$_2$	—	9.80	[102]	1956	
ThSi	—	9.03	[102]	1956	
ThSi$_2$	7.8	7.79	[846]	1960	[421]
U$_3$Si	—	18.00	[104]	1952	
U$_3$Si$_2$	—	12.20	[104]	1952	
USi	—	10.40	[104]	1952	
α-USi$_2$	9.0	8.98	[846]	1960	[101]
β-USi$_2$	9.2	9.25	[846]	1960	[101]
USi$_3$	—	8.12	[104]	1952	

Phase	Pycnometer	X-Ray	Ref.	Year	Remarks
$NpSi_2$	—	9.08	[101]	1949	
α-$PuSi_2$	—	9.12	[420]	1943	
β-$PuSi_2$	—	9.18	[105]	1955	
Ti_5Si_3	—	4.32	[106]	1951	[1020]
TiSi	—	4.21	[107]	1957	
$TiSi_2$	4.39	4.13	[101]	1949	[117]
Zr_4Si (?)	—	6.04	[428]	1954	
Zr_2Si	5.99	6.04	[117]	1959	[428, 640]
Zr_5Si_3	5.90	6.04	[117]	1959	[428, 640]
ZrSi	5.56	5.94	[428]	1954	
$ZrSi_2$	4.88	4.86	[117]	1959	[428]
Hf_2Si	—	11.69	[418]	1958	
Hf_5Si_3	—	10.84	[418]	1958	
HfSi	—	10.28	[418]	1958	[111]
$HfSi_2$	7.2	8.03	[112]	1956	
V_3Si	5.67	5.74	[116]	1941	[114, 1249]
V_5Si_3	4.80	5.13	[114]	1956	
VSi_2	4.34	4.66	[114]	1956	[117]
Nb_4Si	8.01	—	[117]	1959	[114]
α-Nb_5Si_3	6.56	7.13	[117]	1959	[114]
β-Nb_5Si_3	7.34	7.19	[117]	1959	
$NbSi_2$	5.45	5.66	[114]	1956	
$Ta_{4.5}Si$	12.7	12.86	[117]	1959	
Ta_2Si	12.4	13.54	[117]	1959	
Ta_5Si_3	11.6	13.06	[117]	1959	
$TaSi_2$	8.83	9.1	[117]	1959	
Cr_3Si	—	6.52	[117]	1959	[1020]
Cr_5Si_3	5,6	5.73	[117]	1959	[119]
CrSi	—	5.43	[117]	1959	
$CrSi_2$	—	5.00	[117]	1959	
Mo_3Si	8.4	8.97	[486]	1950	
Mo_5Si_3	7.4	8.24	[117]	1959	
$MoSi_2$	5.9—6.3	6.24	[117]	1959	
W_3Si	—	16.2	[123]	1959	
W_5Si_3	—	12.21	[117]	1959	
WSi_2	—	9.25	[117]	1959	
α-Mn_3Si	—	6.71	[417]	1956	$<$ 600°
β-Mn_3Si	—	6.60	[417]	1956	$>$ 600°
Mn_5Si_3	—	6.02	[125]	1933/34	
MnSi	—	5.85	[125]	1933/34	
Re_5Si_3	—	15.44	[127]	1959	
ReSi	—	13.04	[126]	1955	
$ReSi_2$	—	10.71	[127]	1959	
Fe_3Si	—	7.24	[758]	1953	
FeSi	—	6.16	[58]	1954	

DENSITY, g/cm^3 (continued)

Phase	Pycnometer	X-Ray	Ref.	Year	Remarks
FeSi$_2$	4.75	5.06	[117]	1959	[758]
Co$_2$Si	—	7.46	[58]	1954	
CoSi	—	6.60	[756]	1956	
CoSi$_2$	4.94	4.96	[117]	1959	[426]
Ni$_3$Si	—	7.91	[72]	1951	
Ni$_2$Si	—	7.89	[759]	1952	
Ni$_3$Si$_2$	—	6.72	[935]	1960	
NiSi	—	5.86	[427]	1951	
NiSi$_2$	—	4.84	[759]	1952	
Be$_3$P$_2$	2.055	2.058	[490]	1933	
Mg$_3$P$_2$	2.02	—	[977]	1933	[491]
Ba$_3$P$_2$	3.183	—	[491]	1900	
ScP	—	3.28	[1128]	1963	
YP	—	4.32	[1128]	1963	
LaP	—	5.22	[58]	1954	
CeP	—	5.56	[58]	1954	
PrP	---	5.72	[58]	1954	
NdP	—	5.94	[58]	1954	
SmP	—	6.34	[623]	1956	
ThP$_{0.75}$	—	6.96	[245]	1961	
Th$_3$P$_4$	8.44	8.59	[245]	1961	
UP	9.69	9.68	[245]	1961	
U$_3$P$_4$	—	9.83	[245]	1961	
PuP	—	10.18	[245]	1961	
Ti$_3$P	—	4.64	[129]	1954	
TiP	4.08	4.27	[130]	1954	[1177]
α-ZrP	5.10	5.43	[130]	1954	
β-ZrP	5.35	5.57	[130]	1954	
HfP	—	9.78	[1239]	1962	
VP	4.98	5.00	[1177]	1962	
α-NbP	5.91	6.40	[133]	1937	[130]
β-NbP	6.15	6.54	[130]	1954	[1177]
α-TaP	—	11.04	[130]	1954	
β-TaP	10.3	22.15	[130]	1954	[1177]
Cr$_3$P	6.25	6.51	[130]	1954	[132, 492]
CrP	5.25	5.49	[130]	1954	[132, 492, 1177]
CrP$_2$(?)	4.50	—	[492]	1941	
Mo$_3$P	8.60	9.14	[130]	1954	[132]
MoP	6.58	7.20	[132]	1948	[130, 1177]
MoP$_2$	5.30	5.21	[132]	1948	[492]
WP	—	11.7	[130]	1954	[1177]
WP$_2$	—	9.17	[492]	1941	
Mn$_3$P	—	6.698	[134]	1950	
Mn$_2$P	—	6.333	[134]	1950	
MnP	—	5.706	[245]	1961	[1177]
Re$_2$P	15.50	16.4	[1017]	1961	[495]
ReP	11.99	—	[495]	1935	
ReP$_2$	8.33	—	[495]	1935	

DENSITY, g/cm^3 (continued)

Phase	Pycnometer	X-Ray	Ref.	Year	Remarks
ReP$_3$	7.30	—	[495]	1935	
Fe$_3$P	—	7.21	[245]	1961	
Fe$_2$P	—	6.90	[136]	1953	
FeP	—	6.24	[245]	1961	
FeP$_2$	—	5.12	[245]	1961	
Co$_2$P	—	7.55	[136]	1953	
CoP	—	6.24	[245]	1961	
CoP$_3$	—	4.26	[245]	1961	
Ni$_3$P	—	7.66	[496]	1939	
Ni$_{12}$P$_5$	—	7.64	[245]	1961	
Ni$_2$P	—	7.33	[136]	1953	
NiP$_3$	—	4.16	[496]	1939	
Sc$_2$S$_3$	2.89	—	[497]	1930	
YS	4.51	4.92	[145]	1956	
Y$_5$S$_7$	4.10	4.18	[146]	1956	
Y$_2$S$_3$	3.82	3.87	[497]	1930	[146]
YS$_2$	4.25	4.35	[146]	1956	
Y$_2$O$_2$S	4.86	4.90	[147]	1955	
LaS	5.75	5.86	[150]	1957	[151, 148]
La$_3$S$_4$	5.34	5.44	[159]	1956	
La$_2$S$_3$	4.93	4.98	[159]	1956	[151, 497, 498]
LaS$_2$	4.83	4.90	[1225]	1956	[1112]
La$_2$O$_2$S	5.77	5.81	[153]	1956	[152]
CeS	5.88	5.94	[145]	1956	[155, 156]
Ce$_3$S$_4$	5.51	5.67	[151]	1949	[156, 159]
Ce$_2$S$_3$	5.25	5.19	[156]	1950	[151]
CeS$_2$	4.96	5.07	[157]	1956	
Ce$_2$O$_2$S	—	6.01	[158]	1951	[152]
PrS	—	6.08	[149]	1956	[304]
Pr$_3$S$_4$	5.57	5.77	[159]	1956	
Pr$_2$S$_3$	5.27	5.27	[159]	1956	[497]
PrS$_2$	4.83	4.90	[1225]	1956	[1112]
Pr$_2$O$_2$S	—	6.16	[154]	1958	[153]
NdS	6.24	6.36	[149]	1956	[148]
Nd$_3$S$_4$	5.91	6.02	[159]	1956	
Nd$_2$S$_3$	5.49	5.50	[159]	1956	[500]
NdS$_2$	5.29	5.34	[1225]	1956	[1112]
Nd$_2$O$_2$S	6.22	6.47	[153]	1956	
SmS	5.64	6.01	[149]	1956	
Sm$_3$S$_4$	6.11	6.14	[159]	1956	
Sm$_2$S$_3$	5.87	5.83	[159]	1956	[498]
SmS$_{1.94}$	5.56	5.60	[1227]	1959	[1112]
SmS$_2$	5.66	5.66	[1225]	1956	[1112]
Sm$_2$O$_2$S	6.90	6.87	[153]	1956	[154]
EuS	5.71	5.75	[369]	1959	
Eu$_3$S$_4$	6.26	6.27	[369]	1959	
Eu$_2$S$_{3.81}$	5.70	5.70	[369]	1959	
Eu$_2$O$_2$S	—	7.04	[154]	1958	

Phase	Pycnometer	X-Ray	Ref.	Year	Remarks
GdS	6.83	7.26	[1112]	1961	
γ-Gd$_2$S$_3$	6.06	6.15	[160]	1957	[497]
GdS$_2$	5.90	5.98	[160]	1957	
Gd$_2$O$_2$S	7.30	7.33	[160]	1957	
Tb$_2$O$_2$S	—	7.56	[154]	1958	
Dy$_5$S$_7$	6.14	6.35	[160]	1957	
α-Dy$_2$S$_3$	5.97	—	[160]	1957	
γ-Dy$_2$S$_3$	6.48	6.54	[160]	1957	[502]
δ-Dy$_2$S$_3$	5.75	5.91	[502]	1959	
DyS$_2$	6.48	6.11	[160]	1957	
Dy$_2$O$_2$S	7.84	7.88	[160]	1957	[154]
Ho$_2$O$_2$S	—	8.02	[154]	1958	
ErS	6.75	7.10	[502]	1959	
Er$_5$S$_7$	6.39	6.21	[160]	1957	
δ-Er$_2$S$_3$	6.07	6.21	[160]	1957	
Er$_2$O$_2$S	7.92	8.16	[160]	1957	[154]
Tu$_2$O$_2$S	—	8.59	[154]	1958	
YbS	6.68—6.75	6.74	[1012]	1961	
Yb$_3$S$_4$	6.41	6.72	[1012]	1961	
Yb$_2$S$_3$	6.02	6.04	[1012]	1961	[161]
Yb$_2$O$_2$S	8.59	8.69	[1012]	1961	[154]
Lu$_2$O$_3$S	—	8.89	[154]	1958	
Ac$_2$S$_3$	—	6.75	[151]	1949	
ThS	—	9.56	[155]	1949	
Th$_2$S$_3$	—	7.87	[155]	1949	[503]
Th$_4$S$_7$	6.91	7.65—	[162]	1949	[248]
(Th$_7$S$_{12}$)		7.885			
ThS$_2$	7.3	7.36	[155]	1949	[248, 304]
ThOS	—	8.78	[155]	1949	
PaOS	—	9.44	[579]	1950	
US	10.51	10.87	[155]	1949	[304]
U$_2$S$_3$	8.94	9.01	[164]	1955	[155]
U$_3$S$_5$	8.30	8.34	[164]	1955	
α-US$_2$	7.60	7.57	[165]	1953	
β-US$_2$	8.07	8.09	[165]	1953	[304]
γ-US$_2$	8.12	8.18	[166]	1955	
UOS	—	9.60	[155]	1949	
Np$_2$S$_3$	—	8.8	[155]	1949	
NpOS	—	9.71	[155]	1949	
PuS	—	10.60	[155]	1949	
Pu$_2$S$_3$	—	8.41	[151]	1949	
Pu$_2$O$_2$S	—	9.95	[152]	1949	
Am$_2$S$_3$	—	8.50	[151]	1949	
B$_4$C	3.50	2.52$\pm$0.01	[440]	1960	[178]
B$_{6.5}$C	—	2.44	[168]	1954	

Phase	Pycnometer	X-Ray	Ref.	Year	Remarks
SiC	3.211	3.217	[178]	1959	[117]
α-BN	2.20—2.355	2.29	[278]	1933	[169, 178]
β-BN	3.45	3.49	[170]	1957	Borazon, [1282]
γ-BN	—	3.49	[1282]	1963	Dense hexag. form
α-Si$_3$N$_4$	3.187	3.19	[625]	1957	[505, 854]
β-Si$_3$N$_4$	3.21	3.20	[505]	1957	
B$_3$Si	2.44	2.64	[690]	1960	[173, 440]
B$_4$Si	2.44	2.46	[912]	1960	[764, 833]
B$_6$Si	2.43	2.43	[690]	1960	[392, 440]
B$_{13}$P$_2$	—	2.75	[911]	1961	
BP	2.3—2.34	2.97	[440]	1960	
C	1.80—2.22	—	[944]	1959	Pyrographite, [972]

TEMPERATURE STABILITY RANGES

Phase	Temperature stability range,* °C	Ref.	Year	Remarks
Be$_2$B	1530	[573]	1960	
BeB$_2$	1780	[573]	1960	
CaB$_6$	2230	[25]	1951	
SrB$_6$	2230	[353]	1961	
BaB$_6$	2270	[25]	1951	
AlB$_2$	980	[878]	1961	At > 980° : AlB$_2$ → Al + + AlB$_{12}$
AlB$_{10}$	1660—1850	[1154]	1961	
ScB$_2$	2250	[694]	1960	
YB$_6$	2300	[10]	1958	
LaB$_4$	1800±15	[1030]	1961	At > 1800° : 2LaB$_4$ → LaB$_6$ + (La + 2B)$_{liq}$
LaB$_6$	2530	[474]	1961	
CeB$_6$	2190	[25]	1951	
NdB$_6$	2540	[220]	1960	
SmB$_6$	2540	[18]	1959	
ThB$_4$	2500	[28]	1951	
ThB$_6$	2150	[27]	1956	
UB	1050—1250	[832]	1959	
UB$_{12}$	2235	[974]	1960	[973]
Ti$_2$B	1800—2200 (±50)	[87]	1954	
TiB	680—1900 (±50)	[87]	1954	

*Unless a range of values is actually given, the temperature given is the maximum of the stability range, e.g., Be$_2$B is stable up to 1530°C.

Phase	Temperature stability range,* °C	Ref.	Year	Remarks
TiB$_2$	2980	[87]	1954	
Ti$_2$B$_5$ (?)	1700—2100 (± 50)	[47]	1953	
ZrB	800—1250 (± 50)	[47]	1953	
ZrB$_2$	3040	[36]	1953	
ZrB$_{12}$	1650—2680 (± 50)	[47]	1953	
HfB$_2$	3250	[36]	1953	
VB	2250	[224]	1958	
V$_3$B$_2$	2070	[222]	1959	
V$_3$B$_4$	2350	[222]	1959	
VB$_2$	2400 (± 50)	[222]	1959	
NbB	2260	[224]	1958	
Nb$_3$B$_2$	1950	[224]	1958	
Nb$_3$B$_4$	2700	[224]	1958	
NbB$_2$	3000	[224]	1958	
TaB	2430	[224]	1958	
Ta$_3$B$_2$	2120	[224]	1958	
Ta$_3$B$_4$	2650	[224]	1958	
TaB$_2$	3100	[224]	1958	
Cr$_4$B	1750	[224]	1958	
Cr$_2$B	1840	[224]	1958	
Cr$_5$B$_3$	1890	[224]	1958	
CrB	2050	[224]	1958	
Cr$_3$B$_4$	1900	[224]	1958	
CrB$_2$	2200 (± 50)	[224]	1958	
Mo$_2$B	2000	[49]	1952	
Mo$_3$B$_2$	1850—2070	[49]	1952	
α-MoB	2000	[49]	1952	
β-MoB	2000—2180	[49]	1952	
MoB$_2$	1600—2100	[49]	1952	
Mo$_2$B$_5$	1600	[49]	1952	
MoB$_4$	1600	[1007]	1961	
W$_2$B	2770 (± 80)	[2]	1957	
α-WB	2400 (± 50)	[47]	1953	
W$_2$B$_5$	2300	[47]	1953	
WB$_4$	1600	[1007]	1961	
Fe$_2$B	1389	[1]	1957	
α-FeB	1135	[697]	1959	
β-FeB	1135—1540	[697]	1959	

Phase	Temperature stability range,* °C	Ref.	Year	Remarks
Co_2B	1400	[697]	1959	
NiB	1020	[228]	1915	
Ni_2B	1220	[697]	1959	
$\alpha\text{-}Ni_3B_2$	1050	[697]	1959	
$\beta\text{-}Ni_3B_2$	1050—1160	[697]	1959	
Be_2C	2100	[230]	1958	
MgC_2	570	[467]	1959	
Mg_2C_3	570—610	[467]	1959	
CaC_2	2300	[177]	1950	[842]
SrC_2	1900	[177]	1950	
BaC_2	1770—2300	[515]	1932	
Al_4C_3	2100	[467]	1952	
La_2C_3	1415	[570]	1959	
$\delta\text{-}LaC_2$	1800	[570]	1959	[1005]
$\varepsilon\text{-}LaC_2$	1800—2358	[570]	1959	[1005]
$\alpha\text{-}UC_2$	1820 (± 20)	[1005]	1960	
$\beta\text{-}UC_2$	>1820 (± 20)	[1005]	1960	
TiC	3140	[552]	1947	
ZrC	3530	[177]	1950	
HfC	3890	[231]	1954	
V_2C	2165 ± 25	[1269]	1962	
VC	2650 ± 35	[1269]	1962	
NbC	3760	[230]	1958	
Ta_2C	3400	[234]	1943	
TaC	3880	[234]	1943	
$Cr_{23}C_6$	1518	[235]	1950	
Cr_7C_3	1782	[235]	1950	
Cr_3C_2	1895	[235]	1950	
Mo_2C	2400	[236]	1930	
MoC	2700	[236]	1930	
W_2C	2750	[237]	1930	
WC	2600	[237]	1930	
$\alpha\text{-}Mn_3C$	1037	[1]	1957	
$\beta\text{-}Mn_3C$	1037—1520	[1]	1957	
Fe_3C	1550	[1]	1957	
Co_3C	2300	[1]	1957	
Ni_3C	2100	[467]	1952	
Be_3N_2	2200	[467]	1952	
$\alpha\text{-}Mg_3N_2$	550	[585]	1949	
$\beta\text{-}Mg_3N_2$	550—788	[585]	1949	
$\gamma\text{-}Mg_3N_2$	>788	[585]	1949	
$\alpha\text{-}Ca_3N_2$	780—1195	[467]	1952	
$\beta\text{-}Ca_3N_2$	780	[467]	1952	
Ba_3N_2	2200	[177]	1950	

Phase	Temperature stability range,* °C	Ref.	Year	Remarks
AlN	2230	[230]	1958	
ScN	2650	[611]	1959	
YN	2670	[607]	1957	
ThN	2630	[2]	1957	
Th_3N_4	2100	[553]	1950	
UN	2650 (± 100)	[2]	1957	
TiN	3205	[230]	1958	
ZrN	2980	[592]	1956	
HfN	3310	[613]	1951	
VN	2320	[613]	1951	
NbN	2300	[230]	1958	
δ'-NbN	1230	[955]	1960	At $p_{N_2} > 760$ mm,
δ-NbN	Above 1230	[955]	1960	δ'-phase is converted
TaN	3087	[239]	1954	into ε-phase; in vacuum,
CrN	1500	[697]	1959	$\delta' \to \gamma$; [1083]
Mo_3N	600	[94]	1930	
WN_2	400	[697]	1959	
Fe_4N	350—370	[697]	1959	
Fe_3N	330—350	[697]	1959	
Fe_2N	330	[697]	1959	
Co_2N	276	[697]	1959	
Ni_3N	360	[697]	1959	
α-$ThSi_2$	1400	[117]	1959	
β-$ThSi_2$	1400—1700	[117]	1959	
U_3Si_2	1665	[117]	1959	
α-USi_2	1610—1700	[117]	1959	
β-USi_2	1610	[117]	1959	
Ti_5Si_3	2120	[241]	1951	
TiSi	1760	[241]	1951	
$TiSi_2$	1540	[241]	1951	
Zr_4Si(?)	1610	[117]	1959	
Zr_2Si	2110	[117]	1959	
Zr_5Si_3	2250	[117]	1959	
ZrSi	2095	[117]	1959	
$ZrSi_2$	1520	[117]	1959	
V_3Si	2060	[200]	1956	
V_5Si_3	2150	[200]	1956	
VSi_2	1670	[200]	1956	
Nb_4Si	1950	[147]	1955	
α-Nb_5Si_3	2000	[200]	1956	
β-Nb_5Si_3	2000—2400	[200]	1956	
$NbSi_2$	1950	[114]	1956	
$Ta_{4.5}Si$	2500	[419]	1953	
Ta_2Si	2450	[419]	1953	

Phase	Temperature stability range,* °C	Ref.	Year	Remarks
Ta_5Si_3	2500	[419]	1953	
$TaSi_2$	2200	[419]	1953	
Cr_3Si	1710	[419]	1953	
Cr_5Si_3	1560	[117]	1959	
$CrSi$	1545	[117]	1959	
$CrSi_2$	1500	[117]	1959	
Mo_3Si	2050	[243]	1952	
Mo_5Si_3	2100	[243]	1952	
$MoSi_2$	2030	[243]	1952	
W_5Si_3	2320	[423]	1952	
WSi_2	2165	[423]	1952	
Mn_3Si	1075	[117]	1959	[417]; at 600−650°
Mn_5Si_3	1285	[117]	1959	phase conversion:
$MnSi$	1275	[117]	1959	$\alpha \rightarrow \beta$-Mn_3Si [1036]
Re_5Si_3 $(Re_3Si?)$	1020	[127]	1959	
$ReSi$	1900	[127]	1959	
$ReSi_2$	1930	[127]	1959	
Fe_3Si	1300	[117]	1959	
Fe_5Si_3	1195	[117]	1959	
$FeSi$	1410	[117]	1959	
$FeSi_2$	1210	[117]	1959	
Co_3Si	1208	[117]	1959	
Co_2Si	1332	[117]	1959	
$CoSi$	1415	[117]	1959	
$CoSi_2$	1277	[117]	1959	
$CoSi_3(?)$	1306	[117]	1959	
Ni_3Si	1210	[117]	1959	
α-Ni_2Si	1214	[759]	1952	
β-Ni_2Si	Above 1214	[759]	1952	
Ni_3Si_2	830	[117]	1959	
$NiSi$	1000	[117]	1959	
α-$NiSi_2$	1025	[117]	1959	
β-$NiSi_2$	1025—1280	[117]	1959	
β-La_2S_3	650−1300 ± 100	[1228]	1960	[1112]
γ-La_2S_3	1300 to m.p.	[1228]	1960	[1112]
α-Ce_2S_3	< 1150 ± 50	[1228]	1960	[1112]
β-Ce_2S_3	1150−1450 ± 50	[1228]	1960	[1112]
γ-Ce_2S_3	1450 to m.p.	[1228]	1960	[1112]
α-Pr_2S_3	< 925 ± 75	[1228]	1960	[1112]
β-Pr_2S_3	925−1300 ± 200	[1228]	1960	[1112]
γ-Pr_2S_3	1300 to m.p.	[1228]	1960	[1112]
α-Nd_2S_3	< 1050 ± 50	[1228]	1960	[1112]
β-Nd_2S_3	1050−1300 ± 200	[1228]	1960	[1112]

Phase	Temperature stability range,* °C	Ref.	Year	Remarks
γ-Nd_2S_3	1300 to m.p.	[1228]	1960	[1112]
α-Sm_2S_3	< 1050 ± 50	[1228]	1960	[1112]
γ-Sm_2S_3	1050 to m.p.	[1228]	1960	[1112]
α-Gd_2S_3	< 950 ± 150	[1228]	1960	[1112]
γ-Gd_2S_3	950 to m.p.	[1228]	1960	[1112]
α-Dy_2S_3	< 950 ± 150	[1228]	1960	[1112]
B_4C	2200	[277]	1953	
α-SiC	2100	[117]	1959	
β-SiC	2650	[117]	1959	
α-BN	3000	[288]	1950	Under nitrogen pressure,[1282]
α-Si_3N_4	1900	[230]	1958	,, ,, ,,
β-Si_3N_4	> 1900	[230]	1958	,,
B_4Si	1370	[764]	1960	> 1370° : $B_4Si \rightarrow SiB_6 + Si$ [912, 1021]
C	3652	[944]	1959	Pyrographite

Remarks, 1. Phosphides and sulfides are stable up to the melting point under the pressure of the corresponding vapor of phosphorus and sulfur. 2. For temperature stability ranges of sulfides Me_2S_3 of the rare earth metals, see [1072].

Chapter II

THERMAL AND THERMODYNAMIC PROPERTIES

HEAT EFFECT OF FORMATION FROM THE ELEMENTS
AT CONSTANT PRESSURE (AT 298°K)

Phase	Heat effect $-\Delta H^0$, kcal/ mole	Accuracy (±), kcal/ mole	Ref.	Year	Remarks
SrB_6	~ 50.4	—	[353]	1961	
YB_6	~ 24	—	[10]	1958	
LaB_6	112.3	10	[350]	1961	
CeB_4	< 84	—	[180]	1955	
CeB_6	81	16	[3]	1956	
ThB_4	> 52	—	[180]	1955	
ThB_6	> 66	—	[180]	1955	
TiB_2	70.0	—	[859]	1959	[180, 181, 182, 184, 1050, 1170]
$Ti_2B_5(?)$	> 105	—	[180]	1955	
ZrB	> 39	—	[180]	1955	
ZrB_2	76.7	1.5	[1167]	1962	[180,182,183,1168]
ZrB_{12}	> 120	—	[180]	1955	
NbB_2	> 36	—	[180]	1955	
TaB_2	> 52	—	[180]	1955	[1167]
CrB_2	30	—	[858]	1960	[180, 183]
Mo_2B	~ 25.5	—	[180]	1955	
Mo_3B_2	~ 42.0	—	[180]	1955	
α-MoB	~ 16.3	—	[180]	1955	
MoB_2	~ 23.0	—	[180]	1955	
Mo_2B_5	~ 50.0	—	[180]	1955	
W_2B	20—28	—	[180]	1955	[1167]
α-WB	12—22	—	[180]	1955	
W_2B_5	25—45	—	[180]	1955	
MnB_2	19.0(?)	—	[697]	1959	
Be_2C	21.8	5.0	[431]	1959	
BeC_2	57.4	—	[528]	1958	

Phase	Heat effect $-\Delta H^0$, kcal/ mole	Accuracy ($\pm$), kcal/ mole	Ref.	Year	Remarks
Mg_2C_3	19.0	8.0	[928]	1958	[185, 346]
MgC_2	21.0	5.0	[185]	1954	[346]
CaC_2	14.1	2.0	[185]	1954	[346, 928]
BaC_2	12.1	4.0	[432]	1958	[346]
Al_4C_3	46.7	10	[928]	1958	[177, 187, 346]
LaC_2	42	4	[528]	1958	(2200—2550°C)
CeC_2	43.2	—	[1112]	1961	1600°C, [1324]
ThC_2	44.0	—	[185]	1954	[2, 346, 928]
UC	40.0	—	[2]	1957	
U_2C_3	72.0	—	[2]	1957	[928]
UC_2	36.0	—	[2]	1957	[1169]
TiC	43.85	0.39	[186]	1951	[177,185,946,928,1157,1176]
$TiC_{1.00}$	55.3	0.3	[1176]	1962	Synthesized at >2000°C
$TiC_{1.00}$	46.0	0.5	[1176]	1962	Synthesized at >1900°C
$TiC_{0.91}$	49.4	0.2	[1176]	1962	
$TiC_{0.79}$	43.7	0.5	[1176]	1962	
ZrC	47.7	±5.0	[1031]	1955	[185,187,189,346,1157]
HfC	50.0	—	[1157]	1961	[183,1059]
V_2C	11.5	0.5	[693]	1960	At 973—1273°K
VC	30.2	—	[214]	1957	[185, 346, 928,1157]
NbC	34.0	0.6	[1331]	1963	[189,346,928,1001,1157]
Ta_2C	17.0	—	[190]	1955	
TaC	34.3	1.0	[1331]	1963	[186, 191, 214, 346, 928, 1157]
$Cr_{23}C_6$	25.8	2.0	[188]	1944	[346, 928]
Cr_7C_3	—42.52	2.0	[188]	1944	[346, 928]
Cr_3C_2	21.01	2.0	[188]	1944	[346, 928, 1157]
MoC	2.0	—	[1157]	1961	
Mo_2C	4.2	5.0	[185]	1954	[346, 928]
W_2C	7.09	5.0	[346]	1956	
WC	9.1	2.5	[928]	1958	[185, 346, 1157]
Mn_7C_3	—5.1	—	[1040]	1957	
$Mn_{23}C_6$	—3.3	±1.7	[1039]	1961	
Mn_3C	3.6	5.0	[185]	1954	[1, 346, 928]
Fe_3C	—5.8	0.5	[185]	1954	[192, 346, 928]
Co_3C	—4.0	4.0	[928]	1958	[177, 185, 346], stable between 500—800°
Ni_3C	—9.2	2.0	[185]	1954	[928]
Be_3N_2	134.7	5.0	[185]	1954	[928]
Mg_3N_2	110.3	3.0	[185]	1954	[584, 928]
Ca_3N_2	105.0	3.0	[185]	1954	[928]
Sr_3N_2	97.4	5.0	[928]	1958	[185]
Ba_3N_2	86.9	8.0	[928]	1958	[185, 598]
AlN	76.47	0.20	[612]	1957	[185, 1060, 1170]
ScN	68.0	5.0	[185]	1954	[928,1112]
YN	71.5	5.0	[185]	1954	[928 ,1112]
LaN	71.5	4.0	[928]	1958	[185, 1102, 1112]

HEAT EFFECT OF FORMATION FROM THE ELEMENTS
AT CONSTANT PRESSURE (AT 298°K)(continued)

Phase	Heat effect $-\Delta H^0$, kcal/mole	Accuracy ($\pm$), kcal/mole	Ref.	Year	Remarks
CeN	78.0	6.0	[185]	1954	[928, 1102, 1112]
SmN	75	—	[1103]	1955	
GdN	75	—	[1104]	1955	
DyN	75	—	[1105]	1956	
YbN	75	—	[1105]	1956	
ErN	75	—	[1106]	1956	
Th_3N_4	309.5	4.0	[185]	1954	[2, 928]
UN	68.5	3.0	[185]	1954	[2, 928, 1267]
U_2N_3	256	—	[697]	1959	[2]
PuN	7.8	—	[616]	1949	
TiN	80.5	0.3	[184]	1956	[1, 975, 928]
ZrN	82.2	0.4	[194]	1956	[193, 194, 975, 928]
HfN	88.24	0.34	[387]	1953	[194]
VN	60.0	5.0	[185]	1954	
Nb_2N	61.1	1.0	[619]	1958	
NbN	56.8	1.5	[928]	1958	[185, 194]
Ta_2N	64.7	3.0	[619]	1958	
TaN	60.0	0.6	[194]	1956	[185, 928]
Cr_2N	25.2	3.0	[928]	1958	[601]
Mo_2N	16.6	0.5	[185]	1954	[702, 928]
W_2N	17.2	3.0	[185]	1954	
Mn_4N	30.3	0.4	[600]	1951	[928]
Mn_5N_2	48.2	0.6	[610]	1958	[185, 928]
Re_2N	—1	—	[600]	1951	
Fe_4N	2.6	2.0	[928]	1958	[185]
Fe_2N	0.9	2.0	[928]	1958	[185]
Co_3N	—2.0	5.0	[928]	1958	[185]
Ni_3N	—0.2	0.1	[558]	1951	[185]
Mg_2Si	18.5	1.5	[195]	1949	[928]
Ca_2Si	116	1	[1148]	1962	[195, 928]
CaSi	36.0	2.0	[195]	1949	[928]
$CaSi_2$	36.0	3.0	[195]	1949	[928]
SrSi	112.8	—	[196]	1932	
$SrSi_2$	147.4	—	[196]	1932	
BaSi	181.5	—	[196]	1932	
$BaSi_3$	399.2	---	[196]	1932	
YSi	32.2	—	[1236]	1962	[1235]
LaSi	30.0	—	[354]	1960	[1235, 1236]
$LaSi_2$	44.4	--	[354]	1960	[1235, 1236]
$CeSi_2$	50.0	10	[928]	1958	[180]
$ThSi_2$	42.0	—	[834]	1959	[197, 1274]
$PuSi_2$	211	—	[105]	1955	
Ti_5Si_3	147.0	12.0	[198]	1956	[197, 199, 200, 834,1274]

Phase	Heat effect $-\Delta H^0$, kcal/ mole	Accuracy ($\pm$), kcal/ mole	Ref.	Year	Remarks
TiSi	39.2	3.0	[198]	1956	[197, 199, 200, 834,1274]
TiSi$_2$	42.9	4.5	[198]	1956	[197, 199, 200, 834,1274]
Zr$_4$Si	52	—	[199]	1957	[200,1274]
Zr$_2$Si	35	—	[834]	1959	[928,1274]
Zr$_5$Si$_3$	147	—	[197]	1955	[199, 200, 834, 1274]
ZrSi	35.4	—	[197]	1955	[200, 199, 928, 1274]
Zr$_6$Si$_5$(?)	201	—	[197]	1955	[1274]
Zr$_3$Si$_2$	92	—	[199]	1957	[200, 1274]
ZrSi$_2$	38	—	[199]	1957	[197, 834, 928, 1274]
V$_3$Si	27.9	—	[848]	1960	[197, 199, 200, 834, 928, 1044, 1274]
V$_5$Si$_3$	96	4.5	[848]	1960	[1044, 1274]
VSi$_2$	75	—	[848]	1960	[834, 1044, 1274]
Nb$_2$Si	16.5	6	[1274]	1962	[199, 200]
Nb$_5$Si$_3$	87	30	[1274]	1962	[199, 200]
NbSi$_2$	71.6	4	[1274]	1962	[199, 200, 834]
Ta$_{4.5}$Si	32.2	—	[199]	1957	[200, 1274]
Ta$_2$Si	30.7	—	[199]	1957	[200, 1274]
Ta$_5$Si$_3$	86.7	—	[199]	1957	[197, 834, 200, 928, 1274]
TaSi$_2$	66	12	[1274]	1962	[197,199,200,834,928]
Cr$_3$Si	33.7	5.2	[176]	1959	[1274]
Cr$_5$Si$_3$	77.6	11.2	[176]	1959	[1274]
CrSi	18.4	2.2	[176]	1959	[1274]
CrSi$_2$	28.6	4.2	[176]	1959	[834, 1274]
Mo$_3$Si	23.5	4	[864]	1960	[199, 200, 1274]
Mo$_5$Si$_3$	67.8	15	[864]	1960	[199, 200, 1274]
MoSi$_2$	26.0	10	[864]	1960	[197, 199, 200, 864,1274]
W$_5$Si$_3$	46.5	—	[199]	1957	[200, 1274]
WSi$_2$	22.4	—	[199]	1957	[197, 200, 834, 928, 1274]
Mn$_3$Si	8.2	2.0	[1288]	1963	
Mn$_5$Si$_3$	6.9	2.0	[1288]	1963	
MnSi	11.6	2.0	[1288]	1963	[185, 1002]
MnSi$_{1.7}$	3.1	2.0	[1288]	1963	
Re$_3$Si	12.6	—	[199]	1957	[200]
ReSi	10.2	—	[199]	1957	[200, 1274]
ReSi$_2$	16.6	—	[199]	1957	[200, 928, 1274]
Fe$_3$Si	18.3	—	[1134]	1962	[1135, 1144]
Fe$_5$Si$_3$	38.0	3.0	[185]	1954	[928]
FeSi	19.2	1.5	[185]	1954	[928]
Co$_2$Si	27.6	2.0	[185]	1954	[928]
CoSi	24.0	2.0	[185]	1954	[928]
CoSi$_2$	24.6	2.0	[185]	1954	[928]
CoSi$_3$(?)	25.6	2.0	[185]	1954	
Ni$_3$Si	35.5	3.0	[185]	1954	[928]
Ni$_2$Si	33.5	3.0	[185]	1954	[928]
NiSi	20.5	2.0	[185]	1954	[928]

Phase	Heat effect $-\Delta H^0$, kcal/mole	Accuracy $(\pm)$, kcal/mole	Ref.	Year	Remarks
Mg_3P_2	128	—	[185]	1954	[928]
Ca_3P_2	125	1.0	[201]	1959	[185, 202, 1180]
Sr_3P_2	160.03	—	[1180]	1961	
Ba_3P_2	118	—	[1180]	1961	
AlP	~40	—	[204]	1959	
SiP	~15.0	—	[185]	1954	[928]
GeP	6.0	3.0	[185]	1954	
TiP	63.4	0.5	[203]	1959	
Fe_3P	39.0	2.0	[928]	1958	[185]
Fe_2P	38.5	3.0	[185]	1954	[928]
FeP	29.0	2.0	[185]	1954	[928]
FeP_2	42.0	3.0	[185]	1954	[928, 1180]
Co_2P	46.9	3.5	[195]	1954	[928]
CoP	34.0	4.0	[185]	1954	[928, 1180]
CoP_3	64.0	6.0	[185]	1954	[928, 1180]
Ni_3P	52.4	4.0	[185]	1954	[928]
Ni_5P_2 ($Ni_{12}P_5$)	103.5	5.0	[928]	1958	[185]
Ni_2P	44.0	3.0	[185]	1954	[928]
NiP_2	40.0	3.0	[928]	1958	[185]
NiP_3	48.0	3.0	[928]	1958	[185]
La_2S_3	282.0	10.0	[928]	1958	[177, 205]
LaS_2	145.0	7.0	[928]	1958	[177]
CeS	118.0	2.0	[928]	1958	[156, 177]
Ce_3S_4	421.5	3.5	[156]	1950	[177]
Ce_2S_3	300.5	3.0	[156]	1950	[177]
Ce_2O_2S	430	—	[158]	1950	
Nd_2S_3	265	—	[928]	1958	[177]
ThS	120	5	[177]	1950	[928]
Th_2S_3	258.6	2.5	[928]	1958	[177]
Th_4S_7(Th_7S_{12})	665	35	[177]	1950	
ThS_2	110	20	[928]	1958	[177]
B_4C	13.8	2.7	[207]	1955	[206, 928]
SiC	18.0	4	[1011]	1960	[177, 206, 346, 429 928, 999, 928, 1191]
α-BN	60.7	2.5	[208]	1954	[855, 928, 928, 1170]
Si_3N_4	179.5	8	[209]	1950	[185, 928]
$B_{13}P_2$	118	10	[928]	1958	
BP	49	—	[204]	1959	

Phase	Standard entropy S^0_{298}, cal/ deg·mole	Accuracy (±), cal/ deg·mole	Entropy of formation from its elements S^0_{298}, cal/ deg·mole	Ref.	Year	Remarks
MgB_2	8.60	0.04	—2.17	[559]	1957	
MgB	12.41	0.06	—1.36	[559]	1957	
SrB_6	6.72	—	—15.1	[280]	1953	
TiB	5.8	—	—3.1	[280]	1953	
TiB_2	7.8	—	—2.3	[280]	1953	Calculat-
ZrB_2	10.7	—	—1.8	[280]	1953	ed from
HfB_2	14.2	—	—2.1	[280]	1953	Istmen's
VB_2	7.9	—	—2.3	[280]	1953	formula
NbB_2	10.4	—	—1.3	[280]	1953	
TaB_2	13.9	—	+0.8	[280]	1953	
CrB	5.8	—	—1.5	[280]	1953	
CrB_2	9.4	—	+0.5	[280]	1953	[858]
Mo_2B_5	12.21	—	+2.2	[280]	1953	[858]
W_2B_5	28.3	—	+8.3	[280]	1953	[858]
MgC_2	14.0	2.5	+3.51	[928]	1958	
CaC_2	16.8	0.5	+4.13	[928]	1958	[177]
Al_4C_3	31.3	3.0	+0.14	[928]	1958	[177]
ThC_2	19.2	—	+1.88	[928]	1958	[177]
UC	15.9	—	+3.4	[376]	1956	[177]
UC_3	29.3	—	+3.0	[177]	1950	
UC_2	24.3	—	+2.0	[177]	1950	
TiC	5.8	0.1	—2.92	[918]	1958	[177]
ZrC	8.5	1.5	—2.16	[928]	1958	
V_2C	15.9	—	+0.49[1*]	[693]	1960	
VC	6.77	0.1	—1.59	[928]	1958	[177]
Nb_2C	7.15	—	—11.11	[388]	1960	[1055]
NbC	8.9	0.7	—1.19	[388]	1960	[928]
TaC	10.1	0.2	—1.27	[928]	1958	
$Cr_{23}C_6$	25.3	0.3	—113.50	[188]	1944	
Cr_7C_3	48.0	0.3	+4.16	[188]	1944	
Cr_3C_2	20.4	0.2	+0.64	[210]	1953	[188]
Mo_2C	19.8	3.0	+4.78	[928]	1958	[177]
WC	8.5	1.5	—0.31	[928]	1958	
Mn_3C	23.6	0.3	—0.56	[928]	1958	[187]
Fe_3C	24.2	1.2	+3.39	[928]	1958	
Co_3C	23.5	1.5	+0.60	[185]	1954	
Ni_3C	25.4	1.5	+2.68	[928]	1958	
Be_3N_2	12.0	2.0	—40.61	[185]	1954	

[1*] In the range 973—1273°.

Phase	Standard entropy S_{298}^0, cal/deg·mole	Accuracy (±), cal/deg·mole	Entropy of formation from its elements S_{298}^0, cal/deg·mole	Ref.	Year	Remarks
Mg_3N_2	21.0	2.0	—48.08	[928]	1958	
Ca_3N_2	25.4	1.5	—50.22	[185]	1954	
Sr_3N_2	29.5	2.5	—53.77	[928]	1958	
Ba_3N_2	36.4	2.0	—57.97	[918]	1958	
AlN	5.0	1.0	—21.65	[928]	1958	[1060]
ScN	9.0	2.0	—22.88	[185]	1954	
YN	11.0	2.5	—22.38	[185]	1954	
CeN	11.7	1.5	—27.78	[185]	1954	
Th_3N_4	42.7	2.5	—86.10	[928]	1958	
UN	12.5	1.5	—22.41	[928]	1958	
TiN	7.24	0.1	—22.98	[352]	1951	
ZrN	9.3	0.1	—22.88	[355]	1948	
HfN	13.1	1.5	—22.88	[387]	1953	[928]
VN	8.9	0.1	—20.98	[356]	1926	
NbN	10.5	0.2	—21.11	[352]	1951	
TaN	12.2	1.0	—20.18	[357]	1937	
CrN	8.0	1.3	—19.66	[352]	1951	
Cr_2N	18.0	2.0	—12.24	[928]	1958	
Mn_5N_2	45.9	3.0	—37.86	[185]	1954	
CaSi	15.0	2.0	+0.60	[185]	1954	[928, 1057]
$CaSi_2$	22.0	2.5	+3.05	[185]	1954	[928, 1057]
$MoSi_2$	24.5	0.2	+8.67	[211]	1958	
MnSi	14.1	2.5	+1.90	[185]	1954	
Fe_3Si	24.76	—	—	[1057]	1962	α-phase, [1144]
Fe_5Si_3	49.87	—	—	[1057]	1962	η-phase
FeSi	11.00	—	—	[1057]	1962	ε-phase, [928]
$FeSi_2$	13.26	—	—	[1057]	1962	β-lebauite
$FeSi_{2.33}$	16.58	—	—	[1057]	1962	α-lebauite
CoSi	11.5	2.0	—0.18	[185]	1954	
GeP	14.6	—	—	[1180]	1961	[185]
B_4C	6.47	0.1	—0.98	[212]	1941	
SiC	3.95	0.05	—1.91	[928]	1958	
α-BN	3.67	0.05	—20.77	[208]	1954	
Si_3N_4	23.0	2.5	—81.02	[928]	1958	
B, amorphous	1.585	—	—	[976]	1960	
B, crystalline	1.392	—	—	[976]	1960	

FREE ENERGY OF FORMATION OF REFRACTORY COMPOUNDS

Phase	Reaction	Free energy of formation ΔF, cal	Accuracy ±, kcal	Temperature interval, °K	Ref.	Year	Remarks
SrB_6	$SrO + B_4C + 2B = SrB_6 + CO$	$-78010 + 31.65T$	—	1275—2273	[353]	1961	
TiB_2	$TiO + {}^{1}/{}_{2}B_4C + {}^{1}/{}_{c}C = TiB_2 + CO$	$-58600 + 38.9\ T$	—	1273—2273	[443]	1955	[859]
ZrB_2	$ZrO + {}^{1}/{}_{2}B_4C + {}^{1}/{}_{2}C = ZrB_2 + CO$	$-57600 + 40.7\ T$	—	1273—2273	[443]	1955	[443]
CrB_2	$Cr + 2B = CrB_2$	$-30000 - 0.24\ T$	—	298—2173	[858]	1960	
W_2B_5	${}^{1}/{}_{2}WO_2 + {}^{5}/{}_{16}B_4C + {}^{11}/{}_{16}C = {}^{1}/{}_{4}W_2B_5 + CO$	$-55000 + 40.3\ T$	—	1273—2273	[443]	1955	
Be_2C	$2Be + C = Be_2C$	-7830	—	2400	[346]	1956	
Mg_2C_3	$2Mg + 3C = Mg_2C_3$	$+18000 - 0.0\ T$	—	291—922	[187]	1953	[628, 928]
CaC_2	$Ca + 2C = CaC_2$	$-13600 - 5.9\ T$	3	298—720	[187]	1953	[185,346]
Al_4C_3	$4Al + 3C = Al_4C_3$	$-44000 + 0.0\ T$	8	298—1000	[928]	1958	
CeC_2	$Ce(liq.) + 2C(sol.) = CeC_2(liq.)$	$-43200 + 35.5\ T$	—	1853—1893	[1112]	1961	[928]
ThC_2	$Th + 2C = ThC_2$	$-43800 + 4.0\ T$	10	1500—2100	[185]	1954	
UC	$U + C = UC$	-41000	—	298	[346]	1956	
UC_2	$U + 2C = UC_2$	$-42200 + 3.7\ T$	10	298—1400	[185]	1954	[187,444, 625, 665]
TiC	$Ti + C = TiC$	$-43800 + 6.7\ T$	9	1500—2000	[628]	1952	
		$-43750 + 2.41\ T$	3	298—155	[928]	1958	[187, 444, 628]
ZrC	$Zr + C = ZrC$	$-44600 + 3.16\ T$	3	1155—2000	[928]	1958	
		$-441 + 2.2\ T$	3	298—2200	[929]	1958	
V_2C	$2V + C = V_2C$	$-21550 + 26.16T$	—	973—1273	[693]	1960	[877]
VC	$V + C = VC$	$-12500 + 1.6\ T$	—	298—2000	[187]	1953	[187,638]
NbC	$Nb + C = NbC$	$-38000 + 0.54\ T$	—	—	[444]	1951	
TaC	$Ta + C = TaC$	$-9000 + 40\ T$	10	298—2200	[185]	1954	[187]

85

FREE ENERGY OF FORMATION OF REFRACTORY COMPOUNDS (continued)

Phase	Reaction	Free energy of formation ΔF, cal	Accuracy $\pm$, kcal	Temperature interval, °K	Ref.	Year	Remarks
$Cr_{23}C_6$	$^{23}/_6 Cr + C = \frac{1}{6} Cr_{23}C_6$	$-16380 - 1.54\,T$	3	973—1273	[877]	1961	[185, 187, 444, 928]
Cr_7C_3	$\frac{7}{27} Cr_{23}C_6 + C = \frac{23}{27} Cr_7C_3$	$-10050 - 2.85\,T$	3	298—1673	[187]	1953	[444, 928]
Cr_3C_2	$\frac{3}{5} Cr_7C_3 + C = \frac{7}{5} Cr_3C_2$	$-3200 - 0.20\,T$	3	298—1673	[187]	1953	[187, 628, 928]
Mo_2C	$2Mo + C = Mo_2C$	$-6700 + 0.0\,T$	8	298—1273	[185]	1954	
W_2C	$2W + C = W_2C$	$+4775 - 6.06\,T$	—	—	[444]	1951	
WC	$W + C = WC$	$-9100 + 0.4\,T$	3	298—2000	[185]	1954	[928]
Mn_3C	$3Mn + C = Mn_3C$	$-3300 - 0.26\,T$	3	298—1010	[1]	1957	
Fe_2C	$2Fe + C = Fe_2C$	$+4930 - 2.60\,T$	—	—	[187]	1953	
Fe_3C	$3Fe + C = Fe_3C$	$+6200 - 5.56\,T$	1	298—463	[928]	1958	[185]
Fe_3C	$3Fe + C = Fe_3C$	$+6380 - 5.92\,T$	1	463—1115	[187]	1953	
Fe_3C	$3Fe + C = Fe_3C$	$+2475 - 2.43\,T$	2	1115—1808	[187]	1953	
Mn_7C_3	$^{7}/_3 (\beta\text{-}Mn) + C = ^{1}/_3 Mn_7C_3$	$+5130 - 11.64\,T$	—	1075—1235	[1040]	1957	
$Mn_{23}C_6$	$^{23}/_6 Mn + C = ^{1}/_6 Mn_{23}C_6$	$-3300 - 3.35\,T$	—	973—1173	[1039]	1961	
Co_2C	$2Co + C = Co_2C$	$+2950 - 2.08\,T$	5	298—1200	[187]	1953	
Co_3C	$3Co + C = Co_3C$	$-395 + 1.006\,T \lg T - 343\,T$	1	298—1273	[185]	1958	
Ni_3C	$3Ni + C = Ni_3C$	$+8110 - 1.70\,T$	3	298—1000	[187]	1953	
Be_3N_2	$3Be + N_2 = Be_3N_2$	$-134700 + 40.6\,T$	12	298—1000	[185]	1954	[928]
Mg_3N_2	$3Mg + N_2 = Mg_3N_2$	$-115500 + 48.3\,T$	10	298—923	[185]	1954	[928]

FREE ENERGY OF FORMATION OF REFRACTORY COMPOUNDS (continued)

Phase	Reaction	Free energy of formation ΔF, cal	Accuracy $\pm$, kcal	Temperature interval, °K	Ref.	Year	Remarks
Ca_3N_2 Ba_3N_2	$3Ca + N_2 = Ca_3N_2$ $3Ba + N_2 = Ba_3N_2$	$-103200 + 50.2\,T$ $-87000 + 57.4\,T$	10 9	928—923 298—1000	[185] [928]	1954 1958	[928]
AlN	$Al + \frac{1}{2}N_2 = AlN$	$-77000 + 22.3\,T$	8	298—923	[928]	1958	
LaN	$La + \frac{1}{2}N_2 = LaN$	$-72100 + 25.0\,T$	9	298—1000	[928]	1958	
CeN	$Ce + \frac{1}{2}N_2 = CeN$	$-78000 + 25.0\,T$	17	298—1000	[928]	1958	
AlN	$Al + \frac{1}{2}N_2 = AlN$	$-63500 + 27.5\,T$	8	1800—2200	[628]	1952	[928]
Th_3N_4	$3Th + 2N_2 = Th_3N_4$	$-310400 + 89.7\,T$	20	298—2000	[185]	1954	
UN	$U + \frac{1}{2}N_2 = UN$	$-68500 + 21.5\,T$	10	298—2000	[928]	1958	[185]
TiN	$\alpha\text{-Ti} + \frac{1}{2}N_2 = TiN$	$-80250 + 22.2\,T$	2	298—1155	[928]	1958	[185, 702, 975]
TiN	$\beta\text{-Ti} + \frac{1}{2}N_2 = TiN$	$-80850 + 22.78\,T$	2	1155—1500	[928]	1958	
ZrN	$\alpha\text{-Zr} + \frac{1}{2}N_2 = ZrN$	$-87000 + 22.3\,T$	2	298—1135	[928]	1958	[185, 628, 702, 975]

FREE ENERGY OF FORMATION OF REFRACTORY COMPOUNDS (continued)

Phase	Reaction	Free energy of formation ΔF, cal	Accuracy ±, kcal	Temperature interval, °K	Ref.	Year	Remarks
ZrN	$\beta\text{-Zr} + \frac{1}{2} N_2 = ZrN$	$-87925 + 46.22\,T$	2	1135—1500	[928]	1958	
HfN	$Hf + \frac{1}{2} N_2 = HfN$	-81400	5	298	[387]	1953	
VN	$V + \frac{1}{2} N_2 = VN$	$-60000 - 1.75\,T\lg T + 26.3T$	10	298—2000	[185]	1954	[444, 628]
TaN	$2Ta + N_2 = 2TaN$	$-117800 + 13.8\,T\lg T + 79.7\,T$	7	298—2240	[928]	1958	[185]
Cr$_2$N	$4Cr + N_2 = 2Cr_2N$	$-51900 - 11.5\,T\lg T + 66.0\,T$	5	298—1400	[928]	1958	[185, 601]
CrN	$2Cr_2N + N_2 = 4CrN$	$-64000 - 11.5\,T\lg T + 83.2\,T$	5	298—1400	[928]	1958	[185, 628]
Mo$_2$N	$4Mo + N_2 = 2Mo_2N$	$-34400 - 9.2\,T\lg T + 57.9\,T$	8	298—1300	[928]	1958	[185]
Mo$_2$N	$4Mo + N_2 = 2Mo_2N$	$-33200 + 42.0\,T$	6	1500—2000	[628]	1952	
Fe$_4$N	$4Fe + \frac{1}{2} N_2 = Fe_4N$	$-200 + 11.62\,T\lg T - 24.85\,T$	—	298—950	[185]	1954	
Re$_3$Si	$3Re + Si = Re_3Si$	$-24600 - 5.0\,T$	3	1750—1970	[928]	1958	
ReSi	$Re + Si = ReSi$	$-30000 - 0.5\,T$	3	1750—1970	[928]	1958	
ReSi$_2$	$Re + 2Si = ReSi_2$	$-62100 + 1.7\,T$	3	1750—1970	[928]	1958	[628, 928]
SiC	$Si + C = SiC$	$-12770 + 1.66\,T$	3	298—1683	[187]	1953	
		$-24010 + 8.33\,T$	4	1683—2000	[187]	1953	
BN	$B + \frac{1}{2} N_2 = BN$	$-156900 + 53.5\,T$	2	2500—3000	[893]	1961	
Si$_3$N$_4$	$3Si + 2N_2 = Si_3N_4$	$-188800 + 98.5\,T$	9	1800—2000	[628]	1952	[441]

HEAT CAPACITY

Phase	$C_p = a + bT + cT^2$, cal/mole·deg	Accuracy (±), %	Temperature interval, °C	Ref.	Year	Heat capacity at 20° C_p, cal/mole·deg	Ref.	Year	Remarks
MgB₂	—	—	—	—	—	11.43	[559]	1957	
MgB₄	—	—	—	—	—	16.81	[559]	1957	
LaB₆	$21.73 + 20.4 \cdot 10^{-3}\,T$	—	20—1210	[350]	1961	27.85	[350]	1961	
TiB₂	$7.219 + 1.147 \cdot 10^{-2}\,T$	—	18—800	[252]	1960	10.57	[252]	1960	
ZrB₂	$11.78 + 9.986 \cdot 10^{-3}\,T - 4{,}028 \cdot 10^{-6} \cdot T^2$	—	18—800	[276]	1958	12.00	[276]	1958	
CrB₂	$7.808 + 1.517 \cdot 10^{-2}\,T$	—	18—800	[253]	1959	12.24	[253]	1959	
Be₂C	$10.2 + 5.1 \cdot 10^{-3}\,T$	1.5	20—1100	[497]	1930	11.69	[497]	1930	[346, 928]
α-CaC₂	$16.40 + 2.84 \cdot 10^{-3}\,T - 2{,}07 \cdot 10^5\,T^{-2}$	1	25—447	[185]	1954	14.66	[185]	1954	
β-CaC₂	$15.40 + 2.00 \cdot 10^{-3}\,T$		447—1000	[185]	1954	—	—	—	[928]
Al₄C₃	$24.08 + 31.6 \cdot 10^{-3}\,T$	4	25—327	[444]	1951	33.34	[444]	1951	[185]
UC	—					0.048 (at 125°)	[1014]	1959	Porosity 25%
UC	—					0.053 (at 250°)	[1014]	1959	Porosity 25%
TiC	$11.83 + 0.8 \cdot 10^{-3}\,T - 3.58 \cdot 10^5\,T^{-2}$	1	25—1500	[254]	1946	8.04	[945]	1952	[254, 928]
ZrC	$13.1 + 0.53 \cdot 10^{-3}\,T - 26.4 \cdot 10^5\,T^{-2}$	—	25—3000	×	1960	14.6	×	1960	V. E. Levinskii (calculated)
VC	$9.18 + 3.30 \cdot 10^{-3}\,T - 1.95 \cdot 10^5\,T^{-2}$	1.5	25—1350	[255]	1949	7.97	[945]	1952	[255, 928]
Nb₂C	$7.94 + 1.50 \cdot 10^{-3}\,T - 1.025 \cdot 10^5\,T^{-2}$		25—1530	[388]	1960	7.25	[388]	1960	[867]
NbC	$10.79 + 1.726 \cdot 10^{-3}\,T - 2.15 \cdot 10^5\,T^{-2}$		25—1530	[388]	1960	8.92	[388]	1960	[867, 1001]

HEAT CAPACITY (continued)

Phase	$C_p = a + bT + cT^2$, cal/mole·deg	Accuracy (±), %	Temperature interval, °C	Ref.	Year	Heat capacity at 20° C_p, cal/mole·deg	Ref.	Year	Remarks
TaC	$7.28 + 1.65 \cdot 10^{-3} T$	4	25—1800	[352]	1961	8.79	[945]	1952	[185, 351. 352]
$Cr_{23}C_6$	$29.35 + 7.40 \cdot 10^{-3} T - 5.018 \cdot 10^5 T^{-2}$	1.5	25—1450	[185]	1954	25.89	[185]	1954	[928]
Cr_7C_3	$57.00 + 14.38 \cdot 10^{-3} T - 10.104 \cdot 10^5 T^{-2}$	1	25—1200	[185]	1954	49.92	[185]	1954	[928]
Cr_3C_2	$30.03 + 5.58 \cdot 10^{-3} T - 7.396 \cdot 10^5 T^{-2}$	1	25—1200	[185]	1954	23.38	[185]	1954	[265, 868]
WC	$7.98 + 2.17 \cdot 10^{-3} T$	6	25—1700	[352]	1951	8.53	[352]	1951	
Mn_3C	$23.45 + 7.55 \cdot 10^{-3} T - 2.42 \cdot 10^5 T^{-2}$	2	25—1000	[185]	1954	23.3	[185]	1954	[928]
α-Fe_3C	$19.64 + 20.00 \cdot 10^{-3} T$	3	0—190	[928]	1958	25.50	[928]	1958	
β-Fe_3C	$25.62 + 3.00 \cdot 10^{-3} T$	3	190—753	[928]	1958	—	—	—	
Be_3N_2	$7.32 + 30.8 \cdot 10^{-3} T$	7	0—527	[928]	1958	16.6	[928]	1958	
α-Mg_3N_2	$20.77 + 11.20 \cdot 10^{-3} T$	3	25—550	[928]	1958	24.05	[928]	1958	
β-Mg_3N_2	$20.07 + 10.66 \cdot 10^{-3} T$	3	550—788	[928]	1953	—	—	—	
γ-Mg_3N_2	28.50	3	788—1027	[928]	1958	—	—	—	
Ca_3N_2	$20.44 + 22.0 \cdot 10^{-3} T$	7	20—527	[698]	1949	26.89	[698]	1949	[698]
AlN	$5.47 + 7.80 \cdot 10^{-3} T$	5	20—627	[697]	1959	7.75	[697]	1959	[698]
LaN	—	—	—	—	—	11.0	[1112]	1961	
CeN	—	—	—	—	—	11.1	[1112]	1961	
Th_3N_4	$27.00 + 39.45 \cdot 10^{-3} T - 10.45 \cdot 10^{-6} T^2$	5	0—550	[185]	1954	39.44	[185]	1954	[185, 928, 1060]
TiN	$11.91 + 0.94 \cdot 10^{-3} T - 2.96 \cdot 10^5 T^{-2}$	1	25—1550	[254]	1946	8.86	[945]	1952	[928]
ZrN	$11.10 + 1.68 \cdot 10^{-3} T - 1.72 \cdot 10^5 T^{-2}$	2	25—1550	[185]	1954	10.88	[185]	1954	[254]
VN	$10.94 + 2.10 \cdot 10^{-3} T - 2.21 \cdot 10^5 T^{-2}$	1.5	25—1350	[255]	1949	9.08	[945]	1952	[255]
NbN	$8.69 + 5.40 \cdot 10^{-3} T$	5	0—627	[928]	1958	10.41	[185]	1954	[928]
TaN	$12.50 + 2.05 \cdot 10^{-3} T - 3.90 \cdot 10^5 T^{-2}$	4	25—527	[185]	1954	9.7	[945]	1952	

HEAT CAPACITY (continued)

Phase	$C_p = a + bT + cT^2$, cal/mole·deg	Accuracy (±), %	Temperature interval, °C	Ref.	Year	Heat capacity at 20° C_p, cal/mole·deg	Ref.	Year	Remarks
Cr$_2$N	$15.24 + 6.8 \cdot 10^{-3}\,T$	5	0—527	[185]	1954	17.23	[185]	1954	[698]
CrN	$9.84 + 3.9 \cdot 10^{-3}\,T$	7	0—527	[185]	1954	10.98	[185]	1954	[698, 928]
Mo$_2$N	$8.20 + 26.25 \cdot 10^{-3}\,T - 12.85 \cdot 10^{-6}\,T^2$	5	0—527	[185]	1954	16.99	[185]	1954	[928]
Mn$_4$N	$22.3 + 27.2 \cdot 10^{-3}\,T$	6	0—527	[185]	1954	30.26	[185]	1954	[928]
Mn$_5$N$_2$	$32.5 + 35.0 \cdot 10^{-3}\,T$	6	0—527	[185]	1954	43.0	[185]	1954	
Mn$_3$N$_2$	$22.5 + 22.5 \cdot 10^{-3}\,T$	5	0—527	[185]	1954	29.10	[185]	1954	
Fe$_4$N	$26.84 + 8.16 \cdot 10^{-3}\,T$	4	0—727	[185]	1954	31.55	[185]	1954	
Fe$_2$N	$14.91 + 6.09 \cdot 10^{-3}\,T$	4	0—727	[185]	1854	16.8	[185]	1954	
Ti$_5$Si$_3$	$58.22 + 5.742 \cdot 10^{-3}\,T - 2.646 \cdot 10^6\,T^{-2}$	3	25—900	[213]	1959	33.44	[213]	1959	[1175]
TiSi	$15.43 - 0.8832 \cdot 10^6\,T^{-2}$	1.8	25—1125	[213]	1959	6.63	[213]	1959	[1175]
TiSi$_2$	$14.94 + 8.32 \cdot 10^{-3}\,T - 0.455 \cdot 10^6\,T^{-2}$	1.3	25—900	[213]	1959	12.88	[213]	1959	[1175]
V$_3$Si	$6.634 + 0.697 \cdot 10^{-3}\,T - 125805\,T^{-2}$	2.2	25—1037	[1044]	1962	5.38	[1044]	1962	
V$_5$Si$_3$	$7.203 + 0.0268 \cdot 10^{-3}\,T - 246068\,T^{-2}$	1.9	25—1017	[1044]	1962	4.34	[1044]	1962	
VSi$_2$	$7.757 - 0.723 \cdot 10^{-3}\,T - 497463\,T^{-2}$	1.2	25—1017	[1044]	1962	2.2	[1044]	1962	
Cr$_3$Si	$22.62 + 8.80 \cdot 10^{-3}\,T - 5.31 \cdot 10^5\,T^{-2}$	1.5	25—600	[176]	1959	39.36	[176]	1959	
Cr$_5$Si$_3$	$59.144 + 6.42 \cdot 10^{-3}\,T - 2325360\,T^{-2}$	1.5	25—600	[176]	1959	35.26	[176]	1959	
CrSi	$12.15 + 3.420 \cdot 10^{-3}\,T - 384660\,T^{-2}$	1.5	25—600	[176]	1959	9.35	[176]	1959	
CrSi$_2$	$14.30 + 10.53 \cdot 10^{-3}\,T - 417630\,T^{-2}$	1.5	25—600	[117]	1959	12.64	[176]	1959	
MoSi$_2$	—					13.98	[176]	1959	
Fe$_3$Si	$17.04 + 20.87 \cdot 10^{-3}\,T + 35.57 \cdot 10^2\,T^{-2}$	—	0—527	[1134]	1962	23.5	[1134]	1962	α-phase, [1057, 1135, 1144]
Fe$_3$Si	$11.27 + 24.69 \cdot 10^{-3}\,T + 93.09 \cdot 10^4\,T^{-2}$	—	577—1251	[1134]	1962	—	—	—	
Fe$_3$Si	34.6	—	1263—1527	[1134]	1962	—	—	—	
Fe$_5$Si$_3$	$42.13 + 21.24 \cdot 10^{-3}\,T - 96.91 \cdot 10^3\,T^{-2}$	—	0—1087	[1134]	1962	47.7	[1134]	1962	η-phase, [1057, 1135]

HEAT CAPACITY (continued)

Phase	$C_p = a + bT + cT^2$, cal/mole·deg	Accuracy (±), %	Temperature interval, °C	Ref.	Year	Heat capacity at 20° C_p, cal/mole·deg	Ref.	Year	Remarks
FeSi	$10.66 + 36.60 \cdot 10^{-4} T - 94.94 \cdot 10^2 T^{-2}$	—	0–1387	[1134]	1962	11.6	[1134]	1962	ε-phase, [1057,1135]
FeSi	19.5	—	1402–1652	[1134]	1962	—	—	—	
FeSi$_2$	$14.56 + 4.108 \cdot 10^{-3} T$	—	0–965	[1134]	1962	15.8	[1134]	1962	β-lebauite, [1057,1135]
FeSi$_{2.33}$	$8.932 + 15.88 \cdot 10^{-3} T + 17.24 \cdot 10^4 T^{-2}$	—	0–1207	[1134]	1962	17.6	[1134]	1962	α-lebauite, [1057,1135]
FeSi$_{2.33}$	30.08	—	1212–1450	[1134]	1962	—	—	—	
Mn$_3$Si	$3.218 + 7.194 \cdot 10^{-3} T + 58630 T^{-2}$	1.5	315–968	[1288]	1963	—	[1288]	1963	MnSi$_{0.3223}$
Mn$_3$Si	$11.82 - 0.5502 \cdot 10^{-3} T - 847200 T^{-2}$	1.6	470–1059	[1288]	1963	—	[1288]	1963	MnSi$_{0.5458}$
Mn$_5$Si$_3$	$10.00 + 0.32 \cdot 10^{-3} T$	1.6	470–1059	[1288]	1963	—	[1288]	1963	
Mn$_3$Si$_5$	$3.3407 + 3.0952 \cdot 10^{-3} T - 390340 T^{-2}$	2.8	337–1129	[1288]	1963	—	[1288]	1963	MnSi$_{1.7}$
MnSi	$11.288 - 1.426 \cdot 10^{-3} T - 98200 T^{-2}$	1.1	519–1145	[1288]	1963	—	[1288]	1963	
MnSi$_2$	$3.663 + 2.752 \cdot 10^{-3} T - 324300 T^{-2}$	2.8	337–1129	[1288]	1963	—	[1288]	1963	MnSi$_{2.235}$
B$_4$C	$22.99 + 5.40 \cdot 10^{-3} T - 10.72 \cdot 10^5 T^{-2}$	2	25–1100	[185]	1954	12.55	[945]	1952	[271, 928]
SiC	$8.89 + 2.91 \cdot 10^{-3} T - 2.84 \cdot 10^5 T^{-2}$	3	0–1350	[185]	1954	6.44	[185]	1954	[257, 346, 698, 1051]
α-BN	$1.82 + 3.62 \cdot 10^{-3} T$	3	0–900	[928]	1958	2.88	[928]	1958	[209, 346, 440]
Si$_3$N$_4$	$16.83 + 23.6 \cdot 10^{-3} T$	3	0–727	[928]	1958	23.74	[928]	1958	
C	—	—	—	—	—	2.79	[944]	1959	Pyrographite
B	$1.54 + 4.40 \cdot 10^{-3} T$	5	0–927	[928]	1958	2.83	[928]	1958	

HEAT OF DISSOCIATION

Phase	Heat of dissociation at 25°C, kcal/mole	Ref.	Year	Remarks
LaB_6*	169	[25]	1951	
TiB_2	435	[1292]	1962	[1291,1170]
ZrB_2	416 ± 24	[1168]	1962	
BeC_2	922 ± 2.5	[431]	1959	
BaC_2	53.3 ± 0.4	[432]	1958	[467]
ThC_2	184.7 ± 1.5	[1308]	1962	[1309,1262]
ThC_2*	212.8 ± 1.5	[1308]	1962	[1309,1262]
PaC_2	179.2 ± 8.7	[1309]	1962	
UC*	224.8	[1310]	1962	
UC_2	140.15 ± 1.02	[1169]	1962	[1309,1311,1312]
PuC_2	81.98	[1313]	1962	
TiC	144.76	[946]	1961	
HfC	785.6 ± 3.5		1962	V. V. Fesenko, A. S. Bolgar
VC	144.6 ± 2	[1301]	1960	
NbC	797.9 ± 2.2		1961	V. V. Fesenko, A. S. Bolgar,[430,1291, 1303]
TiN	191.2	[975]	1955	
ZrN	79.5	[975]	1955	
$TaSi_{0.22}$	146.4 ± 2.2	[438]	1957	
Ta_2Si	137.2 ± 1.4	[438]	1957	
$TaSi_{0.6}$	125.6 ± 5.5	[438]	1957	
$TaSi_2$	117.2 ± 1.2	[438]	1957	
Mo_3Si	131.9 ± 1.2	[864]	1960	
Mo_5Si_3	393.3 ± 2.1	[864]	1960	
$MoSi_2$	234.4 ± 1.2	[864]	1960	
SiC	113 ± 3.0	[1011]	1961	
B*	141	[180]	1955	[440,536]

* Heat of sublimation, kcal/mole.

HEAT OF FUSION

Phase	Heat of fusion L_{fus}, kcal/mole	Ref.	Year	Remarks
Fe_3C	18.49	[639]	1948	
Fe_3Si	13.7	[1134]	1962	[1135, 1144]
FeSi	16.8	[1134]	1962	[639, 1135]
$FeSi_{2.33}$	25.1	[1134]	1962	[1135]

Phase	Melting point		Ref.	Year	Remarks
	°C	°K			
Be_5B	1160	1433	[863]	1961	
Be_2B	~1520	~1800	[863]	1961	[573]
BeB_2	~1700	~2000	[863]	1961	Decomposes, [573]
BeB_4	>2000	>2273	[863]	1961	
BeB_6	2300	2573	[865]	1961	[863]
BeB_9	>2000	>2273	[863]	1961	
CaB_6	2230	2503	[539]	1957	[25]
SrB_6	2235	2508	[25]	1951	
BaB_6	2230	2503	[25]	1951	
AlB_{10}	2100	2373	[577]	1948	Decomposes, [1154, 1155]
AlB_{12}	2150	2423	[1154]	1961	Decomposes, [1155]
ScB_2	2250	2523	[694]	1960	
YB_2	2100	2373	[1110]	1959	[1111, 1112]
YB_4	2800	3073	[1110]	1959	[1111, 1112]
YB_6	2300	2573	[10]	1958	[1110]
LaB_4	1800±15	2073	[1030]	1961	
LaB_6	2530	2803	[891]	1961	[25, 219, 1030, 1112]
CeB_6	2190	2463	[25]	1951	
NdB_6	2540	2813	[220]	1960	
SmB_6	2540	2813	[18]	1959	[1112]
GdB_6	>2100	>2373	[285]	1960	
ThB_4	2210	2483	[1097]	1962	[28]
ThB_6	2150	2423	[27]	1956	
UB_2	2385	2658	[974]	1960	[973, 1097]
UB_4	2495	2768	[974]	1960	[973, 1097]
UB_{12}	2235	2508	[974]	1960	[973, 1097]
Ti_2B	2200	2473	[87]	1954	
TiB_2	2980	3253	[87]	1954	
ZrB_2	3040±100	3313	[36]	1953	
ZrB_{12}	2680	2953	[32]	1953	
HfB_2	3250±100	3523	[36]	1953	[291]
VB	2250	2523	[224]	1958	[47, 222]
V_3B_2	2070	2343	[222]	1959	
V_3B_4	2350	2623	[222]	1959	Decomposes [47]
VB_2	2400±50	2673	[222]	1959	
Nb_3B_2	1950	2223	[223]	1959	Decomposes [224]
NbB	2280	2553	[223]	1959	
Nb_3B_4	2900	3173	[223]	1959	Decomposes
NbB_2	3000	3273	[223]	1959	
Ta_3B_2	2120	2393	[233]	1931	Decomposes
TaB	2430	2703	[233]	1931	
Ta_3B_4	2650	2923	[233]	1931	Decomposes, [930]

Phase	Melting point		Ref.	Year	Remarks
	°C	°K			
TaB$_2$	3100	3373	[233]	1931	
Cr$_4$B	1750	2023	[224]	1958	Decomposes, [930]
Cr$_2$B	1890	2163	[224]	1958	Decomposes, [930]
Cr$_5$B$_3$	2000	2273	[224]	1958	Decomposes
CrB	2050	2323	[224]	1958	
Cr$_3$B$_4$	1950	2223	[224]	1958	Decomposes
CrB$_2$	2200±50	2473	[225]	1959	[224, 539, 930]
Mo$_2$B	2140	2413	[226]	1953	
Mo$_3$B$_2$	2250	2523	[226]	1953	
β-MoB	2350	2623	[226]	1953	
MoB$_2$	2100	2373	[226]	1953	
Mo$_2$B$_5$	2100	2373	[49]	1952	Decomposes
W$_2$B	2770±80	3043	[2]	1957	
α-WB	2400±100	2673	[47]	1953	
W$_2$B$_5$	2300±50	2573	[1]	1957	
Fe$_2$B	1389	1662	[1]	1957	Decomposes
FeB	1540	1813	[1]	1957	
Co$_2$B	~1400	~1673	[697]	1959	
Ni$_3$B	1155	1428	[836]	1960	[227]
Ni$_2$B	1100	1373	[836]	1960	[228, 697]
Ni$_3$B$_2$	1160	1433	[218]	1915	
NiB	1020	1293	[228]	1915	
Be$_2$C	2200	2473	[346]	1956	Decomposes, [230, 177,516,613]
CaC$_2$	2300	2373	[177]	1950	Decomposes
SrC$_2$	>1900	>2200	[177]	1950	,,
BaC$_2$	1770—2300	2000—2600	[515]	1932	,,
Al$_4$C$_3$	2100	2373	[467]	1952	Dissociates or sub-limes [346]
ScC	1900 ± 50	2173	x	1963	T. Ya. Kosolapova, G.N. Makarenko
YC	1950±20	2223	[820]	1961	[879]
Y$_2$C$_3$	1800±20	2073	[879]	1961	
YC$_2$	2300±50	2573	[879]	1961	[843]
La$_2$C$_3$	2020	2293	x	1963	Decomposes, T. Ya. Ko-solapova, G.N. Maka-renko,[570]
LaC$_2$	2440	2713	x	1963	Decomposes, T. Ya. Ko-solapova, G.N. Maka-renko,[570, 843]

Phase	Melting point °C	°K	Ref.	Year	Remarks
CeC_2	2540	2813	x	1963	Decomposes, T. Ya. Ko-solapova, G.N. Maka-renko, [843]
PrC_2	2535	2808	x	1963	Decomposes, T. Ya. Ko-solapova, G.N. Maka-renko, [843]
NdC_2	>2000	>2273	[843]	1959	Decomposes
SmC_2	2200	2473	[843]	1959	"
GdC_2	2200	2473	[843]	1959	,,
ThC	2625±25	2898	[531]	1952	,,
ThC_2	2655±25	2928	[531]	1952	
UC	2315	2588	[1096]	1962	[2, 97, 346, 1174]
U_2C_3	2400	2673	[613]	1951	[2]
UC_2	2260	2500	[613]	1951	[2]
TiC	3147±50	3420	[531]	1952	[229, 230, 552]
ZrC	3530	3803	[177]	1950	[1, 230, 531]
HfC	3890±150	4163	[64]	1954	
V_2C	2165 ± 25	2438	[1269]	1962	
VC	2810	3083	[230]	1958	Decomposes, [232, 1269]
NbC	3480	3753	[869]	1960	[230, 233, 613, 870]
Ta_2C	3400	3673	[234]	1943	Decomposes
TaC	3880±150	4153	[234]	1943	[552]
$Cr_{23}C_6$	1550	1823	[346]	1956	Decomposes, [235]
Cr_7C_3	1665	1938	[346]	1956	Decomposes, [235]
Cr_3C_2	1895	2163	[346]	1956	Same
Mo_2C	2410±15	2683	[869]	1960	Decomposes, [230, 236, 552, 910]
MoC	2700	2973	[236]	1930	
W_2C	2730±15	3003	[2]	1957	[257, 346]
WC	2720	2993	[869]	1960	[287, 346, 552]
Mn_3C	1520	1793	[1]	1957	
Fe_3C	1650	1923	[1]	1957	
Co_3C	2300	2573	[1]	1957	Decomposes
Ni_3C	2100	2373	[1]	1957	,,
Be_3N_2	2200	2473	[230]	1958	
Be_3N_4	2205	2478	[230]	1958	
Mg_3N_2	Decomposes	—	[185]	1954	
Ca_3N_2	1195	1468	[467]	1952	
Ba_3N_2	≫ 2220	≫ 2493	[533]	1951	Sublimes, [177, 230]

Phase	Melting point		Ref.	Year	Remarks
	°C	°K			
AlN	>2400	2670	[611]	1959	[177. 230,1186]
ScN	2650	2923	[177]	1950	[1094]
YN	≫ 2670	≫ 2950	[607]	1957	
ThN	2630±50	2903	[2]	1957	
Th₃N₄	2100	2373	[553]	1950	
UN	2650±100	2923	[2]	1957	[177, 1096,1206]
TiN	3205	3478	[230]	1958	[233, 613, 553]
ZrN	2980	3253	[230]	1958	[233, 177, 592]
HfN	2982	3255	[613]	1951	[177]
VN	2360	2633	[230]	1958	[1, 613]
Nb₂N	2420	2693	[929]	1961	
NbN	2300	2573	[238]	1950	Decomposes, [230]
Ta₂N	2050	2323	[929]	1961	
TaN	3087±50	3360	[239]	1954	[264, 613]
Cr₂N	1650	1923	[929]	1961	
CrN	Dissociates at 1500	1773	[1]	1957	
Mo₃N	Dissociates at 600	Dissociates at 873	[94]	1930	
WN	Decomposes at 600	Decomposes at 873	[1]	1957	
Mn₄N	Decomposes between 400 and 600	Decomposes between 673 and 873	[1]	1957	
Fe₄N	670	943	[240]	1950	
Mg₂Si	1070	2343	[566]	1926	[568]
Ca₂Si	∼1000	∼1273	[566]	1926	Decomposes
CaSi	1245	1518	[566]	1926	
CaSi₂	1020	1293	[566]	1926	Decomposes
BaSi₂	1850±50	2123	[1289]	1963	
Y₅Si₃	1850	2123	[1110]	1961	
YSi	1840	2113	[1110]	1961	
Y₃Si₅	1635	1908	[1110]	1961	
YSi₂	1520	1793	[846]	1960	
LaSi₂	1520	1793	[846]	1960	
NdSi₂	1525	1798	[846]	1960	
EuSi₂	1500	1773	[846]	1960	
GdSi₂	1540	1813	[846]	1960	
DySi₂	1550	1823	[846]	1960	[285]

MELTING POINT (continued)

Phase	Melting point		Ref.	Year	Remarks
	°C	°K			
ThSi	>1700	>1973	[117]	1959	
ThSi$_2$	1600	1873	[846]	1960	
U$_3$Si$_2$	1665	1938	[117]	1959	
USi	~1600	~1873	[117]	1959	Decomposes
USi$_2$	1700	1973	[846]	1960	[117]
USi$_3$	1620	1893	[117]	1959	Decomposes
Ti$_5$Si$_3$	2120	2393	[230]	1958	[241, 1020]
TiSi	1920	2193	[241]	1951	Decomposes
TiSi$_2$	1460—1540	1733—1813	[241]	1951	
Zr$_2$Si	2220	2493	[117]	1959	Decomposes
Zr$_5$Si$_3$	2250	2523	[117]	1959	
ZrSi	2150	2423	[117]	1959	Decomposes
ZrSi$_2$	1700	1973	[230]	1958	[117]
HfSi$_2$	1750	2023	[1082]	1961	
V$_3$Si	~1730	~2003	[114]	1956	Decomposes
V$_5$Si$_3$	~2150	~2423	[114]	1956	
VSi$_2$	~1660	~1933	[230]	1958	[114]
Nb$_4$Si	~2580	~2853	[117]	1959	Decomposes [114]
Nb$_5$Si$_3$	2400—2480	2673—2753	[114]	1956	[243]
NbSi$_2$	2150	2423	[117]	1959	[114, 230]
Ta$_{4.5}$Si	2510	2788	[117]	1959	
Ta$_2$Si	~2460	~2733	[419]	1953	Decomposes
Ta$_5$Si$_3$	~2500	~2773	[419]	1953	
TaSi$_2$	2200	2473	[419]	1953	
Cr$_3$Si	1710 ± 50	1983	[419]	1953	[1020, 1265]
Cr$_5$Si$_3$	1600 ± 50	1873	[117]	1959	Decomposes
CrSi	1545 ± 50	1818	[117]	1959	
CrSi$_2$	1500 ± 20	1773	[117]	1959	
Mo$_3$Si	2180 ± 50	2453	[243]	1952	Decomposes, [1265]
Mo$_5$Si$_3$	2100 ± 50	2373	[243]	1952	
MoSi$_2$	2030	2303	[230]	1958	[243, 681]
W$_5$Si$_3$	2320	2593	[423]	1952	Decomposes, [909]
WSi$_2$	2165	2438	[423]	1952	[909]
Mn$_3$Si	1120	1393	[117]	1959	Decomposes
Mn$_5$Si$_3$	1285	1558	[117]	1959	
MnSi	1275	1548	[117]	1959	''
Re$_5$Si$_3$(Re$_3$Si?)	1920	2193	[127]	1959	
ReSi	~1900	~2173	[127]	1959	Decomposes
ReSi$_2$	~1930	~2203	[127]	1959	
Fe$_3$Si	~1300	~1573	[117]	1959	
Fe$_5$Si$_3$	1195	1468	[117]	1959	Decomposes
FeSi	1410	1683	[117]	1959	
FeSi$_2$	1210	1483	[117]	1959	Decomposes
Co$_3$Si	1210	1483	[117]	1959	

Phase	Melting point		Ref.	Year	Remarks
	°C	°K			
Co_2Si	1332	1605	[117]	1959	Decomposes
$CoSi$	1415	1688	[117]	1959	
$CoSi_2$	1277	1550	[117]	1959	
$CoSi_3$	1305	1579	[117]	1959	
Ni_3Si	1250	1523	[117]	1959	Decomposes
Ni_2Si	1290	1563	[117]	1959	
Ni_3Si_2	830	1103	[117]	1959	Decomposes
$NiSi$	1000	1273	[117]	1959	
$NiSi_2$	1280	1553	[117]	1959	Decomposes
Ba_3P_2	3080 (?)	3353	[230]	1958	
TiP	1580	1853	[1177]	1962	Decomposition temp.
VP	1320	1593	[1177]	1962	Same
$\beta-NbP$	1730	2003	[1177]	1962	"
$\beta-TaP$	1660	1933	[1177]	1962	"
CrP	1600	1873	[245]	1961	Decomposes at 1360°C,[1177]
MoP	1480	1753	[1177]	1962	Decomposition temp.
WP	1450	1723	[1177]	1962	Same
Mn_3P	1230	1503	[245]	1961	Decomposes
Mn_2P	1327	1600	[245]	1961	
Mn_3P_2	1200	1473	[245]	1961	
MnP	1147	1420	[245]	1961	Decomposes at 1100°C,[1177]
ReP	1200	1473	[245]	1961	
Fe_3P	~1200	~1473	[245]	1961	Decomposes
Fe_2P	1365	1638	[245]	1961	
Co_2P	1386	1659	[245]	1961	
Ni_3P	1100	1373	[245]	1961	Decomposes
$Ni_{12}P_5$	1115	1388	[245]	1961	
Ni_2P	1100	1373	[245]	1961	
BaS	2205	2478	[230]	1958	
YS	2040	2313	[145]	1956	[1229]
Y_5S_7	1630	1903	[146]	1956	[1229]
Y_2S_3	1600	1873	[146]	1956	
YS_2	1660	1933	[1230]	1959	[246]
Y_2O_2S	2120	2393	[145]	1956	
LaS	2200	2573	[1229]	1958	[149,1230]
La_3S_4	2100	2373	[1229]	1958	
La_2S_3	2100–2150	2373–2423	[246]	1931	[247,1112]
La_2O_2S	1940 ± 20	2213	[153]	1956	
CeS	2450	2723	[230]	1958	[156,145,1229]
Ce_3S_4	2050 ± 75	2323	[156]	1950	[1229]
Ce_2S_3	1840 ± 50	2113	[156]	1950	[1221]
CeS_2	1700	1973	[1225]	1956	

Phase	Melting point		Ref.	Year	Remarks
	°C	°K			
Ce_2O_2S	1950	2223	[147]	1955	
PrS	2230	2503	[1229]	1958	[1221]
Pr_3S_4	2100	2373	[1229]	1958	
Pr_2S_3	1795	2068	[1229]	1958	[1221]
PrS_2	1780	2053	[1225]	1956	
NdS	2140	2413	[149]	1956	[1221, 1225]
Nd_3S_4	2040	2313	[1229]	1958	
Nd_2S_3	2200	2473	[246]	1931	[1221, 1230]
NdS_2	1760	2033	[1225]	1956	
Nd_2O_2S	1990 ± 20	2263	[153]	1956	
SmS	1940	2213	[149]	1956	[1229]
Sm_3S_4	1800	2073	[1229]	1958	
Sm_2S_3	1900	2173	[246]	1931	[1229]
SmS_2	1730	2003	[1225]	1956	
Sm_2O_2S	1980 ± 20	2253	[153]	1956	
Gd_2S_3	1885	2158	[160]	1957	[502]
Dy_5S_7	1540	1813	[160]	1957	
Dy_2S_3	1480	1753	[1112]	1961	
Er_5S_7	1620	1893	[160]	1957	
Er_2S_3	1630	1903	[160]	1957	[1112]
ThS	2400—2450	2673—2723	[249]	1957	[248, 154]
Th_2S	∼2300	∼2573	[177]	1950	
$Th_4S_7(Th_7S_{12}?)$	2300	2573	[177]	1950	
ThS_3	>1905	>2178	[250]	1950	[177]
PaOS	>2000	>2273	[251]	1955	
US	2000	2273	[251]	1955	
β-US_2	1850	2123	[251]	1955	
B_4C	2350	2623	[277]	1953	Decomposes,[1077]
SiC	2827	3100±40	[1011]	1960	Decomposes,[177, 230,613,1077]
α-BN	3000	3273	[278]	1933	Under pressure of N_2
Si_3N_4	1900	2173	[230]	1958	Decomposes,[270]
SiB_6	1950	2223	[690]	1960	[392]
BP	∼1250	∼1523	[245]	1960	Under pressure of phosphorus
B	2075±50	2348	[440]	1960	[392]
C	3870	4143	[230]	1958	[944]

BOILING POINT

Phase	Boiling point		Ref.	Year	Remarks
	°C	°K			
SrB$_6$	5100	5373	[430]	1961	Calculated value, [353]
BeC$_2$	2537	2810	[431]	1959	Calculated value
BaC$_2$	2727	3000	[432]	1958	Calculated value
Al$_4$C$_3$	2600	2873	[1295]	1959	Same
ThC$_2$	3927	4200	[1308]	1962	Calculated value, [467]
UC	2397	2670	[1310]	1962	Calculated value
U$_2$C$_3$	4100	4373	[467]	1952	Calculated value
UC$_2$	4370	4643	[346]	1956	
PuC$_2$	3927	4200	[1313]	1962	Calculated value
TiC	4300	4573	[466]	1951	
ZrC	5100	5373	[346]	1956	
HfC	5400	5673	[430]	1961	Calculated value
VC	3900	4173	[346]	1956	[1301]
NbC	4500	4773	[430]	1961	Calculated value, [2]
TaC	5500	5773	[2]	1957	
Cr$_3$C$_2$	3800	4073	[467]	1952	
W$_2$C	6000	6273	[346]	1956	
WC	6000	6273	[346]	1956	
Mn$_7$C$_3$	2081	2354	[1040]	1957	Calculated value
Ta$_{4.5}$Si	~4000	~4270	[438]	1957	Calculated value
Ta$_2$Si	3727	4000	[438]	1957	Same
TaSi$_2$	5347	5620	[438]	1957	"
SiC	2607	2880	[442]	1959	"
BN	5067	5340	[440]	1960	Calculated value, [439]
C	4200	4473	[944]	1959	Pyrographite

VAPOR PRESSURE AND RATE OF EVAPORATION (TOTAL)

Phase	Temperature, °C	Total rate of evaporation, g/cm²·sec	Vapor pressure, atm	Vapor pressure equation	Source	Year	Remarks
SrB$_6$	1500	$0.34 \cdot 10^{-7}$	—				
SrB$_6$	1600	$2.13 \cdot 10^{-7}$	—	$\lg P_{SrB_6} = 6.43 - \dfrac{21423}{T}$	[353]	1961	Assuming molecular evaporation
SrB$_6$	1700	$3.25 \cdot 10^{-7}$	—	[P, mm Hg]			
SrB$_6$	1800	$17.25 \cdot 10^{-7}$	—				
SrB$_6$	2000	$188.90 \cdot 10^{-7}$	—				
AlB$_{12}$	1100	$0.13 \cdot 10^{-7}$	—				
AlB$_{12}$	1200	$0.61 \cdot 10^{-7}$	—				
AlB$_{12}$	1600	$0.70 \cdot 10^{-7}$	—	—	[430]	1961	
AlB$_{12}$	1700	$4.68 \cdot 10^{-7}$	—				
AlB$_{12}$	1800	$48.3 \cdot 10^{-6}$	—				
LaB$_6$*	1620	$2.59 \cdot 10^{-7}$	$0.25 \cdot 10^{-7}$				[25], assuming molecular evaporation
LaB$_6$	1720	$1.54 \cdot 10^{-6}$	$1.49 \cdot 10^{-7}$	$\lg P_{LaB_6} = 9.5 - \dfrac{28421}{T}$	[1291]	1963	
LaB$_6$	1820	$9.45 \cdot 10^{-6}$	$11.0 \cdot 10^{-7}$	[P, mm Hg]			
LaB$_6$	1920	$1.60 \cdot 10^{-5}$	$16.3 \cdot 10^{-7}$				
UB$_{12}$—B**	1737	—	$3.16 \cdot 10^{-8}$				
UB$_{12}$—B***	1739	—	$3.69 \cdot 10^{-8}$				
UB$_{12}$—B**	1778	—	$6.74 \cdot 10^{-8}$	—	[1293]	1962	
UB$_{12}$—B***	1800	—	$9.14 \cdot 10^{-8}$				
UB$_{12}$—B**	1818	—	$11.30 \cdot 10^{-8}$				
UB$_{12}$—B**	1842	—	$18.60 \cdot 10^{-8}$				

Notes. The following notation is used throughout this table: *Vapor pressure of the metal; **vapor pressure of the nonmetal; ***total pressure over compound; ****vapor pressure of the compound.

VAPOR PRESSURE AND RATE OF EVAPORATION (TOTAL) (continued)

Phase	Temperature, °C	Total rate of evaporation, g/cm²·sec	Vapor pressure atm	Vapor pressure equation	Source	Year	Remarks
$UB_4 - UB_{12}$**	1727	—	$0.94 \cdot 10^{-8}$				
$UB_4 - UB_{12}$**	1742	—	$1.18 \cdot 10^{-8}$				
$UB_4 - UB_{12}$**	1754	—	$1.59 \cdot 10^{-8}$				
$UB_4 - UB_{12}$**	1868	—	$1.81 \cdot 10^{-8}$				
$UB_4 - UB_{12}$**	1781	—	$2.32 \cdot 10^{-8}$	—	[1293]	1962	
$UB_4 - UB_{12}$**	1791	—	$3.05 \cdot 10^{-8}$				
$UB_4 - UB_{12}$**	1807	—	$3.81 \cdot 10^{-8}$				
$UB_4 - UB_{12}$**	1827	—	$5.00 \cdot 10^{-8}$				
$UB_4 - UB_{12}$**	1854	—	$5.76 \cdot 10^{-8}$				
TiB_2*	1919	—	$6.46 \cdot 10^{-8}$				
TiB_2*	1973	—	$7.59 \cdot 10^{-8}$	—	[1292]	1962	[1291, 1170]
TiB_2*	2005	—	$1.62 \cdot 10^{-7}$				
TiB_2*	2080	—	$4.90 \cdot 10^{-7}$				
TiB_2	—	—	—	$\lg P_{TiB_2} = 5,9 - \dfrac{2200}{T}$ [P, mm Hg]	×	1961	V. V. Fesenko, A. S. Bolgar
$B + TiB_2$*	1971	—	$3.39 \cdot 10^{-9}$				
$B + TiB_2$*	2023	—	$3.47 \cdot 10^{-9}$				
$B + TiB_2$*	2074	—	$5.49 \cdot 10^{-9}$	—	[1292]	1962	[1291]
$B + TiB_2$*	2109	—	$1.07 \cdot 10^{-8}$				
$B + TiB_2$*	2135	—	$1.82 \cdot 10^{-8}$				
$B + TiB_2$*	2189	—	$4.17 \cdot 10^{-8}$				

VAPOR PRESSURE AND RATE OF EVAPORATION (TOTAL) (continued)

Phase	Temperature, °C	Total rate of evaporation, $g/cm^2 \cdot sec$	Vapor pressure, atm	Vapor pressure equation	Source	Year	Remarks
Ti + TiB$_2$*	1790	—	$1.20 \cdot 10^{-8}$				
Ti + TiB$_2$*	1834	—	$2.57 \cdot 10^{-8}$				
Ti + TiB$_2$*	1999	—	$3.63 \cdot 10^{-7}$	—	[1292]	1962	[1291]
Ti + TiB$_2$*	2054	—	$4.27 \cdot 10^{-6}$				
Ti + TiB$_2$*	2067	—	$1.09 \cdot 10^{-5}$				
ZrB$_2$*	2051	—	$3.67 \cdot 10^{-8}$				
ZrB$_2$*	2080	—	$4.69 \cdot 10^{-8}$				
ZrB$_2$*	2094	—	$7.43 \cdot 10^{-8}$				
ZrB$_2$*	2106	—	$1.051 \cdot 10^{-7}$				
ZrB$_2$*	2140	—	$1.305 \cdot 10^{-7}$	—	[1168]	1962	[430]
ZrB$_2$*	2147	—	$1.494 \cdot 10^{-7}$				
ZrB$_2$*	2159	—	$2.130 \cdot 10^{-7}$				
ZrB$_2$*	2175	—	$2.084 \cdot 10^{-7}$				
ZrB$_2$*	2193	—	$3.124 \cdot 10^{-7}$				
ZrB$_2$*	2216	—	$3.555 \cdot 10^{-7}$				
CrB$_2$	1200	$2.89 \cdot 10^{-7}$	—				
CrB$_2$	1300	$4.52 \cdot 10^{-7}$	—				
CrB$_2$	1400	$4.72 \cdot 10^{-7}$	—	—	[430]	1961	
CrB$_2$	1600	$7.68 \cdot 10^{-7}$	—				
CrB$_2$	1800	$4.98 \cdot 10^{-7}$	—				

VAPOR PRESSURE AND RATE OF EVAPORATION (TOTAL) (continued)

Phase	Temperature, °C	Total rate of evaporation, g/cm²·sec	Vapor pressure, atm	Vapor pressure equation	Source	Year	Remarks
Be$_2$C*	1627	—	$6.04 \cdot 10^{-3}$	—	[998]	1960	
Be$_2$C*	1827	—	$4.01 \cdot 10^{-2}$				
Be$_2$C*	2127	—	$4.14 \cdot 10^{-1}$				
BeC$_2$*	1157	$6.53 \cdot 10^{-7}$	$1.86 \cdot 10^{-7}$	$\lg P_{Be} = (7.026 \pm 0.347) - \dfrac{19720 \pm 537}{T}$ [P, atm]	[431]	1959	[346,528]
BeC$_2$*	1169	$7.00 \cdot 10^{-7}$	$2.00 \cdot 10^{-7}$				
BeC$_2$*	1242	$4.26 \cdot 10^{-6}$	$1.25 \cdot 10^{-6}$				
BeC$_2$*	1263	$7.06 \cdot 10^{-6}$	$2.08 \cdot 10^{-6}$				
BeC$_2$*	1305	$7.30 \cdot 10^{-6}$	$2.18 \cdot 10^{-6}$				
BeC$_2$*	1307	$1.10 \cdot 10^{-5}$	$3.30 \cdot 10^{-6}$				
BeC$_2$*	1317	$1.27 \cdot 10^{-5}$	$3.80 \cdot 10^{-6}$				
BeC$_2$*	1370	$2.79 \cdot 10^{-5}$	$8.50 \cdot 10^{-6}$				
BeC$_2$*	1374	$3.54 \cdot 10^{-5}$	$1.08 \cdot 10^{-5}$				
BeC$_2$*	1386	$5.13 \cdot 10^{-5}$	$1.33 \cdot 10^{-5}$				
BeC$_2$*	1396	$5.83 \cdot 10^{-5}$	$1.79 \cdot 10^{-5}$				
CaC$_2$*	1850	—	$1.32 \cdot 10^{-3}$	—	[467]	1952	[1294]
CaC$_2$*	2500	—	1				

VAPOR PRESSURE AND RATE OF EVAPORATION (TOTAL) (continued)

Phase	Temperature, °C	Total rate of evaporation, g/cm²·sec	Vapor pressure, atm	Vapor pressure equation	Source	Year	Remarks
BaC$_2$*	1082	$4.59 \cdot 10^{-4}$	$3.254 \cdot 10^{-5}$				
BaC$_2$*	1155	$8.20 \cdot 10^{-4}$	$5.967 \cdot 10^{-5}$				
BaC$_2$*	1176	$8.42 \cdot 10^{-4}$	$6.168 \cdot 10^{-5}$				
BaC$_2$*	1213	$1.94 \cdot 10^{-3}$	$1.437 \cdot 10^{-4}$	$\lg P_{Ba} = (3.1 \pm 0.293) - \dfrac{10680}{T}$ [P, atm]	[432]	1958	[467]
BaC$_2$*	1276	$2.75 \cdot 10^{-3}$	$2.090 \cdot 10^{-4}$				
BaC$_2$*	1288	$3.80 \cdot 10^{-3}$	$2.312 \cdot 10^{-4}$				
BaC$_2$*	1298	$3.88 \cdot 10^{-3}$	$2.962 \cdot 10^{-4}$				
BaC$_2$*	1315	$4.65 \cdot 10^{-3}$	$3.174 \cdot 10^{-4}$				
BaC$_2$*	1393	$1.44 \cdot 10^{-2}$	$8.842 \cdot 10^{-4}$				
Al$_4$C$_3$*	1377	$0.57 \cdot 10^{-4}$	$10.1 \cdot 10^{-6}$				
Al$_4$C$_3$*	1378	$0.65 \cdot 10^{-4}$	$11.15 \cdot 10^{-6}$	$\lg P_{Al} = 6.79 - \dfrac{19500}{T}$ [P, atm]	[1295]	1959	[528]
Al$_4$C$_3$	1408	$0.70 \cdot 10^{-4}$	$14.8 \cdot 10^{-6}$				
Al$_4$C$_3$	1422	$1.11 \cdot 10^{-4}$	$20.2 \cdot 10^{-6}$				

VAPOR PRESSURE AND RATE OF EVAPORATION (TOTAL) (continued)

Phase	Temperature, °C	Total rate of evaporation, g/cm²·sec	Vapor pressure, atm	Vapor pressure equation	Source	Year	Remarks
ThC₂***	2136	—	$1.338 \cdot 10^{-9}$				
ThC₂***	2159	—	$1.757 \cdot 10^{-9}$				
ThC₂***	2165	—	$1.864 \cdot 10^{-9}$				
ThC₂***	2176	—	$2.443 \cdot 10^{-9}$				
ThC₂***	2191	—	$3.006 \cdot 10^{-9}$	$\lg P_{ThC_2} = 7.20 - \dfrac{39364}{T}$	[1308]	1962	[1309, 1262]
ThC₂***	2212	—	$3.803 \cdot 10^{-9}$				
ThC₂***	2233	—	$5.102 \cdot 10^{-9}$	$\lg P_{Th} = 5.74 - \dfrac{36025}{T}$ [P, atm]			
ThC₂***	2249	—	$6.782 \cdot 10^{-9}$				
ThC₂***	2267	—	$8.539 \cdot 10^{-9}$				
ThC₂***	2280	—	$9.826 \cdot 10^{-9}$				
ThC₂***	2299	—	$13.12 \cdot 10^{-9}$				
ThC₂***	2322	—	$18.76 \cdot 10^{-9}$				
ThC₂***	2342	—	$24.51 \cdot 10^{-9}$				
ThC₂***	2369	—	$32.45 \cdot 10^{-9}$				
PaC₂*	2153	—	$0.75 \cdot 10^{-17}$				
PaC₂*	1258	—	$2.10 \cdot 10^{-17}$				
PaC₂*	2310	—	$4.30 \cdot 10^{-17}$				
PaC₂*	2344	—	$13.0 \cdot 10^{-17}$	—	[1309]	1962	
PaC₂*	2400	—	$26.0 \cdot 10^{-17}$				
PaC₂*	2459	—	$37.0 \cdot 10^{-17}$				
PaC₂*	2472	—	$83.0 \cdot 10^{-17}$				
PaC₂*	2506	—	$99.0 \cdot 10^{-17}$				
PaC₂*	2542	—	$270.0 \cdot 10^{-17}$				

VAPOR PRESSURE AND RATE OF EVAPORATION (TOTAL) (continued)

Phase	Temperature, °C	Total rate of evaporation, g/cm².sec	Vapor pressure, atm	Vapor pressure equation	Source	Year	Remarks
UC****	1675	$6.15 \cdot 10^{-7}$	$1.82 \cdot 10^{-7}$	$\lg P_{UC} = 18.426 - \dfrac{49200}{T}$ [P, atm]	[1310]	1962	
UC****	1705	$8.30 \cdot 10^{-7}$	$2.92 \cdot 10^{-7}$				
UC****	1745	$3.02 \cdot 10^{-6}$	$1.55 \cdot 10^{-6}$				
UC****	1823	$2.73 \cdot 10^{-5}$	$9.86 \cdot 10^{-6}$				
UC****	1860	$6.43 \cdot 10^{-5}$	$2.32 \cdot 10^{-5}$				
UC₂*	1656.8	$1.12 \cdot 10^{-9}$	$7.18 \cdot 10^{-11}$	$\lg P_{U} = 12.14 - 0.111\left(\lg T + \dfrac{2000}{T}\right) - \dfrac{67164}{T}$ [P, atm]	[1169]	1962	[1309, 1311, 1312]
UC₂*	1686.7	$1.73 \cdot 10^{-9}$	$1.119 \cdot 10^{-10}$				
UC₂*	1702.7	$2.61 \cdot 10^{-9}$	$1.692 \cdot 10^{-10}$				
UC₂*	1739.4	$4.56 \cdot 10^{-9}$	$2.982 \cdot 10^{-10}$				
UC₂*	1789.9	$5.71 \cdot 10^{-9}$	$3.902 \cdot 10^{-10}$				
UC₂*	1813.2	$6.43 \cdot 10^{-9}$	$4.285 \cdot 10^{-10}$				
UC₂*	1824.0	$1.62 \cdot 10^{-8}$	$1.034 \cdot 10^{-9}$				
UC₂*	1852.0	$2.53 \cdot 10^{-8}$	$1.699 \cdot 10^{-9}$				
UC₂*	1890.6	$3.37 \cdot 10^{-8}$	$2.298 \cdot 10^{-9}$				
UC₂*	1928.4	$5.74 \cdot 10^{-8}$	$3.924 \cdot 10^{-9}$				
UC₂*	1983.4	$1.11 \cdot 10^{-7}$	$6.945 \cdot 10^{-9}$				
UC₂*	2037.0	$2.06 \cdot 10^{-7}$	$1.442 \cdot 10^{-8}$				
UC₂*	2091.2	$4.33 \cdot 10^{-7}$	$3.070 \cdot 10^{-8}$				
PuC₂*	–	–	–	$\lg P_{Pu} = 2.779 \pm 0.11 - \dfrac{17920 \pm 250}{T}$ [P, atm]	[1313]	1962	

VAPOR PRESSURE AND RATE OF EVAPORATION (TOTAL) (continued)

Phase	Temperature, °C	Total rate of evaporation, g/cm²·sec	Vapor pressure, atm	Vapor pressure equation	Source	Year	Remarks
TiC*	2110	$1.23 \cdot 10^{-5}$	$3.9 \cdot 10^{-6}$	$\lg P_{Ti} = 7.65 - \dfrac{30830}{T}$ [P, atm]	[946]	1961	[1291, 528, 463, 430]
TiC*	2162	$1.28 \cdot 10^{-5}$	$6.8 \cdot 10^{-6}$				
TiC*	2194	$2.74 \cdot 10^{-5}$	$14.7 \cdot 10^{-6}$				
TiC*	2215	$3.85 \cdot 10^{-5}$	$21.7 \cdot 10^{-6}$				
TiC*	2246	$6.83 \cdot 10^{-5}$	$26.4 \cdot 10^{-6}$				
TiC*	2268	$8.33 \cdot 10^{-5}$	$32.0 \cdot 10^{-6}$				
TiC*	2320	$16.8 \cdot 10^{-5}$	$65.0 \cdot 10^{-6}$				
CrC*	2500	$2.23 \cdot 10^{-6}$	$2.45 \cdot 10^{-7}$	—	×	1962	V. V. Fesenko, A. S. Bolgar; [430, 1030, 1031, 1302]
CrC*	2600	$7.07 \cdot 10^{-6}$	$7.91 \cdot 10^{-7}$				
CrC*	2700	$3.57 \cdot 10^{-5}$	$4.07 \cdot 10^{-6}$				
CrC*	2800	$6.44 \cdot 10^{-6}$	$7.46 \cdot 10^{-6}$				
CrC*	2900	$1.90 \cdot 10^{-4}$	$2.24 \cdot 10^{-5}$				
HfC*	2500	$9.83 \cdot 10^{-7}$	$8.18 \cdot 10^{-8}$	—	×	1962	V. V. Fesenko, A. S. Bolgar
HfC*	2600	$2.71 \cdot 10^{-6}$	$2.29 \cdot 10^{-7}$				
HfC*	2700	$8.31 \cdot 10^{-6}$	$7.17 \cdot 10^{-7}$				
HfC*	2800	$2.59 \cdot 10^{-5}$	$2.27 \cdot 10^{-6}$				
HfC*	2900	$6.41 \cdot 10^{-5}$	$5.70 \cdot 10^{-6}$				
HfC	—	—	—	$\lg P_{HfC} = 8.7 - \dfrac{30555}{T}$ [P, mm Hg]	[430]	1961	

VAPOR PRESSURE AND RATE OF EVAPORATION (TOTAL) (continued)

Phase	Temperature, °C	Total rate of evaporation, g/cm²·sec	Vapor pressure, atm	Vapor pressure equation	Source	Year	Remarks
VC*	2073	$1.76 \cdot 10^{-7}$	$3.59 \cdot 10^{-6}$				
VC*	2104	$2.42 \cdot 10^{-7}$	$0.498 \cdot 10^{-5}$				
VC*	2125	$2.77 \cdot 10^{-7}$	$5.37 \cdot 10^{-6}$				
VC*	2135	$3.29 \cdot 10^{-7}$	$6.79 \cdot 10^{-6}$				
VC*	2146	$4.39 \cdot 10^{-7}$	$9.11 \cdot 10^{-6}$	$\lg P_V = 7.50 - \dfrac{30400}{T}$ [P, atm]	[1301]	1960	
VC*	2156	$5.48 \cdot 10^{-7}$	$1.14 \cdot 10^{-5}$				
VC*	2167	$7.69 \cdot 10^{-7}$	$1.50 \cdot 10^{-5}$				
VC*	2209	$8.53 \cdot 10^{-7}$	$1.79 \cdot 10^{-5}$				
VC*	2219	$9.89 \cdot 10^{-7}$	$1.97 \cdot 10^{-5}$				
VC*	2240	$10.16 \cdot 10^{-7}$	$2.15 \cdot 10^{-5}$				
VC*	2251	$12.36 \cdot 10^{-7}$	$2.62 \cdot 10^{-5}$				
VC*	2272	$13.77 \cdot 10^{-7}$	$2.93 \cdot 10^{-5}$				
NbC*	2500	$7.17 \cdot 10^{-7}$	$7.75 \cdot 10^{-8}$				
NbC*	2600	$2.18 \cdot 10^{-6}$	$2.44 \cdot 10^{-7}$				V. V. Fesenko, A. S. Bolgar; [430, 1291, 1303]
NbC*	2700	$6.57 \cdot 10^{-6}$	$7.46 \cdot 10^{-7}$	—	×	1961	
NbC*	2800	$2.09 \cdot 10^{-5}$	$2.42 \cdot 10^{-6}$				
NbC*	2850	$5.68 \cdot 10^{-5}$	$6.58 \cdot 10^{-6}$				
NbC	—	—	—	$\lg P_{NbC} = 10.6 - \dfrac{36666}{T}$ [P, mm Hg]	[430]	1961	
TaC*	2700	$1.13 \cdot 10^{-6}$	$9.16 \cdot 10^{-8}$				
TaC*	2800	$1.62 \cdot 10^{-6}$•	$1.42 \cdot 10^{-7}$	—	×	1962	V. V. Fesenko, A. S. Bolgar; [1291,1305]
TaC*	2900	$6.20 \cdot 10^{-6}$	$5.49 \cdot 10^{-7}$				
TaC*	3000	$2.33 \cdot 10^{-5}$	$2.01 \cdot 10^{-6}$				

110

VAPOR PRESSURE AND RATE OF EVAPORATION (TOTAL) (continued)

Phase	Temperature, °C	Total rate of evaporation, g/cm²·sec	Vapor pressure, atm	Vapor pressure equation	Source	Year	Remarks
Cr_3C_2*	1671	$1.28 \cdot 10^{-6}$	$0.35 \cdot 10^{-4}$				
Cr_3C_2*	1723	$2.55 \cdot 10^{-6}$	$0.71 \cdot 10^{-4}$				
Cr_3C_2*	1728	$3.25 \cdot 10^{-6}$	$0.91 \cdot 10^{-4}$				
Cr_3C_2*	1770	$6.05 \cdot 10^{-6}$	$1.71 \cdot 10^{-4}$	$\lg P_{Cr} = 6.525 - \dfrac{21194}{T}$ [P, atm]	[1360]	1961	[430]
Cr_3C_2*	1822	$9.30 \cdot 10^{-6}$	$2.66 \cdot 10^{-4}$				
Cr_3C_2*	1869	$16.52 \cdot 10^{-6}$	$3.68 \cdot 10^{-4}$				
Cr_3C_2*	1895	$25.4 \cdot 10^{-6}$	$5.60 \cdot 10^{-4}$				
Cr_3C_2*	1916	$27.75 \cdot 10^{-6}$	$6.15 \cdot 10^{-4}$				
Cr_3C_2*	1964	$31.6 \cdot 10^{-6}$	$9.35 \cdot 10^{-4}$				
WC**	1983	$0.92 \cdot 10^{-8}$	$0.28 \cdot 10^{-8}$				
WC**	2133	$1.31 \cdot 10^{-7}$	$0.42 \cdot 10^{-7}$	—	[1304]	1955	[1307]
WC**	2483	$24.90 \cdot 10^{-7}$	$8.51 \cdot 10^{-7}$				
Mn_7C_3*	855	—	$0.25 \cdot 10^{-6}$				
Mn_7C_3*	887	—	$0.51 \cdot 10^{-6}$				
Mn_7C_3*	909	—	$0.37 \cdot 10^{-6}$				
Mn_7C_3*	935	—	$1.69 \cdot 10^{-6}$	$\lg P_{Mn} = 6.07 - \dfrac{14290}{T}$ [P, atm]	[1040]	1957	
Mn_7C_3*	951	—	$2.28 \cdot 10^{-6}$				
Mn_7C_3*	952	—	$2.60 \cdot 10^{-6}$				
Mn_7C_3*	955	—	$2.66 \cdot 10^{-6}$				
Mn_7C_3*	956	—	$2.89 \cdot 10^{-6}$				

VAPOR PRESSURE AND RATE OF EVAPORATION (TOTAL) (continued)

Phase	Temperature, °C	Total rate of evaporation, $g/cm^2 \cdot sec$	Vapor pressure, atm	Vapor pressure equation	Source	Year	Remarks
Be_3N_2**	1704	—	$1.32 \cdot 10^{-6}$	—	[434]	1937	[435, 437]
Be_3N_2**	1848	—	$1.32 \cdot 10^{-5}$				
Be_3N_2**	2015	—	$1.32 \cdot 10^{-4}$				
Be_3N_2**	2208	—	$1.32 \cdot 10^{-3}$				
Mg_3N_2**	1260	—	$1.32 \cdot 10^{-6}$	—	[434]	1937	[437]
Mg_3N_2**	1359	—	$1.32 \cdot 10^{-5}$				
Mg_3N_2**	1472	—	$1.32 \cdot 10^{-4}$				
Mg_3N_2**	1602	—	$1.32 \cdot 10^{-3}$				
Ca_3N_2**	1000	—	$1.32 \cdot 10^{-6}$	—	[436]	1923	[437]
Ca_3N_2**	1069	—	$1.32 \cdot 10^{-5}$				
Ca_3N_2**	1147	—	$1.32 \cdot 10^{-4}$				
Ca_3N_2**	1235	—	$1.32 \cdot 10^{-3}$				
Sr_3N_2**	302	—	$1.32 \cdot 10^{-6}$	—	[434]	1937	
Sr_3N_2**	975	—	$1.32 \cdot 10^{-5}$				
Sr_3N_2**	1059	—	$1.32 \cdot 10^{-4}$				
Sr_3N_2**	1154	—	$1.32 \cdot 10^{-3}$				

Phase	Temperature, °C	Total rate of evaporation, $g/cm^2 \cdot sec$	Vapor pressure, atm	Vapor pressure equation	Source	Year	Remarks
Ba$_3$N$_2$**	794	—	$1.32 \cdot 10^{-6}$				
Ba$_3$N$_2$**	855	—	$1.32 \cdot 10^{-5}$	—	[434]	1937	[435, 437]
Ba$_3$N$_2$**	923	—	$1.32 \cdot 10^{-4}$				
Ba$_3$N$_2$**	1002	—	$1.32 \cdot 10^{-3}$				
AlN**	727	—	$5 \cdot 10^{-12}$				
AlN**	927	—	$2 \cdot 10^{-8}$				[434, 435,
AlN**	1127	—	$1 \cdot 10^{-5}$	—	[1184]	1962	437, 1170]
AlN**	1227	—	$1 \cdot 10^{-4}$				
AlN**	1327	—	$9 \cdot 10^{-4}$				
ScN**	1230	—	$1 \cdot 10^{-7}$				
ScN**	1495	—	$1.32 \cdot 10^{-6}$				
ScN**	1607	—	$1.32 \cdot 10^{-5}$	—	[177]	1950	[437]
ScN**	1734	—	$1.32 \cdot 10^{-4}$				
ScN**	1880	—	$1.32 \cdot 10^{-3}$				
YN**	1587	—	$1.32 \cdot 10^{-6}$				
YN**	1704	—	$1.32 \cdot 10^{-5}$	—	[177]	1950	[437]
YN**	1838	—	$1.32 \cdot 10^{-4}$				
YN**	1990	—	$1.32 \cdot 10^{-3}$				

VAPOR PRESSURE AND RATE OF EVAPORATION (TOTAL) (continued)

Phase	Temperature, °C	Total rate of evaporation, g/cm².sec	Vapor pressure, atm	Vapor pressure equation	Source	Year	Remarks
LaN**	1602	—	$1.32 \cdot 10^{-6}$	—	[434]	1937	[437]
LaN**	1721	—	$1.32 \cdot 10^{-5}$				
LaN**	1855	—	$1.32 \cdot 10^{-4}$				
LaN**	2010	—	$1.32 \cdot 10^{-3}$				
CeN**	1756	—	$1.32 \cdot 10^{-6}$	—	[434]	1937	[437]
CeN**	1884	—	$1.32 \cdot 10^{-5}$				
CeN**	2029	—	$1.32 \cdot 10^{-4}$				
CeN**	2196	—	$1.32 \cdot 10^{-3}$				
ThN**	1230	—	$<1 \cdot 10^{-7}$	—	[177]	1950	
ThN**	1730	—	$1 \cdot 10^{-7}$				
ThN**	2230	—	$1 \cdot 10^{-6}$				
Th$_3$N$_4$**	1890	—	$1.32 \cdot 10^{-6}$	—	[434]	1937	[437]
Th$_3$N$_4$**	2037	—	$1.32 \cdot 10^{-5}$				
Th$_3$N$_4$**	2206	—	$1.32 \cdot 10^{-4}$				
Th$_3$N$_4$**	2401	—	$1.32 \cdot 10^{-3}$				
UN**	1230	—	$<1 \cdot 10^{-7}$	—	[177]	1950	[437]
UN**	1730	—	$1 \cdot 10^{-7}$				
UN**	2230	—	$1 \cdot 10^{-6}$				

VAPOR PRESSURE AND RATE OF EVAPORATION (TOTAL) (continued)

Phase	Temperature, °C	Total rate of evaporation, g/cm²·sec	Vapor pressure, atm	Vapor pressure equation	Source	Year	Remarks
UN	–	–	–	$\lg P_{N_2} = 8.193 - 29.54 \cdot 10^{-3} \cdot T^{-1} + 5.57 \cdot 10^{-18} T^5$ [P, atm]	[1206]	1963	
U₃N₄**	1753	–	$1.32 \cdot 10^{-6}$		[434]	1937	
U₃N₄**	1950	–	$1.32 \cdot 10^{-5}$	—			
U₃N₄**	2070	–	$1.32 \cdot 10^{-4}$				
U₃N₄**	2267	–	$1.32 \cdot 10^{-3}$				
TiN*	1713	$1.51 \cdot 10^{-6}$	$1.697 \cdot 10^{-6}$	$\lg P_{Ti} = 8.263 - 0.40 \cdot 10^{-4} T - \dfrac{27859}{T}$	[975]	1955	[430, 434, 177, 1314]
TiN*	1744	$1.97 \cdot 10^{-6}$	$2.234 \cdot 10^{-6}$				
TiN*	1777	$3.129 \cdot 10^{-6}$	$3.573 \cdot 10^{-6}$	$\lg P_{N_2} = 7.963 - 0.40 \cdot 10^{-4} T - \dfrac{27859}{T}$			
TiN*	1785	$3.360 \cdot 10^{-6}$	$3.845 \cdot 10^{-6}$	[P, atm]			
TiN*	1882	$15.963 \cdot 10^{-6}$	$18.685 \cdot 10^{-6}$				
TiN*	1884	$20.51 \cdot 10^{-6}$	$24.016 \cdot 10^{-6}$				
TiN*	1939	$33.254 \cdot 10^{-6}$	$39.435 \cdot 10^{-6}$				
TiN*	1968	$70.555 \cdot 10^{-6}$	$84.221 \cdot 10^{-6}$				
ZrN**	1963	$5.34 \cdot 10^{-7}$	$9.15 \cdot 10^{-7}$	$\lg P_{N_2} = 8.934 + 2.96 \cdot 10^{-4} T - \dfrac{34816}{T}$	[975]	1955	[434, 437, 177, 1314]
ZrN**	1986	$1.299 \cdot 10^{-6}$	$2.397 \cdot 10^{-6}$				
ZrN**	2045	$2.979 \cdot 10^{-6}$	$5.583 \cdot 10^{-6}$	[P, atm]			
ZrN**	2060	$2.714 \cdot 10^{-6}$	$4.477 \cdot 10^{-6}$				
ZrN**	2071	$3.020 \cdot 10^{-6}$	$5.536 \cdot 10^{-6}$				
ZrN**	2178	$14.981 \cdot 10^{-6}$	$28.96 \cdot 10^{-6}$				
ZrN**	2193	$16.127 \cdot 10^{-6}$	$31.04 \cdot 10^{-6}$				

VAPOR PRESSURE AND RATE OF EVAPORATION (TOTAL) (continued)

Phase	Temperature, °C	Total rate of evaporation, g/cm·sec	Vapor pressure, atm	Vapor pressure equation	Source	Year	Remarks
HfN**	1427	—	$3.10 \cdot 10^{-14}$				
HfN**	1527	—	$5.16 \cdot 10^{-13}$				
HfN**	1627	—	$6.30 \cdot 10^{-12}$	—	×	1961	V. V. Fesenko (calculated from data)
HfN**	1727	—	$5.96 \cdot 10^{-11}$				
HfN**	1827	—	$4.54 \cdot 10^{-10}$				
HfN**	1927	—	$2.87 \cdot 10^{-9}$				
NbN**	1727	—	$1.58 \cdot 10^{-4}$				
NbN**	1827	—	$6.02 \cdot 10^{-3}$				
NbN**	1927	—	$1.98 \cdot 10^{-2}$	—	×	1961	V. V. Fesenko (calculated from data)
NbN**	2027	—	$5.90 \cdot 10^{-2}$				
NbN**	2127	—	$1.59 \cdot 10^{-1}$				
NbN**	2227	—	$4.01 \cdot 10^{-1}$				
Ta₂N	1200	$1.89 \cdot 10^{-8}$	—				
Ta₂N	1400	$1.80 \cdot 10^{-7}$	—	—	[430]	1961	
Ta₂N	1600	$4.93 \cdot 10^{-7}$	—				
Ta₂N	1800	$11.39 \cdot 10^{-7}$	—				
Ta₂N	1000–2400	—	—	$\lg P_{N_2} = 8.65 - \dfrac{50000}{T}$ [P, mm Hg]	[1028]	1961	

VAPOR PRESSURE AND RATE OF EVAPORATION (TOTAL) (continued)

Phase	Temperature, °C	Total rate of evaporation, g/cm²·sec	Vapor pressure, atm	Vapor pressure equation	Source	Year	Remarks
TaN**	1527	—	$2.88 \cdot 10^{-7}$				
TaN**	1627	—	$1.49 \cdot 10^{-6}$				
TaN**	1727	—	$6.50 \cdot 10^{-6}$				V. V. Fesenko
TaN**	1827	—	$2.46 \cdot 10^{-5}$		×	1961	(calculated from data), [434, 437]
TaN**	1927	—	$8.19 \cdot 10^{-5}$	—			
TaN**	2027	—	$2.43 \cdot 10^{-4}$				
TaN**	2127	—	$6.61 \cdot 10^{-3}$				
TaN**	2227	—	$1.66 \cdot 10^{-3}$				
TaN**	2327	—	$3.83 \cdot 10^{-3}$				
TaN**	2427	—	$8.32 \cdot 10^{-3}$				
CrN**	1000	—	$6.6 \cdot 10^{-4}$				
CrN**	1100	—	$2.58 \cdot 10^{-3}$				
CrN**	1200	—	$8.22 \cdot 10^{-3}$	—	[1316]	1956	[434, 437]
CrN**	1300	—	$2.32 \cdot 10^{-2}$				
CrN**	1400	—	$5.54 \cdot 10^{-2}$				
Mo₃N**	820	—	$4.76 \cdot 10^{-1}$	—	[703]	1936	[437]
Mo₂N**	203	—	$1.32 \cdot 10^{-6}$				
Mo₂N**	243	—	$1.32 \cdot 10^{-5}$	—	[177]	1950	
Mo₂N**	283	—	$1.32 \cdot 10^{-4}$				
Mo₂N**	329	—	$1.32 \cdot 10^{-3}$				

117

VAPOR PRESSURE AND RATE OF EVAPORATION (TOTAL) (continued)

Phase	Temperature, °C	Total rate of evaporation, g/cm²·sec	Vapor pressure, atm	Vapor pressure equation	Source	Year	Remarks
W_2N**	221	—	$1.32 \cdot 10^{-6}$	—	[177]	1950	
W_2N**	256	—	$1.32 \cdot 10^{-5}$				
W_2N**	296	—	$1.32 \cdot 10^{-4}$				
W_2N**	344	—	$1.32 \cdot 10^{-3}$				
USi**	1452	—	$0.17 \cdot 10^{-7}$	—	[1317]	1962	
USi**	1477	—	$0.30 \cdot 10^{-7}$				
USi**	1502	—	$0.51 \cdot 10^{-7}$				
USi**	1527	—	$0.86 \cdot 10^{-7}$				
USi**	1552	—	$1.46 \cdot 10^{-7}$				
U_3Si_5**	1402	—	$0.24 \cdot 10^{-7}$	—	[1317]	1962	
U_3Si_5**	1427	—	$0.42 \cdot 10^{-7}$				
U_3Si_5**	1452	—	$0.71 \cdot 10^{-7}$				
U_3Si_5**	1477	—	$1.20 \cdot 10^{-7}$				
U_3Si_5**	1502	—	$2.04 \cdot 10^{-7}$				
U_3Si_5**	1527	—	$3.37 \cdot 10^{-7}$				
USi_2**	1402	—	$0.53 \cdot 10^{-7}$	—	[1317]	1962	
USi_2**	1427	[	$0.93 \cdot 10^{-7}$				
USi_2**	1452	—	$1.59 \cdot 10^{-7}$				
USi_2**	1477	—	$2.48 \cdot 10^{-7}$				
USi_2**	1502	—	$4.22 \cdot 10^{-7}$				
USi_2**	1527	—	$7.06 \cdot 10^{-7}$				

VAPOR PRESSURE AND RATE OF EVAPORATION (TOTAL) (continued)

Phase	Temperature, °C	Total rate of evaporation, g/cm²·sec	Vapor pressure, atm	Vapor pressure equation	Source	Year	Remarks
USi$_3$**	1402	—	$0.25 \cdot 10^{-6}$	—	[1317]	1962	
USi$_3$**	1427	—	$0.39 \cdot 10^{-6}$				
USi$_3$**	1452	—	$0.61 \cdot 10^{-6}$				
USi$_3$**	1477	—	$0.99 \cdot 10^{-6}$				
USi$_3$**	1502	—	$1.52 \cdot 10^{-6}$				
TaSi$_{0.22}$**	1960	—	$2.62 \cdot 10^{-7}$	$\lg P_{Si} = 6.6 - \dfrac{28660}{T}$ [P, atm]	[438]	1957	
TaSi$_{0.22}$**	1986	—	$2.90 \cdot 10^{-7}$				
TaSi$_{0.22}$**	2010	—	$4.18 \cdot 10^{-7}$				
TaSi$_{0.22}$**	2036	—	$7.14 \cdot 10^{-7}$				
TaSi$_{0.22}$**	2056	—	$10.5 \cdot 10^{-7}$				
TaSi$_{0.22}$**	2118	—	$34.7 \cdot 10^{-7}$				
TaSi$_{0.22}$**	2128	—	$37.5 \cdot 10^{-7}$				
Ta$_2$Si**	1943	—	$7.57 \cdot 10^{-7}$	$\lg P_{Si} = 7.0 - \dfrac{28000}{T}$ [P, atm]	[438]	1957	
Ta$_2$Si**	1968	—	$11.2 \cdot 10^{-7}$				
Ta$_2$Si**	1989	—	$17.5 \cdot 10^{-7}$				
Ta$_2$Si**	2027	—	$27.5 \cdot 10^{-7}$				
Ta$_2$Si**	2055	—	$37.1 \cdot 10^{-7}$				
Ta$_2$Si**	2073	—	$50.9 \cdot 10^{-7}$				
Ta$_2$Si**	2100	—	$52.1 \cdot 10^{-7}$				
Ta$_2$Si**	2114	—	$92.1 \cdot 10^{-7}$				
Ta$_2$Si**	2147	—	$97.9 \cdot 10^{-7}$				

VAPOR PRESSURE AND RATE OF EVAPORATION (TOTAL) (continued)

Phase	Temperature, °C	Total rate of evaporation, $g/cm^2 \cdot sec$	Vapor pressure, atm	Vapor pressure equation	Source	Year	Remarks
TaSi$_{0.60}$**	1760	—	$4.70 \cdot 10^{-7}$				
TaSi$_{0.60}$**	1833	—	$12.3 \cdot 10^{-7}$				
TaSi$_{0.60}$**	1874	—	$17.9 \cdot 10^{-7}$				
TaSi$_{0.60}$**	1940	—	$47.8 \cdot 10^{-7}$	$\lg P_{Si} = 6.2 - \dfrac{25000}{T}$ [P, atm]	[438]	1957	
TaSi$_{0.60}$**	1965	—	$55.0 \cdot 10^{-7}$				
TaSi$_{0.60}$**	2031	—	$138 \cdot 10^{-7}$				
TaSi$_{0.60}$**	2040	—	$189 \cdot 10^{-7}$				
TaSi$_2$**	1630	—	$1.21 \cdot 10^{-7}$				
TaSi$_2$**	1639	—	$1.80 \cdot 10^{-7}$				
TaSi$_2$**	1731	—	$4.79 \cdot 10^{-7}$				
TaSi$_2$**	1738	—	$5.08 \cdot 10^{-7}$				
TaSi$_2$**	1764	—	$8.19 \cdot 10^{-7}$	—	[438]	1957	
TaSi$_2$**	1777	—	$8.92 \cdot 10^{-7}$				
TaSi$_2$**	1781	—	$9.29 \cdot 10^{-7}$				
TaSi$_2$**	1845	—	$23.1 \cdot 10^{-7}$				
TaSi$_2$**	1891	—	$68.4 \cdot 10^{-7}$				
Mo$_3$Si**	1794	—	$4.74 \cdot 10^{-7}$				
Mo$_3$Si**	1799	—	$6.55 \cdot 10^{-7}$				
Mo$_3$Si**	1825	—	$7.44 \cdot 10^{-7}$	$\lg P_{Si} = 28.94 - 5.75 \lg T - \dfrac{33690}{T}$ [P, atm]	[864]	1960	
Mo$_3$Si**	1860	—	$14.30 \cdot 10^{-7}$				
Mo$_3$Si**	1945	—	$26.8 \cdot 10^{-7}$				
Mo$_3$Si**	1968	—	$31.00 \cdot 10^{-7}$				

VAPOR PRESSURE AND RATE OF EVAPORATION (TOTAL) (continued)

Phase	Temperature, °C	Total rate of evaporation, g/cm²·sec	Vapor pressure, atm	Vapor pressure equation	Source	Year	Remarks
Mo₅Si₃**	1777	–	$4.08 \cdot 10^{-7}$				
Mo₅Si₃**	1811	–	$5.29 \cdot 10^{-7}$				
Mo₅Si₃**	1853	–	$9.87 \cdot 10^{-7}$				
Mo₅Si₃**	1863	–	$12.70 \cdot 10^{-7}$				
Mo₅Si₃**	1877	–	$16.30 \cdot 10^{-7}$	$\lg P_{Si} = 28.67 - 5.75 \lg T - \dfrac{32940}{T}$ [P, atm]	[864]	1960	
Mo₅Si₃**	1914	–	$25.30 \cdot 10^{-7}$				
Mo₅Si₃**	1933	–	$31.0 \cdot 10^{-7}$				
Mo₅Si₃**	1955	–	$43.2 \cdot 10^{-7}$				
Mo₅Si₃**	1982	–	$63.00 \cdot 10^{-7}$				
Mo₅Si₃**	1988	–	$66.90 \cdot 10^{-7}$				
MoSi₂**	1653	–	$2.05 \cdot 10^{-7}$				
MoSi₂**	1697	–	$3.71 \cdot 10^{-7}$				
MoSi₂**	1730	–	$4.96 \cdot 10^{-7}$				
MoSi₂**	1749	–	$7.94 \cdot 10^{-7}$	$\lg P_{Si} = 28.62 - 5.75 \lg T - \dfrac{29800}{T}$ [P, atm]	[864]	1960	[430]
MoSi₂**	1800	–	$12.40 \cdot 10^{-7}$				
MoSi₂**	1809	–	$18.90 \cdot 10^{-7}$				
MoSi₂**	1816	–	$25.40 \cdot 10^{-7}$				
MoSi₂**	1866	–	$49.20 \cdot 10^{-7}$				
MoSi₂**	1833	–	$52.00 \cdot 10^{-7}$				
Re₃Si	–	–	–	$\lg P_{Si} = 5.953 - \dfrac{24040}{T}$ [P, atm]	[416]	1953	

VAPOR PRESSURE AND RATE OF EVAPORATION (TOTAL) (continued)

Phase	Temperature, °C	Total rate of evaporation, g/cm²·sec	Vapor pressure, atm	Vapor pressure equation	Source	Year	Remarks
ReSi	-	-	-	$\lg P_{Si} = 7.444 - \dfrac{25800}{T}$ [P, atm]	[416]	1953	
ReSi₂	-	-	-	$\lg P_{Si} = 7.512 - \dfrac{25610}{T}$ [P, atm]	[416]	1953	
BaS***	1620	-	$1.1 \cdot 10^{-6}$	-	[205]	1939	
Ce₃N₄***	1700	-	10^{-7}	-	[177]	1950	
Ce₂S₃***	1840	-	$\sim 10^{-6}$	-	[177]	1950	
Nd₂S₃***	1900	-	$\sim 10^{-6}$	-	[177]	1950	
Th₂S₃***	1700	-	$< 10^{-7}$	-	[177]	1950	
Th₄S₇***	1700	-	$< 10^{-7}$	-	[177]	1950	
ThS₂***	1700	-	$< 10^{-7}$	-	[177]	1950	
US***	1700	-	$< 10^{-7}$	-	[177]	1950	
SiC**	1849	$2.5 \cdot 10^{-5}$	$9.17 \cdot 10^{-6}$	-	[1296]	1961	[1011,1191, 1297-1300]; reduced to vapor pressure of silicon
SiC**	1851	$3.64 \cdot 10^{-5}$	$11.6 \cdot 10^{-6}$				
SiC**	1897	$5.09 \cdot 10^{-5}$	$18.2 \cdot 10^{-6}$				
SiC**	1915	$2.58 \cdot 10^{-5}$	$9.08 \cdot 10^{-6}$				
SiC	-	-	-	$\lg P_{Si} = 17.414 - \dfrac{52420}{T}$ [P, atm]	[442]	1959	

VAPOR PRESSURE AND RATE OF EVAPORATION (TOTAL) (continued)

Phase	Temperature, °C	Total rate of evaporation, g/cm²·sec	Vapor pressure, atm	Vapor pressure equation	Source	Year	Remarks
BN**	1227	—	$2.62 \cdot 10^{-9}$				
BN**	1327	—	$2.66 \cdot 10^{-8}$				[439, 440,
BN**	1427	—	$1.06 \cdot 10^{-7}$				1170, 1314];
BN**	1527	—	$2.44 \cdot 10^{-6}$	—	[893]	1961	reduced to va-
BN**	1627	—	$1.48 \cdot 10^{-5}$				por pressure of
BN**	1727	—	$6.17 \cdot 10^{-5}$				nitrogen
BN**	1827	—	$3.11 \cdot 10^{-4}$				
BN**	1927	—	$1.16 \cdot 10^{-3}$				
BN**	2027	—	$3.76 \cdot 10^{-2}$				
Si₃N₄**	927	$1.87 \cdot 10^{-8}$	$2.76 \cdot 10^{-9}$				
Si₃N₄**	1027	$3.23 \cdot 10^{-7}$	$4.97 \cdot 10^{-8}$	$\lg P_{N_2} = 16.5 - \dfrac{37806}{T}$ [1200–1700°K]			[177, 441];
Si₃N₄**	1127	$3.09 \cdot 10^{-6}$	$4.93 \cdot 10^{-7}$				reduced to va-
Si₃N₄**	1227	$2.58 \cdot 10^{-5}$	$4.26 \cdot 10^{-6}$	$\lg P_{N_2} = 21.2 - \dfrac{45673}{T}$ [2700–2000°K]	[1315]	1959	por pressure of
Si₃N₄**	1327	$1.64 \cdot 10^{-4}$	$2.80 \cdot 10^{-5}$	[P, atm]			nitrogen
Si₃N₄**	1527	$4.27 \cdot 10^{-3}$	$7.73 \cdot 10^{-4}$				
Si₃N₄**	1727	$7.55 \cdot 10^{-2}$	$1.44 \cdot 10^{-2}$				
B	1420	$1.58 \cdot 10^{-5}$	$3.89 \cdot 10^{-6}$	$\lg P_B = 9.6 - \dfrac{19000}{T}$ [P, mm Hg]	[536]	1960	
B	1491	$5.66 \cdot 10^{-5}$	$1.10 \cdot 10^{-5}$				
SiP	—	—	—	$\lg P_P = 10.90 - \dfrac{15400}{T}$ [P, atm]	[442]	1959	

123

Phase	Thermal conductivity, cal/cm · deg · sec	Accuracy (±), cal/cm · deg · sec	Temperature, °C	Ref.	Year	Remarks
YB_6	0.070	± 0.010	20	[1113]	1963	
LaB_6	0.114	± 0.010	20	[1113]	1963	
CeB_6	0.081	± 0.002	20	[1113]	1963	
PrB_6	0.098	± 0.005	20	[1113]	1963	
NdB_6	0.113	± 0.008	20	[1113]	1963	
SmB_6	0.033	± 0.004	20	[1113]	1963	
EuB_6	0.055	± 0.002	20	[1113]	1963	
GdB_6	0.049	± 0.003	20	[1113]	1963	
GdB_6	0.0466	—	80	[285]	1959	[846]
GdB_6	0.0503	—	131	[285]	1959	[846]
GdB_6	0.0519	—	184	[285]	1959	[846]
GdB_6	0.0536	—	203	[285]	1959	[846]
TbB_6	0.048	± 0.003	20	[1113]	1963	
YbB_6	0.060	± 0.004	20	[1113]	1963	
TiB_2	0.058	—	23	[288]	1950	P^{1*} = [1010]
TiB_2	0.063	—	200	[288]	1950	P^{1*} = 15%, [40, 178, 918, 931]
TiB_2	0.010	—	1500	[918]	1961	[1010]
ZrB_2	0.058	—	23	[288]	1950	[178]
ZrB_2	0.060	—	200	[288]	1950	P = 15% [40]
ZrB_{12}	0.029	—	—	[35]	1952	
NbB_2	0.040	—	23	[288]	1950	[342]
NbB_2	0.047	—	200	[288]	1950	[342]
TaB_2	0.026	—	—	[342]	1950	[3]
TaB_2	0.033	—	200	[342]	1950	
Cr_4B	0.0262	0,0009	20	[931]	1961	P = 0, [1166]
Cr_2B	0.0400	0,0018	20	[931]	1961	P = 0, [1166]
CrB	0.0483	0,0019	20	[931]	1961	P = 0, [1166]
CrB_2	0.0534	0,0037	20	[931]	1961	P = 0, [1166]
Cr_2B_5	0.0430	± 0,0020	20	[1166]	1962	
Mo_2B_5	0.064	—	20	[2]	1957	
W_2B_5	0.076	—	20	[2]	1957	
Be_2C^{2*}	0.123	—	150	[526]	1952	[346]
ThC	0.069	—	25	[1278]	1962	$C_{tot.}$5.03% C_{free}0.25%
ThC_2	0.057	—	25	[1278]	1962	$C_{tot.}$9.85% C_{free}1.07%
UC	0.04—0.06	—	—	[221]	1960	[534, 1095]
UC	0.028	—	150	[856]	1960	Data taken from graph
UC	0.03	—	380	[856]	1960	Same
UC	0.04	—	700	[856]	1960	"

[1*] P = porosity.

[2*] $\lambda_{300-950°} = 0.036 \cdot 10^{-4} t + 87 \cdot 10^{-4}$ (± 200%), cal/cm · sec · deg [705, 998].

Phase	Thermal conduc-tivity, cal/cm · deg · sec	Accuracy (±), cal/cm · deg · sec	Tem-pera-ture, °C	Ref.	Year	Remarks
UC	0.080	0.002	60	[1014]	1959	P = 25%
UC	0.074	0.002	115	[1014]	1959	P = 25%
UC	0.061	0.002	195	[1014]	1959	P = 25%
UC	0.050	0.002	265	[1014]	1959	P = 25%
TiC	0.0869	0.0066	20	[931]	1961	P = 0, [1047]
TiC	~0.09	—	600	[1287]	1961	[178,239,263, 192,380,342]
TiC	~0.095	—	800	[1287]	1961	Same
TiC	0.098	—	966	[1287]	1961	[1286]
TiC	0.10	—	1200	[1287]	1961	
TiC	0.102	—	1400	[1287]	1961	[1286]
TiC	0.108	—	1600	[1287]	1961	
TiC	0.11	—	1800	[1287]	1961	
TiC	0.112	—	2000	[1287]	1961	
ZrC	0.10	—	0	[1047]	1961	[342, 178]
ZrC	0.0750	—	530	[1151]	1962	P = 5%
ZrC	0.104	—	2100	[1151]	1962	P = 5%
HfC	0.070	—	0	[1047]	1961	
VC	0.094	—	0	[1047]	1961	[1, 931]
NbC	0.044	—	0	[1047]	1961	[342, 178]
TaC	0.053	—	—	[178]	1959	[342, 1047]
$Cr_{23}C_6$	0.0437	0.0030	20	[931]	1961	P = 0, [1166]
Cr_7C_3	0.0364	0.0015	20	[931]	1961	P = 0, [1166]
Cr_3C_2	0.0458	0.0002	20	[931]	1961	P = 0, [1166]
Mo_2C	0.016	—	—	[2]	1957	
W_2C	0.07	—	—	[344]	1948	
WC	0.47	—	—	[1047]	1961	[343]
UN	0.02	—	100—700	[1096]	1962	
AlN	0.072	—	200	[674]	1960	
AlN	0.060	—	400	[674]	1960	
AlN	0.053	—	600	[674]	1960	
AlN	0.048	—	800	[674]	1960	
TiN	0.046	0.003	20	[931]	1961	P = 0
TiN	0.070	—	100	[239]	1954	Data taken from graph,[178]
TiN	0.019	—	600	[239]	1954	Same
TiN	0.014	—	950	[239]	1954	"
ZrN	0.049	0.002	20	[997]	1961	P = 0, [178]
ZrN	0.033	—	200	[239]	1954	Data taken from graph
ZrN	0.018	—	490	[239]	1954	Same
ZrN	0.013	—	800	[239]	1954	
VN	0.0270	0.007	20	[931]	1961	P = 0, [997]
Nb_2N	0.0200	0.008	20	[929]	1961	P = 0, [997]
$NbN_{0.75}$	0.0191	0.004	20	[929]	1961	P = 0, [997]
NbN	0.009	0.002	20	[997]	1961	[1]
Ta_2N	0.0240	0.005	20	[929]	1961	[997]

Phase	Thermal conduc- tivity, cal/cm · deg · sec	Accuracy (±), cal/cm · deg · sec	Tem- pera- ture, °C	Ref.	Year	Remarks
TaN	0.0205	0.009	20	[997]	1961	[1]
Cr_2N	0.0519	0.004	20	[931]	1961	P = 0,[997,1166]
CrN	0.0284	0.0023	20	[931]	1961	P = 0,[997,1166]
Mo_2N	0.0427	0.007	20	[929]	1961	P = 0,[997]
Mg_2Si	0.0186	—	27	[1284]	1963	Data taken from graph
Mg_2Si	0.0136	—	127	[1284]	1963	Same
Mg_2Si	0.011	—	227	[1284]	1963	"
Mg_2Si	0.01	—	277	[1284]	1963	"
$BaSi_2$	~0.0037	—	20	[1289]	1963	
$LaSi_2$	~0.0263	—	20	×	1963	V. S. Neshpor
$GdSi_2$	~0.0145	—	20	×	1963	"
U_3Si	0.04	—	—	[690]	1960	[1091]
Ti_5Si_3	0.0363	—	20	×	1963	V. S. Neshpor
TiSi	0.041	—	20	×	1963	"
$TiSi_2$	0.111	—	20	×	1963	"
Zr_5Si_3	0.029	—	20	×	1963	"
$ZrSi_2$	0.0373	—	20	×	1963	"
V_3Si	0.031	—	20	×	1963	"
V_5Si_3	0.03	—	20	×	1963	"
VSi_2	0.0383	—	20	×	1963	"
$NbSi_2$	0.0397	—	20	×	1963	"
Ta_4Si	0.0214	—	20	×	1963	"
Ta_5Si_3	0.0238	—	20	×	1963	"
$TaSi_2$	0.0521	—	20	×	1963	"
Cr_3Si	0.0871	—	20	×	1963	"
Cr_5Si_3	0.026	—	20	×	1963	"
CrSi	0.0284	—	20	×	1963	"
$CrSi_2$	0.0253	—	20	×	1963	"
Mo_3Si	0.095	—	20	[1241]	1963	Si bound 8.6%
Mo_5Si_3	0.052	—	20	[1241]	1963	Si bound 14.2%
$MoSi_2$	0.1165	—	20	[1241]	1963	Si bound 36.9%
WSi_2	0.114	—	20	×	1963	V. S. Neshpor
Mn_3Si	0.087	—	20	×	1963	"
Mn_5Si_3	0.016	—	20	×	1963	"
MnSi	0.0224	—	20	×	1963	V. S. Neshpor, [303, 1037]
$MnSi_2$	0.0202	—	20	×	1963	Same
Re_3Si	0.186	—	20	×	1963	V. S. Neshpor
ReSi	0.059	—	20	×	1963	"
$ReSi_2$	0.0452	—	20	×	1963	"
Fe_3Si	0.0405	—	20	×	1963	"
FeSi	0.0242	—	20	×	1963	"
$FeSi_2$	0.0275	—	20	×	1963	"
Co_3Si	0.031	—	20	×	1963	"
CoSi	0.0492	—	20	×	1963	"
$CoSi_2$	0.0359	—	20	×	1963	"

Phase	Thermal conductivity, cal/cm deg·sec	Accuracy (±), cal/cm ·deg·sec	Temperature, °C	Ref.	Year	Remarks
Ni₃Si	0.0434	—	20		1963	„
Ni₅Si₃ (Ni₃Si₂)	~0.038	—	20		1963	„
NiSi₂	~0.0238	—	20		1963	„
LaS	0.177	0.003	20	[1221]	1963	
La₂S₃	0.0229	0.0001	20	[1221]	1963	
CeS	0.0938	0.005	20	[1221]	1963	
Ce₂S₃	0.0349	0.0002	20	[1221]	1963	[1114]
Pr₂S₃	0.0244	0.0002	20	[1221]	1963	
Nd₂S₃	0.0346	0.0001	20	[1221]	1963	
B₄C	0.29	—	100	[346]	1956	P = 1%
B₄C	0.22	—	300	[346]	1956	P = 1%
B₄C	0.18	—	500	[346]	1956	P = 1%
B₄C	0.155	—	700	[346]	1956	P = 1%
B₄C	0.16	—	100	[346]	1956	P = 24%
B₄C	0.14	—	300	[346]	1956	P = 24%
B₄C	0.115	—	500	[346]	1956	P = 24%
B₄C	0.095	—	700	[346]	1956	P = 24%
SiC	0.02	—	—	[271]	1956	Black
SiC	0.10		—	[705]	1950*	
SiC	0.24	—	200	[674]	1960	
SiC	0.10	—	800	[674]	1960	
SiC	0.04	—	871	[385]	1949	Recrystallized
SiC	0.033	—	1093	[385]	1949	„
SiC	0.028	—	1316	[385]	1949	„
SiC	0.170	—	400	[1000]	1961	P = 4%
SiC	0.110	—	1000	[1000]	1961	Comp., % 96.5 SiC, 2.5 Si$_{free}$, 0.4 C$_{free}$, 0.4 Al, 0.2 Fe
BN	0.036	—	300	[272]	1955	
BN	0.034	—	500	[272]	1955	Parallel to
BN	0.032	—	700	[272]	1955	hot-pressing
BN	0.030	—	900	[272]	1955	direction
BN	0.029	—	1000	[272]	1955	
BN	0.069	—	300	[272]	1955	
BN	0.067	—	500	[272]	1955	Perpendicular
BN	0.065	—	700	[272]	1955	to hot-pressing
BN	0.063	—	900	[272]	1955	direction
BN	0.064	—	1000	[272]	1955	
Si₃N₄	0.041	—	—	[209]	1950	α-, β-phase mixt.
B	0.003	—	20—80	[440]	1960	
C	1.6—3.9	—	20	[944]	1959	Pyrographite

Phase	Coefficient of thermal expansion α, $(1/°C)\cdot10^6$	Accuracy of measurement $(\pm)$, $(1/°C)\cdot10^6$	Temperature interval, °C	Ref.	Year	Remarks
CaB_6	6.5	0.5	20—800	[891]	1961	[3, 5]
SrB_6	6.7	0.5	20—800	[891]	1961	[5]
BaB_6	6.8	0.5	20—800	[891]	1961	[3, 258]
ScB_2	6.8—7.6	—	20—800	×	1962	N. N. Zhuravlev
ScB_4	4.1	—	20—1000	×	1963	T. S. Verkhoglya-dova
YB_6	6.2	0.5	20—800	[891]	1961	
LaB_6	6.4	0.5	20—800	[891]	1961	[3,5,1112]
CeB_6	7.3	0.5	20—800	[891]	1961	[3,5, 1210]
PrB_6	7.5	0.5	20—800	[891]	1961	
NdB_6	7.3	1.0	20—800	[891]	1961	
SmB_6	6.8	0.5	20—800	[891]	1961	[18]
EuB_6	6.9	0.5	20—800	[891]	1961	
GdB_6	8.7	0.5	20—800	[891]	1961	
TbB_6	7.8	1.0	20—800	[891]	1961	
HoB_{12}	3.0	0.2	-160—420	[1283]	1961	
ErB_{12}	3.0	0.2	20—350	[1283]	1961	
YbB_6	5.8	0.5	20—800	[891]	1961	
ThB_6	7.8	0.5	20—800	[891]	1961	
UB_2	8.5	—	20—205	[259]	1956	[1097]
TiB_2	5.5	0.8	17—400	[261]	1959	[393 918, 920]
TiB_2	8.1	—	25—1300	[918]	1961	[1010]
ZrB_2	6.88	—	20—1100	[260]	1957	[261, 391, 393]
HfB_2	5.73	—	20—1100	[290]	1959	[36, 260]
VB_2	5.3	—	20—1100	[372]	1958	
NbB_2	7.9—8.3	—	20—1100	×	1959	É. P. Lapteva
TaB_2	11.6	—	20—870	[1117]	1962	Along axis a
TaB_2	11.4	—	20—870	[1117]	1962	Along axis c
Cr_4B	8.2	—	20—1100		1958	P. S. Kislyi
CrB	9.5	—	20—1100	[225]	1959	
CrB_2	11.1	—	20—1100	[260]	1957	[225, 393]
Be_2C	10.5	—	20—600	[346]	1956	[526]
ScC	11.4	—	20—1100	×	1963	T. Ya. Kosolapova, G. N. Makarenko
La_2C_3	9.9	—	20—1100	×	1963	Same
α-LaC_2	12.1	—	20—1100	×	1963	"
Ce_2C_3	10.4	—	20—1100	×	1963	"
CeC_2	10.1	—	20—1100	×	1963	"
PrC_2	11.4	—	20—1100	×	1963	"
UC	10.4	—	20—1000	[920]	1961	[856, 1014]
TiC	7.74	0.12	12—270	[262]	1958	[178, 263, 279, 554, 920]
ZrC	6.73	—	20—1100	[263]	1950	[262, 554, 920]
HfC	6.59	0.04	25—612	[1004]	1960	[920, 994, 1242]
VC	7.2	0.6	17—190	[262]	1958	
Nb_2C	7.0	0.3	12—190	[261]	1959	
NbC	6.65	—	20—1100	[1237]	1958	[262, 921]

Phase	Coefficient of thermal expansion α, $(1/°C)\cdot10^6$	Accuracy of measurement $(\pm)$, $(1/°C)\cdot10^6$	Temperature interval, °C	Ref.	Year	Remarks
TaC	6.29	—	20–1100	[1237]	1958	[178, 264, 921]
$Cr_{23}C_6$	10.1	—	—	[659]	1953	
Cr_7C_3	10.6	—	—	[659]	1953	
Cr_3C_2	10.3	—	—	[659]	1953	
Mo_2C	7.8	0.5	12–190	[261]	1959	[262, 921]
W_2C	5.8	0.2	17–270	[261]	1959	[264]
WC	3.84	—	22–400	[261]	1959	[264]
Fe_3C	4.6	—	[−187−30]	[1146]	1962	$\alpha_a = 1.2$ $\alpha_b = 1.4$ $\alpha_c = 11.3$
Fe_3C	6.0–6.84	—	20–100	[1147]	1960	
AlN	4.03	—	25–200	[674]	1960	[1184]
LaN	9.0	—	[−183−427]	[1099]	1962	
CeN	30	—	[−183−427]	[1099]	1962	
PrN	13	—	[−183−427]	[1099]	1962	
TiN	9.35	0.04	25–1100	[266]	1955	[262, 920]
ZrN	7.24	—	20–1100	[929]	1961	P = 4.4%, [262, 618]
HfN	6.9	—	20—1100	[929]	1961	P = 3.1%
V_3N	8.1	—	20—1100	[929]	1961	P = 10%
VN	8.1	—	20—1100	[929]	1961	P = 4.5%, [262]
Nb_2N	3.26	—	20—1000	[907]	1961	
NbN	10.1	0.2	20—270	[262]	1958	[929]
Ta_2N	5.2	—	20—1000	[929]	1961	P = 5.4%, [262]
TaN	3.6	—	20—700	[929]	1961	P = 7.8%
Cr_2N	9.41	—	20—1100	[929]	1961	P = 6.1%
CrN	2.3	—	20—800	[929]	1961	P = 9.9%
CrN	7.5	—	850—1040	[929]	1961	P = 9.9%
Mo_2N	4.5	—	20—790	[929]	1961	P = 12.3%
Mo_2N	6.2	—	20—1100	[929]	1961	
Fe_4N	7.9	—	18—386	[699]	1936	
Fe_1N	22.2	—	—	[699]	1936	
Mg_2Si	14.8	—	—	[117]	1959	
$BaSi_2$	8.2	—	90–690	[1131]	1963	[1289]
$BaSi_2$	8.6	—	690–1090	[1131]	1963	[1289]
$LaSi_2$	7.8	—	20–570	[1131]	1963	
$LaSi_2$	11.0	—	570–1070	[1131]	1963	
U_3Si	16.0	—	20–800	[690]	1960	
β-USi_2	57	—	20–205	[259]	1956	
Ti_5Si_3	11.0	—	170–1070	[1131]	1963	
TiSi	8.8	—	20–370	[1131]	1963	
TiSi	10.4	—	370–1070	[1131]	1963	
$TiSi_2$	10.5	—	20–1070	[1131]	1963	
$ZrSi_2$	8.6	—	20–1070	[1131]	1963	
V_3Si	8.0	—	20–620	[1131]	1963	
V_3Si	12.0	—	620–820	[1131]	1963	
V_3Si	14.1	—	820–1070	[1131]	1963	

THERMAL EXPANSION (continued)

Phase	Coefficient of thermal expansion α, $(1/°C)\cdot10^6$	Accuracy of measurement $(\pm)$, $(1/°C)\cdot10^6$	Temperature interval, °C	Ref.	Year	Remarks
V_5Si_3	9.5	—	20—770	[1131]	1963	
V_5Si_3	11.1	—	770—1070	[1131]	1963	
VSi_2	11.2	—	20—770	[1131]	1963	
VSi_2	14.65	—	770—1070	[1131]	1963	
Nb_5Si_3	7.3	—	20—650	[920]	1961	Along axis a
Nb_5Si_3	4.6	—	20—650	[920]	1961	Along axis c
$NbSi_2$	8.4	—	20—370	[1131]	1963	
$NbSi_2$	11.7	—	370—1070	[1131]	1963	
Ta_5Si_3	5.5	—	20—1000	[920]	1961	Along axis a
Ta_5Si_3	8.0	—	20—1000	[920]	1961	Along axis c
Ta_5Si_3	6.3	—	20—1000	[920]	1961	Along axis a, Nowotny phase
Ta_5Si_3	6.6	—	20—1000	[920]	1961	Along axis c, Nowotny phase
$TaSi_2$	8.9	—	20—1000	[920]	1961	Along axis a, [1117]
$TaSi_2$	8.8	—	20—1000	[920]	1962	Along axis c, [1117]
Cr_3Si	10.5	—	20—1000	[1131]	1963	
Cr_5Si_3	6.0	—	20—170	[1131]	1963	[1265]
Cr_5Si_3	10.6	—	170—720	[1131]	1963	
Cr_5Si_3	14.2	—	720—1070	[1131]	1963	
$CrSi$	11.3	—	20—770	[1131]	1963	
Mo_5Si_3	4.3	—	20—270	[1131]	1963	[1265]
Mo_5Si_3	6.7	—	270—1070	[1131]	1963	
$MoSi_2$	8.25	—	20—1070	[1131]	1963	[1117]
WSi_2	6.25	—	20—420	[1131]	1963	
WSi_2	7.90	—	420—1070	[1131]	1963	
$MnSi$	16.3	1.0	20—800	[1150]	1962	
$ReSi$	2.7	—	20—220	[1131]	1963	
$ReSi$	4.9	—	220—1070	[1131]	1963	
$ReSi_2$	6.6	—	20—1070	[1131]	1963	
Fe_3Si	14.4	—	220—870	[1131]	1963	
Fe_3Si	11.0	—	870—1070	[1131]	1963	
$FeSi$	12.7	—	20—1070	[1131]	1963	
$FeSi_2$	6.7	—	20—1070	[1131]	1963	
Co_3Si	13.4	—	20—520	[1131]	1963	
Co_3Si	16.6	—	520—970	[1131]	1963	
$CoSi$	11.1	1.0	20—800	[1150]	1962	
Ni_3Si	9.0	—	20—370	[1131]	1963	
Ni_3Si	11.5	—	370—770	[1131]	1963	
Ni_3Si	14.85	—	770—1070	[1131]	1963	
Ni_2Si	16.5	—	20—870	[1131]	1963	
Ni_2Si	19.0	—	870—1070	[1131]	1963	
LaS	11.62	0.27	20—1000	[1132]	1963	
La_2S_3	9.9	0.67	20—1000	[1132]	1963	
CeS	12.37	0.12	20—1000	[1132]	1963	

THERMAL EXPANSION (continued)

Phase	Coefficient of thermal expansion α, $(1/°C) \cdot 10^6$	Accuracy of measurement $(\pm)$, $(1/°C) \cdot 10^6$	Temperature interval, °C	Ref.	Year	Remarks
Ce_2S_3	10.45	0.67	20—1000	[1132]	1963	
PrS	14.3	0.2	20—1000	[1132]	1963	
Pr_2S_3	11.28	0.40	20—1000	[1132]	1963	
NdS	15.35	0.34	20—1000	[1131]	1963	
Nd_2S_3	12.90	0.5	20—1000	[1131]	1963	
B_4C	4.5	—	—	[269]	1934	[178. 263, 279 554]
SiC	5.68	0.11	100—2400	[279]	1959	[270, 271]
α-BN	7.51	—	25—1000	[272]	1955	
Si_3N_4	2.75	—	20—1000	[270]	1955	[319, 1035]
SiB_6	5.9	—	20—500	[5]	1957	
B	8.3	3	20—750	[440]	1960	
C	0.66	—	—	[944]	1959	Pyrographite

ENERGY OF THE CRYSTAL LATTICE

Phase	U, kcal/mole	Ref.	Year	Remarks
CaB_6	1230	[823] [1*]	1961	
SrB_6	1220	[823] [1*]	1961	
BaB_6	1200	[823] [1*]	1961	
YB_6	1780	[823] [1*]	1961	
LaB_6	1770	[823] [1*]	1961	
CeB_6	2410	[823] [1*]	1961	
PrB_6	1780	[823] [1*]	1961	
NdB_6	1780	[823] [1*]	1961	
PmB_6 [2*]	~1770	[823] [1*]	1961	[880]
SmB_6	1780	[823] [1*]	1961	
EuB_6	1220	[823] [1*]	1961	
GdB_6	1790	[823] [1*]	1961	
TbB_6	1790	[823] [1*]	1961	
DyB_6	1790	[823] [1*]	1961	
HoB_6	1790	[823] [1*]	1961	
TuB_6	1790	[823] [1*]	1961	
ThB_6	2430	[823] [1*]	1961	
TiB_2	3260	$\times$ [3*]	1961	
ZrB_2	2540	[457]	1957	
VB_2	2880	[457]	1957	

ENERGY OF THE CRYSTAL LATTICE (continued)

Phase	U, kcal/mole	Ref.	Year	Remarks
NbB$_2$	3060	[457]	1957	
TaB	1610	× [3*]	1955	
TaB$_2$	2940	[457]	1957	
CrB	2140	× [3*]	1955	
Mo$_2$B	2620	[457]	1957	
MoB	1870	× [3*]	1961	
W$_2$B	2470	[457]		
ThC$_2$	1960	× [3*]	1961	
UC$_2$	1970	× [3*]	1961	
TiC	3890	× [3*]	1961	
ZrC	3470	× [3*]	1961	
HfC	2800	× [3*]	1961	
VC	3900	× [3*]	1961	
NbC	3220	× [3*]	1961	
Nb$_2$C	2570	× [3*]	1961	
TaC	2770	× [3*]	1961	
Ta$_2$C	2390	× [3*]	1961	
Mo$_2$C	2280	× [3*]	1961	
WC	2760	× [3*]	1961	
W$_2$C	2000	× [3*]	1961	
ScN	1062	[988]	1959	
LaN	860	[988]	1959	
TiN	3900	[988]	1959	
ZrN	3540	[988]	1959	
HfN	2840	× [3*]	1961	
VN	3890	[988]	1959	
NbN	3560	[988]	1959	
TaN	3320	×	1961	O. I. Shulishova
CrN	2640	[988]	1959	
Mo$_2$N	2670	×	1961	"
W$_2$N	2390	×	1961	"
WN	3300	×	1961	"
TiSi$_2$	2230	×	1961	"
ZrSi$_2$	2490	×	1961	"
VSi$_2$	2730	×	1955	A. P. Mozharikov
Nb$_5$Si$_3$	3735	×	1960	O. I. Shulishova
NbSi$_2$	2490	×	1955	A. P. Mozharikov
TaSi$_2$	2318	×	1955	"
La$_2$S$_3$	2657	[871]	1961	
CeS$_2$	2326	[871]	1961	

[1*] Calculated from the formula of É.S. Sarkisov [822], modified by O. I. Shulishova [823].
[2*] PmB$_6$ not yet obtained experimentally.
[3*] Calculated from the formula of É.S. Sarkisov [822].

ATOMIZATION[1*] ENERGY

Phase	U_A,[2*] kcal/mole	Ref.	Year	Remarks
TiC	328	[661]	1956	
ZrC	360	[661]	1956	
HfC	380	[661]	1956	
VC	338	[661]	1956	[1157]
NbC	384	[661]	1956	
TaC	391	[661]	1956	
MoC	330.4	[1157]	1961	
WC	376	[661]	1956	
TiN	305	[661]	1956	
ZrN	335	[661]	1956	
HfN	371	[661]	1956	
VN	293	[661]	1956	
NbN	347	[661]	1956	
SiC	300	[661]	1956	Cubic modification

[1*] The term "atomization" is a literal translation of the Russian; apparently there is no perfect English equivalent of the term as used here. It's meaning is made clear by the formula given below—Ed. note.

[2*] $U_A = U - \sum_n I - \sum_n E + \alpha$, where U is the lattice energy, $\sum_n I$ is the total ionization potential, $\sum_n E$ is the total energy of attachment of n electrons, α is the energy of transition from the idealized Born state of the crystal to the actual state [661].

CHARACTERISTIC TEMPERATURE

Phase	Θ, °K	Ref.	Year	Remarks
CaB_6	1085	[474]	1961	
SrB_6	889	[474]	1961	
BaB_6	824	[474]	1961	
YB_6	922	[891]	1961	
LaB_6	885	[474]	1961	
CeB_6	747	[474]	1961	
PrB_6	730	[474]	1961	
NdB_6	752	[474]	1961	
SmB_6	755	[474]	1961	
EuB_6	735	[891]	1961	
GdB_6	745	[474]	1961	
TbB_6	690	[891]	1961	

Phase	Θ, °K	Ref.	Year	Remarks
YbB_6	763	[474]	1961	
ThB_2	600	[891]	1961	
TiB_2	842	[261]	1959	[372]
ZrB_2	742—747	[261]	1959	[372]
$TiC_{0.43}$	840	[468]	1956	
TiC	841	[471]	1956	[262, 372, 468, 1242]
ZrC	490	[372]	1958	[262, 1242]
HfC	357	[294]	1961	[1242]
VC	531	[372]	1958	[262]
Nb_2C	411	[261]	1959	
NbC	470	[372]	1958	[262]
TaC	318	[372]	1958	
Mo_2C	366	[262]	1958	[372]
W_2C	330	[261]	1959	[372]
WC	453	[372]	1958	[262]
Fe_3C	370—450	[1146]	1962	
UN	232	[581]	1958	
TiN	650 ± 23	[372]	1958	[262]
ZrN	288 ± 16	[262]	1958	41.3 at.% N
VN	546	[262]	1958	45 at.% N, traces of C
NbN	309 ± 19	[262]	1958	45.8 at.% N, 3.8 at.% C
Ta_2N	231 ± 10	[272]	1958	
Mg_2Si	398 ± 3	[274]	1957	
Ti_5Si_3	506	[1240]	1963	
$TiSi_2$	1000	[1240]	1963	
$ZrSi_2$	678	[1240]	1963	
Mo_3Si	450	[1240]	1963	Si bound 8.6%
Mo_5Si_3	521	[1240]	1963	Si bound 15.0%
$MoSi_2$	684	[1240]	1963	Si bound 36.9%
WSi_2	715	[1240]	1963	
Fe_3Si	435	[1134]	1962	[1135]
FeSi	520	[1134]	1962	[1135]
$FeSi_{2.33}$	550	[1134]	1962	α-lebauite, [1135]
LaS	699	[1221]	1963	
La_2S_3	913	[1221]	1963	
CeS	677	[1221]	1963	
Ce_2S_3	928	[1221]	1963	
PrS	638	[1221]	1963	
Pr_2S_3	880	[1221]	1963	
NdS	633	[1221]	1963	
Nd_2S_3	840	[1221]	1963	

ROOT-MEAN-SQUARE AMPLITUDE OF THERMAL VIBRATIONS OF ATOM COMPLEXES

Phase	$\sqrt{\overline{U}_{291}^2}$, Å	Ref.	Year	Remarks
CaB_6	0.050	[891]	1961	
SrB_6	0.049	[891]	1961	
BaB_6	0.045	[891]	1961	
YB_6	0.047	[891]	1961	
LaB_6	0.042	[891]	1961	
CeB_6	0.047	[891]	1961	
PrB_6	0.049	[891]	1961	
NdB_6	0.047	[891]	1961	
SmB_6	0.045	[891]	1961	
EuB_6	0.047	[891]	1961	
GdB_6	0.045	[891]	1961	
TbB_6	0.047	[891]	1961	
YbB_6	0.043	[891]	1961	
ThB_6	0.045	[891]	1961	
TiB_2	0.073	[469]	1957	
ZrB_2	0.072	[469]	1957	
VB_2	0.083	[470]	1958	
TaB_2	0.079	[470]	1958	
CrB_2	0.121	[469]	1957	
TiC	0.067	[469]	1957	[471]
ZrC	0.074	[469]	1957	
VC	0.088	[469]	1957	[471]
NbC	0.076	[469]	1957	
TaC	0.082	[469]	1957	
Mo_2C	0.055	[469]	1957	[471]
W_2C	0.062	[469]	1957	[471]
WC	0.058	[469]	1957	[471]
TiN	0.091	[469]	1957	
$TiSi_2$	0.087	[472]	1959	Calculated by the method of [473] without taking into account anisotropy of thermal vibrations of the atoms in crystals of noncubic symmetry
$ZrSi_2$	0.087	[472]	1959	
$TaSi_2$	0.073	[472]	1959	
$CrSi_2$	0.082	[472]	1959	
$MoSi_2$	0.076	[472]	1959	
WSi_2	0.073	[472]	1959	

HEAT OF PHASE CONVERSION

Phase	Heat of phase conversion, kcal/mole	Temperature of conversion, °C	Ref.	Year
$\alpha\text{-}Mn_3C \rightarrow \beta\text{-}Mn_3C$	3.57	1037	[1]	1957
$\alpha\text{-}Mg_3N_2 \rightarrow \beta\text{-}Mg_3N_2$	0.22	550	[585]	1949
$\beta\text{-}Mg_3N_2 \rightarrow \gamma\text{-}Mg_3N_2$	0.26	788	[585]	1949
$\alpha\text{-}Mn_3Si \rightarrow \beta\text{-}Mn_3Si$	87.2	600	[417]	1956

PARAMETERS OF THE DIFFUSION OF NONMETALS INTO METALS AND REACTIONS
WITH THE FORMATION OF REFRACTORY COMPOUNDS

System	Temperature, °C	Phase formed	Temperature function of diffusion coefficient	Activation energy, cal/mole	Ref.	Year	Remarks
B→Ti	1200—1500	TiB_2	$D = 8.9 \cdot 10^{-5} \exp(-15300/T)$	30600±7800	[1326]	1963	[445]*
B→Zr	1200—1500	ZrB_2	$D = 1.26 \cdot 10^{-4} \exp(-17250/T)$	34500±8100	[1327]	1963	[1015]
B→Nb	1100—1400	NbB_2	$D = 2.94 \exp(-29500/T)$	59000±8500	[1327]	1963	[445*, 1015]
B→Ta	1100—1400	TaB_2	$D = 9.44 \cdot 10^{-4} \exp(-23500/T)$	47000±600	[1327]	1963	[445*, 1015]
B→Mo	1100—1400	Mo_2B, Mo_2B_5	$D = 6.96 \cdot 10^{-2} \exp(-22500/T)$	45000±5800	[1327]	1963	[445*, 1015]
B→W	1100—1400	W_2B, W_2B_5	$D = 148 \exp(-32000/T)$	64000±6300	[1327]	1963	[445*, 1015]
B→Re	1400—1500	—	—	—	[1015]	1961	
B→Fe	—	Solid solution in austenite	$D = 2 \cdot 10^{-3} \exp(-10500/T)$	21000	[446]	1953	[1015]
B→Co	900—1000	—	—	—	[1015]	1961	[1015]
C→Ti	900—1300	TiC	$D = 2.04 \cdot 10^{-3} \exp(-16500/T)$	33000±5950	×	1963	A. I. Épik, [445*, 1337]
C→Zr	900—1300	ZrC	$D = 3.44 \cdot 10^{-2} \exp(-20500/T)$	41000±9600	×	1963	A. I. Épik, [445*]
C→Nb	1000—2000	NbC	$D = 1.94 \cdot 10^{-6} \exp(8250/T)$	16500	[445]*	1956	[141,1256,1257,1258]
C→V	—	Solid solution	$D = 0.0047 \exp(-13650/T)$	27300	[532]	1958	
C→Ta	1000—1800	Ta_2C	$D = 3.43 \cdot 10^{-6} \exp(-10500/T)$	21000	[445]*	1956	
C→Cr	1200—1400	Cr_3C_2	—	26100±3200	[457]	1957	
C→Mo	1400—1700	Mo_2C	$D = 1.64 \cdot 10^3 \exp(-41500/T)$	83000±8600	[1326]	1963	[445*]
C→W	1600—1900	WC, W_2C	$D = 1.56 \cdot 10^3 \exp(-52000/T)$	104000±9200	[1326]	1963	[445*, 1171]
C→W	—	$W_2C + (WC)$	$D = 25 \cdot 10^3 \exp(-56000/T)$	112000±3000	[447]	1952	
C→α-Fe	—	Solid solution	$D = 6.0 \cdot 10^{-3} \exp(-9600/T)$	19200	[451]	1955	[448, 449]
C→γ-Fe	—	"	$D = 0.1 \exp(-16200/T)$	32400	[451]	1955	[450]
C→Ni	—	"	$D = 0.051 \exp(-16100/T)$	32200	[451]	1955	

* Recalculated from the data of [445].

136

PARAMETERS OF THE DIFFUSION OF NONMETALS INTO METALS AND REACTIONS
WITH THE FORMATION OF REFRACTORY COMPOUNDS (continued)

System	Temperature, °C	Phase formed	Temperature function of diffusion coefficient	Activation energy, cal/mole	Ref.	Year	Remarks
N→Mg	>500	—	$D = 2.2 \cdot 10^4 \exp(-11650/T)$	23300	[582]	1956	
N→Al	530—625	—	$D = 4.2 \cdot 10^{10} \exp(-11850/T)$	23700	[582]	1956	
N→Th	845—1890	—	$D = 2.1 \cdot 10^{-3} \exp(-11250/T)$	22500	[614]	1954	
N→Th	845—1890	—	$K = 5.9 \exp(-12150/T)$	24300	[614]	1954	
N→U	550—900	UN, U₂N₃, UN₂	$K = 202 \exp(-12750/T)$	25500	[609]	1954	Reaction constant ml² per cm⁴·sec,[1250]
N→α-Ti	—	—	$D = 1.2 \cdot 10^{-2} \exp\left(-\dfrac{22625 \pm 1125}{T}\right)$	45250±2250	[452]	1954	
N→β-Ti	—	—	$D = 3.5 \exp\left(-\dfrac{16900 \pm 200}{T}\right)$	33800±400	[452]	1954	
N→γ-Ti	—	—	$D = 5.4 \cdot 10^{-3} \exp\left(-\dfrac{26000 \pm 1750}{T}\right)$	52000±3500	[452]	1954	
N→Ti	600—900	TiN	$D = 4.5 \cdot 10^{-3} \exp\left(-\dfrac{6700}{T}\right)$	13400	[929]	1961	Diffusion into powder [453, 454, 1023, 1024, 1026, 1202] [454]
N→α-Zr	400—825	Solid solution	—	39200	[454]	1950	
N→β-Zr	862—1073	ZrN	—	52000	[456]	1950	
N→β-Zr	920—1640	ZrN	$\begin{cases} D = 1.5 \cdot 10^{-2} \exp(-15350/T) \\ K = 5.0 \cdot 10^{-3} \exp(-24000/T) \end{cases}$	30700 48000	[620] [620]	1955 1955	[621]

PARAMETERS OF THE DIFFUSION OF NONMETALS INTO METALS AND REACTIONS
WITH THE FORMATION OF REFRACTORY COMPOUNDS (continued)

System	Tempera-ture, °C	Phase formed	Temperature function of diffusion coefficient	Activation energy, cal/mole	Ref.	Year	Remarks
N→Zr	500—600	Solid solution	$D = 7.47 \cdot 10^{-5} \exp(-10000/T)$	20000	[929]	1961	Diffusion into powder
N→Hf	876 – 1034	HfN	—	57000±3000	[556]	1958	
N→Nb	500 800	Solid solution	—	25400	[454]	1950	[453]
N→Nb	500—600	Same	$D = 8.05 \cdot 10^{-4} \exp(-3630/T)$	7260	[929]	1961	Diffusion into powder
N→Nb	600—900	Nb$_2$N	$D = 6.16 \cdot 10^{-4} \exp(-3080/T)$	7060	[929]	1961	Same
N→Nb	900 –1200	NbN	$D = 4.5 \cdot 10^{-3} \exp(-5000/T)$	10000	[929]	1961	"
N→Ta	1800	Solid solution	—	36400	[590]	1958	[453, 591]
N→Ta	500—700	Same	$D = 22.19 \exp(-6600/T)$	13200	[929]	1961	Diffusion into powder [1028]
N→Ta	800—900	Ta$_2$N	$D = 4.914 \cdot 10^3 \exp(-25000/T)$	50000	[929]	1961	"
N→Ta	1000 1200	TaN	$D = 1.22 \exp(-17500/T)$	35000	[929]	1961	"
N→Cr	500—900	CrN	$D = 14.8 \exp(-5210/T)$	10420	[929]	1961	Diffusion into powder
N→γ-Fe	—	Solid solution	$D = 1.07 \cdot 10^{-1} \exp(-17000/T)$	34000	[450]	1947	
Si→Ti	800—1000	TiSi, TiSi$_2$	$D = 2.99 \exp\left(-\dfrac{2608 \pm 1183}{T}\right)$	5216±2367	[459]	1959	[457, 458]

PARAMETERS OF THE DIFFUSION OF NONMETALS INTO METALS AND REACTIONS WITH THE FORMATION OF REFRACTORY COMPOUNDS (continued)

System	Temperature, °C	Phase formed	Temperature function of diffusion coefficient	Activation energy, cal/mole	Ref.	Year	Remarks
Si→Ti	900—1200	TiSi$_2$	$D = 8.1 \cdot 10^2 \exp(-19850/T)$	39700	[461]	1959	Diffusion into powder
Si→Zr	1000—1200	ZrSi$_2$	$D = 1.1 \cdot 10^5 \exp(-27875/T)$	55750	[461]	1959	Same
Si→V	1000—1200	VSi$_2$	$D = 6.2 \cdot 10^5 \exp(-31600/T)$	61200	[461]	1959	Diffusion into powder, [1266]
Si→Nb	900—1100	NbSi$_2$	$D = 56.1 \exp(-18420/T)$	36840	[461]	1959	
Si→Ta	900—1200	TaSi$_2$	$D = 36.23 \cdot 10^{-3} \exp\left(-\dfrac{10756 \pm 835}{T}\right)$	21153 ± 1670	[459]	1959	[1290]
Si→Ta	800—1200	TaSi$_2$	$D = 93.1 \exp(-17300/T)$	34600	[461]	1959	Diffusion into powder
Si→Cr	900—1100	CrSi$_2$	$D = 3.92 \exp(-11380/T)$	22760	[461]	1959	Same
Si→Mo	900—1100	MoSi$_2$	$D = 56.1 \exp(-18420/T)$	36840	[461]	1959	Diffusion into powder, [459, 919, 1290]
Si→W	—	WSi$_2$	—	5780	[458]	1957	Diffusion into powder
Si→W	900—1100	WSi$_2$	$D = 4.4 \cdot 10^6 \exp(-31500/T)$	63000	[461]	1959	
Si→α-Fe	700—800	Solid solution	$D = 2.52 \exp(-777/T)$	1554	[459]	1959	Diffusion into powder

139

PARAMETERS OF THE DIFFUSION OF NONMETALS INTO METALS AND REACTIONS
WITH THE FORMATION OF REFRACTORY COMPOUNDS (continued)

System	Temperature, °C	Phase formed	Temperature function of diffusion coefficient	Activation energy, cal/mole	Ref.	Year	Remarks
Si→α-Fe	1200—1350	Solid solution	$D = 0.44\exp(-24000/T)$	48000	[460]	1952	
Si→γ-Fe	—	FeSi$_2$		20170	[458]	1957	
Si→γ-Fe	900—1100	FeSi$_2$	$D = 14.55 \cdot 10^3 \exp(-11015/T)$	22030	[459]	1959	
Si→Co	—	CoSi$_2$		13090	[458]	1957	
Si→Ni	—	NiSi$_2$		24950	[458]	1957	
B→C	1940—2400	Solid solution	$D = 3.02\exp(-28625/T)$	57250	[462]	1960	
B→Si	—	Same	$D = 10^{-3}\exp(-29000/T)$	58000	[463]	1950	
N→B	600—1200	BN	$D = 30.1 \cdot 10^3 (-30650/T)$	61300	[464]	1959	Defect structure
N→B	1200—1500	BN	$D = 20.3 \cdot 10^{-5}\exp(-2000/T)$	4000	[464]	1959	
P→Si	1000—1200	Solid solution	$D = 10^{-3}\exp(-29000/T)$	58000	[463]	1950	[1254, 1255]

Chapter III

ELECTRICAL AND MAGNETIC PROPERTIES

ELECTRICAL CONDUCTIVITY

Phase	Specific resistance, microohm · cm	Accuracy (±), microohm · cm	Temperature, °C	Conductivity, $ohm^{-1} \cdot cm^{-1}$	Ref.	Year	Remarks
Be_5B	$15 \cdot 10^3$	—	20	66.7	[563]	1960	$P^{1*} = 50\%$
Be_2B	$14 \cdot 10^3$	—	20	71.4	[563]	1960	$P = 50\%$
BeB_2	$2 \cdot 10^4$	—	20	50	[573]	1960	[863, 866]
BeB_4	$18 \cdot 10^3$	—	20	55.6	[563]	1960	$P = 50\%$
BeB_6	$25 \cdot 10^8$	—	20	$0.4 \cdot 10^{-3}$	[863]	1961	
CaB_6	222.0	—	20	4500	[281]	1961	[1115]
SrB_6	191.8	—	20	5240	[353]	1961	[1115, 1160]
BaB_6	77	—	20	13000	[281]	1961	[1115]
AlB_{12}	$2 \cdot 10^{12}$	—	20	$0.5 \cdot 10^{-5}$	[578]	1956	$P = 50\%$
ScB_2	7—15	—	20	143000—67000	[694]	1960	
ScB_4	750	–	20	1330	×	1963	T. S. Verkhoglya-dova
YB_2	39	2.6	20	25600	[1115]	1963	Single crystal in basal plane
YB_4	28.5	1.3	20	35100	[1115]	1963	Same
YB_4	31.3	1	20	32000	[1115]	1963	Parallel to c axis
YB_6	40.5	1.6	20	24700	[1115]	1963	[281, 283]
YB_{12}	94.8	2.7	20	10580	[1115]	1963	
LaB_4	24	±12	20	41800	[1030]	1961	
LaB_4	~12	—	−190	83200	[1030]	1961	
LaB_6	15.0	—	20	66700	[281]	1961	[3, 25, 846]
CeB_6	29.4	—	20	34000	[281]	1961	[3, 284]
PrB_6	19.5	—	20	51400	[281]	1961	[284]
NdB_6	20.0	—	20	50000	[281]	1961	[284, 846]
SmB_6	207	—	20	4800	[281]	1961	
EuB_6	84.7	—	20	11800	[281]	1961	
GdB_6	44.7	—	20	22400	[281]	1961	[285, 284, 846]
TbB_6	37.4	—	20	26750	[281]	1961	[284]
YbB_6	46.6	—	20	21500	[281]	1961	[284]
ThB_6	14.8	—	20	67600	[281]	1961	[27]
TiB	40	—	20	25000	[51]	1952	

[1*] P = porosity

Phase	Specific resistance, microohm · cm	Accuracy (±), microohm · cm	Temperature, °C	Conductivity, ohm^{-1}·cm^{-1}	Ref.	Year	Remarks
TiB$_2$	14.4	—	20	69500	[287]	1960	[286, 293, 312, 918, 1010]
ZrB$_2$	16.6	—	20	62500	[287]	1960	[33, 286, 288, 289, 312]
ZrB$_{12}$	60	—	22	16670	[35]	1952	[32]
HfB$_2$	8.8	—	20	113600	[290]	1959	[36]
VB	35—40	—	20	28600—25000	[38]	1952	
VB$_2$	3.5	—	−140	286000	[291]	1931	
VB$_2$	19	—	20	52600	[287]	1960	[51, 291, 312]
NbB	64.5	—	20	15500	[286]	1956	
NbB$_2$	34.0	—	20	29400	[287]	1960	[286, 312]
TaB	100	—	20	10000	[51]	1952	
TaB$_2$	37.4	—	20	26800	[287]	1960	[51, 232, 286, 312]
Cr$_4$B	176	1	20	5624	[930]	1961	
Cr$_2$B	52	2	20	19230	[930]	1961	
CrB	69	1	20	14550	[51]	1952	
CrB$_2$	84	5	20	11680	[287]	1960	[51, 312]
Cr$_2$B$_5$	73	14	20	13700	[1166	1962	
Mo$_2$B	40	—	20	25000	[51]	1952	
α-MoB	45	—	20	22250	[49]	1952	
β-MoB	25	—	20	40000	[49]	1952	
MoB$_2$	45	—	20	22250	[49]	1952	[382]
Mo$_2$B$_5$	18	—	20	55560	[312]	1958	[49, 293]
W$_2$B$_5$	43	—	20	22300	[287]	1960	[286, 312]
Be$_2$C	1.1 · 10^6	—	20	0.98	[526]	1952	[346]
ScC	274	—	20	3650	×	1963	T.Ya.Kosolapova, G. N. Makarenko
YC	4.54 · 10^4	—	20	22.2	[820]	1961	[879]
Y$_2$C$_3$	3.38 · 10^2	—	20	2350	[879]	1961	
YC$_2$	88.7	—	20	11230	[879]	1961	
La$_2$C$_3$	144	15%	25±5	7000	[570]	1959	
LaC$_2$	68	17%	25±5	14710	[570]	1959	
Ce$_2$C$_3$	202	—	20	4950	×	1963	T.Ya.Kosolapova, G. N. Makarenko
CeC$_2$	58.8	—	20	17100	×	1963	Same "
PrC$_2$	25.7	—	20	39100	×	1963	
ThC	25	—	25	40000	[1278]	1962	C$_{tot.}$ 5.03% ; C$_{free}$ 0.25%
ThC$_2$	30	—	25	33333	[1278]	1962	C$_{tot.}$ 9.85%; C$_{free}$ 1.07%

Phase	Specific resistance, microohm · cm	Accuracy (±), microohm · cm	Temperature, °C	Conductivity, $ohm^{-1} \cdot cm^{-1}$	Ref.	Year	Remarks
UC	100	4	20	10000	[1014]	1959	P = 25%, [1118]
TiC	52.5	—	20	19100	[287]	1960	[232, 286, 291, 292, 293, 465, 664]
ZrC	50.0	—	20	20000	[287]	1960	[232, 286, 288, 291, 465, 1151]
HfC	45.0	—	20	22250	[287]	1960	[291, 294, 465]
VC	65	—	20	15400	[287]	1960	[232, 465]
NbC	51.1	—	20	19600	[287]	1960	[232, 286, 287, 465]
TaC	42.1	—	20	23750	[287]	1960	[51, 286, 232, 295, 465, 922]
$Cr_{23}C_6$	127	2	20	7880	[296]	1961	P = 0, [1166]
Cr_7C_3	109	4	20	9180	[296]	1961	P = 0, [1166]
Cr_3C_2	75	5	20	13330	[296]	1961	P = 0, [1166]
Mo_2C	71.0	—	20	14100	[287]	1960	[286, 232]
WC	19.2	0.3	20	52200	[287]	1960	[232, 465, 831]
W_2C	75.7	0.1	20	13200	[287]	1960	[264]
Mg_3N_2	$2 \cdot 10^{10}$	—	—	$0.5 \cdot 10^{-4}$	[583]	1926	
Ba_2N	10^8	—	20	10^{-2}	[697]	1959	
AlN	$10^{14} - 10^{16}$	—	20	$10^{-8} - 10^{-10}$	[611]	1959	
ScN	25.4	—	20	39400	×	1962	V. S. Neshpor, [597, 697, 1094]
LaN	2.37?	—	20	422000	[1112]	1961	[1099]
CeN	$10^{1.3}$	—	27	$10^{4.7}$	[1099]	1962	[1100], powder
PrN	$10^{1.6}$	—	27	$10^{4.4}$	[1099]	1962	[1100], powder
TiN	25	—	20	40000	[287]	1960	[232, 286, 297, 298, 991, 997]
$TiN_{0.765}$	50.6	—	20	19700	×	1963	S. N. L'vov, V. F. Nemchenko
ZrN	21.1	—	20	74400	[287]	1960	[232, 297, 989, 991, 997]

Phase	Specific resistance, microohm·cm	Accuracy (±), microohm·cm	Temperature, °C	Conductivity, ohm^{-1}·cm^{-1}	Ref.	Year	Remarks
$ZrN_{0.879}$	37	—	20	25100	×	1963	S. N. L'vov, V. F. Nemchenko
HfN	33.0	5	20	30400	[997]	1961	
V_3N	123.0	10	20	8140	[997]	1961	
VN	85.0	4	20	11700	[287]	1960	[232, 287, 997]
Nb_2N	142.0	6	20	7042	[929]	1961	
$NbN_{0.75}$	90.0	8	20	11110	[929]	1961	
$NbN_{0.97}$	85.0	2	20	11764	[929]	1961	
NbN	78.0	4	20	12820	[929]	1961	[287]
Ta_2N	263.0	22	20	3802	[929]	1961	
TaN	128.0	15	20	7812	[929]	1961	[287]
Cr_2N	84	5	20	11900	[287]	1960	[886, 1116]
CrN	640	40	20	1562	[886]	1961	
Mo_2N	19.8	7	20	50500	[929]	1961	
Ni_3N	$2.8 \cdot 10^3$	—	25	357.1	[580]	1956	
Mg_2Si	$\rho = 13 \cdot 10^8 \exp [0.011/2KT]$ $(T = 350 - 1000° K)$				[422]	1957	Impurity conductivity
Mg_2Si	$\rho = \dfrac{10^6}{3150} \exp [0.48/2KT]$ $(T > 450°K)$				[422]	1957	Intrinsic conductivity, [1006, 1046]
$BaSi_2$	$38 \cdot 10^4$	14	20	2635	[299]	1960	[1289]
$LaSi_2$	236	20	20	4240	[299]	1960	[846]
$CeSi_2$	408	20	20	2450	[299]	1960	
$PrSi_2$	202	20	20	4950	[299]	1960	
$NdSi_2$	349	—	20	2870	[846]	1960	
$GdSi_2$	263	—	20	3800	[846]	1960	
$DySi_2$	3020(?)	—	20	330	[285]	1959	[846]
U_3Si	55	—	20	18200	[641]	1958	[690]
Ti_5Si_3	55	4	20	18200	[299]	1960	[300]
TiSi	63	6	20	15900	[299]	1960	[300]
$TiSi_2$	16.9	0.5	20	59250	[299]	1960	[300]
ZrSi	49.4	—	20	13200	[300]	1958	
$ZrSi_2$	75.8	3.1	20	13200	[299]	1960	[300]
V_3Si	203.5	37.5	20	4910	[299]	1960	
V_5Si_3	114.5	8.5	20	8780	[299]	1960	
VSi_2	66.5	2.5	20	15050	[299]	1960	[300, 301]

Phase	Specific resistance, microohm·cm	Accuracy (±), microohm·cm	Temperature, °C	Conductivity, ohm^{-1}·cm^{-1}	Ref.	Year	Remarks
NbSi$_2$	50.4	2.3	20	19900	[299]	1960	[117, 301]
Ta$_{4.5}$Si	174.5	—	20	5740	[299]	1960	
Ta$_2$Si	124	—	20	8070	[299]	1960	
Ta$_5$Si$_3$	108	—	20	9270	[299]	1960	
TaSi$_2$	46.1	1.3	20	21700	[299]	1960	[300, 301]
Cr$_3$Si	35	5	20	28600	[299]	1960	[302, 303]
Cr$_3$Si$_2$	80	5	20	12500	[299]	1960	
Cr$_5$Si$_3$	153	—	20	6540	[299]	1960	[1265]
CrSi	129.5	7.5	20	7730	[299]	1960	[302, 303]
CrSi$_2$	914	74.5	20	1095	[299]	1960	[300, 302, 303, 861]
Mo$_3$Si	21.6	0.7	20	46300	[299]	1960	[1231, 1265]
Mo$_5$Si$_3$	45.9	1.2	20	21800	[299]	1960	[1231, 1265]
MoSi$_2$	21.6	0.9	20	46300	[299]	1960	[117, 861, 1231]
W$_3$Si	93	—	20	10760	[299]	1960	
WSi$_2$	12.5	0.2	20	80000	[299]	1960	[117]
Mn$_3$Si	160	3	20	6250	[299]	1960	[303, 1049, 1138]
Mn$_5$Si$_3$	257	14	20	3900	[299]	1960	[303, 1049, 1138]
MnSi	259	12	20	3860	[299]	1960	[303, 1037, 1049, 1136]
MnSi$_2$	462	63	20	2135	[299]	1960	[303, 1037, 1056, 1136]
Re$_3$Si	129	—	20	7730	[299]	1960	
ReSi	736	36	20	1360	[299]	1960	
ReSi$_2$	7000	1000	20	143	[299]	1960	[916]
Fe$_3$Si	130	19	20	7700	[299]	1960	[303, 425, 860]
Fe$_5$Si$_3$	170	—	20	5900	[425]	1949	
FeSi	271	6	20	3700	[299]	1960	[303, 425]
α-FeSi$_2$	1000	—	20	1000	[860]	1960	[1233, 1234]
β-FeSi$_2$	$4 \cdot 10^6$	—	20	0.25	[914]	1960	[299, 303, 425, 1233, 1234]

ELECTRICAL CONDUCTIVITY (continued)

Phase	Specific resistance, microohm·cm	Accuracy (±), microohm·cm	Temperature, °C	Conductivity, ohm^{-1}·cm^{-1}	Ref.	Year	Remarks
Co$_2$Si	66.2	—	—	15100	[303]	1958	
Co$_3$Si	129	9	20	7760	[299]	1960	
CoSi	86	15.5	20	11620	[299]	1960	[303]
CoSi$_2$	68	6	20	14700	[299]	1960	[303]
CoSi$_3$(?)	404	—	20	2475	[303]	1958	
Ni$_3$Si	93	7.5	20	10770	[299]	1960	[303]
Ni$_5$Si$_2$(?)	149.5	—	20	6700	[303]	1958	
Ni$_3$Si$_2$	79	7	20	12660	[299]	1960	
NiSi	20.2	—	20	49500	[303]	1958	
Ni$_2$Si$_3$(?)	280	—	20	3570	[303]	1958	
NiSi$_2$	118	21	20	8490	[299]	1960	[303]
TiP	75	—	20	13000	[1178]	1963	P = 0
TiP	3400	—	20	293	[1177]	1962	P = 36%
TiP	1300	—	(−197)	771	[1177]	1962	P = 36%
VP	9700	—	20	103.2	[1177]	1962	P = 40%
β-NbP	1700	—	20	590	[1177]	1962	P = 35%
β-NbP	400	—	(−197)	2500	[1177]	1962	P = 35%
β-TaP	23000	—	20	43	[1177]	1962	P = 31%
β-TaP	11600	—	(−197)	85.4	[1177]	1962	P = 31%
CrP	25000	—	20	40	[1177]	1962	P = 34%
CrP	14200	—	(−197)	70.5	[1177]	1962	P = 34%
MoP	900	—	20	1113	[1177]	1962	P = 47%
MoP	1400	—	(−197)	713	[1177]	1962	P = 47%
WP	1800	—	20	558	[1177]	1962	P = 43%
WP	480	—	(−197)	2080	[1177]	1962	P = 43%
MnP	2500	—	20	400	[1177]	1962	P = 22%
MnP	1000	—	(−197)	1000	[1177]	1962	P = 22%
LaS	92.0	—	20	10920	[1218]	1963	
La$_3$S$_4$	24·10^4	—	20	4.17	[159]	1956	
La$_2$S$_3$	2·10^6	—	20	0.5	[1219]	1963	
CeS	170	—	20	5900	[1217]	1963	[1218]
Ce$_3$S$_4$	58·10^4	—	20	1.725	[159]	1956	[156, 1114]
Ce$_2$S$_3$	1.19·10^6	—	20	0.84	[1217]	1963	
PrS	240	—	20	4160	[1218]	1963	
Pr$_2$S$_3$	1.10·10^6	—	20	0.91	[1224]	1963	
NdS	242	—	20	4140	[1218]	1963	
Nd$_3$S$_4$	1.2·10^6	—	20	0.835	[159]	1956	
Nd$_2$S$_3$	0.7·10^6	—	20	1.43	[1224]	1963	
Sm$_3$S$_4$	66.4·10^6	—	20	0.015	[159]	1956	
ThS	20·10^4	—	20	5.0	[305]	1958	
Th$_2$S$_3$	10^7	—	20	10^{-1}	[305]	1958	
Th$_4$S$_7$	25·10^9	—	20	4·10^{-5}	[305]	1958	

ELECTRICAL CONDUCTIVITY (continued)

Phase	Specific resistance, microohm·cm	Accuracy (±), microohm·cm	Temperature, °C	Conductivity, ohm^{-1}·cm^{-1}	Ref.	Year	Remarks
ThS$_2$	10^{16}	—	20	10^{-10}	[305]	1958	
B$_4$C	10^6	—	20	1	[306]	1960	[271, 307, 1164]
B$_4$C	$38 \cdot 10^3$	—	600	26.4	[306]	1960	
B$_4$C	$30 \cdot 10^3$	—	1000	33.3	[306]	1960	
B$_4$C	$22 \cdot 10^3$	—	2000	44.5	[306]	1960	
SiC	$> 0.13 \cdot 10^6$	—	25	< 7.7	[1000]	1961	P = 4%
SiC	$> 0.05 \cdot 10^6$	—	1100	< 20	[1000]	1961	Comp., %: 96.5 SiC, 2.5 Si$_{free}$, 0.4 C$_{free}$, 0.4 Al, 0.2 Fe
α-BN	$1.7 \cdot 10^{19}$	—	25	$5.9 \cdot 10^{-14}$	[272]	1955	
α-BN	$2.3 \cdot 10^{16}$	—	500	$4.35 \cdot 10^{-11}$	[272]	1955	
α-BN	$3.1 \cdot 10^{10}$	—	1000	$3.23 \cdot 10^{-5}$	[272]	1955	
α-BN	$6 \cdot 10^8$	—	1500	$1.67 \cdot 10^{-3}$	[272]	1955	
β-BN	$2 \cdot 10^8 - 10^9$	—	25	$5 \cdot 10^{-3} - 10^{-3}$	[1090]	1962	Borazon p-type
β-BN	$10^{11} - 10^{15}$	—	25	$10^{-5} - 10^{-9}$	[1090]	1962	Borazon n-type
Si$_3$N$_4$	$10^{19} - 10^{20}$	—	20	$10^{-13} - 10^{-14}$	[309]	1960	[172]
Si$_3$N$_4$	10^{15}	—	350	10^{-9}	[309]	1960	
Si$_3$N$_4$	$5 \cdot 10^{12}$	—	600	$2 \cdot 10^{-7}$	[309]	1960	
Si$_3$N$_4$	$2 \cdot 10^9$	—	1000	$5 \cdot 10^{-4}$	[309]	1960	
SiB$_4$	$1.75 \cdot 10^6$	—	20	0.57	[1164]	1959	
SiB$_6$	$0.2 \cdot 10^6$	—	25	5	[392]	1963	
B	$1.7 \cdot 10^{12}$	—	27	$5.9 \cdot 10^{-7}$	[636]	1957	
C	200—250	—	—	$(4-5) \cdot 10^3$	[944]	1959	Pyrographite

THERMAL COEFFICIENT OF ELECTRICAL RESISTANCE

Phase	Coefficient of electrical resistance, deg$^{-1} \cdot 10^3$	Accuracy of measurement (±), deg$^{-1} \cdot 10^3$	Temperature interval, °C	Ref.	Year	Remarks
BeB$_2$	—0.9	—	20— 80	[573]	1960	[863]
BeB$_6$	—0.23	—	20— 80	[863]	1961	

147

Phase	Coefficient of electrical resistance, $\deg^{-1} \cdot 10^3$	Accuracy of measurement $(\pm)$, $\deg^{-1} \cdot 10^3$	Temperature interval, °C	Ref.	Year	Remarks
CaB$_6$	$+1.16$	—	0—100	[281]	1961	
SrB$_6$	$+0.83$	—	0—100	[281]	1961	
BaB$_6$	$+1.08$	—	0—100	[281]	1961	
AlB$_{12}$	-0.3	—	0—100	[578]	1956	
YB$_6$	$+1.24$	—	0—100	[281]	1961	
LaB$_6$	$+2.68$	—	0—100	[281]	1961	[25]
CeB$_6$	$+1.00$	—	0—100	[281]	1961	
PrB$_6$	$+1.92$	—	0—100	[281]	1961	
NdB$_6$	$+1.93$	—	0—100	[281]	1961	
SmB$_6$	-0.42	—	0—100	[281]	1961	
EuB$_6$	$+0.90$	—	0—100	[281]	1961	
GdB$_6$	$+1.40$	—	0—100	[281]	1961	[846]
TbB$_6$	$+1.31$	—	0—100	[281]	1961	
YbB$_6$	$+2.34$	—	0—100	[281]	1961	
ThB$_6$	$+2.31$	—	0—100	[281]	1961	
TiB$_2$	$+2.78$	—	300—2000	[317]	1961	[288, 293, 297, 311]
ZrB$_2$	$+1.76$	—	300—1800	[317]	1961	[288, 297, 311, 318]
ZrB$_{12}$	$+1.62$	—	$-79-+64$	[35]	1952	
HfB$_2$	$+3.6$	—	20—2630	[297]	1931	[318]
VB$_2$	$+3.16$	—	100—1100	[311]	1958	[286, 297]
NbB$_2$	$+1.39$	—	100—1100	[311]	1958	[288]
TaB$_2$	$+1.48$	—	100—1100	[311]	1958	[288]
Cr$_4$B	$+1.13$	0.01	20—100	[930]	1961	[1166]
Cr$_2$B	$+1.95$	0.08	20—100	[930]	1961	[1166]
CrB	$+3.26$	0.03	20—100	[930]	1961	[1166]
CrB$_2$	$+2.61$	0.12	20—100	[930]	1961	[282, 1166]
Cr$_2$B$_5$	$+2.06$	0.01	20—100	[1166]	1962	
Mo$_2$B$_5$	$+3.3$	—	100—1100	[311]	1958	[293]
W$_2$B$_5$	$+4.26$	—	100—1100	[311]	1958	
TiC	$+1.16$	—	300—2000	[317]	1961	[286, 292, 293, 297, 311, 318, 1242]
ZrC	$+0.95$	—	300—2300	[317]	1961	[286, 297, 311, 318, 1242, 1253]
HfC	$+1.42$	—	300—2000	[317]	1961	P=0, [297, 318, 1242]
NbC	$+0.86$	—	300—2300	[317]	1961	P=0, [311]
TaC	$+1.07$	—	400—2000	[317]	1961	[286, 297, 311, 318, 349]
Cr$_{23}$C$_6$	$+1.72$	0.11	0—100	[296]	1961	P=0, [1166]
Cr$_7$C$_3$	$+1.06$	0.05	0—100	[296]	1961	P=0, [1166]
Cr$_3$C$_2$	$+2.33$ —	0.04	0—100	[296]	1961	P=0, [1166]
Mo$_2$C	$+3.78$	—	200—800	[317]	1961	P = 0
W$_2$C	$+1.95$	—	200—2000	[317]	1961	[348]
WC	$+0.495$	—	20—1500	[831]	1960	[348]
TiN	$+2.48$	—	100—1100	[311]	1958	[286, 297, 318, 992]
ZrN	$+4.3$	—	20—2560	[297]	1931	[286, 318, 989]

THERMAL COEFFICIENT OF ELECTRICAL RESISTANCE (continued)

Phase	Coefficient of electrical resistance, $deg^{-1} \cdot 10^3$	Accuracy of measurement $(\pm)$, $deg^{-1} \cdot 10^3$	Temperature interval, °C	Ref.	Year	Remarks
VN	+0.7	—	—	[286]	1956	[297]
TaN	+0.03	—	20—1410	[297]	1931	
LaSi$_2$	+2.18	—	20—500	[1240]	1963	[1216]
Ti$_2$Si$_3$	+1.31	—	20—800	[1240]	1963	[300]
TiSi	+4.13	—	20—120	[300]	1958	
TiSi$_2$	+2.33	—	20—800	[1240]	1963	[293, 300]
ZrSi	+3.52	—	20—120	[300]	1958	
ZrSi$_2$	+2.65	—	20—800	[1240]	1963	[300]
V$_3$Si	+0.563	—	20—200	[1245]	1963	
V$_5$Si$_3$	+1.24	—	20—200	[1245]	1963	
VSi$_2$	+3.51	—	20—120	[300]	1958	[1245]
TaSi$_2$	+3.32	—	20—120	[300]	1958	
CrSi$_2$	+2.93	—	20—120	[300]	1958	
Mo$_3$Si	+6.74	—	20—800	[1240]	1963	[1231]
Mo$_3$Si$_3$	+0.66	—	20—800	[1240]	1963	
MoSi$_2$	+6.38	—	20—120	[300]	1958	[1231, 1240]
WSi$_2$	+2.91	—	20—120	[300]	1958	[1240]
MnSi	+0.758	—	20—200	[1245]	1963	
Fe$_3$Si	+1.41	—	20—200	[1245]	1963	
FeSi	+0.511	—	20—200	[1245]	1963	
Co$_3$Si	+0.85	—	20—200	[1245]	1963	
CoSi$_2$	+2.48	—	20—200	[1245]	1963	
Ni$_3$Si	+1.80	—	20—200	[1245]	1963	
Ni$_2$Si	+3.24	—	20—200	[1245]	1963	
NiSi$_2$	+2.58	—	20—800	[1240]	1963	
LaS	+0.44	0.05	20—1000	[1218]	1963	
CeS	+0.67	0.02	20—1000	[1218]	1963	
PrS	+0.54	0.02	20—1000	[1218]	1963	
NdS	+0.61	0.06	20—1000	[1218]	1963	
B$_4$C	+0.032	—	1000—1450	[306]	1960	
SiC	+0.264	—	900—1500	[307]	1957	
BN	$-20930/T^2$	—	—	[272]	1960	Calculated from the width of the forbidden band
Si$_3$N$_4$	$-6570/T^2$	—	350—700	[309]	1960	
Si$_3$N$_4$	$-22670/T^2$	—	700—1000	[309]	1960	

SUPERCONDUCTIVITY

Phase	Critical temperature T$_c$, °K	Ref.	Year	Remarks
BaB$_6$	<1.28	[361]	1952	
ThB$_2$	< 1.28	[535]	1954	[389]
TiB	< 1.28	[535]	1954	

SUPERCONDUCTIVITY (continued)

Phase	Critical temperature T_c, °K	Ref.	Year	Remarks
TiB_2	< 1.28	[535]	1954	[319, 320, 361, 389]
ZrB	3.30	[535]	1954	[319]
ZrB_2	< 1.80	[535]	1954	[320, 321, 322, 361, 389]
HfB	< 1.80	[535]	1954	
HfB_2	<1.26	[319]	1930	
VB	< 1.28	[535]	1954	
VB_2	<1.9	[320]	1958	
NbB	8.25	[535]	1954	[323, 361]
Nb_3B_4	< 1.28	[535]	1954	[323, 1043]
NbB_2	< 1.28	[535]	1954	[323, 389]
Ta_2B	3.12	[535]	1954	Very contaminated
TaB	< 1.28	[535]	1954	[323]
Ta_3B_4	< 1.28	[535]	1954	[323]
TaB_2	< 1.28	[535]	1954	[320, 323, 389]
Cr_2B	< 1.28	[535]	1954	[1043]
CrB	< 1.28	[535]	1954	[1043]
CrB_2	< 1.20	[535]	1954	[1043]
Mo_2B	4.74	[535]	1954	[361]
MoB	4.40 (?)	[535]	1954	[323, 389]
Mo_2B_5	< 1.28	[535]	1954	[320, 323]
W_2B	3.10	[535]	1954	
WB	< 1.28	[535]	1954	[389]
W_2B_5	< 1.28	[535]	1954	[320]
Re_2B	2.80	[1043]	1961	
CeC_2	<1.28	[361]	1952	
ThC	<1.20	[535]	1954	
UC	<1.20	[535]	1954	
TiC	< 1.20	[535]	1954	[319, 320, 324, 361, 389]
ZrC	< 1.20	[535]	1954	[319, 320, 325, 326, 389]
HfC	< 1.20	[535]	1954	[319]
V_2C	<1.20	[535]	1954	[1332]
VC	< 1.20	[535]	1954	[325, 389, 1211]
Nb_2C	9.18	[535]	1954	[1332]
NbC	~13	[1211]	1962	Extrapolated value, [320,324,325,327, 535]
$NbC_{0.977}$	11.1	[1211]	1962	
$NbC_{0.948}$	10.6	[1211]	1962	
$NbC_{0.918}$	7.3	[1211]	1962	
$NbC_{0.881}$	4.2	[1211]	1962	
$NbC_{0.79}$	<1.05	[1211]	1962	
Ta_2C	3.26	[535]	1954	[1332]
TaC	~11	[1211]	1962	Extrapolated value, [319,324,326,328, 389, 535, 1212]

Phase	Critical temperature T_c, °K	Ref.	Year	Remarks
$TaC_{0.987}$	9.7	[1211]	1962	
$TaC_{0.981}$	9.0	[1211]	1962	
$TaC_{0.958}$	7.5	[1211]	1962	
$TaC_{0.910}$	4.75	[1211]	1962	
$TaC_{0.848}$	2.04	[1211]	1962	
$TaC_{0.754}$	<1.05	[1211]	1962	
$Cr_{23}C_6$	< 1.20	[535]	1954	[1332]
Cr_7C_3	< 1.20	[535]	1954	
Cr_3C_2	< 1.20	[535]	1954	
Mo_2C	2.78	[535]	1954	[323, 326, 361, 1043]
MoC	9.26	[535]	1954	[320, 329, 330]
W_2C	2.74	[535]	1954	[320, 327, 331, 361]
WC	< 1.28	[535]	1954	[320, 327, 328, 361, 389]
LaN	<1.8	[389]	1953	[1043]
CeN	<1.8	[389]	1953	
Th_3N_4	< 1.20	[535]	1954	
UN	< 1.20	[535]	1954	
TiN	4.86—5.6	[535]	1954	[319, 324, 326, 328, 332, 1043]
ZrN	8.9—9.05	[535]	1954	[319, 324, 326, 328]
V_2N	< 1.28	[535]	1954	Very contaminated
VN	7.50—8.2	[535]	1954	[326, 328, 361]
Nb_2N	9.5	[323]	1954	[535, 588, 594, 1043]
NbN	15.2	[333]	1947	[323, 333, 334, 361, 535, 588, 594, 1043]
Ta_2N	9.5	[323]	1954	[535]
TaN	1.88	[323]	1954	[535]
CrN	< 1.28	[535]	1954	[361]
Mo_2N	5.00	[535]	1954	[323]
MoN	12.00	[535]	1954	[323, 361]

Phase	Critical temperature $T_c, °K$	Ref.	Year	Remarks
W_2N	< 1.20	[535]	1954	[361]
$ReN_{0.34}$	4—5	[676]	1958	Re_3N
YSi_2	< 1.00	[1214]	1958	
$LaSi_2$	< 1.00	[1214]	1958	
$CrSi_2$	< 1.00	[1214]	1958	
$PrSi_2$	< 1.00	[1214]	1958	
$NdSi_2$	< 1.00	[1214]	1958	
α-$ThSi_2$	3.16	[535]	1954	
β-$ThSi_2$	2.41	[535]	1954	
U_3Si	< 1.10	[1215]	1958	
Ti_5Si_3	< 1.20	[535]	1954	[335]
$TiSi$	< 1.20	[535]	1954	[335]
$TiSi_2$	< 1.20	[535]	1954	[335]
Zr_4Si	< 1.20	[535]	1954	[335]
Zr_2Si	< 1.20	[535]	1954	[335]
Zr_3Si_2	< 1.20	[535]	1954	[335]
Zr_4Si_3	< 1.20	[535]	1954	[335]
Zr_6Si_5	< 1.20	[535]	1954	[335]
$ZrSi$	< 1.20	[535]	1954	[335]
$ZrSi_2$	< 1.20	[535]	1954	[335]
V_3Si	17.1	[535]	1954	[335, 1043]
V_5Si_3	< 1.20	[535]	1954	
VSi_2	< 1.20	[535]	1954	[335]
$Nb_3S_{12}(Nb_5Si_3)$	< 1.20	[535]	1954	[335]
Nb_2Si	< 1.20	[535]	1954	
$NbSi_2$	< 1.20	[335]	1953	
$Ta_5Si(Ta_{4.5}Si)$	< 1.20	[535]	1954	
$TaSi$	4.25	[326]	1934	[319]
Ta_3Si_2	< 1.20	[535]	1954	
Ta_5Si_3	< 1.20	[535]	1954	
$TaSi_2$	< 1.20	[535]	1954	[320, 335]
Cr_3Si	< 1.20	[535]	1954	
$Cr_3Si_2(Cr_5Si_3)$	< 1.20	[535]	1954	
$CrSi$	< 1.20	[535]	1954	
$CrSi_2$	< 1.20	[535]	1954	
Mo_3Si	< 1.3	[535]	1954	[335]
$Mo_3Si_2(Mo_5Si_3)$	< 1.20	[535]	1954	
$MoSi_2$	"	[535]	1954	[335]
$W_3Si_2(W_5Si_3)$	2.84	[535]	1954	W_5Si_3, [335]
WSi_2	< 1.20	[535]	1954	[320, 336]

SUPERCONDUCTIVITY (continued)

Phase	Critical temperature T_c, °K	Ref.	Year	Remarks
V_3P	< 1.00	[1213]	1956	
VP	< 1.02	[1043]	1961	
CrP	< 1.02	[1043]	1961	
Mo_3P	5.31	[1043]	1961	
MoP	< 1.03	[361]	1952	[1043]
W_3P	2.26	[1043]	1961	
WP	< 1.02	[1043]	1961	
CeS	< 1.28	[361]	1952	
Ce_3S_4	< 1.28	[361]	1952	
B_4C	< 1.28	[361]	1952	[1043]
SiC	< 1.28	[361]	1952	
BN	<1.28	[361]	1952	

THERMOELECTRIC PROPERTIES

Phase	Coefficient of thermo-emf (abs. values), $\mu V/deg$	Accuracy ($\pm$), $\mu V/deg$	Ref.	Year	Remarks
CaB_6	—32.0	—	[281]	1961	[310]
SrB_6	—30.3	—	[281]	1961	
BaB_6	—26.2	—	[281]	1961	[310]
ScB_4	–7	×		1963	T.S.Verkhoglyadova
ScB_6	—7.7	—	[694]	1960	
YB_6	—0.5	—	[281]	1961	
LaB_6	+ 7	—	[1133]	1960	[281, 310]
CeB_6	+2.8	—	[281]	1961	[310]
PrB_6	—0.6	—	[281]	1961	[310]
NdB_6	+0.4	—	[281]	1961	[310]
SmB_6	+7.6	—	[281]	1961	
EuB_6	—17.7	—	[281]	1961	
GdB_6	+0.1	—	[281]	1961	
TbB_6	—1.1	—	[281]	1961	

THERMOELECTRIC PROPERTIES (continued)

Phase	Coefficient of thermo-emf (abs. values), μV/deg	Accuracy (±), μV/deg	Ref.	Year	Remarks
YbB$_6$	−25.5	—	[281]	1961	
ThB$_6$	−0.6	—	[281]	1961	[310]
TiB$_2$	−5.1	—	[287]	1960	[310, 311]
ZrB$_2$	+1.2	—	[287]	1960	[310]
VB$_2$	+9.2	—	[287]	1960	[310, 311]
NbB$_2$	−1.4	—	[287]	1960	[310, 311]
TaB$_2$	−3.1	—	[287]	1960	[310, 311]
Cr$_4$B	−7.7	0.2	[930]	1961	[1166]
Cr$_2$B	−3.6	0.1	[930]	1961	[1166]
CrB	−0.94	0.1	[930]	1961	[1166]
CrB$_2$	−0.05	0.01	[930]	1961	[287, 310, 1166]
Cr$_2$B$_5$	−1.6	0.03	[1166]	1962	
Mo$_2$B$_5$	+3.2	—	[287]	1960	[310, 311]
W$_2$B$_5$	+3.2	—	[287]	1961	[310, 311]
YC	−34.6	—	[820]	1961	[879]
Y$_2$C$_3$	−6.4	—	[879]	1961	
YC$_2$	−0.8	—	[879]	1961	
La$_2$C$_3$	+1.1	—	×	1963	T. Ya. Kosolapova, G. N. Makarenko
LaC$_2$	+9.2	—	×	1963	Same
Ce$_2$C$_3$	+1.3	—	×	1963	"
CeC$_2$	+4.4	—	×	1963	"
PrC$_2$	+13.1	—	×	1963	"
TiC	−11.2	—	[287]	1960	[287, 310, 311, 316, 1242]
TiC	−20	—	×	1961	At 1200°, V. F. Nemchenko
ZrC	−11.3	—	[287]	1960	[310, 311,1242]
HfC	−11.7	—	[287]	1960	
VC	+3.7	—	[287]	1960	[310]
NbC	−4.0	—	[287]	1960	[310, 311]
TaC	−5.0	—	[287]	1960	[310, 311]
Cr$_{23}$C$_6$	+2.76	0.02	[296]	1961	[1166]
Cr$_7$C$_3$	−7.1	0.3	[296]	1961	[1166]
Cr$_3$C$_2$	−6.7	0.5	[296]	1961	[310, 1166]
Mo$_2$C	−1.9	—	[287]	1960	[310]
WC	−23.3	—	[287]	1960	[310, 311]
W$_2$C	−8.17	0.02	[287]	1960	
ScN	−20 to −40	—	×	1963	V. S. Neshpor (100–600°C)
TiN	−7.78	1.1	[997]	1961	
TiN$_{0.935}$	−7.1	—	×	1963	S. N. L'vov, V. F. Nemchenko
TiN$_{0.765}$	−3.6	—	×	1963	Same
ZrN	−4.78	0.5	[997]	1961	[989]
HfN	−2.96	0.6	[997]	1961	
V$_3$N	−5.3	1.2	[997]	1961	
VN	−4.6	0.8	[929]	1961	[281]
Nb$_2$N	−4.6	0.7	[929]	1961	

Phase	Coefficient of thermo-emf (abs. values), μV/deg	Accuracy ($\pm$), μV/deg	Ref.	Year	Remarks
$NbN_{0.97}$	—1.65	0.1	[929]	1961	
NbN	—2.24	0.0	[929]	1961	
Ta_2N	—2.17	0.4	[929]	1961	
TaN	—1.6	0.3	[929]	1961	
Cr_2N	—0.52	0.2	[1166]	1962	[281]
CrN	—92.0	4.0	[886]	1961	[1166]
Mo_2N	+2.18	0.5	[929]	1961	
Mg_2Si	600—2000	—	[1006]	1960	[1046]
$MgSi_2$	+180—+240	—	[117]	1959	
$BaSi_2$	+600	—	[1289]	1963	[917]
$LaSi_2$	—4.2	0.3	[1216]	1960	
$CeSi_2$	+7.8	0.5	[314]	1960	
$PrSi_2$	—3.2	0.4	[314]	1960	
Ti_5Si_3	+2.3	—	[316]	1960	[1240]
TiSi	+2.4	—	[316]	1960	
$TiSi_2$	+5.2	—	[316]	1960	[310, 1240]
$ZrSi_2$	+14.7	—	[314]	1960	[310, 1240, 1245]
V_3Si	~0	—	[1245]	1962	Data taken from graph
V_5Si_3	~+3	—	[1245]	1963	Same
VSi_2	+10.5	—	[314]	1960	[1245]
$NbSi_2$	+14.4	—	[314]	1960	
$TaSi_2$	+14.0	—	[314]	1960	[310, 1245]
Cr_3Si	+16.6	—	[302]	1957	
Cr_2Si	—4.0	—	[303]	1958	
CrSi	+5.0	—	[303]	1958	[302, 1240]
Cr_5Si_3	+0.6	—	[302]	1957	
$CrSi_2$	+86.0	—	[314]	1960	[302, 303]
Mo_3Si	—1.0	—	[314]	1960	[1240]
Mo_5Si_3	+2.0	—	[314]	1960	[1240]
$MoSi_2$	—3.0	—	[314]	1960	[310, 1240]
WSi_2	+0.2	—	[314]	1960	[310, 1240]
Mn_3Si	+18.0	—	[303]	1958	[1049,1070,1138]
Mn_5Si_3	+14.0	—	[303]	1958	[1049,1070,1138]
MnSi	+102	—	[303]	1958	[1037,1049,1136,1245]
$MnSi_2$	+46.0	—	[303]	1958	[1037,1049,1136,1245]
Re_3Si	~0	—	[1245]	1963	Data taken from graph
ReSi	~+31	—	[1245]	1963	Same
$ReSi_2$	+174	—	[314]	1960	[916, 1245]
Fe_3Si	—2.0	—	[303]	1958	[1245]
FeSi	$-23.0+0.43t^{1*}$	—	[860]	1960	[303, 1067,1245]
β-$FeSi_2$	—300 (20°)	—	[914]	1960	[310,303,1232,1233]
β-$FeSi_2$	—670 (220°)	—	[914]	1960	[310,303,1232,1233]
Co_2Si	—8.0	—	[303]	1958	[1245]
CoSi	—46.0	—	[303]	1958	
$CoSi_2$	—8.0	—	[303]	1958	[310, 1245]

[1*] Measured relative to Chromel.

THERMOELECTRIC PROPERTIES (continued)

Phase	Coefficient of thermo-emf (abs.values), μV/deg	Accu-racy ($\pm$), μV/deg	Ref.	Year	Remarks
CoSi$_3$	+14.0	—	[303]	1958	
Ni$_3$Si	—2.0	—	[303]	1958	[1245]
Ni$_5$Si$_2$(?)	—2.0	—	[303]	1958	
Ni$_2$Si$_3$	+9.0	—	[303]	1958	[1245]
NiSi	+8.0	—	[303]	1958	
NiSi$_2$	+7.0	—	[303]	1958	[310,1240]
TiP	—11.1	—	[1178]	1963	
BP	+300	—	[103]	1960	
LaS	—11.98	—	[1218]	1963	
La$_3$S$_4$	—243	7	×	1963	V. I. Marchenko
La$_2$S$_3$	+354	—	[1219]	1963	[304]
CeS	—8.58	—	[1218]	1963	[304]
Ce$_2$S$_3$	+574.4	—	[1217]	1963	[304,1112,1114]
PrS	—19.3	—	[1218]	1963	
NdS	—21.8	—	[1218]	1963	
ThS$_{1.7}$	+200	—	[305]	1958	
ThS$_{1.5}$	+100	—	[305]	1958	
B$_4$C	+80	—	[306]	1960	
α-SiC	+0.3	—	[1191]	1963	
β-SiC	—105	—	[1191]	1963	

THERMIONIC EMISSION PROPERTIES

Phase	Electronic workfunction, eV	Richardson constant, amp/cm^2·deg^2	Coefficient of secondary emission	Ref.	Year	Remarks
CaB$_6$	2.86	2.6	—	[25]	1951	
SrB$_6$	2.67	0.14	—	[25]	1951	
BaB$_6$	3.45	16	—	[25]	1951	[897]
ScB$_2$	3.76	—	0.58	[1334]	1963	[898, 899]
ScB$_6$(?)	2.96	4.6	0.58	[876]	1961	
YB$_6$	2.22	15	—	[25]	1951	[876, 899]
LaB$_6$	2.68	29	0.95	[215]	1958	[25, 337, 876, 899, 900, 901, 1062]
CeB$_6$	2.93	580	0.68	[215]	1958	[25, 876, 900]
PrB$_6$	3.46	300	—	[215]	1958	[337, 876, 901]
NdB$_6$	3.97	420	—	[215]	1958	[337, 876]
SmB$_6$	4.4	—	—	[18]	1959	[876]
EuB$_6$	4.9	—	—	[19]	1958	[902, 876]
GdB$_6$	2.05	0.84	0.8	[215]	1958	[876, 899]
TbB$_6$	3.26	120	0.74	[876]	1961	[22]
DyB$_6$	3.53	25.1	0.8	[215]	1958	

Phase	Electronic work function, eV	Richardson constant, amp/cm^2 ·deg^2	Coefficient of secondary emission	Ref.	Year	Remarks
HoB$_6$	3.42	13.9	0.7	[215]	1958	
ErB$_6$	3.37	9.9	0.7	[215]	1958	[876]
TuB$_4$ + + TuB$_6$	3.38*	—	—	[898]	1961	1000—1800°K, [1334]
YbB$_6$	3.13	2.5	—	[215]	1958	
LuB$_6$	3.0	0.36	0.8	[21]	1959	
ThB$_6$	2.92	0.53	—	[25]	1951	
TiB$_6$	3.88	884	0.825	[273]	1958	1700—1900°K, [338, 903]
ZrB	4.48	35000	—	[339]	1951	
ZrB$_2$	3.67	0.5	0.85	[273]	1958	1700—1900°K, [338]
VB$_2$	3.95	35.5	0.825	[273]	1958	[904]
NbB$_2$	3.65	—	—	[338]	1957	
TaB$_2$	2.89	10	—	[339]	1951	
CrB$_2$	3.36	48	0.775	[273]	1958	1700—1900°K, [338]
Mo$_2$B$_5$	3.38	—	—	[338]	1957	
W$_2$B$_5$	2.62	—	—	[338]	1957	
MnB$_2$	4.14	—	—	[338]	1957	
TiC	3.53	—	—	[1161]	1962	300°K
TiC	3.74	—	—	[1161]	1962	1400°K, [339, 905]
TiC	3.82	—	—	[1161]	1962	2000°K
TiC	—	25	—	[339]	1951	
ZrC	3.20	—	—	[1161]	1962	300°K
ZrC	3.53	—	—	[1161]	1962	1400°K, [339, 340, 903, 905]
ZrC	3.64	—	—	[1161]	1962	2000°K
ZrC	3.8	134	—	[340]	1959	1400—1800°K
HfC	3.47	—	—	[1161]	1962	300°K
HfC	3.65	—	—	[1161]	1962	1400°K, [998, 1334]
HfC	3.76	—	—	[1161]	1962	2000°K
VC	3.85	—	—	[1161]	1962	1300—2100°K
NbC	4.02	—	—	[1161]	1962	300°K
NbC	3.74	—	—	[1161]	1962	1400°K, [898, 1334]
NbC	3.58	—	—	[1161]	1962	2000°K
TaC	3.77	—	—	[1161]	1962	1400°K, [339, 905]
TaC	3.65	—	—	[1161]	1962	2000°K
TaC	—	0.30	—	[339]	1951	
ThC$_2$	3.5	550	—	[339]	1951	
UC	2.70	30	—	[1261]	1963	1300-1900°K, [340, 906]
TiN	3.75	—	—	[1334]	1963	[1], 2000°K
ZrN	3.90	—	—	[1334]	1963	[1, 989], 1900°K
NbN	3.92	—	—	[1334]	1963	1950°K
YSi$_2$	3.48	—	—	×	1961	B. M. Tsarev
ZrSi$_2$	3.25	—	—	[1162]	1962	1400°K
VSi$_2$	3.14	—	—	[1162]	1962	»

THERMIONIC EMISSION PROPERTIES (continued)

Phase	Electronic workfunction, eV	Richardson constant, amp/cm² ·deg²	Coefficient of secondary emission	Ref.	Year	Remarks
$NbSi_2$	3.65	—	—	[1162]	1962	"
$TaSi_2$	3.90	—	—	[1162]	1962	"
Cr_3Si	3.23	—	—	[1162]	1962	"
$CrSi$	3.40	—	—	[1162]	1962	"
$CrSi_2$	3.55	—	—	[1162]	1962	"
WSi_2	3.34	—	—	[1162]	1962	1400°K
$ReSi_2$	3.64	—	—	[1162]	1962	
LaS	3.73	—	—	[1222]	1963	1500°K, [1224]
LaS	4.15	—	—	[1222]	1963	1700°K, [1224]
La_2S_3	3.74	—	—	[1222]	1963	1500°K, [1224]
La_2S_3	4.16	—	—	[1222]	1963	1700°K, [1224]
CeS	3.66	—	—	[1222]	1963	1500°K, [1224]
CeS	3.95	—	—	[1222]	1963	1700°K, [1224]
Ce_2S_3	3.70	—	—	[1222]	1963	1500°K, [1224]
Ce_2S_3	3.95	—	—	[1222]	1963	1700°K, [1224]
PrS	3.72	—	—	[1223]	1963	1500°K, [1224]
PrS	3.90	—	—	[1223]	1963	1700°K, [1224]
Pr_2S_3	3.63	—	—	[1223]	1963	1500°K, [1224]
Pr_2S_3	3.84	—	—	[1223]	1963	1700°K, [1224]
NdS	3.69	—	—	[1223]	1963	1500°K, [1224]
NdS	3.89	—	—	[1223]	1963	1700°K, [1224]
Nd_2S_3	3.64	—	—	[1223]	1963	1500°K, [1224]
Nd_2S_3	3.93	—	—	[1223]	1963	1700°K, [1224]

* Value recalculated from the data of [898] by B. M. Tsarev. According to experimental data of B. M. Tsarev, $\varphi_{TuB_6} = 3.6$ eV.

HALL CONSTANT

Phase	Hall constant R, $(cm^3/coul) \cdot 10^{-4}$	Accuracy (±), $(cm^3/coul) \cdot 10^{-4}$	Ref.	Year	Remarks
CaB_6	—91.0	—	[281]	1961	
SrB_6	—76.3	—	[281]	1961	
BaB_6	—57.5	—	[281]	1961	
YB_2	—3.05	1.15	[1115]	1963	
YB_4	—21.3	0.9	[1115]	1963	
YB_6	—4.56	–	[281]	1961	[1115]
YB_{12}	—10.8	1.54	[1115]	1961	
LaB_6	—4.96	—	[281]	1961	
CeB_6	—4.18	—	[281]	1961	
PrB_6	—4.33	—	[281]	1961	
NdB_6	—4.39	—	[281]	1961	

Phase	Hall constant R, $(cm^3/coul) \cdot 10^{-4}$	Accuracy $(\pm)$, $(cm^3/coul) \cdot 10^{-4}$	Ref.	Year	Remarks
SmB_6	$+1.54$	—	[281]	1961	
EuB_6	-50.2	—	[281]	1961	
GdB_6	-4.39	—	[281]	1961	
TbB_6	-4.57	—	[281]	1961	
YbB_6	-83.6	—	[281]	1961	
ThB_6	-2.19	—	[281]	1961	
TiB_2	-17.8	—	[287]	1960	[312]
ZrB_2	-17.6	—	[287]	1960	[312]
HfB_2	-17	—	[312]	1958	
VB_2	-0.54	—	[287]	1960	[312]
NbB_2	-2.1	—	[287]	1960	[312]
TaB_2	-2.2	—	[287]	1960	[312]
Cr_4B	-1.17	0.05	[1166]	1962	[930]
Cr_2B	-1.01	0.01	[1166]	1962	[930]
CrB	-0.68	0.05	[1166]	1962	[930]
CrB_2	-0.06	0.01	[1166]	1962	[287, 312, 930]
Cr_2B_5	-0.60	0.05	[1166]	1962	
Mo_2B_5	-0.5 to $+0.1$	—	[312]	1958	
W_2B_5	-1.7	—	[287]	1960	[312]
TiC	-6.7	—	[287]	1960	[1242]
ZrC	-9.42	—	[287]	1960	[1242]
HfC	-12.4	—	[287]	1960	[1242]
VC	-0.48	0.21	[287]	1960	
NbC	-1.32	—	[287]	1960	
TaC	-1.1	—	[287]	1960	
$Cr_{23}C_6$	$+1.2$	0.2	[296]	1961	[1166]
Cr_7C_3	-0.38	0.03	[296]	1961	[1166]
Cr_3C_2	-0.47	0.03	[296]	1961	[1166]
Mo_2C	-0.85	—	[287]	1960	
W_2C	-13.1	0.7	[287]	1960	
WC	-21.8	0.3	[287]	1960	
TiN	-0.67	0.0	[997]	1961	[287, 929]
$TiN_{0.765}$	$+0.9$	×		1963	S.N.L'vov, V.F.Nemchenko
ZrN	-1.3	0.2	[997]	1961	[929]
$ZrN_{0.879}$	-1.45	×		1963	S.N.L'vov, V.F.Nemchenko
HfN	-4.2	0.5	[997]	1961	[281, 929]
V_3N	$+0.9$	0.1	[997]	1961	[929]
VN	$+0.42$	0.2	[929]	1961	
Nb_2N	$+1.9$	0.4	[929]	1961	
$NbN_{0.75}$	-0.69	0.1	[929]	1961	
$NbN_{0.97}$	-0.47	0.2	[929]	1961	
NbN	$+0.52$	0.19	[929]	1961	

Phase	Hall constant R, $(cm^3/coul) \cdot 10^{-4}$	Accuracy $(\pm)$, $(cm^3/coul)$ $\cdot 10^{-4}$	Ref.	Year	Remarks
Ta₂N	—0.46	0.1	[929]	1961	
TaN	—3.61	. 0.9	[929]	1961	
Cr₂N	—0.72	0.1	[287]	1960	[886, 1166]
CrN	—264	25	[886]	1961	[287, 1166]
Mo₂N	+2.83	1.2	[929]	1961	
Mg₂Si	—1.68 exp [0.064/2KT] (160 < T < 300°K);		[422]	1957	Intrinsic conductivity
	—11.2 exp [0.0072/2KT] (T < 160°K)		[422]	1957	Impurity conductivity, [1006]
LaSi₂	—17.5	0.4	[1216]	1960	
CeSi₂	+19.3	1.0	[314]	1960	
PrSi₂	—10.7	0.5	[314]	1960	
Ti₅Si₃	—0.27	—	[315]	1960	
TiSi	—0.43	—	[315]	1960	
TiSi₂	—0.63	—	[315]	1960	
Zr₅Si₃	—3.75	—	[315]	1960	
ZrSi₂	—1.46	—	[315]	1960	
V₃Si	—0.17	—	[315]	1960	
V₅Si₃	—1.0	—	[315]	1960	
VSi₂	—1.95	—	[315]	1960	
Ta₄.₅Si	—2.46	—	[315]	1960	
Ta₂Si	—4.76	—	[315]	1960	
Ta₅Si₃	—4.54	—	[315]	1960	
TaSi₂	—0.88	—	[315]	1960	
Cr₃Si	+0.49	—	[315]	1960	
Cr₅Si₃	—0.51	—	[315]	1960	
CrSi	—0.46	—	[315]	1960	
CrSi₂	+66.5	—	[315]	1960	
Mo₃Si	—0.26	—	[315]	1960	[1231]
Mo₅Si₃	—0.42	—	[315]	1960	[1231]
MoSi₂	+12.7	—	[315]	1960	[642, 1231]
W₃Si	+1.19	—	[315]	1960	
W₅Si₃	—0.3	—	[315]	1960	
WSi₂	+841	—	[315]	1960	
Re₃Si	+1.79	—	[315]	1960	
ReSi	+65.41	—	[315]	1960	
ReSi₂	+8700	—	[315]	1960	[916]
β-FeSi₂	—0.3	—	[914]	1960	High-purity alloy, [1233, 1234]
CoSi	—1.73	—	[315]	1960	
CoSi₂	+2.53	—	[315]	1960	
Ni₂Si	+0.35	—	[315]	1960	
NiSi₂	+3.77	—	[315]	1960	
TiP	—3.0	—	[304]	1961	
CeS	+2000	—	×	1960	S. N. L'vov
SiC	+(5—8) · 10⁴	–	[1191]	1963	Technical, n-type
SiC	+(5—10) · 10³	–	[1191]	1963	Technical, p-type

WIDTH OF FORBIDDEN BANDS
OF SEMICONDUCTOR REFRACTORY COMPOUNDS

Phase	Width of forbidden band E_0, eV	Ref.	Year	Remarks
CaB_6	0.40	[1115]	1963	
SrB_6	0.38	[1115]	1963	[1160]
BaB_6	0.12	[1115]	1963	
$BaSi_2$	0.48	[1289]	1963	
$LaSi_2$	0.19	[1216]	1960	
AlN	3.8	[634]	1954	[825]
$TiN_{0.765}$	0.40	×	1963	S.N.L'vov, V.F.Nemchenko
$ZrN_{0.879}$	0.29	×	1963	Same
Mg_2Si	0.75—0.77	[631]	1957	[633]
Ca_2Si	1.9	[631]	1957	
$CrSi_2$	1.3	[302]	1956	
$MnSi$	~0.6	[1136]	1961	
Mn_3Si_5	~0.2	[971]	1961	
$MnSi_2$	~0.9	[1136]	1961	
$ReSi_2$	0.13	[916]	1961	
$\beta\text{-}FeSi_2$	0.8	[914]	1960	High-purity alloy, [860]
AlP	2.5—3.0	[631]	1957	[635, 1285]
La_2S_3	1.33	[1219]	1963	
Ce_2S_3	1.12	[1219]	1963	
Pr_2S_3	1.10	[1224]	1963	
Nd_2S_3	1.06	[1224]	1963	
B_4C	1.64	[889]	1961	[306,630,631,847]
SiC	1.5—3.5	[631]	1957	[632, 1191]
$\alpha\text{-}BN$	4.6	[633]	1957	[1285]
$\beta\text{-}BN$	~3	[862]	1960	Borazon
Si_3N_4	3.9	[724]	1960	
BP	~6	[103]	1960	[1285, 1320]
$3AlB_{12} \cdot 2B_4C$	2.3	[637]	1953	
B	1.55	[636]	1957	[1319]

MAGNETIC PROPERTIES

Phase	Magnetic susceptibility, $\varkappa \cdot 10^6$ (per mole)	Effective magnetic moment μ_{eff} of Bohr magneton	Temperature, °K	Ref.	Year	Remarks
YB_6	—	0	—	[358]	1952	
LaB_6	60	~0	293—673	[359]	1932	
CeB_6	+2260	2.30	293—703	[359]	1932	
PrB_6	+4800	3.37	293—713	[359]	1932	[360,1112]
NdB_6	—	3.82	620—1030	[358]	1952	[359,1112]
SmB_6	+1810	2.52	293	[359]	1932	[1112]
GdB_6	+21000	7.63	623—1033	[358]	1952	[872,1112]
YbB_6	+8740	4.58	623—1033	[358]	1952	[1112]

Phase	Magnetic susceptibility, $\varkappa \cdot 10^6$ (per mole)	Effective magnetic moment μ_{eff} of Bohr magneton	Temperature, °K	Ref.	Year	Remarks
UB_{12}	Diamagnetic	—	—	[15]	1954	
Mo_2B	+1	—	7	[361]	1952	
MnB	—	1.65	293	[841]	1959	
Fe_2B	—	1.91	—	[362]	1955	
Co_2B	Ferromagnetic	—	—	[481]	1938	
CeC_2	~1640*	2.19	293	[840]	1959	
PrC_2	~4500*	3.15	293	[840]	1959	
NdC_2	—	3.53	293	[840]	1959	
SmC_2	~2300*	2.85	293	[840]	1959	
GdC_2	—	7.59	293	[840]	1959	[364]
TbC_2	~28500*	9.57	293	[840]	1959	
DyC_2	~38500*	10.53	293	[840]	1959	
HoC_2	~43500*	10.47	293	[840]	1959	
ErC_2	~33300*	8.75	293	[840]	1959	
YbC_2	~2500*	3.69	293	[840]	1959	
UC	+3.15*	—	300	[1139]	1962	
UC_2	+3.40*	—	300	[1139]	1962	
TiC	+6.7	—	293	[824]	1960	[363, 1238]
ZrC	—23	—	293	[824]	1960	[363]
HfC	—25,5	—	293	[824]	1960	
VC	+26,2	—	193	[824]	1960	
NbC	+15,3	—	293	[824]	1960	
TaC	+9,3	—	293	[824]	1960	[363]
WC	+10	—	293	[363]	1931	
Sr_3N_4	+279	—	293	[596]	1957	[599]
LaN	+60	—	295	[1099]	1962	
CeN	+296	—	295	[1099]	1962	
PrN	+4460	—	295	[1099]	1962	
NdN	+5850	—	295	[1099]	1962	
TbN,HoN	—	—	—	[1330]	1962	
UN	+7.73	—	290	[1139]	1962	
U_2N_3	+13.17	—	290	[1139]	1962	
TiN	+48	—	293	[363]	1931	
ZrN	~+60	—	293	[363]	1931	
CrN	Ferromagnetic	—	—	[990]	1961	
Mn_4N	—	1.2	293	[595]	1957	[873]
Mn_5N_2	—	3.94	93—803	[350]	1952	[873]
Fe_4N	—	2.22	—	[362]	1955	
$ScSi_2$	Paramagnetic	—	—	[1214]	1958	

* Data taken from graph.

Phase	Magnetic susceptibility, $\varkappa \cdot 10^6$ (per mole)	Effective magnetic moment μ_{eff} of Bohr magneton	Temper- ature, °K	Ref.	Year	Remarks
YSi$_2$	"	—	—	[1214]	1958	
LaSi$_2$	"	—	—	[1214]	1958	
CeSi$_2$	"	—	—	[1214]	1958	
NdSi$_2$	"	—	—	[1214]	1958	
Ti$_5$Si$_3$	+810	—	298	[300]	1958	[365]
TiSi	+55	—	298	[300]	1958	
TiSi$_2$	+129	—	298	[300]	1958	
TiSi$_2$	+112	—	773	[300]	1958	
ZrSi	—67	—	298	[300]	1958	[365]
ZrSi$_2$	—103	—	298	[300]	1958	
ZrSi$_2$	—113	—	773	[300]	1958	
VSi$_2$	+161	—	298	[300]	1958	[365]
VSi$_2$	+107	—	773	[300]	1958	[365]
NbSi$_2$	—37	—	298	[300]	1958	[365]
TaSi$_2$	—40	—	298	[300]	1958	[365]
Cr$_3$Si	+704	—	298	[176]	1959	[365]
Cr$_5$Si$_3$	+892	—	298	[176]	1959	[365]
CrSi	+318	—	298	[176]	1959	
CrSi$_2$	+41	—	298	[300]	1958	[176, 366]
CrSi$_2$	+36	—	773	[300]	1958	
MoSi$_2$	—36.5	—	298	[300]	1958	
MoSi$_2$	—67	—	773	[300]	1958	
WSi$_2$	—82	—	298	[300]	1958	
WSi$_2$	—84	—	773	[300]	1958	
Mn$_2$Si	—	3.9	170—500	[367]	1948	
MnSi	+2500	—	290	[367]	1948	
MnSi$_2$	—35	—	298	[366]	1954	[365, 367]
FeSi	+1000	—	293	[367]	1948	[367]
FeSi$_2$	+85	—	293	[366]	1954	[365, 1145]
CoSi	+360	—	300	[366]	1954	
CoSi$_2$	< +92	—	300	[366]	1954	[365]
CoSi$_2$	—	1.2	90—250	[366]	1954	
MnP	Ferromagnetic	—	298	[368]	1957	
MnP	>0	2.85	400—600	[368]	1957	
YS	+100	—	298	[145]	1956	
Y$_5$S$_7$	+39.3	—	298	[146]	1956	
δ-Y$_2$S$_3$	+83.4	—	298	[146]	1956	

Phase	Magnetic susceptibility, $\varkappa \cdot 10^6$ (per mole)	Effective magnetic moment μ_{eff} of Bohr magneton	Temperature, °K	Ref.	Year	Remarks
YS_2	+125	—	298	[1225]	1956	
Y_2O_2S	0	—	293	[147]	1955	
LaS	+281	—	298	[149]	1956	[1224]
La_3S_4	+27.2	—	298	[159]	1956	
$\beta\text{-}La_2S_3$	−30.3	—	298	[1225]	1956	
$\gamma\text{-}La_2S_3$	−27.1	—	298	[1225]	1956	[159, 368, 1112, 1224]
LaS_2	−36.3	—	298	[1225]	1956	[247, 368, 1112]
La_2O_2S	≥ 0	—	298	[368]	1957	
CeS	+2125	2.26	298	[145]	1956	[156, 1112, 1224]
Ce_3S_4	+2160	2.27	298	[159]	1956	
$\alpha\text{-}Ce_2S_3$	+2270	2.33	298	[1225]	1956	
$\beta\text{-}Ce_2S_3$	+2194	2.29	298	[1225]	1956	
$\gamma\text{-}Ce_2S_3$	+2520	2.45	298	[156]	1960	[1224]
CeS_2	+2286	2.34	298	[1225]	1956	[368]
Ce_2O_2S	+2139	2.31	298	[147]	1955	
PrS	+4730	3.36	293	[1218]	1963	[1224]
$\alpha\text{-}Pr_2S_3$	+4770	3.37	298	[1228]	1960	
$\beta\text{-}Pr_2S_3$	+4685	3.34	298	[1225]	1956	[1228]
$\gamma\text{-}Pr_2S_3$	+4640	3.33	298	[1225]	1956	[1228]
PrS_2	+4800	3.38	298	[1225]	1956	[1224, 1227]
NdS	+4370	3.23	298	[149]	1956	[1224]
Nd_3S_4	+4849	3.40	298	[159]	1956	
$\alpha\text{-}Nd_2S_3$	+4600	3.31	298	[1225]	1956	
$\beta\text{-}Nd_2S_3$	+4750	3.37	298	[1225]	1956	
$\gamma\text{-}Nd_2S_3$	+4924	3.44	298	[1225]	1956	[247, 1224]
NdS_2	+5082	3.49	298	[1225]	1956	
Nd_2O_2S	+4846	3.48	298	[153]	1956	
SmS	+5020	3.46	298	[149]	1956	[1228]
Sm_3S_4	+2350	2.37	298	[159]	1956	
$\alpha\text{-}Sm_2S_3$	+1120	1.64	298	[1225]	1956	
$\gamma\text{-}Sm_2S_3$	+1020	1.60	298	[1225]	1956	
SmS_2	+1238	1.72	298	[1225]	1956	
Sm_2O_2S	+993.3	1.57	298	[247]	1930	
EuS	+23200	7.45	298	[369]	1959	[370]
Eu_3S_4	+11500	5.25	298	[369]	1959	[360]
$Eu_2S_{3.81}$	+5800	3.72	298	[369]	1959	
$\gamma\text{-}Gd_2S_3$	+27750	8.15	298	[247]	1930	
$GdS_{1.90}$	+21500	7.15	298	[369]	1959	

MAGNETIC PROPERTIES (continued)

Phase	Magnetic susceptibility, $\varkappa \cdot 10^6$ (per mole)	Effective magnetic moment μ_{eff} of Bohr magneton	Temperature,°K	Ref.	Year	Remarks
γ-Dy_2S_3	+47700	10.6	298	[368]	1957	[247]
γ-Er_2S_3	+38600	9.60	298	[247]	1930	
YbS	+1450	1.86	298	[1012]	1961	
Yb_3S_4	+4740	3.36	298	[1012]	1961	[161]
Yb_2S_3	+7130	4.13	298	[1012]	1961	[318, 247]
ThS	Diamagnetic	—	—	[251]	1955	
Th_2S_3	"	—	—	[251]	1955	
US	+4603	—	—	[371]	1955	[177]
U_2S_3	+5206	—	298	[164]	1955	
U_3S_5	+11220	—	298	[164]	1955	
α-USi_2	+3137	—	298	[165]	1953	
β-USi_2	+3470	—	298	[165]	1953	
SiC	—12.8	—	298	[368]	1957	Powder
Si_3N_4	100—150	0.30—0.35	90	[555]	1953	
Si_3N_4	56—101	0.43—0.49	293	[555]	1953	
Si_3N_4	56—81	0.47—0.56	483	[555]	1953	
Si_3N_4	58.7—70	0.56—0.61	673	[555]	1953	

CURIE TEMPERATURE

Phase	Θ_C, °K	Ref.	Year	Remarks
GeB_6	—344	[368]	1957	
PrB_6	~0	[368]	1957	
NdB_6	—455	[368]	1957	
GdB_6	—60	[872]	1961	[368]
YbB_6	—2	[368]	1957	
MnB	562±4	[841]	1959	
Fe_2B	1012	[1085]	1957	
Co_2B	~783	[481]	1938	
CeC_2	—61	[840]	1959	
PrC_2	5.2	[840]	1959	
NdC_2	40	[840]	1959	
SmC_2	—139	[840]	1959	
GdC_2	41.3	[840]	1959	
TbC_2	—91.2	[840]	1959	
DyC_2	—68.9	[840]	1959	
HoC_2	—25.6	[840]	1959	

CURIE TEMPERATURE (continued)

Phase	Θ_C, °K	Ref.	Year	Remarks
ErC_2	14.7	[840]	1959	
YbC_2	—388	[840]	1959	
Pe_3C	488	[1085]	1959	
Pe_2C	653	[1085]	1957	
TbN	~18	[993]	1957	
HoN	~43	[993]	1960	
U_2N_3	186	[1139]	1962	
Mn_4N	752	[626]	1960	
Mn_3N_2	—1070	[368]	1957	
Mn_2N	743—753	[626]	1957	
Fe_4N	761	[700]	1957	
Fe_3N	548	[1085]	1955	
$PrSi_2$	10.5	[1214]	1958	
Mn_2Si	—5	[368]	1957	
Fe_3Si_2	363	[1085]	1957	
$CoSi_2$	—170	[368]	1957	
MnP	317	[368]	1957	[1085]
Fe_3P	693	[1085]	1957	
Ce_2S_3	—57	[368]	1957	

DIELECTRIC PROPERTIES

Phase	Frequency, cps	Temperature, °C	Dielectric constant	Scattering coefficient	Ref.	Year
BN	10^2	10	4.15	0.00103	[272]	1955
BN	10^2	330	4.4	0.032	[272]	1955
BN	10^2	500	9.0	1.0 (470°)	[272]	1955
BN	10^4	10	4.15	0.00042	[272]	1955
BN	10^4	330	—	0.0043	[272]	1955
BN	10^4	500	4.5	0.1 (470°)	[272]	1955
BN	10^6	10	4.15	0.00020	[272]	1955
BN	10^6	330	—	0.0012	[272]	1955
BN	10^8	10	4.15	0.000095	[272]	1955
BN	10^{10}	10	—	0.0003	[272]	1955
BN	10^{10}	330	—	0.0004	[272]	1955
BN	10^{10}	470	—	0.0005	[272]	1955
Si_3N_4	—	18	9.4	—	[505]	1957

Chapter IV

OPTICAL PROPERTIES

COLOR OF SOME REFRACTORY COMPOUNDS

Phase	Color in the disperse state (powder)	Phase	Color in the disperse state (powder)
Be_2B	Gray with pink tinge	UB_4	Gray-steel
		UB_{12}	Black
BeB_2	Dark gray	TiB_2	Gray
BeB_6	Brick-red	ZrB_2	"
MgB_2	Dark brown	HfB_2	"
MgB_6	" "	VB_2	"
MgB_{12}	" "	NbB_2	"
CaB_6	Black	CrB_2	"
SrB_6	Black with green tinge	Mo_2B_5	Light gray
		W_2B_5	" "
BaB_6	Black with violet tinge	MnB	Reddish-brown
AlB_{12}	Brown	MnB_2	"
ScB_2	Gray	Be_2C	Reddish
YB_4	Grayish brown	YC	Golden
YB_6	Blue-violet	YC_2	Yellow
LaB_6	Violet	LaC_2	"
LaB_{12}	Azure green	CeC_2	Reddish yellow
CeB_4	Grayish brown	ThC	Yellowish gray
CeB_6	Blue-violet	ThC_2	" "
PrB_6	Blue-gray	UC	Gray
NdB_6	"	TiC	Light gray
SmB_6	"	ZrC	Gray
EuB_6	Gray	HfC	"
GdB_4	Grayish brown	VC	"
GdB_6	Blue	NbC	Light brown
TbB_4	Gray-brown	TaC	Golden brown
TbB_6	Blue	$Cr_{23}C_6$	Gray
ErB_4	Gray-brown	Cr_7C_3	"
ErB_6	Blue	Cr_3C_2	"
TuB_4	Gray	Mo_2C	Dark gray
YbB_6	"	W_2C	Gray
ThB_6	Red-violet	WC	Gray

Phase	Color in the disperse state (powder)	Phase	Color in the disperse state (powder)
Be_3N_2	Colorless	YS	Ruby-red
Mg_3N_2	Greenish pink	Y_2S_3	Yellow
Sr_3N_2	Black	YS_2	Brown-violet
Ba_3N_2	"		
AlN	Light gray	Y_2O_2S	Grayish white
ScN	Blue (dark blue)	LaS	Golden yellow
LaN	Black		with greenish
CeN	"		tinge
PrN	"	La_3S_4	Blue-black
NdN	"	La_2S_3	From yellow to
SmN	"		red-black
TiN	Yellow-bronze	LaS_3	Brown-yellow
ZrN	Light yellow with	CeS	Brass-yellow
	greenish tinge	Ce_3S_4	Black
HfN	Yellow-brown	Ce_2S_3	Red
V_3N	Gray-brown	CeS_2	Dark brown
VN	Light brown	Ce_2O_2S	From brown to
Nb_2N	Gray		black
NbN	Light gray with yellow	PrS	Golden with green-
	tinge		ish tinge
Ta_2N	Black	Pr_3S_4	Blue-black
TaN	Gray with blue	Pr_2S_3	Dark brown
	tinge	Pr_2O_2S	Black
Cr_2N	Dark gray	NdS	Golden with green-
CrN	Black		ish tinge
Mo_2N	Dark gray	Nd_2S_3	Olive
W_2N	Black	Nd_2O_2S	Light blue
WN	Brown	SmS	Black
Mn_2N	Gray-blue	Sm_3S_4	"
MnN	Black	Sm_2S_3	From yellow to
Re_3N	Gray		pink
Co_3N	Gray-black	Sm_2O_2S	Light brown
Ni_3N	Dark gray	EuS	Black
Silicides	Gray	Eu_3S_4	"
Be_3P_2	Yellow	GdS	Yellow
Mg_3P_2	Colorless with yellow	α-Gd_2S_3	Brown-red
	tinge	γ-Gd_2S_3	Brown
Ca_3P_2	Brown-red	GdS_2	Brown-violet
BaP_2	Dark gray		
AlP	Peat colored	Gd_2O_2S	Light brown
Th_3P_4	Gray-steel	DyS	Red-violet
V_3P	Gray-black	Dy_5S_7	Black
VP	"	α-Dy_2S_3	Brown-red
VP_2	Black	γ-Dy_2S_3	Black
Sc_2S_3	Yellow	δ-Dy_2S_3	Green

COLOR OF SOME REFRACTORY COMPOUNDS (continued)

Phase	Color in the disperse state (powder)	Phase	Color in the disperse state (powder)
DyS_2	Brown-red	ThOS	Yellow
Dy_2O_2S	Light gray	PaOS	"
ErS	Red-violet	US	Gray
Er_5S_7	Black	U_2S_3	Black
δ-Er_2S_3	Light brown	α-US_2	Grayish black
Er_2O_2S	Light pink	UOS	Black with bluish tinge
Yb_2S_3	Yellow	Np_2S_3	Black
ThS	Silver gray	NpOS	"
ThS_3	Brown	PuS	Golden bronze
Th_4S_3		Pu_2S_3	Black
(Th_7S_{12})	Red "	BP	Chestnut
ThS_2	Purple		

EMISSION COEFFICIENT[1*]

Phase	Emission coefficient	Temperature, °C	Ref.	Year	Remarks
Be_2B	0.6	—	[513]	1960	
Be_5B	0.77	800—1600	[933]	1962	
CaB_6	0.75	800—1800	[933]	1962	
SrB_6	0.79	800—1800	[933]	1962	
BaB_6	0.84	800—1600	[933]	1962	
AlB_{12}	0.76	800—1600	[933]	1962	
ScB_2	0.89	800—1800	[933]	1962	[899, 1334]
YB_6	0.66—0.70	800—1700	[933]	1962	[10, 899]
LaB_6	0.82	800—1700	[933]	1962	[215, 216, 899]
CeB_6	0.72—0.77	800—1800	[933]	1962	[215]
PrB_6	0.76—0.79	800—1900	[933]	1962	
NdB_6	0.51—0.47	800—1600	[933]	1962	[216]
SmB_6	0.77	900—1700	[216]	1960	
EuB_6	0.83	800—1800	[933]	1962	
GdB_6	0.66—0.60	800—1800	[933]	1962	[899, 915, 916]
TbB_6	0.74	900—1800	[933]	1962	
DyB_6	0.8	1600	[215]	1958	
HoB_6	0.7	1600	[215]	1958	
ErB_6	0.7	1600	[215]	1958	
TuB_6	0.57—0.78	800—1900	[933]	1962	[1334]
YbB_6	0.73—0.75	800—1700	[933]	1962	
LuB_6	0.7	1600	[215]	1958	

[1*] Wavelength $\lambda = 650$ mμ.

Phase	Emission coefficient	Temperature, °C	Ref.	Year	Remarks
UB_{12}	0.77	800—1900	[933]	1962	
TiB_2	0.71	800—1700	[933]	1962	[273]
ZrB_2	0.89—0.91	800—1700	[933]	1962	[533]
HfB_2	0.89—0.92	800—1700	[933]	1962	
VB_2	0.72—0.76	800—1700	[933]	1962	[27]
NbB_2	0.77	800—2000	[933]	1962	
TaB_2	0.70	—	[933]	1951	
CrB_2	0.72	800—1700	[933]	1962	[273]
Mo_2B_5	0.80—0.76	800—1700	[533]	1962	[273]
W_2B_5	0.83	800—1900	[933]	1962	
Co_3B	0.82—0.87	800—2000	[933]	1962	
YC	0.81	800—1800	[879]	1961	
Y_2C_3	0.73—0.91	800—1800	[879]	1961	[933]
YC_2	0.87—0.68	1100—2000	[879]	1961	
UC	0.67—0.47	827—1327	[1118]	1963	
TiC	0.90	800—1700	[933]	1962	[216, 1286]
TiC	0.31	966	[1286]	1963	Integral emission coefficient
TiC	0.36	1424	[1286]	1963	Same
ZrC	0.75—0.79	800—2000	[933]	1962	[533, 1253]
ZrC	0.37	966	[1286]	1963	Integral emission coefficient
ZrC	0.40	1424	[1286]	1963	Same
HfC	0.77	800—1600	[933]	1962	[1334]
NbC	0.85	800—1800	[933]	1962	[1334]
TaC	0.62—0.85	800—1700	[933]	1962	[1253]
Cr_7C_3	0.92	800—1400	[933]	1962	[261]
Cr_3C_2	0.62—0.80	800—1500	[933]	1962	
Mo_2C	0.71	800—1500	[933]	1962	
W_2C	0.78	800—1800	[933]	1962	
WC	0.73—0.69	800—1700	[933]	1962	
AlN	0.85	800—1400	[933]	1962	Vacuum
AlN	0.80	800—2000	[933]	1962	Argon
ScN	0.79—0.87	800—1800	[933]	1962	
TiN	0.82—0.79	800—1700	[933]	1962	[1334]
ZrN	0.73—0.76	800—1800	[933]	1962	[1334]
HfN	0.84	800—1900	[933]	1961	
V_3N	0.82	800—1600	[933]	1962	
VN	0.77	800—1800	[933]	1962	
Nb_2N	0.82	800—1700	[933]	1962	
NbN	0.83	800—1700	[933]	1962	[1334]
Ta_2N	0.83	800—1700	[933]	1962	
TaN	0.79	800—1700	[933]	1962	
Cr_2N	0.69	800—1700	[933]	1962	
CrN	0.66—0.40	1200—2000	[933]	1962	Above 1300°C becomes Cr_2N
Mg_2Si	0.67—0.69	800—1000	[933]	1962	Argon
$GdSi_2$	0.80—0.83	800—1600	[933]	1962	
$TiSi_2$	0.82	800—1700	[933]	1962	
Ti_5Si_3	0.74	800—1700	[933]	1962	
$ZrSi_2$	0.72	800—1800	[933]	1962	

Phase	Emission coefficient	Temperature, °C	Ref.	Year	Remarks
VSi_2	0.73—0.89	800—1600	[933]	1962	
$NbSi_2$	0.80	800—1700	[933]	1962	
$TaSi_2$	0.74	800—1800	[933]	1962	
CrSi	0.79	800—1800	[933]	1962	
Cr_3Si_2	0.79	800—1700	[933]	1962	
$CrSi_2$	0.79	800—1600	[933]	1962	
Mo_3Si	0.77	800—1700	[933]	1962	
Mo_5Si_3	0.75	800—1700	[933]	1962	
$MoSi_2$	0.75	800—2000	[933]	1962	
Mn_3Si	0.68—0.78	800—1100	[933]	1962	Argon
$MnSi_2$	0.70—0.83	800—1200	[933]	1962	,,
$ReSi_2$	0.70—0.89	800—1400	[933]	1962	
CoSi	0.67—0.86	800—1300	[933]	1962	''
$NiSi_2$	0.67—0.82	800—1200	[933]	1962	,,
TiP	0.83	800—1300	[933]	1962	
LaS	0.45	800—1800	[1218]	1963	
La_2S_3	0.79	800—1500	[933]	1962	[1224]
CeS	0.56	800—1800	[1218]	1963	
Ce_2S_3	0.78—0.91	800—1800	[933]	1962	[1224]
PrS	0.65	800—1800	[1218]	1963	
Pr_2S_3	0.69	800—1300	[933]	1962	[1224]
NdS	0.79	800—1800	[1218]	1963	
Nd_2S_3	0.68	800—1900	[933]	1962	[1224]
B_4C	0.85	800—1500	[933]	1962	[216]
α-BN	0.64—0.62	800—1700	[933]	1962	
Si_3N_4	0.77	800—1600	[933]	1962	α-, β-phase mixture
$Si_xO_yC_z$ (Siloxicon)	0.80—0.81	800—2000	[933]	1962	
BP	0.63	800—1800	[933]	1962	
C	0.81—0.90	—	[944]	1959	Pyrographite

INFRARED ABSORPTION SPECTRA

Phase	Absorption bands wavelength, mm	Ref.	Year	Remarks
Mo_2B	7.2	[638]	1957	Weak
Mg_3N_2	4.8	[638]	1957	Medium
Mg_3N_2	7.1	[638]	1957	,,
Mg_3N_2	15.2	[638]	1957	,,
AlN	8.45	[638]	1957	Weak
AlN	9.46	[638]	1957	,,
AlN	14.0	[638]	1957	Medium
B_4C	9.5	[638]	1957	,,
B_4C	12.9	[638]	1957	Weak
SiC	12.0	[638]	1957	,,
BN	7.28	[638]	1957	Strong
BN	12.3	[638]	1957	Medium

Chapter V
MECHANICAL PROPERTIES

TENSILE STRENGTH

Phase	Tensile strength, kg/mm²	Temperature, °C	Porosity, %	Ref.	Year	Remarks
Be₂C	9.14—9.83	20	—	[526]	1952	
TiC	56—105	20	~0	[380]	1954	
TiC	38.0	800	~0	[380]	1954	
TiC	28.0	1000	~0	[380]	1954	[385]
TiC	0.25(?)	1300	~0	[380]	1954	[385]
ZrC	11.4	1200	8.8	[385]	1949	
TaC	2—3	20	—	[351]	1948	Determined on annealed TaC filaments produced by carburizing tantalum
Cr₃C₂	5.0	900	—	[140]	1961	Stress-rupture strength after 10 hr (data taken from graph)
Cr₃C₂	3.2	1000	—	[140]	1961	
Cr₃C₂	3.5	900	—	[140]	1961	Stress-rupture strength after 100 hr (data taken from graph)
Cr₃C₂	1.7	1000	—	[140]	1961	
WC	35	20	—	[264]	1934	
AlN	27	25	—	[674]	1960	
AlN	18.95	1000	—	[674]	1960	
AlN	12.7	1400	—	[674]	1960	
TiSi₂	15	20	—	[778]	1955	
MoSi₂	28	980	—	[267]	1956	
MoSi₂	29.4	1200	—	[267]	1956	
MoSi₂	9	1000	—	[543]	1959	1000 hr
Ni₂Si	0.6	20	~0	[1027]	1960	
Ni₂Si	11.2	600	~0	[1027]	1960	For cast alloys
Ni₂Si	14.2	650	~0	[1027]	1960	

Phase	Tensile strength, kg/mm^2	Temperature, °C	Porosity, %	Ref.	Year	Remarks
Ni$_2$Si	5.9	750	~0	[1027]	1960	For cast alloys
NiSi	0.6	20	~0	[1027]	1960	
NiSi	0.8	500	~0	[1027]	1960	
NiSi	2.0	550	~0	[1027]	1960	
NiSi	1.1	650	~0	[1027]	1960	
NiSi	0.53	750	~0	[1027]	1960	
U$_3$Si	70	25	—	[641]	1958	Elongation 1%, limit of proportionality 42.0 kg/mm^2 after 1000 hr, [690]
B$_4$C	7.3	25	~0	[346]	1956	
B$_4$C	16.3(?)	20	1	[385]	1959	
SiC	4.2	800	—	[140]	1961	Taken from graph
SiC	6.2	1000	—	[140]	1961	
SiC	7.5	1200	—	[140]	1961	
SiC	6.8	1300	—	[140]	1961	
SiC	2.8	900	—	[140]	1961	Stress-rupture strength after 10 hr (data taken from graph)
SiC	2.4	1000	—	[140]	1961	
SiC	2.3	900	—	[140]	1961	Stress-rupture strength after 100 hr (data taken from graph)
SiC	1.5	1000	—	[140]	1961	
BN	11.12	25	4—5	[272]	1955	Parallel to the hot-pressing direction
BN	10.60	350	4—5	[272]	1955	
BN	2.70	700	4—5	[272]	1955	
BN	1.53	1000	4—5	[272]	1955	
BN	5.10	25	4—5	[272]	1955	Perpendicular to the hot-pressing direction
BN	4.90	350	4—5	[272]	1955	
BN	1.33	700	4—5	[272]	1955	
BN	0.76	1000	4—5	[272]	1955	
Si$_3$N$_4$	1.5—2.75	20	20—25	×	1960	A.G. Dobrovol'skii. Specimens prepared by cold-pressing followed by sintering, [1035]
C	12—14.6	—	—	[944]	1959	Pyrographite
C	>42	2800	—	[972]	1960	Pyrographite (data taken from graph)

Phase	Shear strength, kg/mm^2	Temperature, °C	Porosity, %	Ref.	Year	Remarks
CaB$_6$	14.1	20	~0	[381]	1953	
GdB$_6$	21.1	20	8.5	[846]	1960	[285]
TiB$_2$	24.5	20	1.0	[1010]	1961	
ZrB$_2$	39.1	1000	—	[382]	1954	
ZrB$_2$	9.3	20	22—24	×	1962	
ZrB$_2$	9.6	800	22—24	×	1962	
ZrB$_2$	6.6	1000	22—24	×	1962	
ZrB$_2$	3.4	1100	22—24	×	1962	
ZrB$_2$	2.1	1200	22—24	×	1962	[1107]
ZrB$_2$	2.4	1300	22—24	×	1962	
ZrB$_2$	0.8	1500	22—24	×	1962	
ZrB$_2$	1.0	1670	22—24	×	1962	
ZrB$_2$	0.7	1750	22—24	×	1962	
CrB$_2$	62.0	20	—	×	1960	E. P. Lapteva, [393]
Mo$_2$B, MoB, Mo$_2$B$_5$	17.53—35.1	20	10—35	[49]	1952	
TiC	28.0 —39.9	20	—	[373]	1950	
TiC	51.6	20	1.4	[292]	1952	From powder, particle size from 44 to 74µ
TiC	64.0	20	0.5	[292]	1952	From powder, particle size from 37 to 44µ
TiC	70.3	20	~0	[292]	1952	From powder, particle size from 8 to 37µ
TiC	87.1	20	~0	[292]	1952	From powder, particle size from 2 to 8µ
TiC	10.2	1000	3.5	[293]	1950	Stress-rupture strength after 12.5 hr
TiC	5.6	1220	3.5	[293]	1950	Stress-rupture strength after 4 hr
TiC	62	20	~0	×	1960	E. P. Lapteva
TiC[1*]	5.5	20	18.6	×	1961	L. I. Struk; from powder, particle size: 40%(320µ)+10%(127µ)+50%(75µ)
TiC[1*]	4.2	1000	17.6	×	1961	
TiC[1*]	5.4	1400	17	×	1961	
TiC[1*]	5.9	20	19.0	×	1961	L. I. Struk; from powder, particle size: 10%(320µ)+30%(127µ)+60%(75µ)
TiC[1*]	6.0	1000	16.4	×	1961	
TiC[1*]	5.2	1400	11	×	1961	

SHEAR STRENGTH (continued)

Phase	Shear strength, kg/mm²	Temperature, °C	Porosity, %	Ref.	Year	Remarks
TiC¹*	5.9	20	11.6	×	1961	L. I. Struk; from powder, particle size below 75 μ
TiC¹*	4.9	1000	10.5	×	1961	
TiC¹*	5.2	1400	11	×	1961	
TiC¹*	9.1	1500	15.8	×	1961	
TiC¹*	9.9	1650	17	×	1961	
TiC¹*	13.4	1800	17.6	×	1961	
TiC¹*	10.3	1900	16.7	×	1961	
TiC¹*	1.5	20	21—25	×	1962	[1107]
TiC¹*	2.5	800	21—25	×	1962	
TiC¹*	0.8	1000	21—25	×	1962	
TiC¹*	0.6	1200	21—25	×	1962	
TiC¹*	1.4	1400	21—25	×	1962	
TiC¹*	0.8	1600	21—25	×	1962	
TiC¹*	4.0	1800	21—25	×	1962	
TiC¹*	10.4	1900	21—25	×	1962	
TiC¹*	5.7	2000	21—25	×	1962	
TiC¹*	3.6	2200	21—25	×	1962	
TiC¹*	1.3	2450	21—25	×	1962	
ZrC	7.51	1000	2.2	[263]	1950	Stress-rupture strength after 13 hr
ZrC	8—10	1220	2.2	[263]	1950	Stress-rupture strength after 4 hr
Mo₂C	5.0	20	26—28	×	1962	[1107]
Mo₂C	4.8	1000	26—28	×	1962	
Mo₂C	14.8	1300	26—28	×	1962	
Mo₂C	21.4	1600	26—28	×	1962	
Mo₂C	11.7	1800	26—28	×	1962	
WC	3.0	20	14—16	×	1962	
WC	6.3	1000	14—16	×	1962	
WC	1.6	1500	14—16	×	1962	
WC	6.9	1800	14—16	×	1962	
WC	13.5	2000	14—16	×	1962	
WC	35	20	—	[2]	1957	[269]
LaSi₂	27.2	20	—	[846]	1960	
NdSi₂	6.18	20	—	[846]	1960	
GdSi₂	4.45	20	—	[846]	1960	

1* Specimens prepared by cold-pressing followed by sintering.

175

SHEAR STRENGTH (continued)

Phase	Shear strength, kg/mm²	Temperature, °C	Porosity, %	Ref.	Year	Remarks
$DySi_2$	6.9	20	—	[846]	1960	[285]
$TiSi_2$	21.0	20	—	[778]	1955	
$MoSi_2$	35.1	20	—	[777]	1959	
$MoSi_2$	21	980	—	[267]	1956	Stress-rupture strength after 100 hr
$MoSi_2$	10.6	1040	—	[267]	1956	
$MoSi_2$	6.0	1100	—	[267]	1956	
B_4C	31	20	—	[269]	1934	
B_4C	28.1	20	—	[381]	1953	
B_4C	34.0	20	—	[383]	1952	
B_4C	24.6	870	—	[383]	1952	
B_4C	20.9	1093	—	[383]	1952	
B_4C	19.5	1316	—	[383]	1952	
B_4C	14.45	1000	0,8	[263]	1950	Stress-rupture strength after 13.5 hr
SiC	16.9	25	~4	[1000]	1961	Comp. %: 96,5 SiC, 2.5 Si_{free}, 0.4 C_{free}, 0.4 Al, 0.2 Fe; [206, 777]
SiC	17.6	1200	~4	[1000]	1961	
SiC	12.6	1500	~4	[1000]	1961	
SiC	15.5	20	—	[206]	1952	
SiC	20.9	1200	—	[777]	1959	
Si_3N_4	16.0	20	32.6%	[384]	1957	
Si_3N_4	15.2	600	30.6%	[384]	1957	
Si_3N_4	14.5	900	30.4%	[384]	1957	
Si_3N_4	14.7	1200	32.0%	[384]	1957	

COMPRESSIVE STRENGTH

Phase	Compressive strength, kg/mm²	Temperature, °C	Porosity, %	Ref.	Year	Remarks
TiB_2	135.0	20	~0	[378]	1960	
TiB_2	22.7	1000	~0	[378]	1960	
TiB_2	25.8	1200	~0	[378]	1960	
TiB_2	18.3	1400	~0	[378]	1960	
TiB_2	11.0	1600	~0	[378]	1960	

Phase	Compressive strength, kg/mm²	Temperature, °C	Porosity, %	Ref.	Year	Remarks
ZrB₂	158.7	20	∼0	[378]	1960	
ZrB₂	30.6	1000	∼0	[378]	1960	
ZrB₂	24.1	1200	∼0	[378]	1960	
ZrB₂	24.4	1400	∼0	[378]	1960	
ZrB₂	47.1	1600	∼0	[378]	1960	
CrB₂	127.9	20	∼0	[378]	1960	
CrB₂	86.8	1000	∼0	[378]	1960	
CrB₂	40.2	1200	∼0	[378]	1960	
CrB₂	58.1	1400	∼0	[378]	1960	
Be₂C	73.9	20	∼0	[526]	1952	
UC	30.1±4	20	20—25	[1014]	1959	Parallel to applied pressure
UC	12.6±2.2	20	20—25	[1014]	1959	Perpendicular to applied pressure
TiC	138.0	20	∼0	[378]	1960	[269, 301, 380, 664]
TiC	87.5	1000	∼0	[378]	1960	
TiC	51.0	1200	∼0	[378]	1960	
TiC	35.0	1400	∼0	[378]	1960	
TiC	23.0	1600	∼0	[378]	1960	
TiC	31.0	1800	∼0	[378]	1960	
TiC	16.4	2000	∼0	[378]	1960	
TiC	9.45	2200	∼0	[378]	1960	
ZrC	83.4	20	∼0	×	1961	
ZrC	49.7	1000	∼0	×	1961	
ZrC	26.4	1200	∼0	×	1961	
VC	62	20	—	[1]	1957	
NbC	242.3(?)	20	∼0	×	1961	L. I. Struk, [1]
Cr₃C₂	104.8	20	∼0	×	1961	
Cr₃C₂	94.9	1000	∼0	×	1961	
Cr₃C₂	57.2	1100	∼0	×	1961	
Cr₃C₂	57.1	1200	∼0	×	1961	
Cr₃C₂	42.1	1400	∼0	×	1961	
WC	360	20	—	[1]	1957	[2]
WC	272.1	20	∼0	×	1961	L. I. Struk

Phase	Compressive strength, kg/mm²	Temperature, °C	Porosity, %	Ref.	Year	Remarks
WC	141.0	1000	~0	×	1961	} L. I. Struk
WC	76.4	1100	~0	×	1961	
TiN	129.8	20	3.4	[1]	1957	
ZrN	100	20	~0	[1]	1957	
U₃Si	35.0	600	—	[690]	1960	Reduction in height 20%
U₃Si	5.5	800	—	[690]	1960	
TiSi₂	117.9	20	~0	×	1961	} L. I. Struk
TiSi₂	39.7	1000	~0	×	1961	
TiSi₂	10.5	1100		×	1961	
TiSi₂	5.5	1200	~0	×	1961	
MoSi₂	113.0	20	~0	[378]	1960	[267]
MoSi₂	40.5	1000	~0	[378]	1960	
MoSi₂	35.0	1200	~0	[378]	1960	
MoSi₂	39.0	1400	~0	[378]	1960	
MoSi₂	4.5	1600	~0	[378]	1960	
WSi₂	126.9	20	~0	×	1961	} L. I. Struk
WSi₂	59.5	1000	~0	×	1961	
CoSi	3.8	20	~0	[1027]	1960	
CoSi	6.3	500	~0	[1027]	1960	
CoSi	34.0	750	~0	[1027]	1960	
CoSi₂	10.0	20	~0	[1027]	1960	
CoSi₂	15.2	500	~0	[1027]	1960	
CoSi₂	60.0	750	~0	[1027]	1960	
Ni₂Si	31.6	20	~0	[1027]	1960	
Ni₂Si	57.9	500	~0	[1027]	1960	For cast alloys
Ni₂Si	76.0	600	~0	[1027]	1960	
NiSi	62.5	750	~0	[1027]	1960	
NiSi	15.8	20	~0	[1027]	1960	
NiSi	46.7	500	~0	[1027]	1960	
NiSi	50.7	600	~0	[1027]	1960	
NiSi	32.0	750	~0	[1027]	1960	
B₄C	180	20	—	[269]	1934	
SiC	58 ?	20	—	[269]	1934	
SiC	150	25	~4	[1000]	1961	Comp. %: 96.5 SiC, 2.5 Si free, 0.4 C free, 0.4 Al, 0.2 Fe; [206, 269]
BN	24—32	20	—	[272]	1955	

Phase	Modulus of elasticity, kg/mm²	Temperature, °C	Ref.	Year	Remarks
CaB_6	46000	20	[372]	1958	
BaB_6	39300	20	[372]	1958	
LaB_6	48800	20	[372]	1958	
CeB_6	38600	20	[372]	1958	
ThB_4	15120	20	[1097]	1962	
UB_4	45000	20	[1097]	1962	
TiB_2	54000	20	[1003]	1961[1]*	[372, 1010]
ZrB_2	35000	20	[372]	1958	
VB_2	27300	20	[282]	1960	
TaB_2	26200	20	[1003]	1961	
CrB_2	21500	20	[372]	1958	
Be_2C	32000	20	[526]	1952	
Be_2C	32000	540	[526]	1952	
Be_2C	24600	830	[526]	1952	
Be_2C	21050	1100	[526]	1952	
TiC	46000	20	[1003]	1961[1]*	[372—375]
ZrC	35500	20	[372]	1958	
HfC	35900	20	[1003]	1961	
VC	43000	20	[1003]	1961[1]*	[375]
NbC	34500	20	[375]	1948	
TaC	29100	20	[374]	1953	[375]
Cr_3C_2	38000	20	[1003]	1961[1]*	
Mo_2C	54400	20	[372]	1958	[375]
W_2C	42800	20	[375]	1948	
WC	71000	20	[1003]	1961	E is given equal to $f(T°)$, [344, 372, 375]; [375] gives E = $f(T°)$
AlN	35050	25	[674]	1960	
AlN	32300	1000	[674]	1960	
AlN	28100	1400	[674]	1960	Density 3.03 g/cm³
TiN	8060(?)	20	[375]	1948	Density 5.03 g/cm³
TiN	25600	20	[372]	1958	
Mg_2Si	5430	20	[274]	1957	
U_3Si	19300	20	[690]	1960	
$TiSi_2$	26400	20	[1003]	1961[1]*	
$ZrSi_2$	26800	20	[1003]	1961	
Mo_3Si	30000	20	[1003]	1961[1]*	[117, 543]
$MoSi_2$	43000	20	[1003]	1961[1]*	
SiC	39400	20	[376]	1959	
SiC	39300	200	[376]	1959	
SiC	38900	400	[376]	1959	
SiC	38300	600	[376]	1959	
SiC	37850	800	[376]	1959	
SiC	37700	800	[376]	1959	Single crystal
SiC	37000	1000	[376]	1959	
SiC	36850	1000	[376]	1959	Single crystal

179

Phase	Modulus of elasticity, kg/mm^2	Temperature, °C	Ref.	Year	Remarks
SiC	36700	1100	[376]	1959	
SiC	36200	1200	[376]	1959	
SiC	36100	1200	[376]	1959	
SiC	35600	1250	[376]	1959	Single crystal
SiC	35000	1300	[376]	1959	
SiC	33000	1350	[376]	1959	
SiC	48100	25	[1000]	1961	Porosity 4%
SiC	43200	1200	[1000]	1961	Comp. %: 96.5 SiC, 2.5 Si_{free},
SiC	34700	1500	[1000]	1961	0.4 C_{free} 0.4 Al, 0.2 Fe.
α-BN	8650	25	[272]	1955	
α-BN	6150	350	[272]	1955	Parallel to hot-pressing direction
α-BN	1080	700	[272]	1955	
α-BN	1160	1000	[272]	1955	
α-BN	3440	25	[272]	1955	
α-BN	2430	350	[272]	1955	Perpendicular to hot-pressing direction
α-BN	360	700	[272]	1955	
Si_3N_4	4700	20	[377]	1960	Mixture of α- and β-phases
Si_3N_4	4860	300	[377]	1960	
Si_3N_4	4830	550	[377]	1960	
Si_3N_4	4760	850	[377]	1960	
Si_3N_4	4720	950	[377]	1960	[1035]
Si_3N_4	4600	1100	[377]	1960	

[1*] Determined on specimens of the composition, %: TiC (80 Ti, 20.4 C_{tot}, 0.4 C_{free}), VC (81.7 V, 18.0 C_{tot}, 0.3 C_{free}), Cr_3C_2 (86.5 Cr, 13.3 C_{tot}, 0.3 C_{free}), TiB_2 (69.06 Ti, 30.2 B, 0.3 C), $TiSi_2$ (46.3 Ti, 53.37 Si), Mo_3Si (91.46 Mo, 8.05 Si_{tot}, 0.12 Si_{free}), $MoSi_2$ (64.4 Mo, 34.9 Si).

IMPACT TOUGHNESS

Phase	Impact toughness, $kg \cdot m/cm^2$	Ref.	Year	Remarks
TiC	9.9(?)	[690]	1960	
$MoSi_2$	1.1	[690]	1961	
$MoSi_2$	1.66	[777]	1959	Hot-pressing
SiC	1.12—1.59	[777]	1959	
Si_3N_4	0.77—1.02	[777]	1959	Nitrided briquet pressed from silicon powder

Phase	Hardness number, arb. units	Ref.	Year	Remarks
TiB_2	>9	[394]	1948	[395]. Scratches corundum and silicon carbide
ZrB_2	~8	[394]	1948	[395]
VB	7	[550]	1923	
VB_2	8—9	[2]	1957	[217, 395]
NbB_2	>8	[394]	1948	Scratches quartz and topaz
Cr_3B_2	9	[396]	1929	
CrB	8.5	[45]	1949	
Mo_2B	8—9	[397]	1946	
MoB	8	[397]	1946	
WB	9	[397]	1946	
Mn_3B_4	8	[440]	1960	
UB_4, UB_{12}	>8	[697]	1959	
Be_2C	9	[346]	1956	
UC_2	7	[346]	1956	
TiC	8—9	[395]	1926	[232, 346]
ZrC	8—9	[395]	1926	[232, 346]
VC	>9	[232]	1925	[395]
NbC	>9	[2]	1957	Scratches corundum
TaC	9	[395]	1926	[238, 346]
$Cr_{23}C_6$	>9	[232]	1925	Scratches corundum
Cr_3C_2	~7	[346]	1956	
Mo_2C	~7	[397]	1946	[346, 395]
MoC	7—8	[395]	1926	[346]
W_2C	9—10	[395]	1926	[232, 346]
WC	>9	[2]	1957	[346, 395]
Fe_3C	7—8	[346]	1956	
ScN	7—8	[1094]	1962	
AlN	9	[697]	1959	
TiN	9—10(?)	[2]	1957	[264, 393]
ZrN	8	[2]	1957	[264]
VN	9—10(?)	[264]	1934	
NbN	>8	[2]	1957	[264]
TaN	>8	[2]	1957	[264]
$ZrSi_2$	~6	[2]	1957	Hardness of glass
VSi_2	6—7	[2]	1957	
Cr_3Si	>6	[2]	1957	Scratches glass
Cr_3Si_2	>6	[2]	1957	
Cr_2Si	~9	[2]	1957	Scratches quartz and corundum
B_4C	9.3	[364]	1956	
BN	2	[278]	1933	
B	9.3	[364]	1956	
C	1—4.5	[944]	1949	Pyrographite

Phase	RHA	Ref.	Year	Remarks
LaB_6	83	×	1960 ⎫	É. P. Lapteva,
GdB_6	86	×	1960 ⎬	[285, 846]
TiB_2	86	×	1960 ⎫	É. P. Lapteva,
ZrB_2	84	×	1960 ⎭	[32]
ZrB	69—72	[32]	1953	
ZrB_{12}	92—92.5	[35]	1952	
VB_2	83	×	1960 ⎫	
CrB_2	84	×	1960 ⎭	É. P. Lapteva
$Mo_2B, MoB,$ ⎫	~90	[49]	1952	
MoB_2, Mo_2B_5 ⎭				
Co_2B_5	82	×	1960	É. P. Lapteva
Co_2B	~90	[481]	1938	
La_2C_3	~77	[570]	1959	
TiC	92.5—93.5	[292]	1952	[374, 380, 399]
ZrC	87	×	1960	
HfC	84	×	1960	
NbC	83	×	1960	
TaC	82	×	1960	
$Cr_{23}C_6$	83	×	1960	É. P. Lapteva
Cr_7C_3	67	×	1960	
Cr_3C_2	81	×	1960	
Mo_2C	74	×	1960	
W_2C	80	×	1960	É. P. Lapteva
WC	81	×	1960	
TiN	75	×	1960	É. P. Lapteva,
ZrN	84	×	1960	[402]
NbN	86	×	1960	
CrN	78	×	1960	
YSi_2	32	×	1960	É. P. Lapteva,
$LaSi_2$	31	[846]	1960	[402]
$GdSi_2$	80	[543]	1959	
$EuSi_2$	80	[543]	1959	
$DySi_2$	80	[846]	1960	[285, 543]
U_3Si	~23	[690]	1960	
$TiSi_2$	81	×	1960	
V_3Si	78	×	1960	
V_5Si_3	79	×	1960	
Cr_3Si	85	×	1960	É. P. Lapteva
$CrSi$	82	×	1960	
Mo_5Si_3	74	×	1960	
$MoSi_2$	74	×	1960	É. P. Lapteva, [267]
SiC	70?	[270]	1955	
Si_3C_4	99?	[270]	1955	

VICKERS HARDNESS

Phase	VH, kg/mm^2	Ref.	Year	Remarks
SmB$_6$	1391$\pm$159	[846]	1960	
YbB$_6$	1538$\pm$33	[846]	1960	
UC	700$\pm$150	[1014]	1959	Porosity 20.4%
UC	550$\pm$50	[1014]	1959	Porosity 25%
TiC	3200	[399]	1948	
HfC	3202—2533	[1093]	1954	
WC	1620	[401]	1951	
Zr$_2$Si	1180—1280	[428]	1954	
Zr$_5$Si$_3$	1280—1390	[428]	1954	
ZrSi	1020—1180	[428]	1954	
ZrSi$_2$	830—980	[428]	1954	
Nb$_4$Si	470—550	[114]	1956	
Nb$_5$Si$_3$	400—600	[114]	1956	
NbSi$_2$	600—700	[114]	1956	
Ta$_{4.5}$Si	1000—1200	[419]	1953	
Ta$_2$Si	1200—1500	[419]	1953	$p = 40$ kg, 30 sec
Ta$_5$Si$_3$	1200—1500	[419]	1953	
TaSi$_2$	1000—1200	[419]	1953	
Cr$_3$Si	900—980	[414]	1953	
Cr$_3$Si$_2$	1050—1200	[414]	1953	
CrSi	950—1050	[414]	1953	
CrSi$_2$	880—1100	[414]	1953	
Mo$_3$Si	1320—1550	[117]	1959	
Mo$_5$Si$_3$	1200—1320	[117]	1959	
MoSi$_2$	1320—1550	[117]	1959	
B$_4$C	2250—2260	[1093]	1954	

MICROHARDNESS

Phase	MH, kg/mm^2	Accuracy $\pm$, kg/mm^2	Load, g	Ref.	Year	Remarks
CaB$_6$	2700	220	30	[3]	1956	[970]
SrB$_6$	2920	90	30	[353]	1961	[970]
BaB$_6$	3000	290	30	[3]	1956	[970]
AlB$_{12}$	3694	174	30	[578]	1956	
ScB$_2$	1780	276	200	[694]	1960	
ScB$_4$	4540	—	50	x	1963	T. S. Verkho-glyadova
YB$_6$	3264	21	50	[10]	1958	[970]
LaB$_6$	2770	60	30	[3]	1956	[970]
CeB$_6$	3140	190	30	[3]	1956	[970]
PrB$_6$	2470	—	100	[970]	1961	
NdB$_6$	2540	170	70	[220]	1960	[970]

Phase	MH, kg/mm^2	Accuracy $\pm$, kg/mm^2	Load, g	Ref.	Year	Remarks
SmB_6	2500	300	100	[18]	1959	
EuB_6	2660	—	100	[970]	1961	
GdB_6	2300	—	100	[970]	1961	
TbB_6	2300	—	100	[970]	1961	
YbB_6	2660	—	100	[970]	1961	
ThB_6	1740	123	20	[27]	1956	[970]
TiB	2700—2800	—	30	[677]	1959	
TiB_2	3370	60	30	[404]	1952	[403]
TiB_2	3300	—	—	[918]	1961	Knoop tester, [1010]
ZrB	3500—3600	—	30	[677]	1959	
ZrB_2	2252	22	30	[404]	1959	[403, 405]
HfB_2	2900	500	30	[290]	1959	
V_3B_2	2280	—	50	[222]	1959	
V_3B_4	2350	—	50	[222]	1959	
VB_2	2800	13	30	[404]	1952	[405]
Nb_3B_2	2290	—	50	[223]	1959	
NbB	2195	—	50	[223]	1959	
Nb_3B_4	2290	—	30	[223]	1959	
NbB_2	2600	—	30	[223]	1959	[372, 404]
Ta_3B_2	2770	—	50	[223]	1959	
TaB	3130	—	50	[223]	1959	
Ta_3B_4	3350	—	50	[223]	1959	
TaB_2	2500	42	30	[372]	1958	[404]
Cr_4B	1240	60	50	[930]	1961	
Cr_2B	1350	100	50	[930]	1961	
CrB	1200—1300	—	100	[46]	1958	
Cr_3B_4	1400—1500		100	[46]	1958	
CrB_2	2100	80	50	[930]	1961	[46, 372, 403, 404]
Mo_2B	2500	—	50	[51]	1952	[403]
$\alpha\text{-}MoB$	2350	—	50	[49]	1952	[403]
$\beta\text{-}MoB$	2500	—	50	[49]	1952	[403]
MoB_2	1200	—	50	[49]	1952	[403]
Mo_2B_5	2350	—	50	[49]	1952	[372]
W_2B	2420	120	50	[406]	1957	
WB	3700	—	50	[406]	1957	
WB_2	2660	12	30	[404]	1952	
W_2B_5	2663	12	30	[372]	1958	[406]
Co_3B	1150	—	50	[56]	1959	
Co_2B	1150	—	50	[56]	1959	
CoB	1150	—	50	[56]	1959	
CoB_2	2575	—	50	[56]	1959	
Ni_3B	1145	—	50	[227]	1958	
NiB_2	2575	—	50	[227]	1958	
Be_2C	2690	—	—	[526]	1952	Knoop tester
ScC	2720	—	20	$\times$	1963	T. Ya. Kosasatova, G. N. Makarenko
YC	120	—	5	[820]	1961	[879]

Phase	MH, kg/mm²	Accuracy ±, kg/mm²	Load, g	Ref.	Year	Remarks
Y_2C_3	910	—	50	[879]		
YC_2	708	—	50	[879]	1961	
ThC	850	—	200	[1278]	1962	C_{tot} 5.03%, C_{free} 0.25%
ThC_2	600	—	200	[1278]	1962	C_{tot} 9.85%, C_{free} 1.07%
UC	923	56	50	[856]	1960	[1281]
TiC	2988	125	30	[372]	1958	[407—409]
TiC	2470	—	100	[664]	1950	Knoop tester
TiC	3000	—	—	[690]	1960	20°
TiC	2600	—	—	[690]	1960	200°
TiC	1700	—	—	[690]	1960	400° ⎫ Data taken
TiC	900	—	—	[690]	1960	600° ⎬ from graph
TiC	500	—	—	[690]	1960	800° ⎭
ZrC	2925	184	30	[372]	1958	[407, 409]
HfC	2913	300	50	[410]	1954	[64, 1163]
VC	2094	58	50	[408]	1951	[409, 411]
Nb_2C	2123	199	30	[372]	1958	
NbC	1961	96	30	[372]	1958	[407, 409, 412]
Ta_2C	1714	159	30	[372]	1958	
TaC	1599	49	30	[372]	1958	[408, 409, 411, 412]
$Cr_{23}C_6$	1650	—	50	[413]	1953	[179]
Cr_7C_3	1336	—	50	[179]	1960	[413]
Cr_3C_2	1350	—	50	[179]	1960	[409, 413]
Mo_2C	1499	130	30	[372]	1958	[2, 408, 409]
W_2C	3000(?)	—	50	[409]	1949	Measured on relit * (WC + W_2C)
WC	1780	44	30	[372]	1958	
AlN	1225—1230	—	100	[674]	1960	Knoop tester
TiN	1994	137	50	[908]	1961	[372]
ZrN	1520	±85	50	[929]	1961	
HfN	1640	±161	50	[929]	1961	
V_3N	1900	102	50	[908]	1961	8.5% N
VN	1520	115	50	[908]	1961	16.8% N
Nb_2N	1720	100	50	[908]	1961	6.6% N
$NbN_{0.75}$	1780	—	50	[923]	1961	
$NbN_{0.97}$	1525	136	50	[929]	1961	
NbN	1396	26	50	[929]	1961	12.7% N
Ta_2N	1220	120	50	[929]	1961	4.3% N
TaN	1060	72	50	[929]	1961	7.3% N
Cr_2N	1571	49	50	[929]	1961	11.2% N
CrN	1093	93	50	[929]	1961	21.0% N
Mo_2N	630	86	20	[929]	1961	
Mg_2Si	457	—	—	[488]	1957	20°, [1027]
Mg_2Si	320	—	—	[488]	1957	300°
Mg_2Si	180	—	—	[488]	1957	600°, [1027]
$BaSi_2$	930	50	50	[1289]	1963	
$LaSi_2$	340	—	50	[1216]	1960	
$CeSi_2$	540	—	50	[1216]	1960	
$ThSi_2$	1120	—	100	[846]	1960	[2]

*Transliteration of the Russian. Apparently a trade name, the correct English spelling may differ from that given here.

Phase	MH, kg/mm^2	Accuracy ±, kg/mm^2	Load, g	Ref.	Year	Remarks
Ti$_5$Si$_3$	986	—	100	[241]	1951	
TiSi	1039	—	100	[241]	1951	
TiSi$_2$	892	—	50	[117]	1959	[2, 241]
ZrSi$_2$	1063	—	50	[117]	1959	[2]
HfSi$_2$	930	—	50	[112]	1956	
V$_3$Si	1430—1560	—	50	[114]	1956	[1249]
V$_5$Si$_3$	1350—1510	—	50	[114]	1956	
VSi$_2$	890—960	—	50	[114]	1956	[2]
Nb$_4$Si	690—820	—	100	[117]	1959	
α-Nb$_5$Si$_3$	700	—	100	[117]	1959	
NbSi$_2$	1050	—	50	[118]	1955	[2, 117]
TaSi$_2$	1407	—	50	[117]	1959	[2]
Cr$_3$Si	1005	—	50	[414]	1953	
Cr$_3$Si$_2$	1280	—	50	[414]	1953	
CrSi	1005	—	50	[414]	1953	
CrSi$_2$	1131	—	50	[117]	1959	[414]
CrSi$_2$	704	—	50	[861]	1960	Cast, [1141]
CrSi$_2$	798	—	50	[861]	1960	Annealed
Mo$_3$Si	1310	—	100	[243]	1952	
Mo$_5$Si$_3$	1170	—	100	[243]	1952	
MoSi$_2$	1200	—	50	[117]	1959	[243]
MoSi$_2$	707	—	50	[861]	1960	Cast
MoSi$_2$	735	—	50	[861]	1960	Annealed
W$_3$Si$_2$	770	—	50	[423]	1952	[1265]
WSi$_2$	1074	—	50	[117]	1959	[423]
ReSi$_2$	1500	40	50	[916]	1961	
FeSi$_{2.33}$	870	—	70	[1141]	1962	α-lebauite
CoSi	1000	—	—	[1027]	1960	20°
CoSi	300	—	—	[1027]	1960	500°
CoSi	115	—	—	[1027]	1960	1000°
CoSi$_2$	552	—	—	[1027]	1960	20°, [117]
CoSi$_2$	322	—	—	[1027]	1960	500°
CoSi$_2$	77	—	—	[1027]	1960	1000°
Ni$_3$Si	400	—	—	[117]	1959	
Ni$_2$Si	440	—	—	[488]	1957	20°, [1027]
Ni$_2$Si	320	—	—	[488]	1957	500°, [1027]
Ni$_2$Si	120	—	—	[1027]	1960	750°
NiSi	400	—	—	[488]	1957	20°, [1027]
NiSi	256	—	—	[1027]	1960	500°, [488]
NiSi$_2$	1019	—	50	[117]	1959	
LaP	158	14	1	[1179]	1963	
TiP	1300	—	100	[1178]	1963	
TiP	718	—	—	[1177]	1962	Porosity 36%
VP	541	—	—	[1177]	1962	Porosity 40%
β-NbP	599	—	—	[1177]	1962	Porosity 35%
β-TaP	374	—	—	[1177]	1962	Porosity 31%
CrP	632	—	—	[1177]	1962	Porosity 34%
MnP	633	—	—	[1177]	1962	Porosity 22%
ThS	363	40	30	[249]	1957	

MICROHARDNESS (continued)

Phase	MH, kg/mm²	Accuracy ±, kg/mm²	Load, g	Ref.	Year	Remarks
Th₂S₃	227	28	30	[249]	1957	
B₄C	4950	—	30	[277]	1953	[392]
B₁₂C₂	5600—5800	—	30	[168]	1954	
B₁₂C	4100	400	50	[305]	1960	
SiC	3340	—	—	[206]	1952	
SiC	2500—3000	—	100	[1000]	1961	Knoop tester: Comp. %: 96.5 SiC, 2.5 Si free, 0.4 C free, 0.4 Al, 0.2 Fe
Si₃N₄	3337	120	50	[415]	1957	
B₆Si	2470—2810	—	100	[912]	1960	Knoop tester, [392, 690, 1091]
B₄Si	1830—2240	—	100	[912]	1960	Knoop tester
B₃Si	5352	167	30	[173]	1955	
B₃Si	3000—4000	—	—	[690]	1960	[1091]
BP	3200	—	100	[103]	1960	Knoop tester
B	2410	—	100	[392]	1959	" "
B	3400	—	50	[440]	1960	

COMPRESSIBILITY

Phase	Coefficient of compressibility $\Delta V / V_0$	Temperature, °C	Ref.	Year
TiC	$4.72 \cdot 10^{-7} p - 2.16 \cdot 10^{-12} p^2$	30	[390]	1952
TiC	$4.78 \cdot 10^{-7} p - 2.19 \cdot 10^{-12} p^2$	75	[390]	1952
TiN	$3.32 \cdot 10^{-7} p - 2.13 \cdot 10^{-12} p^2$	30	[390]	1952
TiN	$3.51 \cdot 10^{-7} p - 2.13 \cdot 10^{-12} p^2$	75	[390]	1952
BN	$34 \cdot 10^{-7} p - 54.10^{-12} p^2$	45	[821]	1960, [1282]

Chapter VI

CHEMICAL PROPERTIES

RESISTANCE OF POWDERS OF REFRACTORY COMPOUNDS TO THE ACTION OF CHEMICAL REAGENTS[1*]

Phase	Reagent[2*]	Tempera-ture, °C	Duration of treat-ment, hr	Insoluble residue, %	Ref.	Year	Remarks
MgB$_2$	H$_2$O	20	—	—	[697]	1959	Gradual de-composition
MgB$_2$	HCl(1.19)	20	—	cs	[697]	1959	
MgB$_2$	HNO$_3$(1.43)	20	—	cs	[697]	1959	
MgB$_2$	H$_2$SO$_4$(1.84)	20	—	cs	[697]	1959	
MgB$_{12}$	HCl(1.19)	20	—	i	[697]	1959	
MgB$_{12}$	HCl(1.19)	Boiling	—	i	[697]	1959	
MgB$_{12}$	HNO$_3$(1.43)	20	—	i	[697]	1959	
MgB$_{12}$	HNO$_3$(1.43)	Boiling	—	i	[697]	1959	
MgB$_{12}$	H$_2$SO$_4$(1.84)	20	—	i	[697]	1959	

[1*] Notation: cs—completely soluble; ps—partly soluble; csh—completely soluble with hydrolysis; lss—largely soluble with salt precipitation; i—insoluble.
[2*] The concentration of the reagent or its specific gravity is given in parentheses.

RESISTANCE OF POWDERS OF REFRACTORY COMPOUNDS TO CHEMICAL REAGENTS   (continued)

Phase	Reagent[2*]	Temperature, °C	Duration of treatment, hr	Insoluble residue, %	Ref.	Year	Remarks
MgB$_{12}$	H$_2$SO$_4$(1.84)	Boiling	—	i	[697]	1959	Decomposes completely on long boiling
MgB$_{12}$	HNO$_3$(1.43) + H$_2$O$_2$ (30%)	"	—	cs	[697]	1959	
Mg$_3$B$_2$	H$_2$O	20	—	cs	[697]	1959	
Mg$_3$B$_2$	HCl(1.19)	20	—	cs	[697]	1959	
Mg$_3$B$_2$	HNO$_3$(1.43)	20	—	cs	[697]	1959	
Mg$_3$B$_2$	H$_2$SO$_4$(1.84)	20	—	cs	[697]	1959	
CaB$_6$	HCl(1.19)	20	1	99.5	[1038]	1961	
CaB$_6$	HCl(1.19)	20	2	99.5	[1038]	1961	
CaB$_6$	HCl(1.19)	20	240	98.5	[1038]	1961	
CaB$_6$	H$_2$SO$_4$(1.84)	20	1—240	i	[1038]	1961	
CaB$_6$	HNO$_3$(1.42)	20	1	8.5	[1038]	1961	
CaB$_6$	HNO$_3$(1.42)	20	2	2.7	[1038]	1961	
CaB$_6$	HNO$_3$(1.42)	20	24	cs	[1038]	1961	
CaB$_6$	NaOH(50%)	20	1	97.8	[1038]	1961	
CaB$_6$	NaOH(50%)	20	2	97.8	[1038]	1961	
CaB$_6$	NaOH(50%)	20	240	97.4	[1038]	1961	
CaB$_6$	Na$_2$CO$_3$(50%)	20	1	99.7	[1038]	1961	
CaB$_6$	Na$_2$CO$_3$(50%)	20	2	99.2	[1038]	1961	
CaB$_6$	Na$_2$CO$_3$(50%)	20	240	99.5	[1038]	1961	
SrB$_6$	HCl(1.19)	20	1	99.3	[1038]	1961	
SrB$_6$	HCl(1.19)	20	24	98.6	[1038]	1961	
SrB$_6$	HCl(1.19)	20	240	98.5	[1038]	1961	

RESISTANCE OF POWDERS OF REFRACTORY COMPOUNDS TO CHEMICAL REAGENTS   (continued)

Phase	Reagent[2]*	Temperature, °C	Duration of treatment, hr	Insoluble residue, %	Ref.	Year	Remarks
SrB6	H_2SO_4(1.84)	20	1—240	i	[1038]	1961	
SrB6	HNO_3(1.42)	20	1	1.5	[1038]	1961	[697]
SrB6	HNO_3(1.42)	20	2	cs	[1038]	1961	
SrB6	NaOH(50%)	20	1	99.2	[1038]	1961	
SrB6	NaOH(50%)	20	2	98.8	[1038]	1961	
SrB6	NaOH(50%)	20	240	98.1	[1038]	1961	
SrB6	Na_2CO_3(50%)	20	24	98.7	[1038]	1961	
SrB6	Na_2CO_3(50%)	20	240	98.5	[1038]	1961	
SrB6	H_2O	Boiling	—	i	[697]	1959	
SrB6	H_2O	Boiling	—	i	[697]	1959	
BaB6	HCl(1.19)	20	1	99.2	[1038]	1961	
BaB6	HCl(1.19)	20	2	98.9	[1038]	1961	
BaB6	HCl(1.19)	20	24	98.4	[1038]	1961	
BaB6	HCl(1.19)	20	240	97.0	[1038]	1961	
BaB6	H_2SO_4(1 84)	20	1—240	i	[1038]	1961	
BaB6	HNO_3(1.42)	20	1	6.0	[1038]	1961	
BaB6	HNO_3(1.42)	20	2	1.2	[1038]	1961	
BaB6	HNO_3(1.42)	20	24	cs	[1038]	1961	
BaB6	NaOH(50%)	20	1	98.8	[1038]	1961	
BaB6	NaOH(50%)	20	24	98.1	[1038]	1961	
BaB6	NaOH(50%)	20	240	98.1	[1038]	1961	
BaB6	Na_2CO_3(50%)	20	1	99.4	[1038]	1961	
BaB6	Na_2CO_3(50%)	20	24	98.6	[1038]	1961	
BaB6	Na_2CO_3(50%)	20	240	98.6	[1038]	1961	
YB6	HCl(1:1)	20	2	77—78	[474]	1961	[1066]

RESISTANCE OF POWDERS OF REFRACTORY COMPOUNDS TO CHEMICAL REAGENTS (continued)

Phase	Reagent[2*]	Temperature, °C	Duration of treatment, hr	Insoluble residue, %	Ref.	Year	Remarks
YB_6	$HNO_3(1:1)$	Gentle heat	5 min	cs	[474]	1961	[1066]
YB_6	$H_2SO_4(1:1)$	Same	2	71—72	[474]	1961	
YB_6	3p. $HCl(1.19) + 1p. HNO_3(1.43)$	20	5 min	cs	[474]	1961	
YB_6	$H_2SO_4(1:1)$ with addition of $HNO_3(1.43)$	Gentle heat	5 min	cs	[474]	1961	
LaB_6	$HCl(1:1)$	Same	2	93—94	[474]	1961	[1066]
LaB_6	$HNO_3(1:1)$	"	5 min	cs	[474]	1961	
LaB_6	$H_2SO_4(1:1)$	*	2	89—92	[474]	1961	
LaB_6	3p. $HCl (1.19) + 1p. HNO_3 (1.43)$	20	5 min	cs	[474]	1961	
LaB_6	$H_2SO_4(1:1)$ with addition of $HNO_3(1.43)$	Gentle heat	5 min	cs	[474]	1961	
LaB_6	$NaOH (15\%)$	106	1	99.4	[474]	1961	[1066]
LaB_6	50 ml $NaOH + 25$ ml H_2O_2	103	1	99.9	[474]	1961	
CeB_6	$HCl(1:1)$	103	2	84—86	[474]	1961	
CeB_6	$HNO_3(1:1)$	103	5 min	cs	[474]	1961	
CeB_6	$H_2SO_4(1:1)$	103	2	83—84	[474]	1961	
CeB_6	3p. $HCl (1.19) + 1p. HNO_3 (1.43)$	20	5 min	cs	[474]	1961	
CeB_6	$H_2SO_4(1:1)$ with addition of $HNO_3(1.43)$	Gentle heat	5 min	cs	[474]	1961	
CeB_6	$NaOH(15\%)$	106	1	'99.4	[474]	1961	

RESISTANCE OF POWDERS OF REFRACTORY COMPOUNDS TO CHEMICAL REAGENTS   (continued)

Phase	Reagent [2]*	Temperature, °C	Duration of treatment, hr	Insoluble residue, %	Ref.	Year	Remarks
CeB_6	50 ml NaOH + 25 ml H_2O_2	103	1	99.9	[474]	1961	[1066]
PrB_6	HCl(1:1)	Gentle heat	2 min	90—94	[474]	1961	
PrB_6	HNO_3(1:1)	Same	5 min	cs	[474]	1961	
PrB_6	H_2SO_4(1:1)	*	2	27—30	[474]	1961	
PrB_6	3 p. HCl(1.19) + 1 p. HNO_3(1.43)	20	5 min	cs	[474]	1961	[1066]
PrB_6	H_2SO_4(1:1) with addition of HNO_3(1.43)	Gentle heat	5 min	cs	[474]	1961	
PrB_6	NaOH (15%)	106	1	99.4	[474]	1961	
PrB_6	50 ml NaOH + 25 ml H_2O_2	103	1	99.9	[474]	1961	
NdB_6	HCl (1:1)	103	2	87—88	[474]	1961	
NdB_6	HNO_3(1:1)	Gentle heat	5 min	cs	[474]	1961	
NdB_6	H_2SO_4 (1:1)	Same	2	78	[474]	1961	
NdB_6	3 p. HCl(1.19) + 1 p. HNO_3(1.43)	20	5 min	cs	[474]	1961	[1066]
NdB_6	H_2SO_4(1:1) with addition of HNO_3 (1.43)	Gentle heat	2	cs	[474]	1961	
NdB_6	NaOH (15%)	106	1	99.4	[474]	1961	
NdB_6	50 ml NaOH + 25 ml H_2O_2	103	1	99.6	[474]	1961	
SmB_6	HCl (1:1)	Gentle heat	5 min	78—80	[474]	1961	
SmB_6	HNO_3 (1:1)	Same	5 min	cs	[474]	1961	

RESISTANCE OF POWDERS OF REFRACTORY COMPOUNDS TO CHEMICAL REAGENTS   (continued)

Phase	Reagent [2*]	Temperature, °C	Duration of treatment, hr	Insoluble residue, %	Ref.	Year	Remarks
SmB6	H_2SO_4 (1:1)	Gentle heat	2	77	[474]	1961	[1066]
SmB6	3p.HCl(1.19) + 1p. HNO_3(1.43)	20	5 min	cs	[474]	1961	
SmB6	H_2SO_4(1:1) with addition of HNO_3 (1.43)	Gentle heat	5 min	cs	[474]	1961	
SmB6	NaOH (15%)	106	1	99.4	[474]	1961	
SmB6	50 ml NaOH + 25 ml H_2O_2	103	1	99.6	[474]	1961	
GdB6	HCl(1:1)	103	20	91—93	[474]	1961	
GdB6	HNO_3(1:1)	103	5 min	cs	[474]	1961	
GdB6	H_2SO_4(1:1)	103	2	87	[474]	1961	
GdB6	3p. HCl (1.19) + 1p. HNO_3 (1.43) H_2SO_4 (1:1) with addition of HNO_3 (1.43)	Gentle heat	5 min	cs	[474]	1961	
GdB6	NaOH (15%)	106	1	99.4	[474]	1961	
GdB6	50 ml NaOH + 25 ml H_2O_2	103	1	99.9	[474]	1961	
ThB4	H_2O	20	—	i	[697]	1959	[1066]
ThB4	H_2O	Boiling	—	i	[697]	1959	
ThB4	HNO_3 (1:1)	20	—	cs	[697]	1959	
ThB4	HCl (1.19)	Boiling	—	cs	[697]	1959	
ThB4	H_2SO_4 (1.82)	20	—	cs	[697]	1959	
ThB4	H_2O	Boiling	—	i	[697]	1959	
ThB4	H_2O	20	—	i	[697]	1959	
ThB4	HCl (1.19)	Boiling	—	i	[697]	1959	
ThB4	HNO_3 (1.43)	20*	—	i	[697]	1959	
ThB4	H_2SO_4 (1.82)	Boiling	—	i	[697]	1959	
ThB6	H_2SO_4 (1.82)	Boiling	—	i	[697]	1959	

193

RESISTANCE OF POWDERS OF REFRACTORY COMPOUNDS TO CHEMICAL REAGENTS (continued)

Phase	Reagent[2]*	Temperature, °C	Duration of treatment, hr	Insoluble residue, %	Ref.	Year	Remarks
UB_2	HCl (1.19)	Boiling	—	i	[697]	1959	
UB_2	HNO₃ (1.43)	20	—	cs	[697]	1959	
UB_2	H₂SO₄ (1.82)	Boiling	—	i	[697]	1959	
UB_2	HF (1.15)	20	—	cs	[697]	1959	
UB_2	NaOH (50%)	Boiling	—	i	[697]	1959	
UB_4	HCl (1.19)	20	—	cs	[697]	1959	
UB_4	HCl (1.19)	Boiling	—	cs	[697]	1959	
UB_4	HNO₃ (1.43)	20	—	cs	[697]	1959	
UB_4	HNO₃ (1.43)	Boiling	—	cs	[697]	1959	
UB_4	H₂SO₄ (1.82)	20	—	i	[697]	1959	
UB_4	H₂SO₄ (1.82)	Boiling	—	cs	[697]	1959	Decomposes on long boiling
UB_4	HF (1.15)	"	—	cs	[697]	1959	
UB_4	H₂O₂ (30%)	20	—	cs	[697]	1959	
UB_4	Na₂O₂ (20%)	20	—	cs	[697]	1959	
UB_{12}	HCl (1.19)	20	—	i	[697]	1959	
UB_{12}	HCl (1.19)	Boiling	—	i	[697]	1959	
UB_{12}	HNO₃ (1.43)	20	—	cs	[697]	1959	
UB_{12}	HNO₃ (1.43)	Boiling	—	cs	[697]	1959	
UB_{12}	H₂SO₄ (1.82)	20	—	i	[697]	1959	
UB_{12}	H₂SO₄ (1.82)	Boiling	—	i	[697]	1959	
UB_{12}	HF (1.15)	20	—	i	[697]	1959	
UB_{12}	HF (1.15)	Boiling	—	i	[697]	1959	
UB_{12}	H₂O₂ (30%)	20	—	i	[697]	1959	

RESISTANCE OF POWDERS OF REFRACTORY COMPOUNDS TO CHEMICAL REAGENTS   (continued)

Phase	Reagent[2*]	Temperature, °C	Duration of treatment, hr	Insoluble residue, %	Ref.	Year	Remarks
UB$_{12}$	Na$_2$O$_2$ (20%)	20	—	i	[697]	1959	
TiB$_2$	H$_2$SO$_4$ (1.84)	Boiling	96	5.5	[645]	1960	Particle size 200 μ
TiB$_2$	H$_2$SO$_4$ (1.84)	"	1	43.3	[645]	1960	Particle size 200 μ
TiB$_2$	H$_2$SO$_4$ (1.84)	20	24	89	[644]	1959	
TiB$_2$	H$_2$SO$_4$ (1.84)	Boiling	2	58	[644]	1959	
TiB$_2$	H$_3$PO$_4$ (1:3)	20	24	98	[644]	1959	
TiB$_2$	H$_3$PO$_4$ (1:3)	Boiling	2	65	[644]	1959	
TiB$_2$	H$_3$PO$_4$ (1.21)	20	24	90	[644]	1959	
TiB$_2$	HCl (1.19)	20	24	94	[644]	1959	
TiB$_2$	HCl (1.19)	Boiling	2	58	[644]	1959	
TiB$_2$	HCl (1.19)	20	24	12	[645]	1960	Particle size 200 μ
TiB$_2$	HCl (1.19)	Boiling	1	5.5	[645]	1960	Particle size 200 μ
TiB$_2$	HNO$_3$ (1:10)	20	96	97.5	[645]	1960	
TiB$_2$	HNO$_3$ (1:10)	Boiling	1	95.5	[645]	1960	
TiB$_2$	HNO$_3$ (1:1)	20	24	31	[644]	1959	
TiB$_2$	HNO$_3$ (1:1)	Boiling	2	csh	[644]	1959	Particle size 200 μ
TiB$_2$	HNO$_3$ (1.43)	20	24	97	[645]	1960	Particle size 200 μ
TiB$_2$	HNO$_3$ (1.43)	20	24	28	[644]	1959	Particle size 200 μ
TiB$_2$	HNO$_3$ (1.43)	Boiling	2	csh	[644]	1959	
TiB$_2$	H$_2$SO$_4$ (1:10)	20	168	45.7	[645]	1960	Particle size 200 μ
TiB$_2$	H$_2$SO$_4$ (1:4)	20	24	96	[644]	1960	

Phase	Reagent[2*]	Temperature, °C	Duration of treatment, hr	Insoluble residue, %	Ref.	Year	Remarks
TiB₂	H₂SO₄ (1:4)	Boiling	2	68	[644]	1959	
TiB₂	HNO₃ (1.43) + HF (1.15)	"	15 min	1	[644]	1959	
TiB₂	3 p. HCl (1.19) + 1 p. HNO₃ (1.43)	20	24	9	[644]	1959	
TiB₂	3 p. HCl (1.19) + 1 p. HNO₃ (1.43)	Boiling	2	lss	[644]	1959	
TiB₂	H₃PO₄ (1.21)	"	2	lss	[644]	1959	
TiB₂	HClO₄ (1:3)	20	2	87	[644]	1959	
TiB₂	HClO₄ (1:3)	Boiling	24	28	[644]	1959	
TiB₂	HClO₄ (1.35)	20	24	30	[644]	1959	
TiB₂	HClO₄ (1.35)	Boiling	15 min	cs	[644]	1959	
TiB₂	H₂C₂O₄ (1:3)	20	24	89	[644]	1959	
TiB₂	H₂C₂O₄ (1:3)	Boiling	2	csh	[644]	1959	
TiB₂	H₂C₂O₄ (satur.)	20	24	94	[644]	1959	
TiB₂	H₂C₂O₄ (satur.)	Boiling	2	51	[644]	1959	
TiB₂	HF (1:10)	20	27	15.6	[645]	1960	Particle size 200 μ
TiB₂	HF (1.15)	20	96	16.6	[645]	1960	
TiB₂	HF (1.15)	Boiling	2	64	[644]	1959	
TiB₂	HCl (1:10)	20	96	3.9	[645]	1960	
TiB₂	HCl (1:10)	Boiling	1	12	[645]	1960	
TiB₂	HCl (1:2)	20	0.5	94	[645]	1960	Particle size 200 μ
TiB₂	HCl (1:2)	"	24	93.5	[644]	1959	
TiB₂	HCl (1:2)	Boiling	2	61	[644]	1959	
TiB₂	3 p. HCl (1:1) + 1 p. HNO₃ (1:1)	20	24	30	[644]	1959	
TiB₂	3 p. HCl (1:1) + 1 p. HNO₃ (1:1)	Boiling	2	lss	[644]	1959	
TiB₂	30 ml C₂H₂O₄ (satur.) + 10 ml 30% H₂O₂ + 10 ml HNO₃ (1.43)	20	24	1	[644]	1959	

RESISTANCE OF POWDERS OF REFRACTORY COMPOUNDS TO CHEMICAL REAGENTS   (continued)

Phase	Reagent[2*]	Tempera-ture, °C	Duration of treat-ment, hr	Insoluble residue, %	Ref.	Year	Remarks
TiB_2	30 ml $C_2H_2O_4$ (satur.) $+ 10$ ml 30% $H_2O_2 + 10$ ml HNO_3 (1.43)	Boiling	2	6	[644]	1959	
TiB_2	30 ml $C_2H_2O_4$ (satur.) $+ 20$ ml H_2SO_4 (1.84)	20	24	87	[644]	1959	
TiB_2	30 ml $C_2H_2O_4$ (satur.) $+ 20$ ml H_2SO_4 (1.84)	Boiling	2	50	[644]	1959	
TiB_2	35 ml HCl (1.19) $+ 15$ ml bromine water	11	2	35	[644]	1959	
TiB_2	35 ml $HClO_4$ (1.35) $+ 15$ ml HCl (1.19)	20	24	27	[644]	1959	
TiB_2	35 ml $HClO_4$ (1.35, $+ 15$ ml HCl (1.19)	Boiling	2	1ss	[644]	1959	
TiB_2	1 p. H_2SO_4 (1.84) $+ 4$ p. $H_3PO_4 + 2$ p. H_2O	20	24	91	[644]	1959	
TiB_2	1 p. H_2SO_4 (1.84) $+ 4$ p. $H_3PO_4 + 2$ p. H_2O	Boiling	2	48	[644]	1959	
TiB_2	50 ml H_2SO_4 (1.84) $+ 5$ p. K_2SO_4	"	2	6	[644]	1959	
TiB_2	35 ml H_2SO_4 (1.84) $+ 15$ ml HNO_3 (1.43)	"	2	1	[644]	1959	

RESISTANCE OF POWDERS OF REFRACTORY COMPOUNDS TO CHEMICAL REAGENTS   (continued)

Phase	Reagent²*	Temperature, °C	Duration of treatment, hr	Insoluble residue, %	Ref.	Year	Remarks
TiB₂	30 ml H₂O₂ (1:3) + 5 drops H₂SO₄ (1.84)	Boiling	1	ps	[940]	1961	
TiB₂	30 ml H₂O₂ (1:3) + 10 drops HNO₃(1.43)	"	1	ps	[940]	1961	
TiB₂	KNO₃ — 1% sulfuric acid solution	"	1	ps	[940]	1961	Precipitates TiO₂
TiB₂	HCl (1:1) + H₂C₂O₄	"	1	71	[940]	1961	
TiB₂	H₂C₂O₄ (6%)	"	1	81	[940]	1961	
TiB₂	HCl (1:1) + Trilon B	"	1	84	[940]	1961	
TiB₂	30 ml HCl (1:1) + C₆H₈O₇	" 20	1	83	[940]	1961	
TiB₂	NaOH (30%)	20	24	92	[644]	1959	
TiB₂	NaOH (30%)	Boiling	2	cs	[644]	1959	
TiB₂	NaOH (10%)	" 20	2	cs	[644]	1959	
ZrB₂	HCl (1.19)	20	24	91	[644]	1959	
ZrB₂	HCl (1.19)	Boiling	2	6	[644]	1959	
ZrB₂	HCl (1.19)	" 20	1	25.4	[645]	1960	Particle size 200 μ
ZrB₂	HCl (1.19)	20	24	2	[645]	1960	
ZrB₂	HCl (1:1)	20	24	93	[644]	1959	
ZrB₂	HCl (1:1)	Boiling	2	7	[644]	1959	
ZrB₂	HCl (1:2)	"	2	30.3	[645]	1960	Particle size 200 μ

RESISTANCE OF POWDERS OF REFRACTORY COMPOUNDS TO CHEMICAL REAGENTS   (continued)

Phase	Reagent[2]*	Temperature, °C	Duration of treatment, hr	Insoluble residue, %	Ref.	Year	Remarks
ZrB_2	HCl (1:2)	Boiling	1.5	47.4	[645]	1960	
ZrB_2	HCl (1:2)	"	2.5	27.7	[645]	1960	
ZrB_2	HCl (1:2)	"	1.25	44	[645]	1960	
ZrB_2	HCl (1:2)	"	4	3.02	[645]	1960	Particle size 200 μ
ZrB_2	HCl (1:2)	"	0.5	20.8	[645]	1960	
ZrB_2	HCl (1:2)	"	0.5	20.5	[645]	1960	
ZrB_2	HCl (1:10)	20	1	7	[645]	1960	
ZrB_2	HCl (1:10)	20	16	25.4	[645]	1960	
ZrB_2	H_2O	20	24	5.75	[645]	1960	
ZrB_2	H_2O	Boiling	1	0.94	[645]	1960	
ZrB_2	HNO_3 (1.43)	20	24	12	[644]	1959	
ZrB_2	HNO_3 (1.43)	Boiling	2	cs	[644]	1959	
ZrB_2	HNO_3 (1:1)	20	24	23	[644]	1959	
ZrB_2	HNO_3 (1:1)	Boiling	2	4	[644]	1959	
ZrB_2	HNO_3 (1:10)	20	0.5	cs	[645]	1960	Particle size 200 μ
ZrB_2	HNO_3 (1:10)	Boiling	1	14	[645]	1960	
ZrB_2	HNO_3 (1.43)	20	96	74.5	[645]	1960	Particle size 200 μ
ZrB_2	HNO_3 (1.43)	Boiling	1	93.1	[645]	1960	
ZrB_2	H_2SO_4 (1.84)	20	24	65	[644]	1959	
ZrB_2	H_2SO_4 (1.84)	2	2	1	[644]	1959	
ZrB_2	H_2SO_4 (1:4)	Boiling	24	51	[644]	1959	
ZrB_2	H_2SO_4 (1:4)	20	2	5	[644]	1959	
ZrB_2	H_2SO_4 (1:10)	Boiling	0.5	cs	[645]	1960	Particle size 200 μ
ZrB_2	H_2SO_4 (1:10)	20	1	27	[645]	1960	
ZrB_2	H_2SO_4 (1.84)	Boiling	96	3.99	[645]	1960	
ZrB_2	H_2SO_4 (1.84)	20	1	cs	[645]	1960	

RESISTANCE OF POWDERS OF REFRACTORY COMPOUNDS TO CHEMICAL REAGENTS (continued)

Phase	Reagent[2*]	Temperature, °C	Duration of treatment, hr	Insoluble residue, %	Ref.	Year	Remarks
ZrB_2	H_3PO_4 (1.21)	20	24	63	[644]	1959	
ZrB_2	H_3PO_4 (1.21)	Boiling	2	1ss	[644]	1959	
ZrB_2	H_3PO_4 (1:3)	20	24	89	[644]	1959	
ZrB_2	H_3PO_4 (1:3)	Boiling	2	1ss	[644]	1959	
ZrB_2	$HClO_4$ (1.35)	20	24	10	[644]	1959	
ZrB_2	$HClO_4$ (1.35)	Boiling	25	4	[644]	1959	
ZrB_2	$HClO_4$ (1:3)	20	24	71	[644]	1959	
ZrB_2	$HClO_4$ (1:3)	Boiling	2	48	[644]	1959	
ZrB_2	$H_2C_2O_4$ (satur.)	20	24	55	[644]	1959	
ZrB_2	$H_2C_2O_4$ (satur.)	Boiling	2	5	[644]	1959	
ZrB_2	$H_2C_2O_4$ (6%)	"	1	67	[940]	1961	
ZrB_2	$H_2C_2O_4$ (1:3)	20	24	38	[644]	1959	
ZrB_2	$H_2C_2O_4$ (1:3)	Boiling	2	1ss	[644]	1959	
ZrB_2	HF (1.15)	"	2	25	[644]	1959	
ZrB_2	HF (1.15)	20	24	84.4	[645]	1960	
ZrB_2	HF (1:10)	20	24	77	[645]	1960	
ZrB_2	HF (1:10)	Boiling	1	86.2	[645]	1960	
ZrB_2	3 p. HCl (1.19) + 1 p.HNO_3 (1.43)	20	24	7	[644]	1959	
ZrB_2	3 p. HCl (1.19) + 1 p.HNO_3 (1.43)	Boiling	2	6	[644]	1959	
ZrB_2	3 p. HCl (1:1) + 1p. HNO_3 (1:1)	20	24	16	[644]	1959	
ZrB_2	30 ml $H_2C_2O_4$ (satur.) + 10 ml H_2O_2(30%) + 10 ml HNO_3(1.43)	20	24	6	[644]	1959	
ZrB_2	30 ml $H_2C_2O_4$ (satur.) + 20 ml H_2SO_4 (1.84)	20	24	59	[644]	1959	
ZrB_2	30 ml $H_2C_2O_4$ (satur.) + 20 ml H_2SO_4	Boiling	2	10	[644]	1959	

RESISTANCE OF POWDERS OF REFRACTORY COMPOUNDS TO CHEMICAL REAGENTS (continued)

Phase	Reagent²*	Temperature, °C	Duration of treatment, hr	Insoluble residue, %	Ref.	Year	Remarks
ZrB_2	35 ml HCl (1.19) + 15 ml bromine water	Boiling	2	18	[644]	1959	
ZrB_2	35 ml HClO₄ (1.35) + 15 ml HCl (1.19)	20	24	90	[644]	1959	
ZrB_2	35 ml HClO₄ (1.35) + 15 ml HCl (1 19)	Boiling	2	8	[644]	1959	
ZrB_2	H₂SO₄ 1 p. (1.84) + 4 p. H₃PO₄ (1.21) + 2 p. H₂O	20	24	1ss	[644]	1959	
ZrB_2	H₂SO₄ 1 p. (1.84) + 4 p. H₃PO₄ (1.21) + 2 p. H₂O	Boiling	2	1ss	[644]	1959	
ZrB_2	H₂SO₄ 50 ml (1.84) + 5 p. K₂SO₄	"	2	6	[644]	1959	
ZrB_2	H₂SO₄ 50 ml (1.84) + 5 p. K₂S₂O₈	"	2	7	[644]	1959	
ZrB_2	H₂SO₄ 35 ml (1.84 + 15 ml HNO₃ (1.43)	"	15 min	4	[644]	1959	
ZrB_2	HNO₃ (1.43) + HF (1.15)	"	15 min	4	[644]	1959	
ZrB_2	HCl (1:1) + H₂C₂O₄	"	1	71	[896]	1961	
ZrB_2	30 ml HCl (1:1) + C₆H₈O₇	"	1	76	[896]	1961	
ZrB_2	HCl (1:1) + Trilon B	"	1	75	[896]	1961	
ZrB_2	KNO₃, 1% sulfuric acid solution	"	1	52	[896]	1961	
ZrB_2	30 ml H₂O₂ (1:3) + 5 drops H₂SO₄ (1.84)	"	1	97	[896]	1961	
ZrB_2	30 ml H₂O₂ (1:3) + 10 drops HSO₃ (1.43)	"	1	i	[896]	1961	
ZrB_2	NaOH (30%)	20	24	cs	[644]	1959	
ZrB_2	NaOH (30%)	Boiling	2	cs	[644]	1959	

RESISTANCE OF POWDERS OF REFRACTORY COMPOUNDS TO CHEMICAL REAGENTS (continued)

Phase	Reagent²*	Temperature, °C	Duration of treatment, hr	Insoluble residue, %	Ref.	Year	Remarks
ZrB₂	NaOH (10%)	Boiling	2	98	[644]	1959	
HfB₂	HCl (1.19)	"	1	21	[896]	1961	
HfB₂	HCl (1:1)	"	1	24	[896]	1961	
HfB₂	HNO₃ (1.43)	"	1	13	[896]	1961	
HfB₂	HNO₃ (1:1)	"	1	18	[896]	1961	
HfB₂	H₂SO₄ (1:4)	"	1	21	[896]	1961	
VB₂	HCl (1.19)	20	24	63	[644]	1959	
VB₂	HCl (1.19)	Boiling	2	3	[644]	1959	
VB₂	HCl (1:1)	20	24	62	[644]	1959	
VB₂	HCl (1:1)	Boiling	2	10	[644]	1959	
VB₂	HNO₃ (1.43)	20	24	1	[644]	1959	
VB₂	HNO₃ (1.43)	Boiling	2	2	[644]	1959	
VB₂	HNO₃ (1:1)	20	24	3	[644]	1959	
VB₂	HNO₃ (1:1)	Boiling	2	2	[644]	1959	
VB₂	H₂SO₄ (1.84)	20	24	49	[644]	1959	
VB₂	H₂SO₄ (1.84)	Boiling	2	13	[644]	1959	
VB₂	H₂SO₄ (1:4)	20	24	60	[644]	1959	
VB₂	H₂SO₄ (1:4)	Boiling	2	7	[644]	1959	
VB₂	H₃PO₄ (1.21)	20	24	66	[644]	1959	
VB₂	H₃PO₄ (1.21)	Boiling	2	1ss	[644]	1959	
VB₂	H₃PO₄ (1:3)	20	24	62	[644]	1959	
VB₂	H₃PO₄ (1:3)	Boiling	2	24	[644]	1959	
VB₂	HClO₄ (1:3)	20	24	4	[644]	1959	
VB₂	HClO₄ (1.35)	Boiling	2	0	[644]	1959	
VB₂	HClO₄ (1:3)	20	24	47	[644]	1959	

RESISTANCE OF POWDERS OF REFRACTORY COMPOUNDS TO CHEMICAL REAGENTS (continued)

Phase	Reagent[2]*	Temperature, °C	Duration of treatment, hr	Insoluble residue, %	Ref.	Year	Remarks
VB_2	$HClO_4$ (1:3)	Boiling	2	2	[644]	1959	
VB_2	$H_2C_2O_4$ (satur.)	20	24	60	[644]	1959	
VB_2	$H_2C_2O_4$ (satur.)	Boiling	2	17	[644]	1959	
VB_2	$H_2C_2O_4$ (6%)	"	1	29	[940]	1961	
VB_2	$H_2C_2O_4$ (1:3)	20	24	24	[644]	1959	
VB_2	$H_2C_2O_4$ (1:3)	Boiling	2	37	[644]	1959	
VB_2	HF (1.15)	"	2	13	[644]	1959	
VB_2	30 ml H_2O_2 (1:3) + 5 drops H_2SO_4 (1.82)	"	1	lss	[940]	1961	
VB_2	30 ml H_2O_2 (1:3) + 10 drops HNO_3 (1.43)	"	1	27	[940]	1961	
VB_2	KNO_3 1% sulfuric acid solution	"	1	28	[940]	1961	
VB_2	HCl (1:1) + $H_2C_2O_4$	"	1	34	[940]	1961	
VB_2	30 ml HCl (1:1) + $C_6H_8O_7$ Trilon B	"	1	36	[940]	1961	
VB_2	HCl (1:1) + Trilon B	20	24	64	[644]	1959	
VB_2	$NaOH$ (30%)	Boiling	2	61	[644]	1959	
VB_2	$NaOH$ (30%)	"	2	56	[644]	1959	
VB_2	$NaOH$ (10%)	20	24	98	[644]	1959	
NbB_2	HCl (1.19)	Boiling	2	91	[644]	1959	
NbB_2	HCl (1.19)	20	24	99	[644]	1959	
NbB_2	HCl (1:1)	Boiling	2	95	[644]	1959	
NbB_2	HCl (1:1)	20	24	94	[644]	1959	
NbB_2	HNO_3 (1.4)	Boiling	2	100	[644]	1959	

RESISTANCE OF POWDERS OF REFRACTORY COMPOUNDS TO CHEMICAL REAGENTS (continued)

Phase	Reagent[2]*	Temperature, °C	Duration of treatment, hr	Insoluble residue, %	Ref.	Year	Remarks
NbB_2	HNO_3 (1:1)	20	24	99	[644]	1959	
NbB_2	HNO_3 (1:1)	Boiling	2	100	[644]	1959	
NbB_2	H_2SO_4 (1.84)	20	24	100	[644]	1959	
NbB_2	H_2SO_4 (1.84)	Boiling	2	3	[644]	1959	
NbB_2	H_2SO_4 (1:4)	20	24	100	[644]	1959	
NbB_2	H_2SO_4 (1:4)	Boiling	2	22	[644]	1959	
NbB_2	H_3PO_4 (1.35)	20	24	100	[644]	1959	
NbB_2	H_3PO_4 (1.35)	Boiling	2	ps	[644]	1959	
NbB_2	H_3PO_4 (1:3)	20	24	100	[644]	1959	
NbB_2	H_3PO_4 (1:3)	Boiling	2	24	[644]	1959	
NbB_2	$HClO_4$ (1.35)	20	24	98	[644]	1959	
NbB_2	$HClO_4$ (1.35)	Boiling	2	98	[644]	1959	
NbB_2	$HClO_4$ (1:3)	20	24	98	[644]	1959	
NbB_2	$H_2C_2O_4$ (satur.)	20	24	97	[644]	1959	
NbB_2	$H_2C_2O_4$ (satur.)	Boiling	24	50	[644]	1959	
NbB_2	$H_2C_2O_4$ (6%)	"	1	94	[940]	1961	
NbB_2	$H_2C_2O_4$ (1:3)	20	24	93	[644]	1959	
NbB_2	$H_2C_2O_4$ (1:3)	Boiling	2	98	[644]	1959	
NbB_2	HF (1.15)	"	2	44	[644]	1959	
NbB_2	3 p. HCl (1.19) + 1 p. HNO_3 (1.43)	20	24	71	[644]	1959	
NbB_2	3 p. HCl (1.19) + 1 p. HNO_3 (1.43)	Boiling	2	80	[644]	1959	
NbB_2	3 p. HCl (1:1) + 1 p. HNO_3 (1:1)	20	24	96	[644]	1959	
NbB_2	3 p. HCl (1:1) + 1 p. HNO_3 (1:1)	Boiling	2	cs	[644]	1959	
NbB_2	30 ml $H_2C_2O_4$ (satur.) + 10 ml HNO_3 (1.43)	20	24	5	[644]	1959	

RESISTANCE OF POWDERS OF REFRACTORY COMPOUNDS TO CHEMICAL REAGENTS (continued)

Phase	Reagent[2]*	Temperature, °C	Duration of treatment, hr	Insoluble residue, %	Ref.	Year	Remarks
NbB_2	30 ml $H_2C_2O_4$ (satur.) + 10 ml (30%) H_2O_2 + 10 ml HNO_3 (1.43)	Boiling	2	26	[644]	1959	
NbB_2	30 ml $H_2C_2O_4$ (satur.) + 20 ml H_2SO_4 (1.84)	20	24	86	[644]	1959	
NbB_2	30 ml $H_2C_2O_4$ (satur.) + 20 ml H_2SO_4 (1.84)	Boiling	2	86	[644]	1959	
NbB_2	35 ml HCl (1.19) + 15 ml bromine water	"	2	58	[644]	1959	
NbB_2	35 ml $HClO_4$ (1.35) + 15 ml HCl (1.19)	20	24	73	[644]	1959	
NbB_2	35 ml $HClO_4$ (1.35) + 15 ml HCl (1.19)	Boiling	2	60	[644]	1959	
NbB_2	1 p. H_2SO_4 (1.84) + 4 p. H_3PO_4 (1.21) + 2 p. H_2O	"	2	10	[644]	1959	
NbB_2	1 p. H_2SO_4 (1.84) + 4 p. H_3PO_4 (1.21) + 2 p. H_2O	20	15 min	1ss	[644]	1959	
NbB_2	50 ml H_2SO_4 (1.84) + 5 p. K_2SO_4	Boiling	2	3	[644]	1959	
NbB_2	H_2SO_4 (1.84) + HNO_3 (1.43)	"	2	1ss	[644]	1959	
NbB_2	HNO_3 (1.43) + HF (1.15)	"	15 min	4	[644]	1959	
NbB_2	30 ml H_2O_2 (1 : 3) + 5 drops H_2SO_4 (1.82)	"	1	ps	[940]	1961	
NbB_2	30 ml H_2O_2 (1 : 3) + 10 drops HNO_3 (1.43)	"	1	ps	[940]	1961	
NbB_2	KNO_3, 1% sulfuric acid solution	"	1	98	[940]	1961	NbO_2 precipitated

RESISTANCE OF POWDERS OF REFRACTORY COMPOUNDS TO CHEMICAL REAGENTS (continued)

Phase	Reagent²*	Temperature, °C	Duration of treatment, hr	Insoluble residue, %	Ref.	Year	Remarks
NbB₂	HCl (1:1) + H₂C₂O₄	Boiling	1	95	[940]	1961	
NbB₂	30 ml HCl (1:1) + C₆H₈O₇	"	1	98	[940]	1961	
NbB₂	NaOH (30%)	20	24	95	[644]	1959	
NbB₂	NaOH (30%)	Boiling	2	cs	[644]	1959	
NbB₂	NaOH (10%)	"	2	95	[644]	1959	
TaB₂	HCl (1.19)	20	24	100	[644]	1959	
TaB₂	HCl (1.19)	Boiling	2	99	[644]	1959	
TaB₂	HCl (1:1)	20	24	100	[644]	1959	
TaB₂	HCl (1:1)	Boiling	2	98	[644]	1959	
TaB₂	HNO₃ (1.43)	20	24	100	[644]	1959	
TaB₂	HNO₃ (1.43)	Boiling	2	100	[644]	1959	
TaB₂	HNO₃ (1:1)	20	24	100	[644]	1959	
TaB₂	HNO₃ (1:1)	Boiling	2	100	[644]	1959	
TaB₂	H₂SO₄ (1.84)	20	24	99	[644]	1959	
TaB₂	H₂SO₄ (1.84)	Boiling	2	3	[644]	1959	
TaB₂	H₂SO₄ (1:4)	20	24	100	[644]	1959	
TaB₂	H₂SO₄ (1:4)	Boiling	2	99	[644]	1959	
TaB₂	H₃PO₄ (1.21)	20	24	100	[644]	1959	
TaB₂	H₃PO₄ (1:3)	Boiling	2	100	[644]	1959	
TaB₂	H₃PO₄ (1:3)	20	24	100	[644]	1959	
TaB₂	H₃PO₄ (1:4)	Boiling	2	100	[644]	1959	
TaB₂	HClO₄ (1.35)	20	24	100	[644]	1959	
TaB₂	HClO₄ (1:3)	Boiling	2	100	[644]	1959	
TaB₂	HClO₄ (1:3)	20	24	99	[644]	1959	
TaB₂	H₂C₂O₄ (satur.)	Boiling	24	100	[644]	1959	

RESISTANCE OF POWDERS OF REFRACTORY COMPOUNDS TO CHEMICAL REAGENTS (continued)

Phase	Reagent[2]*	Temperature, °C	Duration of treatment, hr	Insoluble residue, %	Ref.	Year	Remarks
TaB_2	$H_2C_2O_4$ (satur.)	Boiling	2	94	[644]	1959	
TaB_2	$H_2C_2O_4$ (1:3)	20	24	99	[644]	1959	
TaB_2	$H_2C_2O_4$ (1:3)	Boiling	2	99	[644]	1959	
TaB_2	HF (1.15)	"	2	20	[644]	1959	
TaB_2	3 p. HCl (1.19) + 1 p. HNO_3 (1.43)	20	24	99	[644]	1959	
TaB_2	3 p. HCl (1.19) + 1 p. HNO_3 (1.43)	Boiling	2	csh	[644]	1959	
TaB_2	3 p. HCl (1:3) + 1 p. HNO_3 (1:1)	20	24	99	[644]	1959	
TaB_2	3 p. HCl (1:3) + 1 p. HNO_3 (1:1)	Boiling	2	csh	[644]	1959	
TaB_2	30 ml $H_2C_2O_4$ (satur.) + 10 ml H_2O_2 (30%) + 10 ml HNO_3 (1.43)	20	24	32	[644]	1959	
TaB_2	30 ml $H_2C_2O_4$ (satur.) + 10 ml 30% H_2O_2 + 10 ml HNO_3 (1.43)	Boiling	2	csh	[644]	1959	
TaB_2	30 ml $H_2C_2O_4$ + 20 ml H_2SO_4 (1.84)	20	24	99	[644]	1959	
TaB_2	30 ml $H_2C_2O_4$ (satur.) + 20 ml H_2SO_4 (1.84)	Boiling	2	99	[644]	1959	
TaB_2	35 ml HCl (1.19) + 15 ml bromine water	"	2	99	[644]	1959	
TaB_2	35 ml $HClO_4$ (1.35) + 15 ml HCl (1.19)	"	2	100	[644]	1959	
TaB_2	35 ml $HClO_4$ (1.35) + 15 ml HCl (1.19)	20	24	99	[644]	1959	
TaB_2	1 p. H_2SO_4 (1.84) + 4 p. H_3PO_4 (1.21) + 2 p. H_2O	20	24	lss	[644]	1959	

207

RESISTANCE OF POWDERS OF REFRACTORY COMPOUNDS TO CHEMICAL REAGENTS (continued)

Phase	Reagent²*	Temperature, °C	Duration of treatment, hr	Insoluble residue, %	Ref.	Year	Remarks
TaB_2	1 p. H_2CO_4 (1.84) + 4 p. H_3PO_4 (1.21) + 2 p. H_2O	Boiling	2	23	[644]	1959	
TaB_2	50 ml H_2SO_4 (1.84) + 5 p. K_2SO_4	"	2	5	[644]	1959	
TaB_2	35 ml H_2SO_4 (1.84) + 15 ml HNO_3 (1.43)	"	2	cs	[644]	1959	
TaB_2	HNO_3 (1.43) + HF (1.15)	"	1	32	[644]	1959	
TaB_2	KNO_3, 1% sulfuric acid solution	"	1	i	[940]	1961	
TaB_2	30 ml H_2O_2 (1:3) + 15 drops H_2SO_4 (1.82)	"	1	ps	[940]	1961	
TaB_2	30 ml H_2O_2 (1:3) + 10 drops HNO_3 (1.43)	"	1	ps	[940]	1961	
TaB_2	HCl (1:1) + $H_2C_2O_4$	"	1	91	[940]	1961	
TaB_2	30 ml HCl (1:1) + $C_6H_8O_7$	"	1	98	[940]	1961	
TaB_2	HCl (1:1) + Trilon B	20	1	98	[940]	1961	
TaB_2	NaOH (30%)	Boiling	24	cs	[644]	1959	
TaB_2	NaOH (30%)	"	2	cs	[644]	1959	
TaB_2	NaOH (10%)	20	24	45	[644]	1959	
CrB_2	HCl (1.19)	Boiling	2	36	[644]	1959	Particle size 200 μ [1140]. See [1140] for data on Cr_2B, Cr_5B_3, CrB
CrB_2	HCl (1.19)	20	24	3	[644]	1959	
CrB_2	HCl (1:1)	Boiling	2	51	[644]	1959	
CrB_2	HCl (1:1)	"	0.5	6	[645]	1960	
CrB_2	HCl (1:2)	"	0.5	10.2	[645]	1960	
CrB_2	HCl (1:2)	20	24	8.5	[645]	1960	
CrB_2	HNO_3 (1.43)	"	24	99	[644]	1959	
CrB_2	HNO_3 (1.43)	Boiling	2	22	[644]	1959	

RESISTANCE OF POWDERS OF REFRACTORY COMPOUNDS TO CHEMICAL REAGENTS (continued)

Phase	Reagent[2]*	Temperature, °C	Duration of treatment, hr	Insoluble residue, %	Ref.	Year	Remarks
CrB_2	HNO_3 (1:1)	20	24	99	[644]	1959	
CrB_2	HNO_3 (1:1)	Boiling	2	41	[644]	1959	
CrB_2	H_2SO_4 (1.84)	20	24	99	[644]	1959	
CrB_2	H_2SO_4 (1:4)	20	24	9	[644]	1959	
CrB_2	H_2SO_4 (1:4)	Boiling	2	3	[644]	1959	
CrB_2	H_3PO_4 (1.21)	20	24	100	[644]	1959	
CrB_2	H_3PO_4 (1.21)	Boiling	2	1ss	[644]	1959	
CrB_2	H_3PO_4 (1:3)	20	24	100	[644]	1959	
CrB_2	H_3PO_4 (1:3)	Boiling	2	18	[644]	1959	
CrB_2	$HClO_4$ (1.35)	20	24	96	[644]	1959	
CrB_2	$HClO_4$ (1.35)	Boiling	2	0	[644]	1959	
CrB_2	$HClO_4$ (1:3)	20	24	100	[644]	1959	
CrB_2	$HClO_4$ (1:3)	Boiling	2	4	[644]	1959	
CrB_2	$H_2C_2O_4$ (satur.)	20	24	44	[644]	1959	
CrB_2	$H_2C_2O_4$ (satur.)	Boiling	2	2	[644]	1959	
CrB_2	$H_2C_2O_4$ (6%)	"	1	ps	[940]	1961	
CrB_2	$H_2C_2O_4$ (1:3)	20	24	97	[644]	1961	
CrB_2	$H_2C_2O_4$ (1:3)	Boiling	2	75	[644]	1959	
CrB_2	HF (1.15)	"	2	2	[644]	1959	
CrB_2	3 p. HCl (1.19) + 1 p. HNO_3 (1.43)	20	24	80	[644]	1959	
CrB_2	3 p. HCl (1.19) + 1 p. HNO_3 (1.43)	Boiling	2	29	[644]	1959	
CrB_2	3 p. HCl (1:3) + 1 p. HNO_3 (1:1)	20	24	95	[644]	1959	
CrB_2	3 p. HCl (1:3) + 1 p. HNO_3 (1:1)	Boiling	2	27	[644]	1959	
CrB_2	30 ml $H_2C_2O_4$ (satur.) + 10 ml H_2O_2 (30%) + 10 ml HNO_3 (1.43)	20	24	99	[644]	1959	

RESISTANCE OF POWDERS OF REFRACTORY COMPOUNDS TO CHEMICAL REAGENTS (continued)

Phase	Reagent²*	Temperature, °C	Duration of treatment, hr	Insoluble residue, %	Ref.	Year	Remarks
CrB₂	30 ml $H_2C_2O_4$ (satur.) + 10 ml H_2O_2 (30%) + 10 ml HNO_3 (1.43)	Boiling	2	89	[644]	1959	
CrB₂	30 ml $H_2C_2O_4$ (satur.) + 20 ml H_2SO_4 (1.84)	20	24	31	[644]	1959	
CrB₂	30 ml $H_2C_2O_4$ (satur.) + 20 ml H_2SO_4 (1.84)	Boiling	2	3	[644]	1959	
CrB₂	HCl (1.19) 35 + 15 ml bromine wat.	"	2	2	[644]	1959	
CrB₂	$HClO_4$ (1.35) 35 ml + 15 ml HCl (1.19)	"	2	3	[644]	1959	
CrB₂	$HClO_4$ (1.35) 35 ml + 15 ml HCl (1.19)	20	24	49	[644]	1959	
CrB₂	1 p. H_2SO_4 (1.84) + 4 p. H_3PO_4 (1.21) + 2 p. H_2O	20	24	86	[644]	1959	
CrB₂	1 p. H_2SO_4 (1.84) + 4 p. H_3PO_4 (1.21) + 2 p. H_2O	Boiling	2	6	[644]	1959	
CrB₂	50 ml H_2SO_4 (1.84) + 5 g K_2SO_4	"	2	cs	[644]	1959	
CrB₂	50 ml H_2SO_4 (1.84) + 5 g $K_2S_2O_8$	"	2	cs	[644]	1959	
CrB₂	36 ml H_2SO_4 (1.84) 215 ml HNO_3 (1.43)	"	2	2	[644]	1959	
CrB₂	HNO_3 (1.43) + HF (1.15)	"	1	4	[644]	1959	
CrB₂	HCl (1:1) + $H_2C_2O_4$ (satur.)	"	1	ps	[940]	1961	
CrB₂	30 ml H_2O_2 (1:3) + 5 drops H_2SO_4 (1.84)	"	1	97	[940]	1961	
CrB₂	30 ml H_2O_2 (1:3) + 10 drops HNO_3	"	1	ps	[940]	1961	

Phase	Reagent[2]*	Temperature, °C	Duration of treatment, hr	Insoluble residue, %	Ref.	Year	Remarks
CrB_2	KNO_3 1% sulfuric acid solution	Boiling	1	98	[940]	1961	
CrB_2	NaOH (30%)	20	20	99	[644]	1959	
CrB_2	NaOH (30%)	Boiling	2	88	[644]	1959	
CrB_2	NaOH (10%)	20	2	98	[644]	1959	
Mo_2B_5	HCl (1.19)	20	24	95	[644]	1959	
Mo_2B_5	HCl (1.19)	Boiling	2	73	[644]	1959	
Mo_2B_5	HCl (1:1)	20	24	94	[644]	1959	
Mo_2B_5	HCl (1:1)	Boiling	2	85	[644]	1959	
Mo_2B_5	HNO_3 (1.43)	20	24	9	[644]	1959	
Mo_2B_5	HNO_3 (1.43)	Boiling	2	3	[644]	1959	
Mo_2B_5	HNO_3 (1:1)	20	24	9	[644]	1959	
Mo_2B_5	HNO_3 (1:1)	Boiling	2	9	[644]	1959	
Mo_2B_5	H_2SO_4 (1.84)	20	24	95	[644]	1959	
Mo_2B_5	H_2SO_4 (1.84)	Boiling	2	7	[644]	1959	
Mo_2B_5	H_2SO_4 (1:4)	20	24	97	[644]	1959	
Mo_2B_5	H_2SO_4 (1:4)	Boiling	2	65	[644]	1959	
Mo_2B_5	H_3PO_4 (1.21)	20	24	93	[644]	1959	
Mo_2B_5	H_3PO_4 (1.21)	Boiling	2	1ss	[644]	1959	
Mo_2B_5	H_3PO_4 (1:3)	20	24	93	[644]	1959	
Mo_2B_5	H_3PO_4 (1:3)	Boiling	2	77	[644]	1959	
Mo_2B_5	$HClO_4$ (1.35)	20	24	8	[644]	1959	
Mo_2B_5	$HClO_4$ (1.35)	Boiling	2	9	[644]	1959	
Mo_2B_5	$HClO_4$ (1:3)	20	24	90	[644]	1959	
Mo_2B_5	$HClO_4$ (1:3)	Boiling	2	16	[644]	1959	
Mo_2B_5	$H_2C_2O_4$ (satur.)	20	24	91	[644]	1959	

RESISTANCE OF POWDERS OF REFRACTORY COMPOUNDS TO CHEMICAL REAGENTS (continued)

Phase	Reagent²*	Temperature, °C	Duration of treatment, hr	Insoluble residue, %	Ref.	Year	Remarks
Mo_2B_5	$H_2C_2O_4$ (satur.)	Boiling	2	88	[644]	1959	
Mo_2B_5	$H_2C_2O_4$ (6:3)	"	1	75	[940]	1961	
Mo_2B_5	$H_2C_2O_4$ (1:3)	20	24	92	[644]	1959	
Mo_2B_5	$H_2C_2O_4$ (1:3)	Boiling	2	88	[644]	1959	
Mo_2B_5	HF (1.15)	"	2	60	[644]	1959	
Mo_2B_5	3 p. HCl (1.19) + 1 p. HNO_3 (1.43)	20	24	0.5	[644]	1959	
Mo_2B_5	3 p. HCl (1.19) + 1 p. HNO_3 (1.43)	Boiling	2	lss	[644]	1959	
Mo_2B_5	3 p. HCl (1:1) + 1 p. HNO_3 (1:1)	20	24	0.3	[644]	1959	
Mo_2B_5	3 p. HCl (1:1) + 1 p. HNO_3 (1:1)	Boiling	2	cs	[644]	1959	
Mo_2B_5	30 ml $H_2C_2O_4$ (satur.) + 10 ml H_2O_2 (30%) + 10 ml HNO_3 (1.43)	20	24	0.3	[644]	1959	
Mo_2B_5	H_2O_2 + 10 ml HNO_3 (1.43)	Boiling	2	lss	[644]	1959	
Mo_2B_5	30 ml $H_2C_2O_4$ (satur.) + 20 ml H_2SO_4 (1.84)	20	24	63	[644]	1959	
Mo_2B_5	30 ml $H_2C_2O_4$ (satur.) + 20 ml H_2SO_4 (1.84)	Boiling	2	63	[644]	1959	
Mo_2B_5	35 ml HCl (1.19) + 15 ml bromine water	"	2	23	[644]	1959	
Mo_2B_5	35 ml HCl (1.19) + 35 ml bromine water	20	24	61	[644]	1959	
Mo_2B_5	1 p. H_2SO_4 (1.84) + 4 p. H_3PO_4 (1.21) + 2 p. H_2O	20	24	58	[644]	1959	
Mo_2B_5	1 p. H_2SO_4 (1.84) + 4 p. H_3PO_4 (1.21) + 2 p. H_2O	Boiling	2	67	[644]	1959	

RESISTANCE OF POWDERS OF REFRACTORY COMPOUNDS TO CHEMICAL REAGENTS (continued)

Phase	Reagent[2*]	Temperature, °C	Duration of treatment, hr	Insoluble residue, %	Ref.	Year	Remarks
Mo$_2$B$_5$	50 ml H$_2$SO$_4$ (1.84) + 5 g K$_2$SO$_4$	Boiling	2	8	[644]	1959	
Mo$_2$B$_5$	50 ml H$_2$SO$_4$ (1.84) + 5 g K$_2$S$_2$O$_8$	"	2	1	[644]	1959	
Mo$_2$B$_5$	HNO$_3$ (1.43) + HF (1.15)	"	1	1	[644]	1959	
Mo$_2$B$_5$	30 ml H$_2$O$_2$ (1:3) + 5 drops H$_2$SO$_4$	"	1	cs	[940]	1961	
Mo$_2$B$_5$	30 ml H$_2$O$_2$ (1:3) + 10 drops HNO$_3$	"	1	cs	[940]	1961	
Mo$_2$B$_5$	KNO$_3$, 1% sulfuric acid solution	"	1	45	[940]	1961	
Mo$_2$B$_5$	HCl (1:1) + H$_2$C$_2$O$_4$	"	1	75	[940]	1961	
Mo$_2$B$_5$	30 ml HCl (1:1) + C$_6$H$_8$O$_7$	"	1	95	[940]	1961	
Mo$_2$B$_5$	HCl (1:1) + Trilon B	"	1	96	[940]	1961	
Mo$_2$B$_5$	NaOH (30%)	20	24	68	[644]	1959	
Mo$_2$B$_5$	NaOH (30%)	Boiling	2	67	[644]	1959	
				Loss of weight, mg/cm^2·hr			
α-MoB	KCl (0.1 N)	30.2±0.2°	10	3.96·10^{-2}	[923]	1961	Data taken from graph
α-MoB	KCl (0.1 N)	30.2±0.2°	100	1.25·10^{-2}	[923]	1961	
α-MoB	KCl (0.1 N)	30.2±0.2°	500	0.83·10^{-2}	[923]	1961	
β-MoB	KCl (0.1 N)	30.2±0.2°	10	2.75·10^{-2}	[923]	1961	
β-MoB	KCl (0.1 N)	30.2±0.2°	100	0.83·10^{-2}	[923]	1961	
β-MoB	KCl (0.1 N)	30.2±0.2°	500	0.5 ·10^{-2}	[923]	1961	
W$_2$B$_5$	HCl (1.19)	20	2	96	[644]	1959	

RESISTANCE OF POWDERS OF REFRACTORY COMPOUNDS TO CHEMICAL REAGENTS (continued)

Phase	Reagent[2]*	Temperature, °C	Duration of treatment, hr	Insoluble residue, %	Ref.	Year	Remarks
W_2B_5	HCl (1.19)	Boiling	24	96	[644]	1959	
W_2B_5	HCl (1:1)	20	2	95	[644]	1959	
W_2B_5	HCl (1:1)	Boiling	24	97	[644]	1959	
W_2B_5	HNO_3 (1.43)	20	24	csh	[644]	1959	
W_2B_5	HNO_3 (1.43)	Boiling	2	9	[644]	1959	
W_2B_5	HNO_3 (1:1)	20	24	csh	[644]	1959	
W_2B_5	HNO_3 (1:1)	Boiling	2	11	[644]	1959	
W_2B_5	H_2SO_4 (1.84)	20	24	100	[644]	1959	
W_2B_5	H_2SO_4 (1.84)	Boiling	2	2	[644]	1959	
W_2B_5	H_2SO_4 (1:1)	20	24	96	[644]	1959	
W_2B_5	H_2SO_4 (1:1)	Boiling	2	97	[644]	1959	
W_2B_5	H_3PO_4 (1.21)	20	24	96	[644]	1959	
W_2B_5	H_3PO_4 (1.21)	Boiling	2	1ss	[644]	1959	
W_2B_5	H_3PO_4 (1:3)	20	24	89	[644]	1959	
W_2B_5	H_3PO_4 (1:3)	Boiling	2	93	[644]	1959	
W_2B_5	$HClO_4$ (1.35)	20	24	94	[644]	1959	
W_2B_5	$HClO_4$ (1.35)	Boiling	2	3	[644]	1959	
W_2B_5	$HClO_4$ (1:3)	20	24	96	[644]	1959	
W_2B_5	$HClO_4$ (1:3)	Boiling	2	100	[644]	1959	
W_2B_5	$H_2C_2O_4$ (satur.)	20	24	92	[644]	1959	
W_2B_5	$H_2C_2O_4$ (satur.)	Boiling	2	87	[644]	1959	
W_2B_5	$H_2C_2O_4$ (1:3)	20	24	91	[644]	1959	
W_2B_5	$H_2C_2O_4$ (1:3)	Boiling	2	88	[644]	1959	
W_2B_5	HF (1.15)	"	2	75	[644]	1959	
MnB	HCl (1.19)	"	—	cs	[697]	1959	[1075,1140]

RESISTANCE OF POWDERS OF REFRACTORY COMPOUNDS TO CHEMICAL REAGENTS (continued)

Phase	Reagent²*	Temperature, °C	Duration of treatment, hr	Insoluble residue, %	Ref.	Year	Remarks
MnB	HCl (1.19)	20	—	cs	[697]	1959	[1075]
MnB	HNO₃ (1.43)	20	—	cs	[697]	1959	
MnB	HNO₃ (1.43)	Boiling	—	cs	[697]	1959	
MnB	H₂SO₄ (1.84)	20	—	cs	[697]	1959	
MnB	H₂SO₄ (1.84)	Boiling	—	cs	[697]	1959	
Mn₃B	H₂O	20	—	i	[1]	1957	[1140]. See [1140] also for data on Mn₃B₄, MnB₂
Mn₃B	H₂O	Boiling	—	i	[1]	1957	
Mn₃B	HCl (1.19)	20	—	cs	[1]	1957	
Mn₃B	HNO₃ (1.43)	20	—	cs	[1]	1957	
Mn₃B	H₂SO₄ (1.84)	20	—	cs	[1]	1957	
Mn₄B	H₂O	20	—	i	[1]	1957	
Mn₄B	H₂O	Boiling	—	cs	[1]	1957	[217, 1075]
Mn₄B	HCl (1.19)	20	—	cs	[1]	1957	
Mn₄B	HNO₃ (1.4)	20	—	cs	[1]	1957	
Mn₄B	H₂SO₄ (1.84)	20	—	cs	[1]	1957	
FeB	HCl (1.19)	Boiling	—	i	[1]	1957	
FeB	HCl (1.19)	20	—	i	[1]	1957	[1140]. See [1140] also for data on Fe₂B
FeB	HNO₃ (1.43)	Boiling	—	cs	[1]	1957	
FeB	HNO₃ (1.43)	20	—	cs	[1]	1957	
FeB	H₂SO₄ (1.84)	Boiling	—	i	[1]	1957	
FeB	H₂SO₄ (1.84)	20	—	i	[1]	1957	
Co₂B	HCl (1.19)	20	—	i	[1]	1957	
Co₂B	HCl (1.19)	Boiling	—	cs	[1]	1957	Dissolves completely on long boiling. [1140]

RESISTANCE OF POWDERS OF REFRACTORY COMPOUNDS TO CHEMICAL REAGENTS (continued)

Phase	Reagent[2]*	Temperature, °C	Duration of treatment, hr	Insoluble residue, %	Ref.	Year	Remarks
Co_2B	HNO_3 (1.43)	20	—	cs	[1]	1957	[217, 1075, 1140]
CoB	HCl (1.19)	20	—	i	[1]	1957	
CoB	HCl (1.19)	Boiling	—	i	[1]	1957	
NiB	H_2O	20	—	cs	[697]	1959	[1075, 1140]
NiB	HNO_3 (1.43)	20	—	cs	[697]	1959	
NiB	3 p. HCl (1.19) + 1 p. HNO_3 (1.43)	20	—	cs	[697]	1959	
BeC_2	H_2O	20	—	cs	[467]	1952	C_2H_2 liberated on decomposition
BeC_2	HCl (1:1)	20	—	cs	[467]	1952	
BeC_2	H_2SO_4 (1:1)	20	—	cs	[467]	1952	
BeC_2	HNO_3 (1:1)	20	—	cs	[467]	1952	
Be_2C	H_2O	20	—	cs	[467]	1952	CH_4 liberated
Be_2C	HCl (1:1)	20	—	cs	[467]	1952	
Be_2C	H_2SO_4 (1:1)	20	—	cs	[467]	1952	
Be_2C	HNO_3 (1:1)	20	—	cs	[467]	1952	
Be_2C	H_2O	20	—	cs	[467]	1952	CH_4 liberated
Be_3C_2	HCl (1:1)	20	—	cs	[467]	1952	
Be_3C_2	HNO_3 (1:1)	20	—	cs	[697]	1959	
Be_3C_2	H_2SO_4 (1:1)	20	—	cs	[697]	1959	
MgC_2	H_2O	20	—	cs	[697]	1959	
Mg_2C_3	H_2O	20	—	cs	[467]	1952	
CaC_2	H_2O	20	—	cs	[467]	1952	C_2H_2 liberated
CaC_2	HCl (1.19)	20	—	cs	[346]	1956	
SrC_2	H_2O	20	—	cs	[467]	1952	
BaC_2	H_2O	20	—	cs	[467]	1952	

RESISTANCE OF POWDERS OF REFRACTORY COMPOUNDS TO CHEMICAL REAGENTS (continued)

Phase	Reagent[2]*	Temperature, °C	Duration of treatment, hr	Insoluble residue, %	Ref.	Year	Remarks
BaC$_2$	HCl (1:1)	20	—	cs	[346]	1956	
Al$_4$C$_3$	HCl (1:1)	20	—	i	[467]	1952	
Al$_4$C$_3$	H$_2$SO$_4$ (1:)	20	—	i	[467]	1952	
Al$_4$C$_3$	HNO$_3$ (1:1)	20	—	i	[467]	1952	
Al$_4$C$_3$	HCl (1:1)	Boiling	—	cs	[467]	1952	
Al$_4$C$_3$	H$_2$SO$_4$ (1:1)	"	—	cs	[467]	1952	
Al$_4$C$_3$	HNO$_3$ (1:1)	"	—	cs	[467]	1952	
ScC	H$_2$O	"	—	100	[467]	1952	On long boiling
YC	H$_2$O	20	5 min	cs	[879]	1961	
YC	HCl (1.19)	20	1 min	cs	[879]	1961	
YC	HCl (1:1)	20	1 min	cs	[879]	1961	
YC	HNO$_3$ (1.43)	20	15 min	ps	[879]	1961	
YC	HNO$_3$ (1:1)	20	5 min	cs	[879]	1961	
YC	H$_2$SO$_4$ (1.84)	20	15	ps	[879]	1961	
YC	H$_2$SO$_4$ (1:1)	20	2	ps	[879]	1961	
YC	NaOH (25%)	20	15 min	cs	[879]	1961	
Y$_2$C$_3$	H$_2$O	20	3 min	cs	[879]	1961	
Y$_2$C$_3$	HCl (1.19)	20	1 min	cs	[879]	1961	
Y$_2$C$_3$	HCl (1:1)	20	1 min	cs	[879]	1961	
Y$_2$C$_3$	HNO$_3$ (1.43)	20	20	ps	[879]	1961	
Y$_2$C$_3$	HNO$_3$ (1:1)	20	5 min	cs	[879]	1961	
Y$_2$C$_3$	H$_2$SO$_4$ (1.84)	20	20	ps	[879]	1961	
Y$_2$C$_3$	H$_2$SO$_4$ (1:1)	20	5 min	cs	[879]	1961	
Y$_2$C$_3$	NaOH (25%)	20	10 min	cs	[879]	1961	

RESISTANCE OF POWDERS OF REFRACTORY COMPOUNDS TO CHEMICAL REAGENTS (continued)

Phase	Reagent[2]*	Temperature, °C	Duration of treatment, hr	Insoluble residue, %	Ref.	Year	Remarks
YC_2	H_2O	20	5 min	cs	[879]	1961	
YC_2	HCl (1.19)	20	20	ps	[879]	1961	
YC_2	HCl (1:1)	20	20	ps	[879]	1961	
YC_2	HNO_3 (1.43)	20	20	i	[879]	1961	
YC_2	HNO_3 (1:1)	20	20	ps	[879]	1961	
YC_2	H_2SO_4 (1.84)	20	20	i	[879]	1961	
YC_2	H_2SO_4 (1:1)	20	20	ps	[879]	1961	
YC_2	NaOH (25%)	20	15	cs	[879]	1961	
CeC_2	Inorganic acids	20	—	cs	[346]	1956	
PrC_2	H_2O, dilute inorganic acids	20	—	cs	[346]	1956	
NdC_2	H_2O	20	—	cs	[467]	1952	
SmC_2	H_2O	20	—	cs	[467]	1952	[1278]
SmC_2	Inorganic acids	20	—	cs	[346]	1956	
ThC	H_2O	100	2	csh	[977]	1960	
ThC	HCl (1:1)	20	1	csh	[977]	1960	
ThC	HCl (1:1)	110	0.5	csh	[977]	1960	
ThC	H_2SO_4 (1:1)	20	0.5	csh	[977]	1960	
ThC	H_2SO_4 (1:1)	135	0.15	csh	[977]	1960	
ThC	HNO_3 (1:1)	20	2	100	[977]	1960	
ThC	HNO_3 (1:1)	115	1	csh	[977]	1960	
ThC	Tartaric acid	120	1	csh	[977]	1960	
ThC	NaOH (25%)	20		csh	[977]	1960	
ThC	NaOH (25%)	110	0.5	csh	[977]	1960	
ThC	Moist air	20	12		[977]	1960	ThO_2 formed

RESISTANCE OF POWDERS OF REFRACTORY COMPOUNDS TO CHEMICAL REAGENTS (continued)

Phase	Reagent[2]*	Temperature, °C	Duration of treatment, hr	Insoluble residue, %	Ref.	Year	Remarks
ThC_2	Concentrated inorganic acids						
U_2C_3	HCl (1.19)	20	—	ps	[346]	1956	[1278]
U_2C_3	HNO_3 (1.4)	Boiling	—	cs	[467]	1952	[346]
U_2C_3	H_2SO_4 (1.84)	"	—	cs	[467]	1952	[346]
UC_2	HCl (1.4)	"	—	cs	[467]	1952	
UC_2	HNO_3 (1.4)	"	—	cs	[467]	1952	
UC_2	H_2SO_4 (1.84)	"	—	cs	[467]	1952	
TiC	HCl (1.19)	20	24	99	[667]	1958	
TiC	HCl (1.19)	Boiling	2	100	[667]	1958	
TiC	HCl (1:1)	20	24	100	[667]	1958	
TiC	HCl (1:1)	Boiling	2	97	[667]	1958	
TiC	HNO_3 (1.43)	20	24	cs	[667]	1958	
TiC	HNO_3 (1.43)	Boiling	2	lss	[667]	1958	
TiC	HNO_3 (1:1)	20	24	cs	[667]	1958	
TiC	HNO_3 (1:1)	Boiling	2	cs	[667]	1958	
TiC	H_2SO_4 (1.84)	20	24	i	[667]	1958	
TiC	H_2SO_4 (1.84)	Boiling	2	88	[667]	1958	
TiC	H_2SO_4 (1:4)	20	24	100	[667]	1958	
TiC	H_2SO_4 (1:4)	Boiling	2	97	[667]	1958	
TiC	H_3PO_4 (1.21)	20	24	99	[667]	1958	
TiC	H_3PO_4 (1.21)	Boiling	2	98	[667]	1958	
TiC	H_3PO_4 (1:3)	20	24	98	[667]	1958	
TiC	H_3PO_4 (1:3)	Boiling	2	99	[667]	1958	
TiC	$HClO_4$ (1.35)	20	24	100	[667]	1958	

RESISTANCE OF POWDERS OF REFRACTORY COMPOUNDS TO CHEMICAL REAGENTS (continued)

Phase	Reagent[2]*	Temperature, °C	Duration of treatment, hr	Insoluble residue, %	Ref.	Year	Remarks
TiC	$HClO_4$ (1.35)	Boiling	2	cs	[667]	1958	
TiC	$HClO_4$ (1:3)	20	24	100	[667]	1958	
TiC	$HClO_4$ (1:3)	Boiling	2	cs	[667]	1958	
TiC	$H_2C_2O_4$ (satur.)	20	24	100	[667]	1958	
TiC	$H_2C_2O_4$ (satur.)	Boiling	2	100	[667]	1958	
TiC	3 p. HCl (1.19) + 1 p. HNO_3 (1.43)	20	24	4	[667]	1958	
TiC	3 p. HCl (1.19) + 1 p. HNO_3 (1.43)	Boiling	2	cs	[667]	1958	
TiC	2 p. H_2SO_4 (1.84) + 1 p. HNO_3 (1.43)	20	24	cs	[667]	1958	
TiC	2 p. H_2SO_4 (1.84) + 1 p. HNO_3 (1.43)	Boiling	2	cs	[667]	1958	
TiC	4 p. HNO_3 (1.4) + 1 p. HF (1.15)	20	24	cs	[667]	1958	
TiC	H_2SO_4 (1.81) + H_3PO_4 (1:4)	20	24	98	[667]	1958	
TiC	H_2SO_4 (1.84) + H_3PO_4 (1:3)	20	24	100	[667]	1958	
TiC	H_2SO_4 (1.84) + H_3PO_4 (1:3)	Boiling	2	100	[667]	1958	
TiC	H_2SO_4 (1.84) + $H_2C_2O_4$ (satur.)	20	24	99	[667]	1958	
TiC	H_2SO_4 (1.84) + $H_2C_2O_4$ (satur.)	Boiling	2	84	[667]	1958	
ZrC	HNO_3 (1:1)	2U	24	76	[667]	1958	
ZrC	HNO_3 (1:1)	Boiling	2	6	[667]	1958	
ZrC	H_2SO_4 (1.84)	20	24	97	[667]	1958	
ZrC	H_2SO_4 (1.84)	Boiling	2	cs	[667]	1958	
ZrC	H_2SO_4 (1:4)	20	24	98	[667]	1958	
ZrC	H_2SO_4 (1:4)	Boiling	2	76	[667]	1958	
ZrC	H_3PO_4 (1.21)	20	24	98	[667]	1958	
ZrC	H_3PO_4 (1.21)	Boiling	2	1ss	[667]	1958	
ZrC	H_3PO_4 (1:3)	20	24	96	[667]	1958	
ZrC	H_3PO_4 (1:3)	Boiling	2	88	[667]	1958	

RESISTANCE OF POWDERS OF REFRACTORY COMPOUNDS TO CHEMICAL REAGENTS (continued)

Phase	Reagent[2]*	Temperature, °C	Duration of treatment, hr	Insoluble residue, %	Ref.	Year	Remarks
ZrC	HClO₄ (1.35)	20	24	97	[667]	1958	
ZrC	HClO₄ (1.35)	Boiling	2	2	[667]	1958	
ZrC	HClO₄ (1:3)	20	24	99	[667]	1958	
ZrC	HClO₄ (1:3)	Boiling	2	84	[667]	1958	
ZrC	H₂C₂O₄ (satur.)	20	24	98	[668]	1958	
ZrC	H₂C₂O₄ (satur.)	Boiling	2	92	[667]	1958	
ZrC	3 p. HCl (1.19) + 1 p. HNO₃ (1.43)	20	24	14	[667]	1958	
ZrC	3 p. HCl (1.19) + 1 p. HNO₃ (1.43)	Boiling	2	6	[667]	1958	
ZrC	1 p. H₂SO₄ (1.84) + 1 p. H₂C₂O₄ (satur.)	"	2	cs	[667]	1958	
ZrC	1 p. H₂SO₄ (1:4) + 1 p. H₂C₂O₄ (satur.)	20	20	96	[667]	1958	
ZrC	1 p. H₂SO₄ (1:4) + 1 p. H₂C₂O₄ (satur.)	Boiling	2	91	[667]	1958	
ZrC	4 p. HNO₃ (1.43) + 1 p. HF (1.15)	20	24	cs	[667]	1958	
ZrC	H₂SO₄ (1.84) + H₃PO₄ (1:1)	Boiling	2	97	[667]	1958	
ZrC	H₂SO₄ (1.84) + H₃PO₄ (1:1)	20	2	ps	[667]	1958	
ZrC	H₂SO₄ (1:4) + H₃PO₄ (1:3)	Boiling	24	ps	[667]	1958	
ZrC	H₂SO₄ (1:4) + H₃PO₄ (1:3)	20	2	ps	[667]	1958	
ZrC	1 p. H₂SO₄ (1.84) + 1 p. H₂C₂O₄ (satur.)	20	24	96	[667]	1958	
ZrC	1 p. H₂SO₄ (1.84) + 1 p. H₂C₂O₄ (satur.)	Boiling	2	cs	[667]	1958	
ZrC	1 p. H₂SO₄ (1:4) + H₂C₂O₄ (satur.)	20	20	96	[667]	1958	

RESISTANCE OF POWDERS OF REFRACTORY COMPOUNDS TO CHEMICAL REAGENTS (continued)

Phase	Reagent[2*]	Temperature, °C	Duration of treatment, hr	Insoluble residue, %	Ref.	Year	Remarks
ZrC	1p. H_2SO_4 (1 : 4) + $H_2C_2O_4$ (satur.)	Boiling	2	91	[667]	1958	
ZrC	NaOH (10%)	20	24	100	[667]	1958	
ZrC	NaOH (10%)	Boiling	2	100	[667]	1958	
ZrC	NaOH (20%)	20	24	100	[667]	1958	
ZrC	NaOH (20%)	Boiling	2	100	[667]	1958	
ZrC	40 ml NaOH (20%) + 10 ml bromine water	20	24	93	[667]	1958	
ZrC	40 ml NaOH (20%) + 10 ml bromine water	Boiling	2	87	[667]	1958	
ZrC	40 ml NaOH (20%) + 10 ml H_2O_2 (30%)	20	24	53	[667]	1958	
ZrC	40 ml NaOH (20%) + 10 ml H_2O_2 (30%)	Boiling	2	3	[667]	1958	
HfC	HCl (1.19)	20	24	100	[668]	1961	
HfC	HCl (1.19)	120	2	100	[668]	1961	
HfC	HCl (1:1)	20	24	96	[668]	1961	
HfC	H_2SO_4 (1.84)	20	24	100	[668]	1961	
HfC	H_2SO_4 (1.84)	280	2	cs	[668]	1961	
HfC	H_2SO_4 (1 : 4)	116	2	88	[668]	1961	
HfC	HNO_3 (1.43)	20	24	60	[668]	1961	
HfC	HNO_3 (1.43)	112	2	cs	[668]	1961	
HfC	H_3PO_4 (1.21)	20	24	97	[668]	1961	
HfC	H_3PO_4 (1.21)	115	2	cs	[668]	1961	
HfC	H_3PO_4	110	2	90	[668]	1961	

RESISTANCE OF POWDERS OF REFRACTORY COMPOUNDS TO CHEMICAL REAGENTS (continued)

Phase	Reagent 2*	Temperature, °C	Duration of treatment, hr	Insoluble residue, %	Ref.	Year	Remarks
HfC	$HClO_4$ (1.35)	20	24	97	[668]	1961	
HfC	$HClO_4$ (1.35)		2	2	[668]	1961	
HfC	$H_2C_2O_4$ (satur.)	20	24	98	[668]	1961	
HfC	$H_2C_2O_4$ (satur.)	104	2	98	[668]	1961	
HfC	3 p. HCl (1.19) + 1 p. HNO_3 (1.43)	20	24	14	[668]	1961	
HfC	3 p. HCl (1.19) + 1 p. HNO_3 (1.43)	106	2	6	[668]	1961	
HfC	H_2SO_4 (1:1) + H_3PO_4 (1:1)	20	24	2	[668]	1961	
HfC	H_2SO_4 (1:1) + H_3PO_4 (1:1)	160	2	cs	[668]	1961	
HfC	H_2SO_4 (1.84) + H_3PO_4 (1.21)	20	24	97	[668]	1961	
HfC	H_2SO_4 (1.84) + H_3PO_4 (1.21)	250	2	1ss	[668]	1961	
HfC	NaOH (20%)	110	2	i	[668]	1961	
HfC	NaOH + bromine water	20	24	81	[668]	1961	
HfC	NaOH + H_2O_2 (30%)	20	24	53	[668]	1961	
HfC	NaOH + H_2O_2 (30%)	110	2	csh	[868]	1961	
HfC	$K_3[Fe(CN)_6]$ (10%) + NaOH (20%)	100	2	37	[668]	1961	
HfC	$K_3[Fe(CN)_6]$ (10%) + NaOH (20%)	100	2	37	[668]	1961	
NbC	HCl (1.19)	20	24	83	[668]	1961	
NbC	HCl (1.19)	115	2	100	[668]	1961	
NbC	HCl (1:1)	20	24	96	[668]	1961	
NbC	HCl (1:1)	108	2	100	[668]	1961	
NbC	HNO_3 (1.43)	20	24	99	[668]	1961	
NbC	HNO_3 (1.43)	120	2	100	[668]	1961	
NbC	HNO_3 (1:1)	105	2	ps	[668]	1961	

RESISTANCE OF POWDERS OF REFRACTORY COMPOUNDS TO CHEMICAL REAGENTS (continued)

Phase	Reagent[2]*	Temperature, °C	Duration of treatment, hr	Insoluble residue, %	Ref.	Year	Remarks
NbC	H_2SO_4 (1.84)	20	24	i	[668]	1961	
NbC	H_2SO_4 (1.84)	275	2	cs	[668]	1961	
NbC	H_2O_4 (1 : 4)	112	2	98	[668]	1961	
NbC	H_3PO_4 (1.21)	120	24	100	[668]	1961	
NbC	H_3PO_4 (1.21)	104	2	i	[668]	1961	
NbC	$H_2C_2O_4$ (satur.)	20	2	99	[668]	1961	
NbC	3 p. HCl (1.19) + 1 p. $H.NO_3$ (1.43)	20	24	92	[668]	1961	
NbC	3 p. HCl (1.19) + 1 p. $H.NO_3$ (1.43)	105	2	csh	[668]	1961	
NbC	1 p. H_2SO_4 (1.84) + 1 p. HNO_3 (1.43)	20	24	100	[668]	1961	
NbC	1 p. H_2SO_4 (1.84) + 1 p. H_3PO_4 (1.21)	140	2	22	[668]	1961	
NbC	1 p. H_2SO_4 (1.84) + 1 p. H_3PO_4 (1.21)	20	24	91	[668]	1961	
NbC	1 p. H_2SO_4 (1.84) + 1 p. H_2C_2O	240	2	cs	[668]	1961	
NbC	1 p. H_2SO_4 (1.84) + 1 p. $H_2C_2O_4$ (satur.)	20	24	100	[668]	1961	
NbC	NaOH (20%)	180	2	95	[668]	1961	
NbC	NaOH (20%)	20	24	99	[668]	1961	
NbC	NaOH + bromine water	110	2	100	[668]	1961	
NbC	NaOH + bromine water	20	24	100	[668]	1961	
NbC	NaOH + bromine water	105	2	84	[668]	1961	

RESISTANCE OF POWDERS OF REFRACTORY COMPOUNDS TO CHEMICAL REAGENTS (continued)

Phase	Reagent[2]*	Temperature, °C	Duration of treatment, hr	Insoluble residue, %	Ref.	Year	Remarks
NbC	NaOH (20%) + H$_2$O$_2$ (30%)	20	24	71	[668]	1961	
NbC	NaOH (20%) + H$_2$O$_2$ (30%)	112	2	88	[668]	1961	
NbC	K$_3$[Fe(CN)$_6$] + NaOH (20%)	20	24	ps	[668]	1961	
NbC	K$_3$[Fe(CN)$_6$] + NaOH (20%)	110	2	ps	[668]	1961	
TaC	HCl (1.19)	20	24	100	[668]	1961	
TaC	HCl (1.19)	120	2	98	[668]	1961	
TaC	HCl (1:1)	112	2	98	[668]	1961	
TaC	HNO$_3$ (1.43)	20	24	100	[668]	1961	
TaC	HNO$_3$ (1.43)	114	2	99	[668]	1961	
TaC	HNO$_3$ (1:1)	105	2	98	[668]	1961	
TaC	H$_2$SO$_4$ (1.84)	20	24	100	[668]	1961	
TaC	H$_2$SO$_4$ (1.84)	260	2	0	[668]	1961	
TaC	H$_2$SO$_4$ (1:4)	115	2	93	[668]	1961	
TaC	H$_3$PO$_4$ (1.21)	20	24	98	[668]	1961	
TaC	H$_3$PO$_4$ (1.21)	Boiling	2	lss	[668]	1961	
TaC	NaOH (20%)	20	24	99	[668]	1961	
TaC	NaOH	108	2	100	[668]	1961	
TaC	NaOH + bromine water	20	24	100	[668]	1961	
TaC	NaOH + bromine water	110	2	lss	[668]	1961	
TaC	NaOH (20%) + H$_2$O$_2$ (30%)	20	24	62	[668]	1961	
TaC	NaOH (20%) + H$_2$O$_2$ (30%)	105	2	lss	[668]	1961	

RESISTANCE OF POWDERS OF REFRACTORY COMPOUNDS TO CHEMICAL REAGENTS (continued)

Phase	Reagent[2]*	Temperature, °C	Duration of treatment, hr	Insoluble residue, %	Ref.	Year	Remarks
TaC	K₃[Fe(CN)₆] + NaOH (20%)	100	2	57	[668]	1961	
TaC	H₂C₂O₄ (satur.)	20	24	97	[668]	1961	
TaC	H₂C₂O₄ (satur.)	105	2	98	[668]	1961	
TaC	3 p. HCl (1.19) + 1 p. HNO₃ (1.43)	20	24	99	[668]	1961	
TaC	3 p. HCl (1.19) + 1 p. HNO₃ (1.43)	115	2	98	[668]	1961	
TaC	HF (1.15) + HNO₃ (1.43)	20	—	cs	[346]	1956	
TaC	1 p. H₂SO₄ (1.84) + 1 p. HNO₃ (1.43)	20	24	91	[668]	1961	
TaC	1 p. H₂SO₄ (1.84) + 1 p. HNO₃ (1.43)	150	2	96	[668]	1961	
TaC	1 p. H₂SO₄ (1.84) + 1 p. H₃PO₄ (1.21)	20	24	98	[668]	1961	
TaC	1 p. H₂SO₄ (1.84) + 1 p. H₃PO₄ (1.21)	180	2	lss	[668]	1961	
TaC	1 p. H₂SO₄ (1.84) + 1 p. H₂C₂O₄ (satur.)	20	24	97	[668]	1961	
TaC	1 p. H₂SO₄ (1.84) + 1 p. H₂C₂O₄ (satur.)	—	2	97	[668]	1961	
Ta₂C	HF (1.15) + HNO₃ (1.43)	20	—	cs	[346]	1956	
Cr₃C₂	H₂O	20	100	97	[668]	1961	
Cr₃C₂	HCl (1.19)	Boiling	—	cs	[346]	1956	
Cr₃C₂	H₂SO₄ (1:1)	20	48	100	[669]	1958	[346]
Cr₃C₂	H₂SO₄ (1.84)	280	1	cs	[669]	1958	[346]
Cr₃C₂	H₂SO₄ (1:1)	136	1	65.1	[669]	1958	[346]
Cr₃C₂	H₂SO₄ (1:4)	105	1	95.3	[669]	1958	[346]

226

RESISTANCE OF POWDERS OF REFRACTORY COMPOUNDS TO CHEMICAL REAGENTS (continued)

Phase	Reagent[2]*	Temperature, °C	Duration of treatment, hr	Insoluble residue, %	Ref.	Year	Remarks
Cr_3C_2	1 p. H_2SO_4 (1.84) + 1 p. HNO_3 (1.43)	120	1	83.5	[669]	1958	
Cr_3C_2	H_2SO_4 (1.84) + tartaric acid	134	1	74.13	[669]	1958	
Cr_3C_2	H_2SO_4 (1.84) + CrO_3	120	1	33.2	[669]	1958	
Cr_3C_2	H_2SO_4 (1.84) + Trilon B	126	1	10.36	[669]	1958	
Cr_3C_2	H_3PO_4 (1.21)	20	48	100	[669]	1958	
Cr_3C_2	4 p. H_3PO_4 (1.21) + 1 p. H_2SO_4 (1.84) + 2 p. H_2O	119	1	94.1	[669]	1958	
Cr_3C_2	4 p. H_3PO_4 (1.21) + 1 p. H_2SO_4 (1.84)	20	48	100	[669]	1958	
Cr_3C_2	3 p. HCl (1.19) + 1 p. HNO_3 (1.43)	106	1	90.9	[669]	1958	
Cr_3C_2	3 p. HCl (1.19) + 1 p. HNO_3 (1.43)	20	48	98.7	[669]	1958	
Cr_3C_2	$H_2C_2O_4$ (satur.)	100	1	98.5	[669]	1958	
Cr_3C_2	$H_2C_2O_4$ (satur.)	20	48	100	[669]	1958	
Cr_3C_2	Tartaric acid, 30% solution	100	1	100	[669]	1958	
Cr_3C_2	NaOH (50%)	20	48	100	[669]	1958	
Cr_3C_2	NaOH (30%)	110	1	99.8	[669]	1958	
Cr_3C_2	NaOH(20%) + H_2O_2(30%)	100	1	95.5	[669]	1958	
Cr_3C_2	NaOH + bromine water	106	1	88.1	[669]	1958	
Cr_3C_2	$K_3[Fe(CN)_6]$, alkaline solution	100	1	61.5	[669]	1958	
Cr_3C_2	Ethyl alcohol	20	48	99.0	[669]	1958	
Cr_3C_2	Methyl alcohol	20	48	99.4	[669]	1958	
Cr_3C_2	Toluene	20	48	99.5	[669]	1958	
Cr_3C_2	Benzene	20	48	99.4	[669]	1958	
Cr_3C_2	Dichloroethane	20	48	99.6	[669]	1958	

RESISTANCE OF POWDERS OF REFRACTORY COMPOUNDS TO CHEMICAL REAGENTS (continued)

Phase	Reagent[2]*	Temperature, °C	Duration of treatment, hr	Insoluble residue, %	Ref.	Year	Remarks
Cr_3C_2	Acetone	20	48	99.6	[669]	1958	
Cr_3C_2	Chloroform	20	48	99.8	[669]	1958	
Cr_7C_3	HCl (1.19)	20	48	92.3	[669]	1958	
Cr_7C_3	HCl (1:1)	20	48	99.9	[669]	1958	
Cr_7C_3	HCl (1:1)	110	1	3.49	[669]	1958	
Cr_7C_3	H_2SO_4 (1:1)	20	48	99.8	[669]	1958	
Cr_7C_3	H_2SO_4 (1.84)	265	1	cs	[669]	1958	
Cr_7C_3	H_2SO_4 (1:1)	137	1	1.52	[669]	1958	
Cr_7C_3	H_2SO_4 (1.84) + HNO_3 (1.43)	125	1	90.6	[669]	1958	
Cr_7C_3	H_3PO_4 (1.21)	20	48	100	[669]	1958	
Cr_7C_3	4 p. H_3PO_4 (1.21) + 1 p. H_2SO_4 (1.84) + 2 p. H_2O	127	1	3.93	[669]	1958	
Cr_7C_3	4 p. H_3PO_4 (1.21) + 1 p. H_2SO_4 (1.84) + 2 p. H_2O	20	48	100	[669]	1958	
Cr_7C_3	3 p.HCl (1.19) + 1 p. HNO_3 (1.43)	20	48	94.9	[669]	1958	
Cr_7C_3	3 p. HCl (1.19) + 1 p. HNO_3 (1.43)	106	48	93.8	[669]	1958	
Cr_7C_3	HCl (1.19) + H_2O_2 (30%)	105	1	5.55	[669]	1958	
Cr_7C_3	$H_2C_2O_4$ (satur.)	104	1	95.47	[669]	1958	
Cr_7C_3	$H_2C_2O_4$ (satur.)	20	48	100	[669]	1958	
Cr_7C_3	Tartaric acid	102	1	99.7	[669]	1958	
Cr_7C_3	NaOH (50%)	20	48	100	[669]	1958	
Cr_7C_3	NaOH (30%)	110	1	96.1	[669]	1958	
Cr_7C_3	NaOH' + H_2O_2 (30%)	100	1	96.3	[669]	1958	
Cr_7C_3	NaOH (20%) + bromine water	102	1	85.9	[669]	1958	
Cr_7C_3	$K_3[Fe(CN)_6]$, alkaline solution	100	1	53.2	[669]	1958	

RESISTANCE OF POWDERS OF REFRACTORY COMPOUNDS TO CHEMICAL REAGENTS (continued)

Phase	Reagent[2]*	Temperature, °C	Duration of treatment, hr	Insoluble residue, %	Ref.	Year	Remarks
Cr_7C_3	Ethyl alcohol	20	48	99.6	[669]	1958	
Cr_7C_3	Methyl alcohol	20	48	99.7	[669]	1958	
Cr_7C_3	Toluene	20	48	99.8	[669]	1958	
Cr_7C_3	Benzene	20	48	99.6	[669]	1958	
Cr_7C_3	Dichloroethane	20	48	99.8	[669]	1958	
Cr_7C_3	Acetone	20	48	99.6	[669]	1958	
Cr_7C_3	Chloroform	20	24	99.8	[669]	1958	
Mo_2C	HCl (1.19)	Boiling	2	80	[667]	1958	
Mo_2C	HCl (1.19)	20	24	89	[667]	1958	
Mo_2C	HCl (1:1)	Boiling	2	88	[667]	1958	
Mo_2C	HCl (1:1)	20	24	83	[667]	1958	
Mo_2C	HNO_3 (1.43)	Boiling	2	cs	[667]	1958	
Mo_2C	HNO_3 (1.43)	20	24	cs	[667]	1958	
Mo_2C	HNO_3 (1:1)	Boiling	2	cs	[667]	1958	
Mo_2C	HNO_3 (1:1)	20	24	cs	[667]	1958	
Mo_2C	H_2SO_4 (1.84)	Boiling	2	89	[667]	1958	
Mo_2C	H_2SO_4 (1.84)	20	24	cs	[667]	1958	
Mo_2C	H_2SO_4 (1:4)	Boiling	2	90	[667]	1958	
Mo_2C	H_2SO_4 (1:4)	20	24	83	[667]	1958	
Mo_2C	H_3PO_4 (1.21)	Boiling	2	93	[667]	1958	
Mo_2C	H_3PO_4 (1.21)	20	24	76	[667]	1958	
Mo_2C	H_3PO_4 (1:3)	Boiling	2	92	[667]	1958	
Mo_2C	$HClO_4$ (1.35)	20	24	cs	[667]	1958	
Mo_2C	$HClO_4$ (1.35)	Boiling	2	73	[667]	1958	
Mo_2C	$HClO_4$ (1:3)	20	24	89	[667]	1958	

229

RESISTANCE OF POWDERS OF REFRACTORY COMPOUNDS TO CHEMICAL REAGENTS (continued)

Phase	Reagent² *	Temperature, °C	Duration of treatment, hr	Insoluble residue, %	Ref.	Year	Remarks
Mo_2C	$HClO_4$ (1:3)	Boiling	2	58	[667]	1958	
Mo_2C	$H_2C_2O_4$ (satur.)	20	24	89	[667]	1958	
Mo_2C	$H_2C_2O_4$ (satur.)	Boiling	2	90	[667]	1958	
Mo_2C	3 p. HCl (1.19) + 1 p. HNO_3 (1.43)	20	24	cs	[667]	1958	
Mo_2C	3 p. HCl (1.19) + 1 p. HNO_3 (1.43)	Boiling	2	cs	[667]	1958	
Mo_2C	2 p. H_2SO_4 (1.84) + 1 p. HNO_3 (1.43)	20	24	1	[667]	1958	
Mo_2C	4 p. HNO_3 (1.4) +1p. HF (1.15)	Boiling	2	cs	[667]	1958	
Mo_2C	H_2SO_4 (1.84) + H_3PO_4 (1.21)	20	24	cs	[667]	1958	
Mo_2C	H_2SO_4 (1.84) + H_3PO_4 (1.21)	Boiling	24	90	[667]	1958	
Mo_2C	H_2SO_4 (1.84) + H_3PO_4 (1:3)	20	2	cs	[667]	1958	
Mo_2C	H_2SO_4 (1.84) + H_3PO_4 (1:3)	Boiling	24	92	[667]	1958	
Mo_2C	H_2SO_4 (1.84) + $H_2C_2O_4$ (satur.)	20	2	88	[667]	1958	
Mo_2C	H_2SO_4 (1.84) + $H_2C_2O_4$ (satur.)	Boiling	24	80	[667]	1958	
Mo_2C	H_2SO_4 (1:4) + $H_2C_2O_4$ (satur.)	20	2	73	[667]	1958	
Mo_2C	H_2SO_4 (1:4) + $H_2C_2O_4$ (satur.)	Boiling	24	89	[667]	1958	
Mo_2C	NaOH (20%)	20	2	88	[667]	1958	
Mo_2C	NaOH (20%)	Boiling	24	90	[667]	1958	
Mo_2C	NaOH (10%)	20	24	90	[667]	1958	
Mo_2C	NaOH (10%)	Boiling	2	94	[667]	1958	
Mo_2C	40 ml NaOH (20%) + 10 ml bromine water	20	24	65	[667]	1958	

230

RESISTANCE OF POWDERS OF REFRACTORY COMPOUNDS TO CHEMICAL REAGENTS (continued)

Phase	Reagent²*	Temperature, °C	Duration of treatment, hr	Insoluble residue, %	Ref.	Year	Remarks
Mo₂C	40 ml NaOH (20%) + 10 ml bromine water	Boiling	2	60	[667]	1958	
Mo₂C	40 ml NaOH (20%) + 10 ml H₂O₂ (30%)	20	24	31	[667]	1958	
Mo₂C	40 ml NaOH (20%) + 10 ml H₂O₂ (30%)	Boiling	2	36	[667]	1958	
Mo₂C	10 ml K₃[Fe(CN)₆] + 40 ml NaOH (20%)	20	24	69	[667]	1958	
Mo₂C	10 ml K₃[Fe(CN)₆] + 40 ml NaOH (20%)	Boiling	2	69	[667]	1958	
Mo₂C	HNO₃ (1.43) + HF (1.15)	"	—	cs	[346]	1956	
WC	HCl (1.19)	20	24	97	[667]	1958	
WC	HCl (1.19)	Boiling	2	48	[667]	1958	
WC	HCl (1:1)	20	24	96	[667]	1958	
WC	HCl (1:1)	Boiling	2	92	[667]	1958	
WC	HNO₃ (1.43)	20	24	63	[667]	1958	
WC	HNO₃ (1.43)	Boiling	2	1	[667]	1958	
WC	HNO₃ (1:1)	20	24	72	[667]	1958	
WC	HNO₃ (1:1)	Boiling	2	10	[667]	1958	
WC	H₂SO₄ (1.84)	20	24	91	[667]	1958	
WC	H₂SO₄ (1.84)	Boiling	2	1	[667]	1958	
WC	H₂SO₄ (1 : 4)	20	24	96	[667]	1958	
WC	H₂SO₄ (1 : 4)	Boiling	2	95	[667]	1958	
WC	H₂PO₄ (1.21)	20	24	91	[667]	1958	
WC	H₂PO₄ (1.21)	Boiling	2	93	[663]	1958	

RESISTANCE OF POWDERS OF REFRACTORY COMPOUNDS TO CHEMICAL REAGENTS (continued)

Phase	Reagent²*	Temperature, °C	Duration of treatment, hr	Insoluble residue, %	Ref.	Year	Remarks
WC	H_2PO_4 (1:3)	20	24	96	[667]	1958	
WC	H_2PO_4 (1:3)	Boiling	2	90	[667]	1958	
WC	$HClO_4$ (1.35)	20	24	98	[667]	1958	
WC	$HClO_4$ (1.35)	Boiling	2	40	[667]	1958	
WC	$HClO_4$ (1:3)	20	24	98	[667]	1958	
WC	$HClO_4$ (1:3)	Boiling	2	93	[667]	1958	
WC	$H_2C_2O_4$ (satur.)	20	24	95	[667]	1958	
WC	$H_2C_2O_4$ (satur.)	Boiling	2	95	[667]	1958	
WC	3 p. HCl (1.19) + 1 p. HNO_3 (1.43)	20	24	28	[667]	1958	
WC	3 p. HCl (1.19) + 1 p. HNO_3 (1.43)	Boiling	2	3	[667]	1958	
WC	2 p. H_2SO_4 (1.84) + 1 p. HNO_3 (1.43)	20	24	92	[667]	1958	
WC	4 p. HNO_3 (1.4) + 1 p. HF (1.15)	Boiling	2	42	[667]	1958	
WC	H_2SO_4 (1.84) + H_3PO_4 (1:4)	20	24	60	[667]	1958	
WC	H_2SO_4 (1.84) + H_3PO_4 (1:4)	Boiling	2	96	[667]	1958	
WC	H_2SO_4 (1.84) + H_3PO_4 (1:3)	20	24	cs	[667]	1958	
WC	H_2SO_4 (1.84) + H_3PO_4 (1:3)	Boiling	2	96	[667]	1958	
WC	H_2SO_4 (1.84) + $H_2C_2O_4$ (satur.)	20	24	93	[667]	1958	
WC	H_2SO_4 (1.84) + $H_2C_2O_4$ (satur.)	Boiling	2	95	[667]	1958	
WC	H_2SO_4 (1:4) + $H_2C_2O_4$ (satur.)	20	24	70	[667]	1958	
WC	H_2SO_4 (1:4) + $H_2C_2O_4$ (satur.)	Boiling	2	94	[667]	1958	
WC	NaOH (20%)	20	24	95	[667]	1958	
WC	NaOH (20%)	Boiling	2	97	[667]	1958	
WC				98	[667]	1958	

RESISTANCE OF POWDERS OF REFRACTORY COMPOUNDS TO CHEMICAL REAGENTS (continued)

Phase	Reagent[2]*	Temperature, °C	Duration of treatment, hr	Insoluble residue, %	Ref.	Year	Remarks
WC	NaOH (10%)	20	24	98	[667]	1958	
WC	NaOH (10%)	Boiling	2	98	[667]	1958	
WC	40 ml NaOH (20%) + 10 ml bromine water	20	24	70	[667]	1958	
WC	40 ml NaOH (20%) + 10 ml bromine water	Boiling	2	60	[667]	1958	
WC	40 ml NaOH (20%) + 10 ml 30% H_2O_2	20	24	88	[667]	1958	
WC	40 ml NaOH (20%) + 10 ml 30% H_2O_2	Boiling	2	87	[667]	1958	
WC	10 ml $K_3[Fe(CN)_6]$ (10%) + 40 ml NaOH (20%)	20	24	68	[667]	1958	
WC	10 ml $K_3[Fe(CN)_6]$ (10%) + 40 ml NaOH (20%)	Boiling	2	58	[667]	1958	
Mn_3C, Mn_4C	HCl (1.19)	"	—	i	[943]	1929	
Mn_5C_2	HCl (1.19)	"	—	i	[943]	1929	
Mn_7C_3	HNO_3 (1.43)	20	—	i	[943]	1929	[346]
$Mn_{23}C_6$	HNO_3 (1.4)	Boiling	—	i	[943]	1929	
$Mn_{23}C_6$	H_2SO_4 (1.84)	20	—	i	[943]	1929	
$Mn_{23}C_6$	H_2SO_4 (1.84)	Boiling	—	i	[943]	1929	
$Mn_{23}C_6$	HF (1.15) + HNO_3 (1.43)	20	—	cs	[943]	1929	
Mn_3C	Concentrated inorganic acids	20	—	cs	[346]	1956	
Fe_2C	HCl (1.19)	Boiling	—	i	[1088]	1932	
Fe_2C	HNO_3 (1.43)	"	—	cs	[1088]	1932	

RESISTANCE OF POWDERS OF REFRACTORY COMPOUNDS TO CHEMICAL REAGENTS (continued)

Phase	Reagent[2]*	Temperature, °C	Duration of treatment, hr	Insoluble residue, %	Ref.	Year	Remarks
Fe₂C	HCl (1.19)	Boiling	—	i	[1088]	1932	
Fe₃C	HNO₃ (1.43)	"	—	cs	[1088]	1932	
Fe₃C	Dilute inorganic acids	20	—	cs	[346]	1956	
Co₂C	HCl (1.19)	Boiling	—	i	[1089]	1932	
Co₂C	HNO₃ (1.43)	"	—	i	[1089]	1932	
Co₂C	H₂SO₄ (1.84)	20	—	i	[1089]	1932	
Co₂C	3 p. HCl (1.19) + 1 p. HNO₃ (1.43)	20	—	cs	[1089]	1932	
Co₂C	HF (1.15) + HNO₃ (1.43)	20	—	cs	[1089]	1932	
Co₂C	HCl (1.19) + H₂SO₄ (1.84)	20	—	cs	[1089]	1932	
Co₃C	HCl (1.19)	Boiling	—	i	[1089]	1932	
Co₃C	HNO₃ (1.43)	"	—	i	[1089]	1932	
Co₃C	H₂SO₄ (1.84)	20	—	i	[1089]	1932	
Co₃C	3 p. HCl (1.19) + 1 p. HNO₃ (1.43)	20	—	cs	[1089]	1932	
Co₃C	HF + HNO₃ (1.43)	20	—	cs	[1089]	1932	
Co₃C	HCl (1.19) + H₂SO₄ (1.84)	20	—	cs	[1089]	1932	
Mg₃N₂	H₂O	100	—	i	[697]	1959	
Mg₃N₂	HCl (1.19)	Boiling	—	cs	[697]	1959	
Mg₃N₂	HCl (1.19)	Boiling	—	cs	[792]	1959	
Mg₃N₂	HNO₃ (1.4)	20	—	cs	[792]	1959	
Mg₃N₂	C₂H₅OH	20	—	i	[697]	1959	
Ca₃N₂	HCl (1:1)	20	—	cs	[697]	1959	
Ca₃N₂	H₂SO₄ (:1)	20	—	cs	[697]	1959	
Ca₃N₂	C₂O₅OH	20	—	i	[697]	1959	
SrN, Sr₂N	H₂O	20	—	cs	[697]	1959	

RESISTANCE OF POWDERS OF REFRACTORY COMPOUNDS TO CHEMICAL REAGENTS (continued)

Phase	Reagent[2]*	Temperature, °C	Duration of treatment, hr	Insoluble residue, %	Ref.	Year	Remarks
Sr_3N_2	H_2O	20	—	ps	[697]	1959	$Ba(OH)_2$ precipitated
Ba_3N_2	H_2O	20	—	i	[611]	1959	
InN	H_2O	100	—	i	[611]	1959	
InN	HCl	20	—	i	[611]	1959	
InN	HNO_3	20	—	i	[611]	1959	
InN	H_2SO_4	20	—	i	[611]	1959	
InN	HF	20	—	i	[611]	1959	
InN	Na_2CO_3 (30%)	80	—	cs	[611]	1959	[231]
AlN	Na_2CO_3 (30%)	80	—	cs	[611]	1959	
AlN	Na_2CO_3	100	—	i	[611]	1959	
AlN	HCl (1.19)	20	—	i	[611]	1959	
AlN	HNO_3 (1.43)	20	—	i	[611]	1959	
AlN	H_2SO_4 (1.84)	20	—	i	[611]	1959	
AlN	HF (1.15)	20	—|	i	[611]	1959	
AlN	NaOH (25%)	20	—	i	[611]	1959	
AlN	Dry air	700	—	Increase in weight	[611]	1959	Slowly decomposes
AlN	Moist air	700	—	0.2	[611]	1959	
AlN	Dry air	1000	—	0.7	[611]	1959	Converted to Al_2O_3
AlN	Moist air	1000	—	2.7	[611]	1959	
AlN	Dry air	1200	—	i	[611]	1959	
ScN	H_2O	20	—	i	[697]	1959	
ScN	H_2O	100	—	csh	[697]	1959	
ScN	HCl (1.19)	20	—	cs	[697]	1959	

RESISTANCE OF POWDERS OF REFRACTORY COMPOUNDS TO CHEMICAL REAGENTS (continued)

Phase	Reagent[2]*	Temperature, °C	Duration of treatment, hr	Insoluble residue, %	Ref.	Year	Remarks
ScN	HNO_3 (1.43)	20	—	cs	[697]	1959	
GaN	H_2SO_4 (1.84)	20	—	cs	[697]	1959	
GaN	NaOH (20%)	80	—	cs	[611]	1959	
LaN	H_2O	80	—	csh	[697]	1959	
NdN	H_2O	80	—	cs	[467]	1952	
PrN	H_2O	80	—	cs	[467]	1952	Dissolves with formation of ThO_2
Th_3N_4	H_2O	20	—	cs	[467]	1952	
UN	H_2O	20	—	cs	[467]	1952	
UN	H_3PO_4 (1.21)	20	—	lss	[467]	1952	
U_2N_3 } UN_2 }	HCl (1.19)	20	—	i	[467]	1952	
UN_2	H_2SO_4 (1.82)	20	—	i	[467]	1952	
U_2N_3	HNO_3 (1.43)	20	—	ps	[697]	1959	
U_2N_3	NaOH (25%)	20	—	i	[697]	1959	
NpN	H_2O	20	—	i	[697]	1959	
NpN	HCl (1:1)	20	—	cs	[697]	1959	
NpN	HCl (1:1)	20	—	cs	[697]	1959	
TiN	HCl (1:1)	Boiling	24	99	[670]	1958	
TiN	HCl (1:1)	20	2	98	[670]	1958	
TiN	HCl (1.19)	Boiling	24	89	[670]	1958	
TiN	HCl (1.19)	20	2	98	[670]	1958	
TiN	H_2SO_4 (1:4)	Boiling	24	98	[670]	1958	
TiN	H_2SO_4 (1:4)	Boiling	2	95	[670]	1958	

RESISTANCE OF POWDERS OF REFRACTORY COMPOUNDS TO CHEMICAL REAGENTS (continued)

Phase	Reagent²*	Temperature, °C	Duration of treatment, hr	Insoluble residue, %	Ref.	Year	Remarks
TiN	H_2SO_4 (1.84)	20	24	97	[670]	1958	
TiN	H_2SO_4 (1.84)	Boiling	2	24	[670]	1958	
TiN	HNO_3 (1:1)	20	24	11	[670]	1958	
TiN	HNO_3 (1:1)	Boiling	2	5	[670]	1958	
TiN	HNO_3 (1.43)	20	24	10	[670]	1958	
TiN	HNO_3 (1.43)	Boiling	2	ps	[670]	1958	
TiN	$HClO_4$ (1:3)	20	24	98	[670]	1958	
TiN	$HClO_4$ (1:3)	Boiling	2	94	[670]	1958	
TiN	$HClO_4$ (1.35)	20	24	99	[670]	1958	
TiN	$HClO_4$ (1.35)	Boiling	2	ps	[670]	1958	
TiN	$HClO_4$ (1:4)	20	24	97	[670]	1958	
TiN	H_3PO_4 (30%)	Boiling	2	2	[670]	1958	
TiN	NaOH (1%)	"	2	ps	[670]	1958	
TiN	NaOH (10%)	"	2	ps	[670]	1958	
TiN	NaOH (40%)	"	2	ps	[670]	1958	
TiN	$NaOH + H_2O_2$ (1%)	"	2	9	[670]	1958	
TiN	$NaOH + H_2O_2$ (10%)	"	2	16	[670]	1958	
TiN	$NaOH + H_2O_2$ (40%)	"	2	43	[670]	1958	
GeN	H_2O	80	—	i	[611]	1959	
GeN	HCl (1.19)	100	—	i	[611]	1959	
GeN	HCl (1:1)	100	—	i	[611]	1959	
GeN	HNO_3 (1.43)	100	—	i	[611]	1959	
GeN	H_2SO_4 (1.84)	100	—	i	[611]	1959	
GeN	HF (1.15)	100	—	i	[611]	1959	
GeN	Na_2CO_3 (30%)	80	—	cs	[611]	1959	

RESISTANCE OF POWDERS OF REFRACTORY COMPOUNDS TO CHEMICAL REAGENTS (continued)

Phase	Reagent 2*	Temperature, °C	Duration of treatment, hr	Insoluble residue, %	Ref.	Year	Remarks
ZrN	HNO_3 (1.43)	20	24	98	[670]	1958	
ZrN	HNO_3 (1.43)	Boiling	2	84	[670]	1958	
ZrN	$HClO_4$ (1 : 3)	20	24	100	[670]	1958	
ZrN	$HClO_4$ (1 : 3)	Boiling	2	98	[670]	1958	
ZrN	$HClO_4$ (1.35)	20	24	99	[670]	1958	
ZrN	$HClO_4$ (1.35)	Boiling	2	98	[670]	1958	
ZrN	H_3PO_4 (1 : 4)	Boiling	24	1ss	[670]	1958	
ZrN	H_2O_2 (30%)	"	2	100	[670]	1958	
ZrN	NaOH (1%)	"	2	100	[670]	1958	
ZrN	NaOH (10%)	"	2	100	[670]	1958	
ZrN	NaOH (40%)	"	2	42	[670]	1958	
ZrN	$NaOH + H_2O_2$ (1%)	"	2	99	[670]	1958	
ZrN	$NaOH + H_2O_2$ (10%)	"	2	87	[670]	1958	
ZrN	$NaOH + H_2O_2$ (40%)	"	2	48	[670]	1958	
ZrN	3 p. HCl (1.19) + 1 p. HNO_3 (1.43)	20	24	82	[670]	1958	
ZrN	3 p. HCl (1.19) + 1 p. HNO_3 (1.43)	Boiling	2	25	[670]	1958	
ZrN	1 p. $HClO_4$ (1.35) + 1 p. HCl (1.19)	20	24	76	[670]	1958	
ZrN	1 p. $HClO_4$ (1.35) + 1 p. HCl (1.19)	Boiling	2	12	[670]	1958	
ZrN	1 p. HNO_3 (1.43) + 1 p. H_2O_2	20	24	94	[670]	1958	
ZrN	1 p. HNO_3 (1.43) + 1 p. H_2O_2	Boiling	24	65	[670]	1958	
ZrN	HNO_3 (1.43) + HF (1.15)	"	5min	cs	[670]	1958	
ZrN	3 p. $H_2C_2O_4$ (satur.) + 1 p. H_2SO_4 (1.84)	20	24	90	[670]	1958	
ZrN	2 p. H_2SO_4 + 1 p. H_2O_2	20	24	25	[670]	1958	

RESISTANCE OF POWDERS OF REFRACTORY COMPOUNDS TO CHEMICAL REAGENTS (continued)

Phase	Reagent²*	Temperature, °C	Duration of treatment, hr	Insoluble residue, %	Ref.	Year	Remarks
ZrN	1p. H_2SO_4 (1.84) + 1p. HNO_3 (1.43) + 4 p. H_2O	20	24	81	[670]	1958	
ZrN	10 p. K_2SO_4 + 10 ml H_2SO_4 (1.84)	Boiling	2	cs	[670]	1958	
HfN	HCl (1.19)	20	—	i	[697]	1959	
VN	HCl (1.19)	20	—	i	[697]	1959	
VN	H_2SO_4 (1.82)	Boiling	—	cs	[697]	1959	
VN	HNO_3 (1.4)	"	—	cs	[697]	1959	
NbN	HCl (1:3)	20	24	99	[670]	1958	
NbN	HCl (1:3)	Boiling	2	94	[670]	1958	
NbN	HCl (1.19)	20	24	100	[670]	1958	
NbN	HCl (1.19)	Boiling	2	99	[670]	1958	
NbN	H_2SO_4 (1:4)	20	24	99	[670]	1958	
NbN	H_2SO_4 (1:4)	Boiling	2	84	[670]	1958	
NbN	H_2SO_4 (1.84)	20	24	100	[670]	1958	
NbN	H_2SO_4 (1.84)	Boiling	2	0	[670]	1958	
NbN	HNO_3 (1:1)	20	24	98	[670]	1958	
NbN	HNO_3 (1:1)	Boiling	2	100	[670]	1958	
NbN	HNO_3 (1.43)	20	24	100	[670]	1958	
NbN	HNO_3 (1.43)	Boiling	2	100	[670]	1958	
NbN	$HClO_4$ (1:3)	20	24	100	[670]	1958	
NbN	$HClO_4$ (1:3)	Boiling	2	98	[670]	1958	
NbN	$HClO_4$ (1.35)	20	24	100	[670]	1958	
NbN	$HClO_4$ (1.35)	Boiling	2	16	[670]	1958	
NbN	H_2O_2 (30%)	"	2	96	[670]	1958	
NbN	NaOH (1%)	"	2		[670]	1958	

RESISTANCE OF POWDERS OF REFRACTORY COMPOUNDS TO CHEMICAL REAGENTS (continued)

Phase	Reagent[2*]	Temperature, °C	Duration of treatment, hr	Insoluble residue, %	Ref.	Year	Remarks
NbN	NaOH (10%)	Boiling	2	87	[670]	1958	
NbN	NaOH (40%)	"	2	87	[670]	1958	
NbN	NaOH (1%) + H_2O_2	"	2	17	[670]	1958	
NbN	NaOH (10%) + H_2O_2	"	2	cs	[670]	1958	
NbN	NaOH (40%) + H_2O_2	20	24	cs	[670]	1958	
NbN	3 p. HCl (1.19) + 1 p. HNO_3 (1.43)	Boiling	2	99	[670]	1958	
NbN	3 p. HCl (1.19) + 1 p. HNO_3 (1.43)	20	24	99	[670]	1958	
NbN	1 p. $HClO_4$ (1.35) + 1 p. HCl (1.19)	Boiling	2	98	[670]	1958	
NbN	1 p. $HClO_4$ (1.35) + 1 p. HCl (1.19)	20	24	95	[670]	1958	
NbN	1 p. HNO_3 (1.4) + 1 p. H_2O	Boiling	2	26	[670]	1958	
NbN	1 p. HNO_3 (1.43) + 1 p. H_2O	Boiling	5 min	15	[670]	1958	
NbN	HNO_3 (1.43) + HF (1.15)	"	2	cs	[670]	1958	
NbN	10g K_2SO_4 + 10 ml H_2SO_4 (1.84)	20	24	cs	[670]	1958	
TaN	HCl (1:1)	Boiling	2	98	[670]	1958	
TaN	HCl (1:1)	20	24	99	[670]	1958	
TaN	HCl (1.19)	Boiling	2	99	[670]	1958	
TaN	HCl (1.19)	20	24	98	[670]	1958	
TaN	H_2SO_4 (1:4)	Boiling	2	100	[670]	1958	
TaN	H_2SO_4 (1:4)	20	24	100	[670]	1958	
TaN	H_2SO_4 (1.84)	Boiling	2	100	[670]	1958	
TaN	H_2SO_4 (1.84)	20	24	77	[670]	1958	
TaN	HNO_3 (1:1)	Boiling	2	99	[670]	1958	
TaN	HNO_3 (1:1)	20	24	98	[670]	1958	
TaN	HNO_3 (1.43)	Boiling	2	98	[670]	1958	
TaN	HNO_3 (1.43)	20	24	98	[670]	1958	

RESISTANCE OF POWDERS OF REFRACTORY COMPOUNDS TO CHEMICAL REAGENTS (continued)

Phase	Reagent²*	Temperature, °C	Duration of treatment, hr	Insoluble residue, %	Ref.	Year	Remarks
TaN	$HClO_4$ (1:3)	20	24	100	[670]	1958	
TaN	$HClO_4$ (1.35)	20	24	100	[670]	1958	
TaN	$HClO_4$ (1.35)	Boiling	2	98	[670]	1958	
TaN	H_3PO_4 (1:4)	20	24	96	[670]	1958	
TaN	H_2O_2 (1:4)	Boiling	2	100	[670]	1958	
TaN	H_2O_2 (30%)	"	2	41	[670]	1958	
TaN	NaOH (1%)	"	2	93	[670]	1958	
TaN	NaOH (10%)	"	2	ps	[670]	1958	
TaN	NaOH (40%)	"	2	ps	[670]	1958	
TaN	NaOH (1%) + H_2O_2	"	2	84	[670]	1958	
TaN	NaOH (10%) + H_2O_2	"	2	39	[670]	1958	
TaN	NaOH (40%) + H_2O_2	"	2	lss	[670]	1958	
TaN	3 p. HCl (1.19) + 1 p. HNO_3 (1.43)	"	2	100	[670]	1958	
TaN	HNO_3 (1.43) + HF (1.15)	"	5 min	0	[670]	1958	
TaN	2 p. H_2SO_4 (1.82) + 1 p. H_2O_2	"	2	93	[670]	1958	
TaN	10 g K_2SO_4 + 10 ml H_2SO_4 (1.84)	"	—	—	[670]	1958	Specimen dissolves completely in 5-6 hr
Cr$_2$N	HCl (1.19)	"	—	cs	[697]	1959	
CrN	HCl (1.19)	20	—	i	[697]	1959	
CrN	3 p, HCl (1.19) + 1 p. HNO_3 (1.43)	Boiling	—	cs	[697]	1959	
CrN	NaOH (20%)	20	—	i	[697]	1959	
W$_2$N	H_2O	20	—	cs	[697]	1959	Dissolves with liberation of NH_3

RESISTANCE OF POWDERS OF REFRACTORY COMPOUNDS TO CHEMICAL REAGENTS (continued)

Phase	Reagent²*	Temperature, °C	Duration of treatment, hr	Insoluble residue, %	Ref.	Year	Remarks
W_2N_3	HCl (1.19)	Boiling	—	i	[697]	1959	
Mn_3N_2	H_2O	"	—	i	[697]	1959	
Mn_3N_2	HCl (1.19)	"	—	i	[697]	1959	
Mn_3N_2	HNO_3 (1.4)	"	—	i	[697]	1959	
Mn_3N_2	H_2SO_4 (1.82)	"	—	i	[697]	1959	
Mn_3N_2	3 p. HCl (1.19) + 1 p. HNO_3 (1.43)	"	—	cs	[697]	1959	
Mn_5N_2	HCl (1.19)	"	—	cs	[467]	1952	
Mn_5N_2	HNO_3 (1:1)	"	—	cs	[467]	1952	
Mn_5N_2	H_2SO_4 (1.82)	"	—	cs	[467]	1952	
Mn_5N_2	3 p. HCl (1.19) + 1 p.HNO_3	"	—	cs	[467]	1952	
Mn_2N	HCl (1.19)	"	—	i	[943]	1929	
Mn_2N	HNO_3 (1.43)	"	—	i	[943]	1929	
Mn_2N	H_2SO_4 (1.82)	"	—	i	[943]	1929	
Mn_2N	HF (1.15) +HNO_3 (1.43)	"	—	cs	[943]	1929	Dissolves with liberation of NH_3
Mn_2N	HCl (1.19) + H_2SO_4 (1.82)	"	—	cs	[943]	1929	
Mn_2N	NaOH (25%)	"	—	cs	[943]	1929	
FeN	H_2O	100	—	cs	[697]	1959	
Fe_2N	HCl (1.19)	20	—	cs	[697]	1959	
Fe_3N_2	HNO_3 (1.4)	20	—	cs	[697]	1959	
Fe_3N	H_2SO_4 (1.82)	20	—	cs	[697]	1959	
Co_3N	HCl (1:1)	Boiling	—	cs	[1089]	1932	
Co_2N	HCl (1:1)	20	—	cs	[1089]	1932	
Co_3N	H_2SO_4 (1.82)	20	—	cs	[1089]	1932	
Co_3N	HCl (1.19)	20	—	cs	[1089]	1932	

RESISTANCE OF POWDERS OF REFRACTORY COMPOUNDS TO CHEMICAL REAGENTS (continued)

Phase	Reagent²*	Temperature, °C	Duration of treatment, hr	Insoluble residue, %	Ref.	Year	Remarks
Co_2N	HNO_3 (1.4)	20	—	cs	[1089]	1932	Dissolves slowly
Co_3N	HCl (1:1)	20	—	cs	[1089]	1932	Dissolves slowly
Co_3N	HCl (1:1)	Boiling	—	cs	[1089]	1932	Dissolves slowly
Co_3N	HNO_3 (1:1)	20	—	cs	[1089]	1932	
Co_3N	HNO_3 (1:1)	Boiling	—	cs	[1089]	1932	Dissolves slowly
Co_3N	HCl (1.19)	20	—	cs	[1089]	1932	
Co_3N	HNO_3 (1.43)	20	—	cs	[1089]	1932	
Co_3N	H_2SO_4 (1.82)	20	—	cs	[1089]	1932	
Co_3N	H_2SO_4 (1.82)	Boiling	—	cs	[1089]	1932	
Ni_3N	HCl (1:1)	20	—	cs	[943]	1929	
Ni_3N	HNO_3 (1:1)	20	—	cs	[943]	1929	
Ni_3N	H_2SO_4 (1:1)	20	—	cs	[943]	1929	
Ni_3N	NaOH (25%)	20	—	i	[943]	1929	
$MgSi_2$	HCl (1:1)	20	—	cs	[117]	1959	
CaSi, $CaSi_2$	H_2SO_4 (1:1)	20	—	cs	[117]	1959	
CaSi, $CaSi_2$	HCl (1.19)	20	—	cs	[117]	1959	} Dissolves very slowly
CaSi, $CaSi_2$	HNO_3 (1.43)	20	—	cs	[117]	1959	}
CaSi, $CaSi_2$	H_2SO_4 (1.82)	20	—	cs	[117]	1959	
$SrSi_2$ } $BaSi_2$ }	H_2O	Boiling	—	cs	[117]	1959	Readily decomposes with liberation of H_2

243

RESISTANCE OF POWDERS OF REFRACTORY COMPOUNDS TO CHEMICAL REAGENTS (continued)

Phase	Reagent²*	Temperature, °C	Duration of treatment, hr	Insoluble residue, %	Ref.	Year	Remarks
BaSi₂	HCl (1:1)	20	—	cs	[1]	1957	Dissolves rapidly
BaSi₂	HCl (1.19)	20	—	cs	[1]	1957	Dissolves slowly
LaSi₂ and silicides of other rare earth metals	HCl (1.19)	20	—	cs	[117]	1959	
DySi₂	H₂O	Boiling	8—16	i	[285]	1956	
				Loss of weight, mg/cm²			
ThSi	Moist air	20	—	i			
ThSi₂	HCl (1.19)	20	—	73	[1]	1957	
ThSi₂	HJ(1.47)	20	—	lss	[102]	1957	
ThSi₂	HF (1.15)	20	—	lss	[102]	1956	
ThSi₂	3 p. HCl (1.19) + 1 p. HNO₃ (1.43)	20	—	lss	[102]	1956	
ThSi, ThSi₂	NaOH (20%)	20	—	lss	[102]	1956	
ThSi, ThSi₂	H₂O₂	20	—	i	[102]	1956	
Th₃Si₂	HCl (1.19)	20	—	i	[102]	1956	
ThSi₂	HNO₃ (1.43)	20	—	i	[251]	1955	
ThSi₂	H₂SO₄ (1.82)	20	—	i	[251]	1955	

RESISTANCE OF POWDERS OF REFRACTORY COMPOUNDS TO CHEMICAL REAGENTS (continued)

Phase	Reagent[2]*	Temperature, °C	Duration of treatment, hr	Insoluble residue, %	Ref.	Year	Remarks
$ThSi_2$	H_2SO_4 (1:1)	20	—	lss / Loss of weight, mg/cm²	[251]	1955	
U_3Si	H_2O	260	720	0.05—1	[641]	1958	$p = 46$ atm
U_3Si	H_2O	340	720	0.05—1	[641]	1958	$p = 150$ atm
$NpSi_2$	H_2O	20	—	i	[251]	1955	
$NpSi_2$	H_2O	Boiling	—	lss	[251]	1955	
$TiSi_2$	HCl (1.19)	20	—	lss	[251]	1955	
$TiSi_2$	HCl (1.19)	Boiling	2	99.7	[67]	1958	
$TiSi_2$	HCl (1.19)	"	1	i	[67]	1958	
$TiSi_2$	HCl (1:1)	"	2	99.8	[67]	1958	
$TiSi_2$	HCl (1:1)	"	1	i	[67]	1958	
$TiSi_2$	H_2SO_4 (1.84)	"	2	99.6	[67]	1958	
$TiSi_2$	H_2SO_4 (1.84)	"	3.5	i	[67]	1958	
$TiSi_2$	H_2SO_4 (1:1)	"	2	99.6	[67]	1958	
$TiSi_2$	H_2SO_4 (1:1)	"	3.5	i	[67]	1958	
$TiSi_2$	H_2SO_4 (1:10)	"	2	99.8	[67]	1958	
$TiSi_2$	H_2SO_4 (1:10)	"	3.5	i	[67]	1958	
$TiSi_2$	H_3PO_4 (1.21)	"	2	99.7	[67]	1958	
$TiSi_2$	HF (1.15)	"	2.5	lss	[67]	1958	
$TiSi_2$	$KHSO_4$	"	1	i	[67]	1958	
$TiSi_2$	HF (1.15) $2HNO_3$ (1.43)	"	2	cs	[67]	1958	
$TiSi_2$	HCl (1.19) + HNO_3 (1.43)	"	2	99.5	[67]	1958	

RESISTANCE OF POWDERS OF REFRACTORY COMPOUNDS TO CHEMICAL REAGENTS (continued)

Phase	Reagent[2*]	Temperature, °C	Duration of treatment, hr	Insoluble residue, %	Ref.	Year	Remarks
TiSi$_2$	4 p. H$_3$PO$_4$ (1.21) + 1 p. H$_2$SO$_4$ (1.82) + 2 p. H$_2$O	Boiling	2	lss	[671]	1958	Insoluble residue appears when solution is evaporated until SO$_3$ fumes begin to appear
TiSi$_2$	H$_2$C$_2$O$_4$ (satur.) + H$_2$O	"	2	86.4	[671]	1958	
TiSi$_2$	1 p. H$_2$C$_2$O$_4$ + 2 p. H$_2$SO$_4$ (1.82)	"	2	85.5	[671]	1958	
TiSi$_2$	NaOH (1%)	"	2	i	[671]	1958	
TiSi$_2$	NaOH (20%)	"	30 min	cs	[671]	1958	
TiSi$_2$	Na$_2$O$_2$ (30%)	"	15 min	cs	[671]	1958	
ZrSi$_2$	HCl (1.19)	"	1	i	[671]	1958	
ZrSi$_2$	HCl (1:1)	"	1	i	[671]	1958	
ZrSi$_2$	H$_2$SO$_4$ (1.84)	"	3	i	[671]	1958	
ZrSi$_2$	H$_2$SO$_4$ (1:1)	"	3	i	[671]	1958	
ZrSi$_2$	H$_2$SO$_4$ (1:10)	"	3	i	[671]	1958	
ZrSi$_2$	H$_3$PO$_4$ (1.21)	"	2	99.9	[671]	1958	
ZrSi$_2$	HF (1:15)	"	2	lss	[671]	1958	
ZrSi$_2$	KHSO$_4$	"	2	i	[671]	1958	
ZrSi$_2$	HF (1.15) + HNO$_3$ (1.43)	"	15 min	cs	[671]	1958	
ZrSi$_2$	KHSO$_4$ + KHF$_2$	"		cs	[671]	1958	
ZrSi$_2$	KHF + H$_2$SO$_4$ (1.84)	"	4	cs	[671]	1958	
ZrSi$_2$	KHSO$_4$ + H$_2$SO$_4$ + SiOCl$_2$	"	7	i	[671]	1958	

RESISTANCE OF POWDERS OF REFRACTORY COMPOUNDS TO CHEMICAL REAGENTS (continued)

Phase	Reagent2*	Temperature, °C	Duration of treatment, hr	Insoluble residue, %	Ref.	Year	Remarks
$ZrSi_2$	NaOH (1%)	Boiling	—	i	[671]	1958	
$ZrSi_2$	NaOH (20%)	"	30min	cs	[671]	1958	
$ZrSi_2$	Na_2O_2 (30%)	"	15min	cs	[671]	1958	
VSi_2	HCl (1 : 1)	"	1	i	[671]	1958	
VSi_2	HCl (1 19)	"	1	i	[671]	1958	
VSi_2	H_2SO_4 (1.84)	"	3	i	[671]	1958	
VSi_2	H_2SO_4 (1 : 1)	"	3	i	[671]	1958	
VSi_2	H_2SO_4 (1 : 10)	"	3	i	[671]	1958	
VSi_2	H_3PO_4 (1.21)	"	2	99.5	[671]	1958	
VSi_2	HF (1.15)	"	3	iss	[671]	1958	
VSi_2	HF (1.15) + HNO_3 (1.43)	"	2	cs	[671]	1958	
VSi_2	4 p.H_3PO_4 + 1 p.H_2SO_4 + 2 p.H_2O		—	cs	[671]	1958	Insoluble residue appears when solution is evaporated until SO_3 fumes begin to appear
VSi_2	$KHSO_4$ + KHF_2	"	1.5min	cs	[671]	1958	
VSi_2	KHF + H_2SO_4 (1.82)	"	3	cs	[671]	1958	
VSi_2	$KHSO_4$ + H_2SO_4 + CrO_3	"	5	i	[671]	1958	
VSi_2	$KHSO_4$ + H_2SO_4 + CrO_2Cl_2	"	4.5	i	[671]	1958	
VSi_2	$KHSO_4$ + H_2SO_4 + $SiOCl_2$	"	3	i	[671]	1958	
VSi_2	NaOH (1%)	"	—	i	[671]	1958	

RESISTANCE OF POWDERS OF REFRACTORY COMPOUNDS TO CHEMICAL REAGENTS (continued)

Phase	Reagent² *	Temperature, °C	Duration of treatment, hr	Insoluble residue, %	Ref.	Year	Remarks
VSi₂	NaOH (20%)	Boiling	30 min	cs	[671]	1958	
VSi₂	Na₂CO₃	"	1	i	[671]	1958	
VSi₂	Na₂O₂ (30%)	"	15 min	cs	[671]	1958	
NbSi₂	HCl (1:1)	"	1	i	[671]	1958	
NbSi₂	HCl (1.19)	"	1	i	[671]	1958	
NbSi₂	H₂SO₄ (1.84)	"	3	i	[671]	1958	
NbSi₂	H₂SO₄ (1:1)	"	3	i	[671]	1958	
NbSi₂	H₂SO₄ (1:10)	"	3	i	[671]	1958	
NbSi₂	HF (1.15)	"	1	lss	[671]	1958	
NbSi₂	HF (1.15) + HNO₃ (1.43)	"	2	cs	[671]	1958	
NbSi₂	HCl (1.19) + HNO₃ (1.43)	"	2	95.4	[671]	1958	
NbSi₂	H₃PO₄ + H₂SO₄ + H₂O (4:1:2)	"	—	cs	[671]	1958	Insoluble residue appears when solution is evaporated until SO₃ fumes begin to appear
NbSi₂	KHSO₄ + H₂SO₄ + SiOCl₂	"	2	96.5	[671]	1958	
NbSi₂	NaOH (1%)	"	5	i	[671]	1958	
NbSi₂	NaOH (20%)	"	5	i	[671]	1958	
NbSi₂	Na₂O₂ (30%)	"	—	cs	[671]	1958	
TaSi₂	HF (1.15) + HNO₃ (1.43)	"	15 min	cs	[671]	1958	
TaSi₂	HCl (1.19) + HNO₃ (1.43)	"	2	95.5	[671]	1958	

RESISTANCE OF POWDERS OF REFRACTORY COMPOUNDS TO CHEMICAL REAGENTS (continued)

Phase	Reagent[2]*	Temperature, °C	Duration of treatment, hr	Insoluble residue, %	Ref.	Year	Remarks
$TaSi_2$	4 p. H_3PO_4 (1.21) + 1 p. H_2SO_4 (1.84) + 2 p. H_2O	Boiling	—	cs	[671]	1958	Insoluble residue appears when solution is evaporated until SO_3 fumes begin to appear
$TaSi_2$	$H_2C_2O_4 + H_2O$	"	2	96.5	[671]	1958	
$TaSi_2$	1 p. $H_2C_2O_4$ + 2 p. H_2O + 2 p. H_2SO_4 (1.84)	"	2	96.6	[671]	1958	
$TaSi_2$	$KHSO_4 + H_2SO_4 + SiOCl_2$	"	10	i	[671]	1958	
$TaSi_2$	NaOH (1%)	"	—	i	[671]	1958	
$TaSi_2$	NaOH (1%)	"	30 min	cs	[671]	1958	
$TaSi_2$	Na_2O_2 (30%)	"	15 min	cs	[671]	1958	
$CrSi_2$	HCl (1.19)	"	2	44.5	[671]	1958	
$CrSi_2$	HF (1.15)	"	1	ps	[671]	1958	
$CrSi_2$	$HF + HNO_3$	"	2	cs	[671]	1958	
$CrSi_2$	$HCl + HNO_3$	"	2	91.6	[671]	1958	

RESISTANCE OF POWDERS OF REFRACTORY COMPOUNDS TO CHEMICAL REAGENTS (continued)

Phase	Reagent²*	Temperature, °C	Duration of treatment, hr	Insoluble residue, %	Ref.	Year	Remarks
$CrSi_2$	4 p. H_3PO_4 + 1 p. H_2SO_4 (1.84) + 2 p. H_2O	Boiling	—	cs	[671]	1958	Insoluble residue appears when solution is evaporated until SO_3 fumes begin to appear
$CrSi_2$	$H_2C_2O_4$ + H_2O	"	2	41.4	[671]	1958	
$CrSi_2$	$H_2C_2O_4$ + H_2O	"	2	62.8	[671]	1958	
$CrSi_2$	NaOH (1%)	"	—	i	[671]	1958	
$CrSi_2$	NaOH (20%)	"	30 min	cs	[671]	1958	
$CrSi_2$	Na_2O_2 (30%)	"	15 min	cs	[671]	1958	
$MoSi_2$	HCl (1.19)	"	2	99.4	[671]	1958	
$MoSi_2$	HCl (1:1)	"	2	99.6	[671]	1958	
$MoSi_2$	H_2SO_4 (1.18)	"	2	99.2	[671]	1958	
$MoSi_2$	H_2SO_4 (1:1)	"	2	99.8	[671]	1958	
$MoSi_2$	H_3PO_4 (1.21)	"	2	96.7	[671]	1958	
$MoSi_2$	HF (1.15)	"	1	ps	[671]	1958	
$MoSi_2$	HCl (1.19)	"	3	i	[671]	1958	
$MoSi_2$	$KHSO_4$	"	1	i	[671]	1958	
$MoSi_2$	HF (1.15) + HNO_3 (1.43)	"	2	cs	[671]	1958	
$MoSi_2$	HCl (1.19) + HNO_3 (1.43)	"	2	99.0	[671]	1958	

RESISTANCE OF POWDERS OF REFRACTORY COMPOUNDS TO CHEMICAL REAGENTS (continued)

Phase	Reagent² *	Temperature, °C	Duration of treatment, hr	Insoluble residue, %	Ref.	Year	Remarks
MoSi₂	4 p. H₃PO₄ (1.21) + 1 p. H₂SO₄ (1.84) + 2 p. H₂O	Boiling	—	cs	[671]	1958	Insoluble residue appears when solution is evaporated until SO₃ fumes begin to appear
MoSi₂	H₂C₂O₄ + H₂O	"	2	99.2	[671]	1958	
MoSi₂	1 p. H₂C₂O₄ + 2 p. H₂SO₄ (1.84)	"	2	95.2	[671]	1958	
MoSi₂	KHSO₄ + KHF₂	"	1.2	cs	[671]	1958	
MoSi₂	KHF + H₂SO₄ (1.84)	"	3	ps	[671]	1958	
MoSi₂	KHSO₄ + H₂SO₄ + CuO	"	5	i	[671]	1958	
MoSi₂	KHSO₄ + H₂SO₄ + CrO₂Cl₂	"	4.5	i	[671]	1958	
MoSi₂	KHSO₄ + H₂SO₄ + SiOCl₂	"	9.5	i	[671]	1958	
MoSi₂	NaOH (1%)	"	—	i	[671]	1958	
MoSi₂	NaOH (20%)	"	30 min	cs	[671]	1958	
MoSi₂	Na₂CO₃ (20%)	"	1	i	[671]	1958	
MoSi₂	Na₂O₂ (30%)	"	15 min	cs	[671]	1958	
WSi₂	HCl (1.19)	"	2	99.2	[671]	1958	
WSi₂	HCl (1:1)	"	1.5	i	[671]	1958	
WSi₂	H₂SO₄ (1.82)	"	4	i	[671]	1958	
WSi₂	H₂SO₄ (1:1)	"	4	i	[671]	1958	
WSi₂	H₂SO₄ (1:10)	"	4	i	[671]	1958	
WSi₂	HF (1.15)	"	2.5	cs	[671]	1958	

RESISTANCE OF POWDERS OF REFRACTORY COMPOUNDS TO CHEMICAL REAGENTS (continued)

Phase	Reagent[2]*	Temperature, °C	Duration of treatment, hr	Insoluble residue, %	Ref.	Year	Remarks
WSi₂	HJ (1.47)	Boiling	3	i	[671]	1958	
WSi₂	KHSO₄	,,	1	ps	[671]	1958	
WSi₂	HF (1.15) + HNO₃ (1.43)	,,	2	cs	[671]	1958	
WSi₂	H₂C₂O₄ + H₂O	,,	2	93.4	[671]	1958	
WSi₂	KHSO₄ + KHF₂	,,	20 min	cs	[671]	1958	
WSi₂	KHF + H₂SO₄ (1.82)	,,	3.5	ps	[671]	1958	
WSi₂	KHSO₄ + H₂SO₄ + CuO	,,	5	i	[671]	1958	
WSi₂	KHSO₄ + H₂SO₄ + CrO₂Cl₂	,,	4.5	i	[671]	1958	
WSi₂	KHSO₄ + H₂SO₄ + SiOCl₂	,,	3	i	[671]	1958	
WSi₂	NaOH (20%)	,,	30 min	cs	[671]	1958	
WSi₂	Na₂CO₃ (20%)	,,	1	i	[671]	1958	
WSi₂	Na₂O₂ (30%)	,,	15 min	i	[671]	1958	
Mn₂Si MnSi MnSi₂	HCl (1.19)	20	—	i	[117]	1959	
Mn₂Si MnSi	HNO₃ (1.43)	20	—	i	[117]	1959	
MnSi₂	H₂SO₄ (1.82)	20	—	i	[117]	1959	
MnSi₂	HF (1.15)	20	—	cs	[117]	1959	
MnSi₂	3 p. HCl (1.19) + 1p. HNO₃ (1.43)	20	—	cs	[117]	1959	
MnSi₂	HF (1.15) + HNO₃ (1.43)	20	—	cs	[117]	1959	
FeSi	HCl (1.19)	20	—	cs	[117]	1959	
FeSi	HNO₃ (1.43)	20	—	cs	[117]	1959	

RESISTANCE OF POWDERS OF REFRACTORY COMPOUNDS TO CHEMICAL REAGENTS (continued)

Phase	Reagent[2]*	Temperature, °C	Duration of treatment, hr	Insoluble residue, %	Ref.	Year	Remarks
FeSi	H_2SO_4 (1.82)	20	—	cs	[117]	1959	
CoSi CoSi₂ Co₂Si	HCl (1.19)	20	—	cs	[117]	1959	
CoSi₂	HNO_3 (1.43)	20	—	cs	[117]	1959	
Ni₂Si	H_2SO_4 (1.82)	20	—	cs	[117]	1959	
Ni₂Si	HCl (1.19)	20	—	cs	[117]	1959	
NiSi	HNO_3 (1.43)	20	—	cs	[117]	1959	
NiSi	H_2SO_4 (1.82)	20	—	cs	[117]	1959	
Mg₃P₂	H_2O	20	—	cs	[245]	1961	
Mg₃P₂	HCl (1:1)	20	—	cs	[245]	1961	
Mg₃P₂	HCl (1.19)	20	—	cs	[245]	1961	Phosphine liberated
Mg₃P₂	H_2SO_4 (1.82)	20	—	cs	[245]	1961	
Mg₃P₂	HNO_3 (1.43)	20	—	cs	[245]	1961	
Mg₃P₂	HF	20	—	cs	[245]	1961	
Mg₃P₂	HF + HNO_3 (1:1)	20	—	cs	[245]	1961	
Mg₃P₂	3 p. HCl (1.19) + 1 p. HNO_3 (1.43)	20	—	cs	[245]	1961	
Mg₃P₂	KBr + Br_2	20	—	cs	[245]	1961	
Ca₃P₂	H_2O	20	—	cs	[245]	1961	
Ca₃P₂	HCl (1:1)	20	—	cs	[245]	1961	
Ca₃P₂	HCl (1.19)	20	—	cs	[245]	1961	Phosphine liberated
Ca₃P₂	H_2SO_4 (1.82)	20	—	cs	[245]	1961	
Ca₃P₂	HNO_3 (1.43)	20	—	cs	[245]	1961	
Ca₃P₂	HF (40%)	20	—	cs	[245]	1961	
Ca₃P₂	HF + HNO_3 (1:1)	20	—	cs	[245]	1961	

RESISTANCE OF POWDERS OF REFRACTORY COMPOUNDS TO CHEMICAL REAGENTS (continued)

Phase	Reagent[2]*	Temperature, °C	Duration of treatment, hr	Insoluble residue, %	Ref.	Year	Remarks
Ca₃P₂	3 p. HCl (1.19) + 1 p. HNO₃ (1.43)	20	—	cs	[245]	1961	
Sr₃P₂	H₂O	20	—	cs	[245]	1961	
Sr₃P₂	HCl (1.19)	20	—	98	[245]	1961	
Sr₃P₂	H₂SO₄ (1.82)	20	—	cs	[245]	1961	
AlP	H₂O	20	—	cs	[683]	1956	
AlP	HCl (1:1)	20	—	cs	[683]	1956	
AlP	HCl (1.19)	20	—	cs	[683]	1956	
AlP	H₂SO₄	20	—	cs	[683]	1956	
AlP	HNO₃ (1.43)	20	—	cs	[683]	1956	
AlP	HF (40%)	20	—	cs	[683]	1956	
SmP	H₂O	20	—	cs	[683]	1956	
SmP	HCl (1.19)	Gentle heat	4,5	cs	[245]	1961	
SmP	HCl (1:1)	Same	—	cs	[245]	1961	
TiP	H₂O	Boiling	8—10	cs	[245]	1961	
TiP	HCl (1:1)	"	6	100	[245]	1961	
TiP	HCl (1.19)	"	6	100	[245]	1961	
TiP	HNO₃ (1.43)	"	6	100	[245]	1961	
TiP	H₂SO₄ (1.82)	"	6	100	[245]	1961	
TiP	HF (40%)	"	6	100	[245]	1961	
TiP	HF (40%) + HNO₃ (1:1)	20	—	cs	[245]	1961	
TiP	3 p. HCl (1.19) + 1 p. HNO₃ (1.43)	Gentle heat	—	cs	[245]	1961	Phosphine liberated

RESISTANCE OF POWDERS OF REFRACTORY COMPOUNDS TO CHEMICAL REAGENTS (continued)

Phase	Reagent²*	Temperature, °C	Duration of treatment, hr	Insoluble residue, %	Ref.	Year	Remarks
V₃P	H_2O	Boiling	—	100	[684]	1942	
V₃P	HCl (1:1)	"	—	i	[684]	1942	
V₃P	HCl (1.19)	"	—	i	[684]	1942	
V₃P	HNO_3 (1.43)	"	—	i	[684]	1942	Phosphine liberated
V₃P	H_2SO_4 (1.82)	"	—	cs	[684]	1942	
V₃P	3 p. HCl (1.19) + 1 p. HNO_3 (1.43)	"	—	cs	[684]	1942	
VP	H_2O	"	—	100	[684]	1942	
VP	HCl (1:1)	"	—	i	[684]	1942	
VP	HCl (1.19)	"	—	i	[684]	1942	
VP	H_2SO_4 (1.82)	"	—	cs	[684]	1942	Phosphine liberated
VP	HNO_3 (1.43)	"	—	cs	[684]	1942	
VP	3 p. HCl (1.19) + 1 p. HNO_3 (1.43)	"	—	cs	[684]	1942	
VP₂	H_2O	"	—	i	[684]	1942	
VP₂	HCl (1:1)	"	—	i	[684]	1942	
VP₂	HCl (1.19)	"	—	i	[684]	1942	
VP₂	H_2SO_4 (1.82)	"	—	cs	[684]	1942	Phosphine liberated
VP₂	HNO_3 (1.43)	"	—	cs	[684]	1942	
VP₂	3 p. HCl (1.19) + 1 p. HNO_3 (1.43)	"	—	cs	[684]	1942	
NbP	H_2O	"	—	i	[684]	1942	
NbP	HCl (1:1)	"	—	i	[684]	1942	
NbP	HCl (1.19)	"	—	i	[684]	1942	
NbP	H_2SO_4 (1.82)	"	—	cs	[684]	1942	Phosphine liberated
NbP	HNO_3 (1.43)	"	—	cs	[684]	1942	
NbP	3 p. HCl (1.19) + 1 p. HNO_3 (1.43)	"	—	cs	[684]	1942	

RESISTANCE OF POWDERS OF REFRACTORY COMPOUNDS TO CHEMICAL REAGENTS (continued)

Phase	Reagent[2]*	Temperature, °C	Duration of treatment, hr	Insoluble residue, %	Ref.	Year	Remarks
TaP	H_2O	Boiling	—	i	[684]	1942	
TaP	HCl (1:1)	"	—	i	[684]	1942	
TaP	HCl (1.19)	"	—	i	[684]	1942	
TaP	H_2SO_4 (1.82)	"	—	cs	[684]	1942	
TaP	HNO_3 (1.43)	"	—	cs	[684]	1942	Phosphine liberated
TaP	3 p. HCl (1.19) + 1 p. HNO_3 (1.43)	"	—	cs	[684]	1942	
CrP	H_2O	"	—	i	[245]	1961	
CrP	HCl (1.19)	"	—	i	[881]	1961	
CrP	HCl (1:1)	"	—	i	[881]	1961	
CrP	H_2SO_4 (1.84)	"	—	cs	[881]	1961	
CrP	H_2SO_4 (1:4)	"	—	i	[881]	1961	
CrP	H_2SO_4 (1.84) + HNO_3 (1.43)	"	—	cs	[881]	1961	
CrP	H_2SO_4 (1:1) + HNO_3 (1.43)	"	—	i	[881]	1961	
CrP	HNO_3 (1.43) + HF (40%)	"	—	i	[881]	1961	
CrP	3 p. HCl (1.19) + 1 p. HNO_3 (1.43)	"	—	i	[881]	1961	
CrP	H_2SO_4 (1:4) + $(NH_4)_2S_2O_8$	"	—	i	[881]	1961	
CrP	HNO_3 (1.43) + $H_2C_2O_4$ (35%)	"	—	i	[881]	1961	
CrP	HNO_3 (1.43) + H_2O_2 (30%)	"	—	i	[881]	1961	
CrP	HNO_3 (1.43) + H_2SO_4 (1:1) + $H_2C_2O_4$ (35%)	"	—	i	[881]	1961	
CrP	NaOH (20%) + bromine water	"	—	i	[881]	1961	
CrP	NaOH (20%) + H_2O_2 (30%) + $C_2H_2O_4$ (35%)	"	—	i	[881]	1961	
MoP	H_2O	"	—	i	[881]	1961	
MoP	H_2O	"	—	i	[685]	1941	

RESISTANCE OF POWDERS OF REFRACTORY COMPOUNDS TO CHEMICAL REAGENTS (continued)

Phase	Reagent[2]*	Temperature, °C	Duration of treatment, hr	Insoluble residue, %	Ref.	Year	Remarks
MoP	HCl (1:1)	Boiling	—	i	[685]	1941	
MoP	HCl (1.19)	"	—	i	[685]	1941	
MoP	H_2SO_4 (1.82)	"	—	i	[685]	1941	
MoP	HNO_3 (1.43)	"	—	cs	[685]	1941	
MoP_2	H_2O	"	—	i	[685]	1941	
MoP_2	HCl (1.19)	"	—	i	[685]	1941	
MoP_2	HCl (1.19)	"	—	i	[685]	1941	
MoP_2	H_2SO_4 (1.82)	"	—	i	[685]	1941	
MoP_2	HNO_3 (1.43)	"	—	cs	[685]	1941	
FeP_2	H_2O	"	—	i	[245]	1961	
FeP_2	HCl (1.19)	"	—	i	[245]	1961	
FeP_2	H_2SO_4 (1.82)	"	—	i	[245]	1961	
FeP_2	3 p. HCl (1.19) + 1 p. HNO_3 (1.43)	"	—	cs	[245]	1961	
Fe_3P	H_2O	"	—	i	[245]	1961	
Fe_3P	HCl (1.19)	"	—	i	[245]	1961	
Fe_3P	H_2SO_4 (1.82)	"	—	i	[245]	1961	
Fe_3P	3 p. HCl (1.19) + 1 p. HNO_3 (1.43)	"	—	cs	[245]	1961	
FeP	H_2O	"	—	i	[245]	1961	
FeP	HCl (1.19)	"	—	i	[245]	1961	
FeP	H_2SO_4 (1.82)	"	—	i	[245]	1961	
U_3P_4	H_2O	"	—	ps	[686]	1941	
U_3P_4	HCl (1.19)	"	—	cs	[685]	1941	
U_3P_4	HNO_3 (1.43)	"	—	cs	[686]	1941	
U_3P_4	H_2SO_4 (1.82)	"	—	cs	[686]	1941	
U_3P_4	HF (40%)	"	—	cs	[686]	1941	

RESISTANCE OF POWDERS OF REFRACTORY COMPOUNDS TO CHEMICAL REAGENTS (continued)

Phase	Reagent[2]*	Temperature, °C	Duration of treatment, hr	Insoluble residue, %	Ref.	Year	Remarks
U_3P_4	$HF + HNO_3$ (1:1)	Boiling	—	cs	[686]	1941	
U_3P_4	3 p. HCl (1.19) + 1 p. HNO_3 (1.43) H_2O	"	—	cs	[686]	1941	
Np_3P_4	$HF + HNO_3$ (1:1)	20	—	i	[687]	1953	
Np_3P_4	HCl (1.19)	20	—	cs	[687]	1953	
Np_3P_4	Dilute inorganic acids	20	—	cs	[687]	1953	
YS	CH_3COOH (1:10)	20	—	cs	[145]	1956	Dissolves completely in the cold
YS	I_2 (solution)	20	—	cs	[145]	1956	Oxidized by iodine solution
YS	$KMnO_4$	20	—	—	[304]	1961	Oxidized by $KMnO_4$ solution
Y_2O_2S	HCl (1.19)	20	—	—	[689]	1958	
LaS	HCl (1:5)	20	—	cs	[304]	1961	Dissolves completely in the cold
LaS	HNO_3 (1:5)	20	—	cs	[304]	1961	
LaS	H_2SO_4 (1:5)	20	—	cs	[304]	1961	
LaS	CH_3COOH (1:1)	20	—	cs	[304]	1961	
LaS_3	H_2O	100	1	99.9	[304]	1961	
La_2S_3	HCl (1:5)	20	—	cs	[304]	1961	Dissolves completely in the cold
La_2S_3	HNO_3 (1:5)	20	—	cs	[304]	1961	
La_2S_3	H_2SO_4 (1:5)	20	—	cs	[304]	1961	Dissolves completely in the cold
La_2S_3	H_3PO_4 (1.21)	20	—	cs	[304]	1961	
La_2S_3	CH_3COOH (1:1)	20	—	cs	[304]	1961	

RESISTANCE OF POWDERS OF REFRACTORY COMPOUNDS TO CHEMICAL REAGENTS (continued)

Phase	Reagent[2]*	Temperature, °C	Duration of treatment, hr	Insoluble residue, %	Ref.	Year	Remarks
La_2S_3	$H_6C_4O_6$ (50%)	100	—	cs	[304]	1961	Dissolves completely on boiling
La_2S_3	NaOH (20%)	100	1	100	[304]	1961	
La_2S_3	Dilute inorganic acids	20	—	cs	[153]	1956	Dissolves completely in the cold
La_2O_2S	CH_3COOH (1:10)	20	—	cs	[153]	1956	Dissolves completely
CeS	HCl (1:5)	20	—	cs	[304]	1961	Dissolves completely in the cold
CeS	HNO_3 (1:5)	20	—	cs	[304]	1961	
CeS	H_2SO_4 (1:5)	20	—	cs	[304]	1961	
CeS	CH_3COOH (1:1)	20	—	cs	[304]	1961	
CeS	H_2O	100	1	100	[304]	1961	
Ce_2S_3	HCl (1:5)	20	—	cs	[304]	1961	Dissolves completely in the cold
Ce_2S_3	HNO_3 (1:5)	20	—	cs	[304]	1961	
Ce_2S_3	H_2SO_4 (1:5)	20	—	cs	[304]	1961	
Ce_2S_3	H_3PO_4 (1.21)	20	—	cs	[304]	1961	
Ce_2S_3	CH_3COOH (1:1)	20	—	cs	[304]	1961	Dissolves completely in the cold
Ce_2S_3	$H_6C_4O_6$ (50%)	100	—	cs	[304]	1961	
Ce_2S_3	NaOH (20%)	100	1	99.9	[304]	1961	Not decomposed

RESISTANCE OF POWDERS OF REFRACTORY COMPOUNDS TO CHEMICAL REAGENTS (continued)

Phase	Reagent[2]*	Temperature, °C	Duration of treatment, hr	Insoluble residue, %	Ref.	Year	Remarks
Ce_2S_3	H_2O_2 (30%)	100	1	21.0	[304]	1961	Slowly decomposed on boiling under reflux condenser
Ce_2S_3	Dilute inorganic acids	20	—	cs	[689]	1959	Dissolves completely with liberation of H_2S and H_2 in volume ratio of 2:1
PrS	HCl (1:1)	20	—	cs	[149]	1956	
PrS	HNO_3 (1.43)	20	—	cs	[149]	1956 }	Dissolves completely with precipitation of S
PrS	H_2SO_4 (1.82)	20	—	cs	[149]	1956 }	
Pr_3S_4	Dilute inorganic acids and CH_3COOH	20	—	cs	[159]	1956	Dissolves completely with liberation of H_2S and H_2 in volume ratio of 8:1

RESISTANCE OF POWDERS OF REFRACTORY COMPOUNDS TO CHEMICAL REAGENTS (continued)

Phase	Reagent[2]*	Temperature, °C	Duration of treatment, hr	Insoluble residue, %	Ref.	Year	Remarks
Pr_2S_3	HCl (1 : 5)	20	—	cs	[304]	1961	Dissolves completely in the cold
Pr_2S_3	HNO_3 (1 : 5)	20	—	cs	[304]	1961	
Pr_2S_3	H_2SO_4 (1 : 5)	20	—	cs	[304]	1961	Dissolves completely in the cold
Pr_2S_3	CH_3COOH (1 : 1)	20	—	cs	[304]	1961	
Pr_2O_2S	Dilute inorganic acids	20	—	cs	[153]	1956	
Pr_2O_2S	CH_3COOH (1 : 10)	20	—	cs	[153]	1956	
NdS	HCl (1 : 1)	20	—	cs	[149]	1956	Dissolves completely with liberation of H_2S and H_2 in volume ratio of 2:1
NdS	HNO_3 (1.43)	20	—	cs	[149]	1956	Dissolves completely with precipitation of S
NdS	H_2SO_4 (1.82)	20	—	cs	[149]	1956	

261

RESISTANCE OF POWDERS OF REFRACTORY COMPOUNDS TO CHEMICAL REAGENTS (continued)

Phase	Reagent[2]*	Tempera-ture, °C	Duration of treat-ment, hr	Insoluble residue, %	Ref.	Year	Remarks
Nd₃S₄	Dilute inorganic acids and CH₃COOH	20	—	cs	[159]	1956	Dissolves completely with liberation of H₂S and H₂ in volume ratio of 2:1
Nd₂S₃	HCl (1:5)	20	—	cs	[304]	1961	Dissolves completely in the cold
Nd₂S₃	HNO₃ (1:5)	20	—	cs	[304]	1961	
Nd₂S₃	H₂SO₄ (1:5)	20	—	cs	[304]	1961	
Nd₂S₃	CH₃COOH (1:1)	20	—	cs	[304]	1961	
Nd₂O₂S	Dilute inorganic acids	20	—	cs	[153]	1956	Dissolves completely with liberation of H₂S and H₂ in volume ratio of 2:1
Nd₂O₂S	CH₃COOH (1:10)	20	—	cs	[153]	1956	
SmS	HCl (1:1)	20	—	cs	[153]	1956	
SmS	HNO₃ (1.43)	20	—	cs	[153]	1956	Dissolves completely with precipitation of S
SmS	H₂SO₄ (1.82)	20	—	cs	[153]	1956	

RESISTANCE OF POWDERS OF REFRACTORY COMPOUNDS TO CHEMICAL REAGENTS (continued)

Phase	Reagent² *	Temperature, °C	Duration of treatment, hr	Insoluble residue, %	Ref.	Year	Remarks
Sm_3S_4	Dilute inorganic acids and CH_3COOH	20	—	cs	[153]	1956	Dissolves completely with liberation of H_2S and H_2 in volume ratio of 8:1
Sm_2O_2S	Dilute inorganic acids	20	—	cs	[153]	1956	Dissolves completely in the cold
Yb_2O_2S	HCl (1.19)	20	—	ps	[689]	1958	
Yb_2O_2S	3 p. HCl (1.19) + 1 p. HNO_3 (1.43)	20	—	cs	[689]	1958	
ThS	HCl (1:1)	20	—	cs	[251]	1955	
Th_2S_3	HCl (1:1)	20	—	cs	[251]	1955	
Th_2S_3	HCl (1:1)	20	—	cs	[251]	1955	
Th_2S_3	HNO_3 (1.43)	20	—	cs	[251]	1955	
US	Inorganic acids	20	—	cs	[688]	1955	
US	Alkalis and NH_4OH (solution)	20	—	i	[688]	1955	
U_2S_3	Inorganic acids	20	—	cs	[164]	1955	
U_2S_3	CH_3COOH (1:1)	100	—	i	[164]	1955	
U_2S_5	Inorganic acids	20	—	i	[164]	1955	
U_3S_5	CH_3COOH (1:1)	100	—	i	[164]	1955	

RESISTANCE OF POWDERS OF REFRACTORY COMPOUNDS TO CHEMICAL REAGENTS (continued)

Phase	Reagent²*	Temperature, °C	Duration of treatment, hr	Insoluble residue, %	Ref.	Year	Remarks
US₃	Inorganic acids and CH₃COOH						
B₄C	HCl (1.19)	20	—	cs	[164]	1955	
B₄C	HCl (1.19)	20	24	98	[672]	1959	
B₄C	HCl (1:1)	115	1	98	[672]	1959	
B₄C	HCl (1:1)	105	30 min	97.8	[672]	1959	
B₄C	HCl (1:1)	105	2	97.8	[672]	1959	
B₄C	H₂SO₄ (1.82)	20	24	98	[672]	1959	
B₄C	H₂SO₄ (1.82)	130	1	98	[672]	1959	
B₄C	H₂SO₄ (1:1)	130	30 min	98	[672]	1959	
B₄C	H₂SO₄ (1:1)	130	2	97.7	[672]	1959	
B₄C	H₂SO₄ (1:1)	130	4	98	[672]	1959	
B₄C	HNO₃ (1.43)	20	24	97	[672]	1959	
B₄C	HNO₃ (1.43)	130	1	97	[672]	1959	
B₄C	HNO₃ (1:1)	105	30 min	96.9	[672]	1959	
B₄C	HNO₃ (1:1)	105	1	96.5	[672]	1959	
B₄C	HNO₃ (1:1)	105	2	96.1	[672]	1959	
B₄C	HClO₄ (1.35)	110	4	96.9	[672]	1959	
B₄C	HClO₄ (1.35)	20	24	98	[672]	1959	
B₄C	HClO₄ (1.35)	115	1	98	[672]	1959	
B₄C	3 p. HCl (1.19) +1p. HNO₃ (1.43)	20	24	97	[672]	1959	
B₄C	H₂SO₄ (1.82) + HNO₃ (1.43)	230	4	91.2	[672]	1959	
B₄C	NaOH (50%)	20	40	98.3	[672]	1959	
B₄C	NaOH (25%)	20	40	99.2	[672]	1959	
B₄C	NaOH (12%)	20	40	99	[672]	1959	
B₄C	NaOH (6%)	20	40	98.6	[672]	1959	

RESISTANCE OF POWDERS OF REFRACTORY COMPOUNDS TO CHEMICAL REAGENTS (continued)

Phase	Reagent²*	Temperature, °C	Duration of treatment, hr	Insoluble residue, %	Ref.	Year	Remarks
B₄C	NaOH (3%)	20	40	98.8	[672]	1959	
B₄C	NaOH (1%)	20	40	99	[672]	1959	
B₄C	NaOH (1%)	100	1	99	[672]	1959	
B₄C	NaOH (1%)	100	2	98.5	[672]	1959	
B₄C	NaOH (25%)	100	2	99	[672]	1959	
B₄C	NaOH (12%)	100	2	98.5	[672]	1959	
B₄C	HF (1.15) + HNO₃ (1.43)	180	2	90.8	[672]	1959	
B₄C	HClO₄ (1:1)	115	30 min	98	[672]	1959	
B₄C	HClO₄ (1:1)	115	1	98	[672]	1959	
B₄C	HClO₄ (1:1)	115	2	96.7	[672]	1959	
B₄C	10% NaOH + 10% H₂O₂	100	1	98	[672]	1959	
B₄C	10% NaOH + 10% H₂O₂	100	2	96	[672]	1959	
B₄C	10% NaOH + Br₂	100	1	99.6	[672]	1959	
B₄C	3 p. HCl (1.19) + 1 p. HNO₃ (1.43)	Boiling	59	i	[942]	1938	
B₄C	50 ml H₂SO₄ (1.84) + K₂Cr₂O₇ (1.5 g)	"	25	i	[942]	1938	
B₄C	30% H₂O₂ + H₂SO₄ (1.84)	"	45	i	[942]	1938	
B₄C	30% H₂O₂ + 0.01 g KNO₃	"	42	i	[942]	1938	
B₄C	2 ml H₂SO₄ (1.84) + 0.6 ml HNO₃ (1.43)	"	6	cs	[942]	1938	[962]
SiC	HCl (1:1)	"	—	100	[673]	1938	
SiC	HCl (1.19)	"	1	100	[673]	1938	
SiC	HNO₃ (1:1)	"	1	100	[673]	1938	
SiC	HNO₃ (1.43)	"	1	100	[673]	1938	
SiC	HF (1.15)	"	1	100	[673]	1938	
SiC	HNO₃ (1.43) + HF (1.15)	"	1	100	[673]	1938	

265

RESISTANCE OF POWDERS OF REFRACTORY COMPOUNDS TO CHEMICAL REAGENTS (continued)

Phase	Reagent[2]*	Temperature, °C	Duration of treatment, hr	Insoluble residue, %	Ref.	Year	Remarks
SiC	H_3PO_4 (1.21)	230	1	ps	[673]	1938	
BN	H_2SO_4 (1.82)	Boiling	6—10	cs	[943]	1925	
BN	H_2SO_4 (1.82)	20	—	—	[272]	1955	With regard to the chemical stability of BN in HCl, H_2SO_4, H_3PO_4 (pure and with addition of $KMnO_4$, $K_2Cr_2O_7$, $KClO_3$) at 190-300°C, see also [1008,1190]
				Loss of weight, mg/cm^2			
BN	H_2SO_4 (20%)	20	—	10.7	[272]	1955	
BN	H_3PO_4 (1.21)	20	—	1.3	[272]	1955	
BN	HNO_3 (1.43)	20	—	8.9	[272]	1955	
BN	HF (1.15)	20	—	17.5	[272]	1955	
BN	NaOH (20%)	Boiling	15—20min	cs	[943]	1925	
BN	NaOH (20%)	20	—	8.9	[272]	1955	
BN	CCl_4	20	—	1.3	[272]	1955	
BN	C_2H_5OH (95%)	20	—	14.6	[272]	1955	
BN	CH_3COCH_3	20	—	13.0	[272]	1955	
Si₃N₄	HCl (20%)	Boiling	500	i	[270]	1955	
Si₃N₄	HNO_3 (65%)	"	500	i	[270]	1955	
Si₃N₄	HNO_3 (65%)	Fuming	500	i	[270]	1955	
Si₃N₄	H_2SO_4 (10%)	70	500	i	[270]	1955	
Si₃N₄	H_2SO_4 (77%)	20	500	i	[270]	1955	

RESISTANCE OF POWDERS OF REFRACTORY COMPOUNDS TO CHEMICAL REAGENTS (continued)

Phase	Reagent[2*]	Temperature, °C	Duration of treatment, hr	Insoluble residue, %	Ref.	Year	Remarks
Si_3N_4	H_2SO_4 (85%)	20	500	i	[270]	1955	
Si_3N_4	H_3PO_4 (1.21)	20	500	i	[270]	1955	
Si_3N_4	$H_4P_2O_7$	20	500	i	[270]	1955	
Si_3N_4	HF (1.15)	Boiling	192	13.9	[941]	1959	
Si_3N_4	NaOH (20%)	20	500	i	[270]	1955	
Si_3N_4	NaOH (50%)	Boiling	115	i	[270]	1955	
Si_3N_4	H_2SO_4 + $CuSO_4$ + $KHSO_4$ (conc.)	"	500	i	[270]	1955	
Si_3N_4	HF (1.15) + HNO_3 (1.43)	"	68	56	[941]	1959	
B_6Si	HNO_3 (1.43)	20	—	cs	[209]	1950	Dissolves with formation of salts
B_6Si	H_2SO_4 (1.82)	Boiling	—	cs	[209]	1950	
B_3Si	HNO_3 (1.43)	"	—	i	[209]	1950	
B_3Si	H_2SO_4 (1.82)	"	—	cs	[209]	1950	
BP	H_2O	"	8	100	[245]	1961	
BP	HCl (1.19)	"	6	100	[245]	1961	
BP	H_2SO_4 (1.82)	"	8	100	[245]	1961	
BP	HNO_3 (1.43)	"	8	100	[245]	1961	
BP	HF + HNO_3 (1:1)	"	6	100	[245]	1961	
BP	HF (40%)	"	8	100	[245]	1961	
BP	HNO_3 + H_2O_2 (1:1)	"	4.5	100	[245]	1961	

RESISTANCE OF COMPACT REFRACTORY COMPOUNDS TO THE ACTION OF CHEMICAL REAGENTS

Phase	Reagent (concentration is given in parentheses)	Temperature, °C	Duration of treatment, hr	Insoluble residue, %	Ref.	Year	Remarks
TiB_2	HCl (1.19)	20	24	99.9	[645]	1960	
TiB_2	HCl (1.19)	20	96	99.6	[645]	1960	
TiB_2	HCl (1.19)	20	240	99.2	[645]	1960	
TiB_2	HNO_3 (1.42)	20	24	93.9	[645]	1960	
TiB_2	HNO_3 (1.42)	20	96	84.9	[645]	1960	
TiB_2	HNO_3 (1.42)	20	240	69.5	[645]	1960	
TiB_2	H_2SO_4 (1.84)	20	24	99.9	[645]	1960	Specimens prepared by hot pressing, $P = 3\text{-}7\%$
TiB_2	H_2SO_4 (1.84)	20	96	99.6	[645]	1960	
TiB_2	H_2SO_4 (1.84)	20	240	99.0	[645]	1960	
ZrB_2	HCl (1.19)	20	24	97.5	[645]	1960	
ZrB_2	HCl (1.19)	20	96	95.1	[645]	1960	
ZrB_2	HCl (1.19)	20	240	89.6	[645]	1960	
ZrB_2	HNO_3 (1.42)	20	24	94.4	[645]	1960	
ZrB_2	HNO_3 (1.42)	20	96	86.4	[645]	1960	
ZrB_2	HNO_3 (1.42)	20	240	66.7	[645]	1960	
ZrB_2	H_2SO_4 (1.84)	20	24	98.8	[645]	1960	
ZrB_2	H_2SO_4 (1.84)	20	96	95.8	[645]	1960	
ZrB_2	H_2SO_4 (1.84)	20	240	94.8	[645]	1960	
UC	H_2O	50	Several hours	100	[1014]	1959	
UC	H_2O	60	1	~100	[1014]	1959	
UC	H_2O	65	1	<100 ($600\ mg/cm^2 \cdot hr$)	[1014]	1959	Specimens with $P = 25\%$
UC	H_2O	100	1	Dissolves rapidly	[1014]	1959	

RESISTANCE OF COMPACT REFRACTORY COMPOUNDS TO CHEMICAL REAGENTS (continued)

Phase	Reagent (concentration is given in parentheses)	Temperature, °C	Duration of treatment, hr	Insoluble residue, %	Ref.	Year	Remarks
UC	Glycerin	100	—	Dissolves rapidly	[1014]	1959	Specimens with P = 25%
UC	Diphenyl	350	5	~100	[1014]	1959	
UC	Triphenyl	350	5	~100	[1014]	1959	
UC	Paratriphenyl	350	5	~100	[1014]	1959	
Cr_3C_2	HCl (1:1)	112	1	0.039	[896]	1961	
Cr_3C_2	HNO_3 (1:1)	112	1	None found	[896]	1961	
Cr_3C_2	H_2SO_4 (1:1)	136	1	0.003	[896]	1961	
Cr_3C_2	$H_2C_2O_4$ (satur.)	130	1	0.015	[896]	1961	
Cr_3C_2	3p. HCl + 1 p. HNO_3	102	1	0.15	[896]	1961	
Cr_3C_2	4 p. H_3PO_4 + 1 p. H_2SO_4 + 2 p. H_2O	189	1	None found	[896]	1961	Specimens prepared by hot pressing, P = 3–7%
Cr_3C_2	NaOH + bromine water	110	1	0.12	[896]	1961	
Cr_7C_3	HCl (1:1)	110	1	7.5	[896]	1961	The fifth column gives the rate of corrosion, $g/m^2 \cdot hr$
Cr_7C_3	HNO_3 (1:1)	112	1	None found	[896]	1961	
Cr_7C_3	H_2SO_4 (1:1)	125	1	26.4	[896]	1961	
Cr_7C_3	$H_2C_2O_4$ (satur.)	135	1	0.036	[896]	1961	

RESISTANCE OF COMPACT REFRACTORY COMPOUNDS TO CHEMICAL REAGENTS (continued)

Phase	Reagent (concentration is given in parentheses)	Temperature, °C	Duration of treatment, hr	Insoluble residue, %	Ref.	Year	Remarks
Cr_7C_3	3 p. HCl (1.19) + 1 p. HNO$_3$ (1.42)	102	1	0.036	[896]	1961	The fifth column gives the rate of corrosion, g/m²·hr
Cr_7C_3	4 p. H$_3$PO$_4$ (1.21) + 1 p. H$_2$SO$_4$ (1.84) + 2 p. H$_2$O	135	1	None found	[896]	1961	
Cr_7C_3	NaOH + bromine water	115	1	None found	[896]	1961	
$MoSi_2$	HCl (1.19)	20	24	99.91	[645]	1960	
$MoSi_2$	HCl (1.19)	20	96	99.84	[645]	1960	
$MoSi_2$	HCl (1.19)	20	240	99.26	[645]	1960	
$MoSi_2$	HNO$_3$ (1.42)	20	24	99.54	[645]	1960	
$MoSi_2$	HNO$_3$ (1.42)	20	96	99.16	[645]	1960	
$MoSi_2$	HNO$_3$ (1.42)	20	240	98.44	[645]	1960	
$MoSi_2$	H$_2$SO$_4$ (1.84)	20	24	99.93	[645]	1960	Specimens prepared by hot pressing, P = 3-7%
$MoSi_2$	H$_2$SO$_4$ (1.84)	20	96	99.99	[645]	1960	
$MoSi_2$	H$_2$SO$_4$ (1.84)	20	240	99.93	[645]	1960	
B_4C	HCl (1.19)	20	24	99.76	[645]	1960	
B_4C	HCl (1.19)	20	96	99.41	[645]	1960	
B_4C	HCl (1.19)	20	240	99.35	[645]	1960	
B_4C	HNO$_3$ (1.42)	20	24	99.59	[645]	1960	
B_4C	HNO$_3$ (1.42)	20	96	99.35	[645]	1960	
B_4C	HNO$_3$ (1.42)	20	240	99.35	[645]	1960	
B_4C	H$_2$SO$_4$ (1.84)	20	24	98.95	[645]	1960	
B_4C	H$_2$SO$_4$ (1.84)	20	96	98.95	[645]	1960	
B_4C	H$_2$SO$_4$ (1.84)	20	240	98.43	[645]	1960	

Phase	Reagent (concentration is given in parentheses)	Temperature, °C	Duration of treatment, hr	Insoluble residue, %	Ref.	Year	Remarks
SiC	HCl (20%)	Boiling	1008	0.3	[1000]	1961	See note at end of table
SiC	HCl (20%)	175	144	—0.3	[1000]	1961	Under pressure
SiC		200	144	0.9	[1000]	1961	
SiC		225	144	1.5	[1000]	1961	
SiC	HCl (37%)	Boiling	144	0.0	[1000]	1961	Under pressure
SiC		200	48	0.0	[1000]	1961	
SiC	HNO₃ (30%)	Boiling	144	0.6	[1000]	1961	Under pressure
SiC	HNO₃ (50%)	"	1008	0.0	[1000]	1961	Under pressure
SiC		200	144	12.2	[1000]	1961	
SiC	HNO₃ (70%)	Boiling	144	0.6	[1000]	1961	Under pressure
SiC		200	144	5.8	[1000]	1961	
SiC		225	144	3.0	[1000]	1961	
SiC	H₂SO₄ (60%)	Boiling	144	0.0	[1000]	1961	Under pressure
SiC		200	144	—0.9	[1000]	1961	
SiC	H₂SO₄ (80%)	Boiling	1008	—0.3	[1000]	1961	
SiC	H₂SO₄ (95%)	"	144	—2.4	[1000]	1961	Under pressure
SiC		200	288	—0.6	[1000]	1961	
SiC		225	144	3.66	[1000]	1961	
SiC	H₃PO₄ (40%)	Boiling	144	0.0	[1000]	1961	
SiC	H₃PO₄ (60%)	"	1008	0.0	[1000]	1961	
SiC		200	144	—0.3	[1000]	1961	
SiC	H₃PO₄ (85%)	Boiling	144	6.9	[1000]	1961	Under pressure
SiC		200	288	1.5	[1000]	1961	
SiC		225	144	0.3	[1000]	1961	

RESISTANCE OF COMPACT REFRACTORY COMPOUNDS TO CHEMICAL REAGENTS (continued)

Phase	Reagent (concentration is given in parentheses)	Temperature, °C	Duration of treatment, hr	Insoluble residue, %	Ref.	Year	Remarks
SiC	40% HF + 10% HNO$_3$	60	24	960.8	[1000]	1961	
SiC		60	144	496	[1000]	1961	
SiC		60	288	369	[1000]	1961	
SiC		60	432	308	[1000]	1961	
SiC		60	576	263	[1000]	1961	
SiC	Na$_2$SO$_4$ (10%)	Boiling	144	0.6	[1000]	1961	
SiC		"	288	0.6	[1000]	1961	
SiC	NaOH (50%)	"	24	6668	[1000]	1961	
SiC	NaOH (25%)	"	144	224	[1000]	1961	
SiC	Na$_2$CO$_3$ (10%)	"	144	84.2	[1000]	1961	
SiC			285	40.6	[1000]	1961	
SiC			432	19.2	[1000]	1961	
SiC			576	8.5	[1000]	1961	
SiC			720	—1.2	[1000]	1961	
SiC			864	—9.6	[1000]	1961	
SiC			1008	—10.2	[1000]	1961	

Remark. For all cases of the determination of the resistance of SiC the column "Insoluble residue" gives, instead of the value of the insoluble residue, the rate of solution (corrosion), mm/min, determined for the time of treatment given in the table. The resistance was determined on specimens of the composition (%): 96.5 SiC, 2.5 Si$_{free}$; 0.4 C$_{free}$; 0.4 Al; 0.2 Fe; P = 0.4. See [673].

RESISTANCE TO OXIDATION

Phase	Temperature, °C	Oxidation time, hr	Change in weight		Ref.	Year	Remarks
			mg/cm²	mg/cm²·hr			
Be₅B	1000	20	+258	+12.9	[651]	1960	
Be₅B	1200	14.5	+120	+8.3	[651]	1960	
Be₂B	1000	20	+132	+6.6	[651]	1960	
Be₂B	1200	14.5	+126	+8.7	[651]	1960	
BeB₂	1000	20	+22	+1.1	[651]	1960	[1140]
BeB₂	1100	20	+34	+1.7	[651]	1960	
BeB₂	1200	14.5	+99.4	+6.8	[651]	1960	
BeB₂	1200	20	+48	+2.4	[651]	1960	
BeB₄	1000	20	+30	+1.5	[651]	1960	
BeB₄	1200	15.5	−2.9	−0.2	[651]	1960	
BeB₆	1000	20	+64	+3.2	[651]	1960	
BeB₆	1200	14.5	−7.25	−0.5	[651]	1960	
CaB₆	900	0.5	+36%		[1038]	1961	
CaB₆	900	1	+37%				
CaB₆	900	2	+38%				
CaB₆	900	10	+38%				
SrB₆	900	0.5	+42%				
SrB₆	900	1	+42%	—			
SrB₆	900	2	+43%				
SrB₆	900	10	+43%				
BaB₆	900	0.5	+44%				
BaB₆	900	1	+45%				
BaB₆	900	2	+46%				
BaB₆	900	10	+46%				
Magnesium borides	400—600	0.5—1.5	—	—	[1076]	1962	Powder

273

RESISTANCE TO OXIDATION (continued)

Phase	Temperature, °C	Oxidation time, hr	Change in weight		Ref.	Year	Remarks
			mg/cm²	mg/cm²·hr			
GdB$_6$	1000	140	+6	+0.043	[285]	1956	
TiB$_2$	450	1	+0.42	+0.42	[653]	1958	
TiB$_2$	500	1	+0.63	+0.63	[653]	1958	
TiB$_2$	550	1	+0.63	+0.63	[653]	1958	
TiB$_2$	600	1	+1.78	+1.78	[653]	1958	P = 1.6%
TiB$_2$	700	1	+2.00	+2.00	[653]	1958	
TiB$_2$	800	1	+7.36	+7.36	[653]	1958	
TiB$_2$	900	1	+20.4	+20.4	[653]	1958	
TiB$_2$	1000	1	+12.0	+12.0	[653]	1958	
TiB$_2$	1000	0.8	+6.8	+8.5	[652]	1958	
TiB$_2$	1000	2.8	+10	+3.6	[652]	1958	
TiB$_2$	1000	9.3	+19	+2.1	[652]	1958	P = 2-3%
TiB$_2$	1000	19	+25	+1.3	[652]	1958	
TiB$_2$	1000	29	+20	+0.7	[652]	1958	
TiB$_2$	1000	40	+24	+0.6	[652]	1958	
TiB$_2$	1000	48	+28	+0.58	[652]	1958	
TiB$_2$	1000	63	+29	+0.45	[652]	1958	
TiB$_2$	1000	82.5	+32	+0.39	[652]	1958	
TiB$_2$	1000	102	+30	+0.29	[652]	1958	P = 2-3%
TiB$_2$	1000	119	+29	+0.24	[652]	1958	
TiB$_2$	1000	147	+29	+0.19	[652]	1958	
TiB$_2$	1000	170	+31	+0.18	[652]	1958	
TiB$_2$	1100	20	+26	+1.3	[651]	1960	
TiB$_2$	1200	2	+10	+5	[270]	1955	

RESISTANCE TO OXIDATION (continued)

Phase	Temperature, °C	Oxidation time, hr	Change in weight		Ref.	Year	Remarks
			mg/cm²	mg/cm²·hr			
TiB₂	1200	5	+24.5	+4.9	[654]	1960	
TiB₂	1200	25	+38.4	+1.54	[654]	1960	
TiB₂	1200	50	+62.0	+1.24	[654]	1960	
TiB₂	1200	75	+68.1	+0.91	[654]	1960	
TiB₂	1200	100	+73.7	+0.74	[654]	1960	
ZrB₂	1000	150	+30	+0.2	[652]	1958	P = 2-3%
ZrB₂	1100	20	+22	+1.1	[651]	1960	
ZrB₂	1150	8	+0.5	+0.06	[292]	1952	
ZrB₂	1150	16	+1.2	+0.08	[292]	1952	
ZrB₂	1150	24	+2	+0.08	[292]	1952	
ZrB₂	1150	32	+3	+0.09	[292]	1952	
ZrB₂	1150	48	+3	+0.06	[292]	1952	
ZrB₂	1150	200	+4	+0.02	[293]	1953	
NbB₂	450	1	+0.25	+0.25	[653]	1958	P = 1.4%
NbB₂	500	1	+0.99	+0.99	[653]	1958	
NbB₂	550	1	+1.74	+1.74	[653]	1958	
NbB₂	600	1	+1.86	+1.86	[653]	1958	
NbB₂	700	1	+4.99	+4.99	[653]	1958	
NbB₂	800	1	+16.2	+16.2	[653]	1958	P = 2-3%
NbB₂	900	1	+28.6	+28.6	[653]	1958	
NbB₂	1000	1	+32.5	+32.5	[653]	1958	

Phase	Temperature, °C	Oxidation time, hr	Change in weight		Ref.	Year	Remarks
			mg/cm²	mg/cm²·hr			
TaB₂	700	2	+1.24	+0.62	×	1954	N. K. Golubeva
TaB₂	700	2	+1.81	+0.9	×	1954	
TaB₂	800	1	+1.69	+0.84	×	1954	
TaB₂	900	1	+2.52	+1.26	×	1954	
TaB₂	900	2	+3.34	+1.67	×	1954	
CrB₂	1000	150	+2.1	+0.014	[652]	1958	P = 2-3%
Mo₂B	1000	1	—5.8%	—5.8%	[49]	1952	
Mo₂B	1000	2	—18%	—9%	[49]	1952	
MoB	1000	3	—2%	—0.66%	[49]	1952	
MoB	1000	10	—4.5%	—0.45%	[49]	1952	Data taken from graph
MoB	1000	20	—9%	—0.45%	[49]	1952	
MoB	1000	40	—20%	—0.5%	[49]	1952	
Mo₂B₅	1000	20	—4%	—0.2%	[49]	1952	
Mo₂B₅	1000	60	—9%	—0.15%	[49]	1952	
Mo₂B₅	1000	90	—18%	—0.20%	[49]	1952	
W₂B₅	1000	150	+0.33	+0.002	[652]	1958	P = 2-3%
W₂B₅	1100	2	+2.2	+1.1	×	1954	N. K. Golubeva
YC	20	1	3.33	3.33	[820]	1961	
YC	20	2	3.74	4.37	[820]	1961	
YC	20	1	+3.33	+3.33	[879]	1961	[820]
YC	20	5	+21.65	+5.73	[879]	1961	
YC	20	15	+123	+8.20	[879]	1961	
YC	20	20	+138	+6.95	[879]	1961	
YC	20	40	+156	+3.90	[879]	1961	[820]
YC	20	50	+162	+3.24	[879]	1961	
Y₂C₃	20	1	+2.51	+2.51	[879]	1961	
Y₂C₃	20	5	+10.65	+2.13	[879]	1961	
Y₂C₃	20	15	+48	+3.20	[879]	1961	

RESISTANCE TO OXIDATION (continued)

Phase	Temperature, °C	Oxidation time, hr	Change in weight mg/cm²	Change in weight mg/cm²·hr	Ref.	Year	Remarks
Y_2C_3	20	30	+81	+2.70	[879]	1961	
Y_2C_3	20	40	+97.2	+2.43	[879]	1961	
Y_2C_3	20	50	+112.5	+2.25	[879]	1961	
YC_2	20	1	+3.52	+3.52	[879]	1961	
YC_2	20	5	+10	+2.00	[879]	1961	
YC_2	20	20	+19.6	+0.98	[879]	1961	
YC_2	20	30	+25.2	+0.84	[879]	1961	
YC_2	20	50	+33	+0.66	[879]	1961	
UC	350	0.5	+2	+4	[1275]	1962	In oxygen; data taken from graph
UC	450	0.25	+14	+56	[1275]	1962	
UC	450	0.5	+17.2	+34.4	[1275]	1962	
UC	500	0.25	+29.6	+118.4	[1275]	1962	
UC	700	0.17	+39	+230	[1275]	1962	
UC	850	0.17	+48.5	+285	[1275]	1962	
UC	1000	0.17	+45	+264	[1275]	1962	
UC	500	2	+1.6	+0.8	[1275]	1962	In CO_2; data taken from graph
UC	500	4	+4.8	+1.2	[1275]	1962	
UC	700	2	+21.8	+10.9	[1275]	1962	
UC	700	4	+43.5	+10.9	[1275]	1962	
UC	850	1	+46.5	+46.5	[1275]	1962	
UC	1000	0.5	+54.5	+109	[1275]	1962	
TiC	600	2	+0.047	+0.023	[658]	1953	
TiC	700	1	+0.532	+0.532	X	1954	N. K. Golubeva; [1123]
TiC	700	2	+1.55	+0.77	X	1954	
TiC	800	1	+0.426	+0.426	X	1954	
TiC	800	2	+1.14	+0.57	X	1954	
TiC	900	1	+1.05	+0.53	X	1954	
TiC	900	2	+2.33	+1.16	X	1954	
TiC	900	–	—	+1.21	[659]	1953	N. K. Golubeva
TiC	1000	1	+1.6	+0.8	X	1954	N. K. Golubeva
TiC	1000	2	+1.85	+0.93	X	1954	

RESISTANCE TO OXIDATION (continued)

Phase	Temperature, °C	Oxidation time, hr	Change in weight		Ref.	Year	Remarks
			mg/cm²	mg/cm²·hr			
TiC	1000	2	+7.15	+3.57	[658]	1953	
TiC	1100	2	+9.83	+4.81	[658]	1953	
TiC	1150	8	+6	+0.75	[292]	1952	
TiC	1150	16	+7	+0.43	[292]	1952	
TiC	1150	24	+7	+0.29	[292]	1952	
TiC	1150	32	+9	+0.29	[292]	1952	
TiC	1150	48	Disintegrates		[292]	1952	
TiC	1200	2	+42.5	+21.2	[270]	1955	N. K. Golubeva
ZrC	450	1	+19.35	+9.7	×	1954	
ZrC	450	2	+61.8	+30.9	×	1954	
ZrC	900	—	—	+46.0	[659]	1953	
ZrC	1100—1400	Active oxidation begins	—	—	[947]	1949	
VC	900	—	—	+73.5	[659]	1953	
Nb₂C	300	3	Compound formed of Nowotny phase type		[1200]	1963	
NbC	450	1	+1.39	+0.69	×	1954	N. K. Golubeva
NbC	450	2	+4.96	+2.48	×	1954	
NbC	600	1	+11.7	+5.8	×	1954	
NbC	900	—	—	+20.5	[659]	1953	
NbC	1100—1400	Active oxidation begins	—	—	[947]	1949	
TaC	800	1	+0.493	+0.25	×	1954	N. K. Golubeva
TaC	800	2	+1.29	+0.65	×	1954	
TaC	900	1	+10.0	+5.0	×	1954	
TaC	900	2	+39.4	+19.7	×	1954	

RESISTANCE TO OXIDATION (continued)

Phase	Temperature, °C	Oxidation time, hr	Change in weight		Ref.	Year	Remarks
			mg/cm²	mg/cm²·hr			
TaC	900	—	—	+20.5	[659]	1953	
TaC	1100—1400	Active oxidation begins	—	—	[947]	1949	[659]
$Cr_{23}C_6$	800	1	0	0	[660]	1961	
$Cr_{23}C_6$	800	2	0	0	[660]	1961	
$Cr_{23}C_6$	900	1	0	0	[660]	1961	
$Cr_{23}C_6$	900	2	0	0	[660]	1961	
$Cr_{23}C_6$	1000	1	0	0	[660]	1961	
$Cr_{23}C_6$	1000	2	0	0	[660]	1961	
$Cr_{23}C_6$	1100	1	0	0	[660]	1961	
$Cr_{23}C_6$	1100	2	0	0	[660]	1961	
Cr_7C_3	800	1	+8.7	+8.7	[660]	1961	
Cr_7C_3	800	2	+12.1	+6.8	[660]	1961	
Cr_7C_3	800	3	+12.8	+6.4	[660]	1961	
Cr_7C_3	800	4	+12.8	+6.4	[660]	1961	
Cr_7C_3	900	1	+28.7	+14.3	[660]	1961	
Cr_7C_3	900	2	+35.4	+17.7	[660]	1961	
Cr_7C_3	900	3	+42.1	+21.0	[660]	1961	
Cr_7C_3	900	4	+47	+23.5	[660]	1961	
Cr_7C_3	900	1	—	+0.11 (?)	[659]	1953	
Cr_7C_3	1000	1	+69.9	+35	[660]	1961	
Cr_7C_3	1000	2	+116.9	+58.4	[660]	1961	
Cr_7C_3	1000	3	+142.5	+71.3	[660]	1961	
Cr_3C_2	800	1	0	0	[660]	1961	

RESISTANCE TO OXIDATION (continued)

Phase	Temperature, °C	Oxidation time, hr	Change in weight		Ref.	Year	Remarks
			mg/cm²	mg/cm²·hr			
Cr_3C_2	800	2	0	0	[660]	1961	
Cr_3C_2	800	3	0	0	[660]	1961	
Cr_3C_2	800	4	0	0	[660]	1961	
Cr_3C_2	900	1	0	0	[660]	1961	
Cr_3C_2	900	2	0	0	[660]	1961	[652]
Cr_3C_2	900	3	0	0	[660]	1961	
Cr_3C_2	900	4	0	0	[660]	1961	
Cr_3C_2	1000	1	0	0	[660]	1961	
Cr_3C_2	1000	2	0	0	[660]	1961	
Cr_3C_2	1000	3	0	0	[660]	1961	
Cr_3C_2	1000	4	0	0	[660]	1961	
Cr_3C_2	1100	1	0	0	[660]	1961	
Cr_3C_2	1100	2	0	0	[660]	1961	
Cr_3C_2	1100	3	0	0	[660]	1961	
Cr_3C_2	1100	4	0	0	[660]	1961	
WC	700	1	+16.5	+8.3	×	1954	
WC	700	2	+18.2	+9.1	×	1954	N. K. Golubeva
WC	1000	1	+27.4	+13.7	×	1954	
WC	1000	2	+37.6	+18.8	×	1954	
WC	900	1	+1.14	+114	[659]	1953	
WC	500—520	Rapid oxidation	—	—	[662]	1955	
WC	530	Complete combustion	—	—	[662]	1955	Fine powder

RESISTANCE TO OXIDATION (continued)

Phase	Temperature, °C	Oxidation time, hr	Change in weight mg/cm²	Change in weight mg/cm²·hr	Ref.	Year	Remarks
WC	565	Oxidation begins	—	—	[662]	1955	Coarse powder
TiN	700	1	+16	+16	[266]	1955	[1053]
TiN	800	1	+17	+17	[266]	1955	[1053]
TiN	900	1	+18	+18	[266]	1955	[1053]
TiN	1000	1	+25	+25	[266]	1955	[1053]
$BaSi_2$	1300	1.0	+4.9	+4.9	[917]	1961	[1289]
$BaSi_2$	1300	2.3	+8.25	+3.58	[917]	1961	
$BaSi_2$	1300	4.5	+10.6	+2.35	[917]	1961	
$BaSi_2$	1300	6.5	+13.0	+2	[917]	1961	$K = 2.2 \cdot 10^{-5}$ g/cm²·hr
$BaSi_2$	1300	10.5	+15.2	+1.45	[917]	1961	(1300°C)
$BaSi_2$	1300	14.7	+16.7	+1.1	[917]	1961	
$LaSi_2$	1300	0.5	+12.7	+25.4	[917]	1961	
$LaSi_2$	1300	1.0	+16.2	+16.2	[917]	1961	
$DySi_2$	1000	0.5	+2.60	+5.20	[285]	1956	
$DySi_2$	1000	1	+4.34	+4.34	[285]	1956	
$DySi_2$	1000	18	+21.4	+1.19	[285]	1956	
$DySi_2$	1000	42	+35.8	+0.85	[285]	1956	
$DySi_2$	1000	66	+44.5	+0.67	[285]	1956	
$DySi_2$	1000	96	+52.1	+0.54	[285]	1956	
$DySi_2$	1000	398	+80.0	+0.20	[285]	1956	
$DySi_2$	1000	782	+90.8	+0.16	[285]	1956	
U_3Si	260	1	-50	-50	[690]	1960	Specimen disintegrated
U_3Si	345	1	+1000	+1000	[690]	1960	

RESISTANCE TO OXIDATION (continued)

Phase	Temperature, °C	Oxidation time, hr	Change in weight		Ref.	Year	Remarks
			mg/cm²	mg/cm²·hr			
$TiSi_2$	1260	100	+2.3	+0.023	[1020]	1961	
$TiSi_2$	1200	2	+0.3	+0.07	[423]	1952	
$TiSi_2$	980	200	−0.02	−0.0001	[778]	1955	
$ZrSi_2$	1100	4	+38.0	+9.5	[428]	1954	
Zr_2Si	1100	4	+29.0	+7.2	[428]	1954	
Zr_5Si_3	1100	4	+26.9	+6.7	[428]	1954	
$ZrSi$	1100	4	+3.0	+0.75	[428]	1954	
$ZrSi$	1200	1	+42	+10.5	[423]	1952	
VSi	1250	1	−63	−63	[650]	1956	
V_5Si_3	1250	1	−7.7	−7.7	[650]	1956	
V_5Si_3	1400	1	−600	−600	[650]	1956	
VSi_2	1200	4	+4.9	+1.2	[423]	1952	
VSi_2	1250	1	+25	+25	[650]	1956	
VSi_2	1400	1	−12	−12	[650]	1956	
$NbSi_2$	1000	1	−3.0	−3.0	[655]	1958	
$NbSi_2$	1000	2	−96	−48	[655]	1958	
$NbSi_2$	1200	4	−54	−13.5	[423]	1952	
$Ta_{4.5}Si$	1500	1	+240	+240	[419]	1953	
Ta_5Si_3	1500	1	+125	+125	[419]	1953	
$TaSi_2$	1500	1	+2	+2	[419]	1953	
$TaSi_2$	1200	4	−51	−12.7	[423]	1952	
Cr_3Si	1300	4	+11.4	+2.9	[414]	1953	
Cr_3Si	1260	100	+7.3	+0.073	[1020]	1961	
Cr_5Si_3	1300	4	+12.5	+3.1	[414]	1953	

RESISTANCE TO OXIDATION (continued)

Phase	Temperature, °C	Oxidation time, hr	Change in weight mg/cm²	Change in weight mg/cm²·hr	Ref.	Year	Remarks
$CrSi_2$	1300	4	+8.7	+2.2	[414]	1953	[917]
$CrSi_2$	1300	0.5	-0.9	-1.8	[917]	1961	
$CrSi_2$	1300	1.8	-0.9	-0.5	[917]	1961	
$CrSi_2$	1300	3.0	-0.7	-0.26	[917]	1961	
$CrSi_2$	1300	12.2	-1.4	-0.11	[917]	1961	
$CrSi_2$	1200	4	+51	+12.7	[423]	1952	
Mo_3Si	500	4	-812	-203	[423]	1952	
Mo_5Si_3	500	4	-67	-16.9	[423]	1952	
$MoSi_2$	1100	20	+1.4	+0.07	[651]	1960	
$MoSi_2$	1100	—	—	+0.07	[681]	1957	
$MoSi_2$	1150	8	+2	+0.25	[292]	1952	
$MoSi_2$	1150	16	+4	+0.25	[292]	1952	
$MoSi_2$	1150	24	+5	+0.21	[292]	1952	
$MoSi_2$	1150	32	+5	+0.15	[292]	1952	
$MoSi_2$	1150	48	+6	+0.13	[292]	1952	
$MoSi_2$	1150	200	+8	+0.04	[292]	1952	
$MoSi_2$	1200	4	+0.3	+0.07	[423]	1952	
$MoSi_2$	1200	20	+0.6	+0.03	[656]	1959	
$MoSi_2$	1200	50	+1.9	+0.037	[656]	1959	
$MoSi_2$	1200	75	+2.1	+0.028	[656]	1959	
$MoSi_2$	1200	100	+2.1	+0.021	[656]	1959	
$MoSi_2$	1500	4	+1.3	+0.32	[423]	1952	

RESISTANCE TO OXIDATION (continued)

Phase	Temperature, °C	Oxidation time, hr	Change in weight		Ref.	Year	Remarks
			mg/cm^2	$mg/cm^2 \cdot hr$			
MoSi$_2$	1095	75	$-2.25 \cdot 10^{-2}$	$-3 \cdot 10^{-4}$	[649]	1955	According to the data of various authors; [649]
MoSi$_2$	1095	150	-60	-0.4	[649]	1955	
MoSi$_2$	1200	200	$+200$	$+1.0$	[649]	1955	
MoSi$_2$	1200	300	$+21$	$+0.7$	[649]	1955	
MoSi$_2$	1320	50	$+250$	$+5.0$	[649]	1955	
MoSi$_2$	1320	100	$+400$	$+4.0$	[649]	1955	
MoSi$_2$	1565	100	-367	-3.67	[649]	1955	
MoSi$_2$	1565	135	-420	-3.10	[649]	1955	
W$_3$Si	1500	4	-445	-111	[423]	1952	
W$_5$Si$_3$	1500	4	-205	-51	[423]	1952	
WSi$_2$	1200	4	-17	-4.2	[423]	1952	
WSi$_2$	1500	4	-23	-5.9	[423]	1952	
MnSi$_2$	1200	0.5	$+2.5$	$+5$	[917]	1961	
MnSi$_2$	1200	1.0	$+3.6$	$+3.6$	[917]	1961	
MnSi$_2$	1200	2.0	$+5.4$	$+2.7$	[917]	1961	
MnSi$_2$	1200	4.0	$+6.8$	$+1.7$	[917]	1961	
MnSi$_2$	1200	6.0	$+7.5$	$+1.25$	[917]	1961	
ReSi$_2$	1400	0.5	$+3.2$	$+6.4$	[917]	1961	
ReSi$_2$	1400	1.0	$+7.15$	$+7.15$	[917]	1961	
ReSi$_2$	1400	3.0	$+7.3$	$+2.43$	[917]	1961	
ReSi$_2$	1400	4.3	$+7.2$	$+1.67$	[917]	1961	
ReSi$_2$	1400	6.0	$+7.2$	$+1.2$	[917]	1961	
ReSi$_2$	1400	8.3	$+7.3$	$+0.9$	[917]	1961	
FeSi$_2$	1200	0.5	-0.5	-1	[917]	1961	

RESISTANCE TO OXIDATION (continued)

Phase	Temperature, °C	Oxidation time, hr	Change in weight		Ref.	Year	Remarks
			mg/cm²	mg/cm²·hr			
FeSi₂	1200	1.0	+0.4	+0.4	[917]	1961	
FeSi₂	1200	2.0	+1.8	+0.9	[917]	1961	
FeSi₂	1200	4.0	+2.7	+0.67	[917]	1961	
FeSi₂	1200	6.0	+3.3	+0.55	[917]	1961	[942]
B₄C	1100	20	−0.8	−0.04	[651]	1960	
B₄C	1200	5	−1.11	−0.22	[654]	1960	
B₄C	1200	25	−3.88	−0.15	[654]	1960	
B₄C	1200	50	−8.1	−0.16	[654]	1960	
B₄C	1200	100	−11.3	−0.11	[654]	1960	
SiC	1400	50	−5.2%	−0.14 %	[429]	1950	
SiC	1400	100	+8.2%	+0.082%	[429]	1950	
SiC	1400	200	+9.2%	+0.041%	[429]	1950	[1000. 1064]
SiC	1400	500	+16.1%	+0.032%	[429]	1950	
SiC	1400	1200	+20.7%	+0.017%	[429]	1950	
BN	700	2	−0.014	−0.007	[272]	1955	
BN	700	10	−0.062	−0.006	[272]	1955	
BN	700	30	−0.138	−0.0046	[272]	1955	
BN	700	60	−0.235	−0.004	[272]	1955	
BN	1000	2	−0.35	−0.175	[272]	1955	
BN	1000	10	−0.85	−0.085	[272]	1955	
BN	1000	30	−4.8	−0.16	[272]	1955	
BN	1000	60	−10.0	−0.167	[272]	1955	
Si₃N₄	1200	80	+5	+0.06	[270]	1955	[942]

RESISTANCE TO THE ACTION OF CHLORINE

Phase	Temperature, °C	Chlorination time, hr	Variation in weight		Ref.	Year	Remarks
			mg/cm²	mg/cm²·hr			
CaB₆, SrB₆ and other alkaline earth borides	900	—	—	—	[957]	1897	[958]
TiC	400—700	—	—	—	[1]	1957	Powder, readily chlorinates, forming chlorides, oxychlorides, sulfochlorides; [400]
ZrC, TaC, NbC	700—800	—	—	—	[1]	1957	Powder readily decomposes
VC	300—500	—	—	—	[1]	1957	[937]
Cr₃C₂	To 900—1000	—	—	—	[1]	1957	Resistant powder
Mo₂C	1000—1200	—	—	—	[1]	1957	Powder decomposes
MoC	900—1000	—	—	—	[791]	1904	Powder decomposes, MoCl₆ and C formed; [921]
W₂C	400	—	—	—	[1]	1957	Powder reacts, WCl₆ and C formed
WC	To 500—700	—	—	—	[968]	1914	Resistant
AlN	1400	—	—	—	[117]	1959	
AlN	To 900	—	—	—	[231]	1947	Compact reacts, AlCl₃ formed
TiSi₂	900	—	—	—	[256]	1954	
ZrSi₂	900	—	—	—	[256]	1954	Decomposes
HfSi₂	900	—	—	—	[256]	1954	

RESISTANCE TO THE ACTION OF CHLORINE (continued)

Phase	Temperature, °C	Chlorination time, hr	Variation in weight mg/cm²	Variation in weight mg/cm²·hr	Ref.	Year	Remarks
MoSi$_2$	1000	2	—1500	—750	[117]	1959	Compact
WSi$_2$	200—300	—	—	—	[647]	1956	Decomposes, WCl$_6$ formed
SiC*	600	—	—	—	[959]	1938	Surface disintegrates
SiC*	900—1000	—	—	—	[959]	1938	SiC + 2Cl$_2$ → SiCl$_4$ + C
SiC*	1100—1200	—	—	—	[959]	1938	SiC + 4Cl$_2$ → SiCl$_4$+CCl$_4$
SiC	200	6	0.0	—	[1000]	1961	
SiC	200	24	8.4	—	[1000]	1961	
SiC	200	48	0.9	—	[1000]	1961	
SiC	200	72	0.6	—	[1000]	1961	
SiC	200	120	0.3	—	[1000]	1961	
SiC	200	192	0.3	—	[1000]	1961	
SiC	200**	6	0.0	—	[1000]	1961	
SiC	200**	24	0.0	—	[1000]	1961	
SiC	200**	48	—1.2	—	[1000]	1961	
SiC	200**	120	—0.3	—	[1000]	1961	** Moist Cl$_2$
SiC	300	6	13.5	—	[1000]	1961	
SiC	300	24	3.9	—	[1000]	1961	
SiC	300	48	1.2	—	[1000]	1961	
SiC	300	120	1.8	—	[1000]	1961	
SiC	300	192	0.9	—	[1000]	1961	
SiC	300	264	0.9	—	[1000]	1961	
SiC	300	336	0.9	—	[1000]	1961	
SiC	400	6	21.7	—	[1000]	1961	
SiC	400	24	8.7	—	[1000]	1961	

RESISTANCE TO THE ACTION OF CHLORINE (continued)

Phase	Temperature, °C	Chlorination time, hr	Variation in weight		Ref.	Year	Remarks
			mg/cm²	mg/cm²·hr			
SiC	400	48	4.5	—	[1000]	1961	
SiC	400	120	2.7	—	[1000]	1961	
SiC	400	192	2.1	—	[1000]	1961	
SiC	400	264	2.1	—	[1000]	1961	
SiC	400	336	2.7	—	[1000]	1961	
SiC	400*	48	7.2	—	[1000]	1961	
SiC	400*	120	4.2	—	[1000]	1961	
SiC	400*	192	3.0	—	[1000]	1961	
SiC	400*	264	1.8	—	[1000]	1961	
SiC	500	6	24	—	[1000]	1961	
SiC	500	24	3040	—	[1000]	1961	
SiC	500	72	1142	—	[1000]	1961	
SiC	500	144	580	—	[1000]	1961	
SiC	500	216	387	—	[1000]	1961	
SiC	500	288	290	—	[1000]	1961	
SiC	500	360	290	—	[1000]	1961	
SiC	500	432	193	—	[1000]	1961	
SiC	600	6	4840	—	[1000]	1961	
SiC	600	24	4444	—	[1000]	1961	
SiC	600	120	1080	—	[1000]	1961	
SiC	600	192	700	—	[1000]	1961	
SiC	600	264	533	—	[1000]	1961	
SiC	600	336	433	—	[1000]	1961	
SiC	600	408	378	—	[1000]	1961	
SiC	600	480	340	—	[1000]	1961	

RESISTANCE TO THE ACTION OF CHLORINE (continued)

Phase	Temperature, °C	Chlorination time, hr	Variation in weight		Ref.	Year	Remarks
			mg/cm²	mg/cm²·hr			
SiC	600**	48	322	—	[1000]	1961	
SiC	600**	120	133	—	[1000]	1961	
SiC	600**	192	85	—	[1000]	1961	
SiC	600**	264	63	—	[1000]	1961	
SiC	800	6	22600		[1000]	1961	
SiC	1000	6	32300		[1000]	1961	Reacts, BCl₃ and C formed; [269]
B₄C	<1000	—	—	—	[257]	1954	
BN	700	3	—0.25	—0.012	[272]	1955	Compact; [943]
BN	700	20	—0.55	—0.014	[272]	1955	
BN	700	40	—2.7	—0.9	[272]	1955	Compact; [943]
BN	1000	3	—17.0	—0.85	[272]	1955	
BN	1000	20			[272]	1955	Compact resistant
Si₃N₄	To 900	500	—0.7	—0.35	[270]	1955	Powder decomposes
Si₃N₄	350—240	2			[270]	1955	Reacts, SiCl₄ and BCl₃ liberated
SiB₃	900				[209]	1950	
SiB₆	900				[209]	1950	

Remarks, 1. For all cases of the determination of the resistance of SiC to the action of chlorine, with the exception of those marked *, the fourth and fifth columns give the rate of decomposition (corrosion, mm/min) determined for the chlorination time given in the column "Chlorination time." Resistance was determined on specimens of the composition (%): 96.5 SiC, 2.5 Si_free, 0.4 C_free, 0.4 Al, 0.2 Fe; P = 4%. 2. For the cases marked **, the resistance was determined in moist chlorine.

289

WETTABILITY BY MOLTEN METALS

Phase	Wetting metal	Temperature, °C	Contact angle, $\Theta°$	Ref.	Year	Remarks
TiB_2	Cu	1100—1500	158—154	[506]	1958	Argon
TiB_2	Ni	1480 (at moment of melting)	100	[506]	1958	Helium
TiB_2	Ni	1480 (20 min)	38.5	[506]	1958	„
ZrB_2	Cu	1160—1400	123—36	[506]	1958	
VB_2	Cu	1100—1400	150—114	[506]	1958	
TaB_2	Ag	1300	118	[506]	1958	Argon
TaB_2	Cu	1100—1400	77—47	[506]	1958	
CrB_2	Cu	1480	50	[506]	1958	
CrB_2	Ni	1480 (1 min)	11	[506]	1958	Helium
Mo_2B_5	Ni	1480 (1 min)	8	[506]	1958	
Be_2C	Si	1450	54	[509]	1954	Hydrogen
Be_2C	Si	1450	63	[509]	1954	Helium
Be_2C	Ni	1500	92	[509]	1954	Vacuum
Be_2C	Ni	1500	90	[509]	1954	Hydrogen
Be_2C	Ni	1500	75	[509]	1954	Helium
Al_4C_3	Al	1000	104	[508]	1952	Vacuum
UC	Zn	High wettability	—	[1014]	1959	
UC	Sn	Does not wet	—	[1014]	1959	
UC	Silumin	Does not wet	—	[1014]	1959	
TiC	Ag	980	108	[507]	1956	Vacuum, [1065. 1086]
TiC	Cu	1100—1300	108—70	[507]	1956	
TiC	Pb	400—1000	152—90	[507]	1956	Argon
TiC	Pb	650	120	[507]	1956	
TiC	Pb + 0.3% Ni	660	98	[507]	1956	Vacuum
TiC	Bi	300—600	138—122	[507]	1956	
TiC	Zn	600	120	[508]	1952	„
TiC	Al	700	118	[508]	1952	Argon
TiC	Fe	1550	39	[509]	1954	Hydrogen
TiC	Fe	1550	36	[509]	1954	Helium
TiC	Fe	1550	41	[509]	1954	Vacuum
TiC	Co	1500	36	[509]	1954	Hydrogen
TiC	Co	1500	39	[509]	1954	Helium
TiC	Co	1500	5	[509]	1954	Vacuum

Phase	Wetting metal	Temperature, °C	Contact angle, Θ°	Ref.	Year	Remarks
TiC	Ni	1450	17	[509]	1954	Hydrogen
TiC	Ni	1450	32	[509]	1954	Helium
TiC	Ni	1450	30	[509]	1954	Vacuum, [1325]
TiC	Ni	1500	0	[506]	1958	Argon
TiC	Ni + 10% Ti		25	[509]	1954	
TiC	Ni + 10% Cr		23	[509]	1954	
TiC	Ni + 10% Mn		23	[509]	1954	
TiC	Ni + 10% Zr		22	[509]	1954	
TiC	Ni + 10% Nb	At the melting point	22	[509]	1954	Vacuum
TiC	Ni + 10% V		21	[509]	1954	
TiC	Ni + 10% Ta		15	[509]	1954	
TiC	Ni + 10% W		14	[509]	1954	
TiC	Ni + 10% Mo		0	[509]	1954	
ZrC	Cu	1100	135	[507]	1956	Argon
ZrC	Cu	1100—1500	140—118	[507]	1956	Vacuum
ZrC	Cu + 0.01% Ni	1200	96	[507]	1956	
ZrC	Cu + 0.05% Ni	1200	70	[507]	1956	
ZrC	Cu + 0.01% Ni	1200	63	[507]	1956	Vacuum
ZrC	Cu + 0.25% Ni	1200	54	[507]	1956	
VC	Ni	200—400	158—90	[507]	1956	Argon, [1325]
VC	Cu	1090—1200	54—39	[507]	1956	
NbC	Ni	1450	21.2	[1325]	1963	Vacuum
TaC	Cu	1100—1250	75—36	[507]	1956	Vacuum
Cr₇C₃	Ni	1500	Instantaneous attack on the surface of carbide	[506]	1958	Argon
Cr₃C₂	Ni	1500	0 (Rapid spreading)	[506]	1958	Argon
WC	Cu	1100	30	[507]	1956	
WC	Cu	1100	20	[510]	1952	
WC	Sn	500—1300	120—30	[507]	1956	Argon
WC	Bi	700—1100	140—52	[507]	1956	
WC	Co	1500	0	[511]	1954	Hydrogen
WC	Co	1500	0	[510]	1952	
UC	Na	240—400	165—141	[507]	1956	Argon
UC	Cu	1100—1260	113—69	[507]	1956	
UC	Cu + 6% Ni	1100—1330	93—45	[507]	1956	Vacuum
UC	Bi	300—700	140—93	[507]	1956	
UC	Bi + 0.3% Ni	350—650	141—52	[507]	1956	
B₄C	Zn	540—620	121.5—119	[440]	1960	

Phase	Wetting metal	Tempera- ture, °C	Contact angle, $\Theta°$	Ref.	Year	Remarks
B_4C	Cu	995—1090	130—17	[440]	1960	Vacuum
B_4C	Al	600—670	117—118	[440]	1960	
B_4C	Pb	225—395	121—113.5	[440]	1960	
B_4C	Fe	1780	Strong reaction	[383]	1952	Helium
B_4C	Co	1780	>90	[383]	1952	
B_4C	Ni	1780	>90	[383]	1952	„
B_4C	Brass	905—950	54.5—30.0	[440]	1960	Vacuum
TiC	Ni	1500	38	[1074]	1962	Vacuum, [1149]
30% TiC + + 70% WC (sol. solution)	Ni	1500	21	[1074]	1962	Same
WC	Ni	1500	~0	[1074]	1962	„
TiC	Ni	1300	~16	[1065]	1960	Taken from graph. [1086]
TiC	Ni + 20% Cu	1300	~12	[1065]	1960	
TiC	Ni + 46% Cu	1300	~10	[1065]	1960	
TiC	Ni + 83% Cu	1300	~13	[1065]	1960	
TiC	Ni + 95% Cu	1300	~32	[1065]	1960	
	Cu	1300	~48	[1065]	1960	

RESISTANCE TO THE ACTION OF MOLTEN SALTS, ALKALIS, OXIDES

Phase	Composition of melt	Temperature, °C	Nature of reaction	Ref.	Year	Remarks
MgB₂	Na_2CO_3	800	Decomposes	[943]	1925	
MgB₂	NaOH	550	"	[943]	1925	
CaB₆	KOH	800	"	[957]	1897	
SrB₆ and other alkaline earth borides	K_2CO_3	800	"	[957]	1897	[958]
	$KHSO_4$	800	"	[957]	1897	
	PbO_2	800	"	[957]	1897	
	KNO_3	800	"	[957]	1897	
AlB₂	Na_2CO_3	800	"	[257]	1954	
AlB₂	NaOH*	550	"	[257]	1954	
TiB₂	NaOH	550	"	[257]	1954	[896]
TiB₂	Na_2CO_3	800	"	[257]	1954	
TiB₂	$KHSO_4$	200—300	"	[257]	1954	
TiB₂	PbO_2	900	Reacts violently	[257]	1954	
TiB₂	Na_2O_2	750	Same	[257]	1954	
ZrB₂	NaOH	550	Decomposes	[257]	1954	[896]
ZrB₂	$KHSO_4$	800	"	[257]	1954	
ZrB₂	PbO_2	300	"	[257]	1954	
ZrB₂	Na_2O_2	750	"	[257]	1954	
VB₂	NaOH	550	"	[257]	1954	
VB₂	K_2CO_3	800—900	"	[257]	1954	Instantaneously
VB₂	Na_2O_2	750	"	[257]	1954	

RESISTANCE TO THE ACTION OF MOLTEN SALTS, ALKALIS, OXIDES (continued)

Phase	Composition of melt	Temperature, °C	Nature of reaction	Ref.	Year	Remarks
NbB_2	NaOH	550	Decomposes	[257]	1954	
NbB_2	Na_2CO_3	800	"	[257]	1954	
TaB_2	NaOH	550	"	[257]	1954	Very quickly [896]
TaB_2	Na_2CO_3	800	"	[257]	1954	
TaB_2	$KHSO_4$	200—300	"	[257]	1954	
TaB_2	Na_2O_2	750	"	[257]	1954	
CrB_2	NaOH	550	"	[257]	1954	
MoB_2	NaOH	550	"	[257]	1954	
WB_2	Na_2CO_3	800	"	[257]	1954	
WB_2	KNO_3	350	"	[257]	1954	
Ni_3B	$NaOH + Na_2O_2$	650—700	"	[257]	1954	
UB_2	NaOH	550	"	[257]	1954	H_2 liberated
UB_4, UB_{12}	PbO_2	900	"	[257]	1954	
UB_4, UB_{12}	Na_2O_2	750	"	[257]	1954	$TiC : NaOH = 1 : 16$
TiC	NaOH	650	"	[943]	1932	
TiC	NaOH	900	"	[943]	1932	
TiC	Na_2O_2	750	"	[943]	1932	
ZrC	NaOH	650	"	[943]	1932	
AlN	KOH	400	"	[943]	1913	
TiN	NaOH	650	"	[943]	1932	Reacts slowly, NH_3 liberated
ZrN	NaOH	650	"	[943]	1932	
VN	NaOH	650	"	[943]	1932	
Nb_2N	NaOH	650	"	[943]	1932	

RESISTANCE TO THE ACTION OF MOLTEN SALTS, ALKALIS, OXIDES (continued)

Phase	Composition of melt	Temperature °C	Nature of reaction	Ref.	Year	Remarks
NbN	NaOH	350	Decomposes	[697]	1959	N_2 liberated
WN	Na_2CO_3	800	"	[1]	1957	NH_3 liberated, Na_2WO_4 formed
LaSi₂	NaOH	650—700	"	[256]	1954	In 20 min
ThSi₂	NaOH	550	"	[256]	1954	Very readily
TiSi₂ ZrSi₂ HfSi₂	NaOH	650—700	"	[256]	1954	In 20 min
HfSi₂	Na_2CO_3	650—700	"	[697]	1959	
HfSi₂	$Na_2B_4O_7$	650—700	"	[256]	1954	
HfSi₂	$KHSO_4$	200—300	Resistant	[697]	1959	
VSi₂	$NaOH + Na_2CO_3$	650—700	"	[256]	1954	
NbSi₂	$NaOH + Na_2CO_3$	650—700	"	[256]	1954	
TaSi₂	$NaOH + Na_2CO_3$	650—700	"	[256]	1954	
CrSi Cr₃Si₂ Cr₂Si CrSi₂	$K_2CO_3 + NaNO_3$	650—700	Decomposes	[117]	1959	Silicates and chromates formed
MoSi₂	$K_2CO_3 + KNO_3$	650—700	Reacts actively	[256]	1954	
MoSi₂	NaOH	400—500	Same	[117]	1959	
WSi₂	$K_2CO_3 + KNO_3$	650—700	Decomposes	[256]	1954	
MnSi MnSi₂	$K_2CO_3 + KNO_3$	650—700	"	[256]	1954	Very readily

RESISTANCE TO THE ACTION OF MOLTEN SALTS, ALKALIS, OXIDES (continued)

Phase	Composition of melt	Temperature °C	Nature of reaction	Ref.	Year	Remarks
FeSi CoSi NiSi	NaOH	550	Decomposes	[256]	1954	
FeSi	K_2CO_3	800—900	"	[256]	1954	
B_4C	$BaCO_3$	600—700	"	[963]	1961	
B_4C	$Na_2CO_3 + NaNO_3$	600—700	"	[964]	1951	
B_4C	$NaOH + NaNO_3$	600—700	"	[257]	1954	
B_4C	$NaOH + NaNO_3$	600—700	"	[964]	1951	
B_4C	CaO, MgO	600—700	"	[963]	1961	
SiC	$Na_2B_4O_7$	750	"	[959]	1938	
SiC	Na_2CO_3	800	"	[673]	1938	
SiC	$K_2Cr_2O_7$	400	"	[959]	1938	[960, 961]
SiC	$PbCrO_4$	850	"	[959]	1938	In 1 hr
SiC	Na_2SO_4	900	"	[959]	1938	[960, 961]
SiC	$NaOH$	550	"	[673]	1938	[960, 961]
SiC	Na_2O_2	750	"	[673]	1938	
SiC	CuO	800	"	[959]	1938	
SiC	CaO	800	"	[959]	1938	Silicates of the alkali metal formed, [960, 961]
SiC	MgO	1000	"	[959]	1938	
SiC	SiO_2	2000—2500	$SiO_2 + SiC = Si + CO_2$	[959]	1938	
SiC	PbO_2	900	Decomposes	[673]	1938	
SiC	Cr_2O_3	1370	"	[959]	1938	
SiC	MnO	1360	"	[959]	1938	
SiC	FeO	1360	"	[959]	1938	
SiC	NiO	1300	"	[959]	1938	

RESISTANCE TO THE ACTION OF MOLTEN SALTS, ALKALIS, OXIDES (continued)

Phase	Composition of melt	Temperature °C	Nature of reaction	Ref.	Year	Remarks
SiC	NaOH	350	1725	[1000]	1961	The fourth column gives the rate of disintegration (corrosion, mm/min). Comp. %: 96.5 SiC, 2.5 Si free, 0.4 C free. 0.4 Al, 0.2 Fe; Porosity 4%
SiC	NaOH	500	33100	[1000]	1961	
SiC	Na_2CO_3	900	>360	[1000]	1961	
SiC	LiCl	900	1038	[1000]	1961	
SiC	NaCl	900	9.3	[1000]	1961	
SiC	KCl	900	322	[1000]	1961	
SiC	$MgCl_2$	900	152	[1000]	1961	
SiC	$CaCl_2$	900	3000	[1000]	1961	
SiC	LiF	900	2360	[1000]	1961	
BN	Sb_2O_3	—	Reacts			
BN	Cr_2O_7					
BN	MoO_3					
BN	As_2O_3	—	Decomposes	[967]	1932	[966]
BN	K_2CO_3 (anhydrous)	800—900	Decomposes	[943]	1925	
BN	NaOH (anhydrous)	400—500	„	[270]	1955	
Si_3N_4	Na_2O_2	—	Partly decomposes	[505]	1957	
Si_3N_4	$PbCrO_4$, PbO_2, PbO		Decomposes	[505]	1957	
Si_3N_4	NaCl + KCl	900	Decomposes after 144 hr	[270]	1955	
Si_3N_4	$NaB(SiO_3)_2$ + V_2O_5	1100	Decomposes after 4 hr	[270]	1955	
Si_3N_4	$NaF + ZrF_4$	850	Decomposes after 100 hr	[270]	1955	Reduced to metal
SiB_6	K_2CO_3 (anhydrous)	800—900	Decomposes	[209]	1950	
SiB_3	K_2CO_3 (anhydrous)	800—900	„	[209]	1950	

RESISTANCE TO THE ACTION OF MOLTEN METALS, ALLOYS, AND SLAGS

Phase	Composition of melt	Temperature, °C	Time of contact, hr	Nature of reaction	Content of components of compound passing into metal	Ref.	Year	Remarks
TiB_2	Zn	550	80		Ti not found	[932]	1961	Air
TiB_2	Zn	940	240		—	[932]	1961	—
TiB_2	Cd	450	10	Does not react	Ti traces	[646]	1960	
TiB_2	Cd	450	40		Ti 0.026%	[646]	1960	Air
TiB_2	Cd	450	80		—	[932]	1961	
TiB_2	Al	1000	0.2	Reacts actively	—	[932]	1961	Argon
TiB_2	Si	1550	0.3		—	[932]	1961	
TiB_2	Sn	350	10		Ti traces	[646]	1960	
TiB_2	Sn	350	40		Ti 0.01%	[646]	1960	
TiB_2	Sn	350	80		—	[932]	1960	
TiB_2	Pb	450	10	Does not react	Ti traces	[646]	1960	Air
TiB_2	Pb	450	40		Ti 0.06%	[646]	1960	
TiB_2	Pb	450	80		—	[932]	1961	
TiB_2	Bi	375	10		Ti traces	[646]	1960	
TiB_2	Bi	375	40		Ti 0.05%	[646]	1960	
TiB_2	Bi	375	80	Does not react	—	[932]	1960	
TiB_2	Cr	1900	0.3	Reacts actively	—	[932]	1960	
TiB_2	Co	1550	0.3		—	[932]	1961	$Co + N_2$
TiB_2	Ni	1500	0.3	Reacts	—	[932]	1961	
TiB_2	Carbon steel	1600	0.1		—	[646]	1960	
TiB_2	Cast iron	1600	0.1		—	[646]	1960	

RESISTANCE TO THE ACTION OF MOLTEN METALS, ALLOYS, AND SLAGS (continued)

Phase	Composition of melt	Temperature, °C	Time of contact, hr	Nature of reaction	Content of components of compound passing into metal	Ref.	Year	Remarks
TiB_2	Cryolite	1050	19.5		—	[932]	1961	In the melt
ZrB_2	Zn	550	80		—	[932]	1961	"
ZrB_2	Zn	940	180	Does not react	Zr traces	[932]	1961	"
ZrB_2	Cu	450	80		Zr absent	[932]	1961	"
ZrB_2	Al	1000	0.2	Reacts slightly	—	[932]	1961	Argon
ZrB_2	Si	1550	0.2		—	[932]	1961	
ZrB_2	Sn	350	80	Does not react	Zr traces	[932]	1961	Air
ZrB_2	Pb	450	80		Zr traces	[932]	1961	
ZrB_2	Bi	375	80	Reacts slightly	Zr absent	[932]	1961	
ZrB_2	Cr	1900	0.2	Reacts	—	[932]	1961	
ZrB_2	Co	1550	0.2		—	[932]	1961	$CO + N_2$
ZrB_2	Ni	1500	0.3		—	[932]	1961	
ZrB_2	Carbon steel	1620	2		—	[932]	1961	$CO + N_2$
ZrB_2	Cast iron	1520	12		—	[932]	1961	Air
ZrB_2	Brass	900	86	Does not react	—	[932]	1961	
ZrB_2	Basic slag	1520	12		—	[932]	1961	$CO + N_2$
ZrB_2	Basic slag	1520	12		—	[932]	1961	
ZrB_2	Acid slag	1520	12		—	[932]	1961	In the melt
ZrB_2	Cryolite	1050	20	Reacts slightly	—	[932]	1961	

RESISTANCE TO THE ACTION OF MOLTEN METALS, ALLOYS, AND SLAGS (continued)

Phase	Composition of melt	Temperature, °C	Time of contact, hr	Nature of reaction	Content of components of compound passing into metal	Ref.	Year	Remarks
CrB_2	Cd	450	10	Does not react	Cr traces	[646]	1960	Air
CrB_2	Cd	450	40		Cr < 0.01%	[646]	1960	
CrB_2	Cd	450	80		Cr traces	[932]	1961	
CrB_2	Al	1000	0.2	Reacts slightly	—	[932]	1961	Argon
CrB_2	Si	1550	0.2	Reacts actively	—	[932]	1961	
CrB_2	Pb	450	10		Cr traces	[932]	1961	Air
CrB_2	Pb	450	40		Cr 0.01%	[646]	1960	
CrB_2	Pb	450	80	Does not react	Cr traces	[646]	1960	
CrB_2	Bi	375	80		Cr traces	[932]	1961	
CrB_2	Zn	940	132		—	[932]	1961	
C_1B_2	Steel KhVG	1620	0.1	Does not react	—	[646]	1960	
CrB_2	Cast iron	1520	0.1		—	[646]	1960	
CrB_2	Basic slag	1520	0.1	Reacts	—	[646]	1960	CO + N₂
CrB_2	Acid slag	1520	—	Reacts slightly	—	[646]	1960	
CrB_2	Cryolite	1050	20	Does not react	—	[932]	1961	
W_2B_5	Zn	940	168		—	[932]	1961	In the melt
W_2B_5	Cryolite	1050	8	Reacts	—	[932]	1961	

RESISTANCE TO THE ACTION OF MOLTEN METALS, ALLOYS, AND SLAGS (continued)

Phase	Composition of melt	Temperature, °C	Time of contact, hr	Nature of reaction	Content of components of compound passing into metal	Ref.	Year	Remarks
MnB	Zn	600	—	Does not react	—	[706]	1955	
NiB	Zn	600	—	Disintegrates	—	[706]	1955	
TiC	Zn	550	10	Does not react	—	[932]	1961	
TiC	Cd	450	10	Does not react	Ti 0.02%	[646]	1960	Air
TiC	Al	1000	0.1	Reacts slightly	Ti 0.01%	[932]	1961	
TiC	Si	1500	0.1	Reacts	—	[932]	1961	
TiC	Sn	350	10	Does not react	—	[932]	1961	
TiC	Pb	450	10	Does not react	—	[646]	1960	
TiC	Bi	375	10	Does not react	—	[646]	1960	
TiC	Co	1550	0.2	Reacts	Ti 0.01%	[932]	1961	CO + N_2
TiC	Ni	1500	0.3	Reacts	Ti 0.018%	[932]	1961	
TiC	Carbon steel	1620	0.3	Reacts	—	[646]	1960	
TiC	Cast iron	1520	0.3	Does not react	—	[646]	1960	Air
TiC	Basic slag	1520	0.1	Does not react	—	[646]	1960	
TiC	Slag	1520	0.1	Reacts slightly	—	[646]	1960	Air
TiC	Cryolite	1050	8	Reacts slightly	—	[932]	1961	
ZrC	Zn	550	6	Does not react	Zr 0.02%	[932]	1961	Argon
ZrC	Cd	450	10	Does not react	Zr 0.01%	[932]	1961	Air

301

RESISTANCE TO THE ACTION OF MOLTEN METALS, ALLOYS, AND SLAGS (continued)

Phase	Composition of melt	Temperature, °C	Time of contact, hr	Nature of reaction	Content of components of compound passing into metal	Ref.	Year	Remarks
ZrC	Al	1000	0.2	Reacts	—	[932]	1961	Argon
ZrC	Si	1500	0.2	Reacts	—	[932]	1961	
ZrC	Sn	350	10	Does not react	Zr not found	[932]	1961	Air
ZrC	Pb	450	10		Zr traces	[932]	1961	
ZrC	Bi	375	10			[932]	1961	
ZrC	Cr	1900	0.2	Reacts		[932]	1961	$CO + N_2$
ZrC	Co	1550	0.2			[932]	1961	
ZrC	Ni	1500	0.2			[932]	1961	
ZrC	Cryolite	1050	20	Dissolves slightly		[932]	1961	
Cr_3C_2	Zn	940	24	Reacts	—	[932]	1961	Air
Mo_2C	Zn	940	168	Dissolves slightly	—	[932]	1961	Argon
WC	Zn	940	144	Reacts slightly	—	[932]	1961	
WC	Cryolite	1050	3.5	Reacts	—	[932]	1961	In the melt
TiN	Cd	450	10	Reacts slightly	Ti 0.20%	[932]	1961	
TiN	Cd	450	40	Reacts slightly	Ti 0.07%	[646]	1960	
TiN	Sn	350	10	Does not react	Ti not found	[646]	1960	
TiN	Sn	350	40	Reacts slightly	Ti 0.26%	[646]	1960	Air
TiN	Pb	450	10	Reacts slightly	Ti 0.04%	[646]	1960	
TiN	Pb	450	40	Reacts slightly	Ti 0.20%	[646]	1960	

RESISTANCE TO THE ACTION OF MOLTEN METALS, ALLOYS, AND SLAGS (continued)

Phase	Composition of melt	Temperature, °C	Time of contact, hr	Nature of reaction	Content of components of compound passing into metal	Ref.	Year	Remarks
TiN	Bi	375	10	Does not react	Ti traces	[646]	1960	Air
TiN	Carbon steel	1620	0.1	Wets slightly	—	[646]	1960	
TiN	Cast iron	1520	0.1	Does not react	—	[646]	1960	CO + N_2
TiN	Basic slag	1520	0.3	Does not react	—	[646]	1960	
TiN	Acid slag	1520	0.1		—	[646]	1960	
TiN	Cryolite	1050	36	Reacts slightly	—	[932]	1961	
ZrN	"	1050	8	Does not react	Th,Si not found	[932]	1961	
ThSi$_2$	Cu	1130	0.5		Complex silicide; solidified melt has a two-phase structure	[834]	1959	
ThSi$_2$	Ni	1500	0.5	Complete solution		[834]	1959	
TiSi$_2$	Cu	1130	0.5	Complete solution	Ti, Si in solid solution	[834]	1959	
TiSi$_2$	Ni	1500	0.5	Complete solution	Ti, Si in solid solution	[834]	1959	
Ti$_5$Si$_3$	Cu	1130	1		—	[834]	1959	
Ti$_5$Si$_3$	Ni	1500	1	Does not react	—	[834]	1959	
ZrSi	Cu	1130	1		—	[834]	1959	
ZrSi	Ni	1500	1		—	[834]	1959	

303

Phase	Composition of melt	Temperature, °C	Time of contact, hr	Nature of reaction	Content of components of compound passing into metal	Ref.	Year	Remarks
$ZrSi_2$	Cu	1130	0.5	Partial reaction	Two-phase structure of solidified melt	[834]	1959	
$ZrSi_2$	Ni	1500	0.5	Complete solution	(Complex silicide with low melting point. Finely disperse precipitations of a new phase	[834]	1959	
$TaSi_2$	Ni	1500	0.5			[834]	1959	
$MoSi_2$	Cu	1130	0.5	Partial reaction	Mo. Si not found	[834]	1959	
$MoSi_2$	Na	—	—	Does not react	—	[834]	1959	
$MoSi_2$	Cu, Fe	—			—	[649]	1955	
$MoSi_2$	Ag	940	204		—	[649]	1955	
$MoSi_2$	Zn	1000	5	Reacts	—	[649]	1955	Vacuum
$MoSi_2$	Al	1550	0.1		—	[932]	1961	
$MoSi_2$	Si	1000	5		—	[932]	1961	
$MoSi_2$	Sn	1000	5	Does not react	—	[932]	1961	
$MoSi_2$	Pb	1000	5		—	[649]	1955	
$MoSi_2$	Ni	1500	0.5	Complete solution	{Single-phase structure	[834]	1959	
WSi_2	Ni	1500	0.5		New phase	[834]	1959	
BaS	Ce	1150	0.1	Does not react	—	[648]	1951	

RESISTANCE TO THE ACTION OF MOLTEN METALS, ALLOYS, AND SLAGS (continued)

Phase	Composition of melt	Temperature, °C	Time of contact, hr	Nature of reaction	Content of components of compound passing into metal	Ref.	Year	Remarks
BaS	Ce	1300	0.5	Reacts	—	[648]	1951	
BaS	U	1400	0.2	Reacts	—	[648]	1951	
BaS	U	1500	—	Reacts	—	[648]	1951	
BaS	U	1900	—	Reacts	—	[648]	1951	Vacuum
CeS	Zn	500	0.1	Does not react	—	[648]	1951	
CeS	Zn	700	0.1	Does not react	—	[648]	1951	
CeS	Mg	900	0.1	Reacts slightly	—	[648]	1951	
CeS	Al	1500	0.2	Reacts slightly	—	[648]	1951	
CeS	Ti	1500	0.2	Does not react	—	[648]	1951	
CeS	Sn	1200	0.1	Does not react	—	[648]	1951	
CeS	Th	1825	0.1	Does not react	—	[648]	1951	
CeS	Ce	1500	0.3	Does not react	—	[648]	1951	
CeS	Bi	1400	0.1	Does not react	—	[648]	1951	Vacuum
CeS	Bi	1500	0.2	Does not react	—	[648]	1951	
CeS	Bi	1400	0.5	Does not react	—	[648]	1951	
CeS	Pt	1900	0.1	Reacts slightly	—	[648]	1951	
CeS	Pt	1900	0.2	Formation of $CePt_2$	—	[648]	1951	
ThS	Ce	1500	0.3	Does not react	—	[648]	1951	
ThS	Th	1825	0.1	Does not react	—	[648]	1951	
ThS	Mg	900	0.1	Does not react	—	[648]	1951	Vacuum, [1091]
ThS	Al	1500	0.2	Does not react	—	[648]	1951	
ThS	Fe	1500	0.2	Does not react	—	[648]	1951	

RESISTANCE TO THE ACTION OF MOLTEN METALS, ALLOYS, AND SLAGS (continued)

Phase	Composition of melt	Temperature, °C	Time of contact, hr	Nature of reaction	Content of components of compound passing into metal	Ref.	Year	Remarks
Th_2S_3	Ce	1500	0.3	Reacts slightly	—	[648]	1951	Porosity 20%
Th_2S_3	Ce	1500	0.3		—	[648]	1951	Porosity 5%
Th_2S_3	Th	1825	0.1		—	[648]	1951	
Th_4S_7	U	1300	0.1	Does not react	—	[648]	1951	
Th_4S_7	U	1475	0.5		—	[648]	1951	Porosity 20%
Th_4S_7	Ce	1500	0.3	Reacts slightly	—	[648]	1951	Porosity 5%
Th_4S_7	Ce	1500	0.3		—	[648]	1951	
SiC	Al	700	72	970.5	—	[1000]	1961	The fifth column gives the rate of disintegration, corrosion, of the SiC specimen, mm/min (porosity 4%); comp.,%: 96.5 SiC, 2.5 Si free, 0.4 C free, 0.4 Al, 0.2 Fe
SiC	Al	900	24	30	—	[1000]	1961	
SiC	Al	900	72	Reacts slightly	—	[1000]	1961	
SiC	Sn	400	24	5.4	—	[1000]	1961	
SiC	Pb	600	24	31.8	—	[1000]	1961	
SiC	Bi	600	22	~100	—	[1000]	1961	
SiC	Mg	750	24	546	—	[1000]	1961	
SiC	Mg	800	24	Reacts slightly	—	[1000]	1961	
SiC	Zn	600	72	14.6	—	[1000]	1961	
Si_3N_4	Cu	1150	7	Reacts	—	[1000]	1961	
Si_3N_4	Zn	550	500	Does not react	—	[270]	1955	
Si_3N_4	Mg	750	20	Reacts slightly	—	[270]	1955	
Si_3N_4	Al	800	900	Does not react	—	[270]	1955	

RESISTANCE TO THE ACTION OF MOLTEN METALS, ALLOYS, AND SLAGS (continued)

Phase	Composition of melt	Temperature, °C	Time of contact, hr	Nature of reaction	Content of components of compound passing into metal	Ref.	Year	Remarks
Si_3N_4	Al	1000	950	Does not react	—	[270]	1955	
Si_3N_4	Sn	300	144		—	[270]	1955	
Si_3N_4	Fe	—		Decomposes	—	[647]	1957	
Si_3N_4	Cast iron	1450	2	Reacts slightly	—	[932]	1961	
Si_3N_4	Brass	950	72	Does not react	—	[932]	1961	
Si_3N_4	Cryolite	1050	36	Reacts slightly	—	[932]	1961	
Si_3N_4 + SiC	"	1050	8	Does not react	—	[932]	1961	} In the melt
Si_3N_4	$NaNH_2$ + $NaNO_3$	350	—	Perfectly resistant	—	[270]	1955	
Si_3N_4	NaCl + KCl	790	—		—	[270]	1955	
Si_3N_4	NaCl + KCl	900	144		—	[270]	1955	
Si_3N_4	NaOH	450	5	Resistant	—	[270]	1955	
Si_3N_4	$NaF + ZrF_4$	800	100		—	[270]	1955	
Si_3N_4	$NaB(SiO_3)_2$ + V_2O_5	1100	4		—	[270]	1955	
Si_3N_4 + SiC	Cryolite	1050	100		—	×	1961	Argon
BN	Fe	1600	0.5		—	×	1961	Air
BN + C	Al	1000	0.2	Does not react	—	×	1961	Argon
BN + C	B + Si	2000	2		—	×	1961	Vacuum
BN + C	KBF_4	900	3		—	×	1961	Air
BN + C	Cryolite	1000	4		—	×	1961	Argon

G. A. Yasin-skaya

RESISTANCE IN REACTIONS IN THE SOLID PHASE AND WITH NITROGEN

Reacting mixture	Temperature, °C	Principal reaction products	Duration of action, hr	Ref.	Year	Remarks
$Be_5B + 25\%$ C	800	$Be_5B + C$	—	[863]	1961	[1140]; see data on interaction of Be_3B, Be_2B, Be_5B, BeB_2, BeB_6 with N_2 at 900–1200°C (up to 12 hr)
$Be_5B + 25\%$ C	900	$Be_2C + C$	—	[863]	1961	
$Be_5B + 25\%$ C	1000	$Be_2C + C$ impurity	—	[863]	1961	
$Be_5B + 25\%$ C	1300	Be_2C	—	[863]	1961	
$Be_2B + 25\%$ C	800	$Be_2B + C$	—	[863]	1961	
$Be_2B + 25\%$ C	900	$Be_2B + C$	—	[863]	1961	
$Be_2B + 25\%$ C	1000	$Be_2C + C$ impurity	—	[863]	1961	
$Be_2B + 25\%$ C	1200	$Be_2C + C$ impurity	—	[863]	1961	
$Be_2B + 25\%$ C	1300	Be_2C	—	[863]	1961	
$BeB_2 + 25\%$ C	800	$BeB_2 + C$	—	[863]	1961	
$BeB_2 + 25\%$ C	900	$BeB_2 + C +$ trace Be_2C	—	[863]	1961	
$BeB_2 + 25\%$ C	1000	$Be_2C + C$	—	[863]	1961	
$BeB_2 + 25\%$ C	1200	$Be_2C + C$ impurity	—	[863]	1961	
$BeB_2 + 25\%$ C	1300	Be_2C	—	[863]	1961	
$BeB_4 + 25\%$ C	800	$BeB_4 + C$	—	[863]	1961	
$BeB_4 + 25\%$ C	900	$BeB_4 + C$	—	[863]	1961	
$BeB_4 + 25\%$ C	1000	$BeB_4 + C$	—	[863]	1961	
$BeB_4 + 25\%$ C	1200	$Be_2C + C +$ trace BeB_4	—	[863]	1961	
$BeB_4 + 25\%$ C	1300	Be_2C	—	[863]	1961	
$LaB_6 + Ta$	1000—1600	$LaB_6 + Ta$	2—5	[895]	1961	
$L_2B_6 + Ta$	1800	Reacts	2	[895]	1961	
$LaB_6 + Ta$	1800	"	5	[895]	1961	

Reacting mixture	Temperature, °C	Principal reaction products	Duration of action, hr	Ref.	Year	Remarks
LaB_6 + Ta	2000	Reacts*	2	[895]	1961	
LaB_6 + Ta	2000	" *	5	[895]	1961	
LaB_6 + Ta	2100	Reacts strongly*		[895]	1961	
LaB_6 + Ta	2100	" *	2	[895]	1961	
LaB_6 + Mo	1200	Reacts slightly*	5	[895]	1961	
LaB_6 + Mo	1400	Reacts*	2	[895]	1961	
LaB_6 + Mo	1600	*	5	[895]	1961	
LaB_6 + Mo	1800	"	5	[895]	1961	
LaB_6 + Mo	1800	"	2	[895]	1961	
LaB_6 + Mo	2000	"	5	[895]	1961	
LaB_6 + Mo	2000	"	2	[895]	1961	
LaB_6 + Mo	2100	"	5	[895]	1961	
LaB_6 + Mo	2100	Reacts strongly*	2	[895]	1961	
LaB_6 + W	1400	"	5	[895]	1961	
LaB_6 + W	1600	LaB_6 + W	2—5	[895]	1961	
LaB_6 + W	1600	Reacts slightly*	5	[895]	1961	
LaB_6 + W	1800	"	2	[895]	1961	
LaB_6 + W	1800	" Reacts*	5	[895]	1961	
LaB_6 + W	2000	*	2	[895]	1961	
LaB_6 + W	2000	"	5	[895]	1961	
LaB_6 + W	2100	"	2	[895]	1961	
LaB_6 + W	2100	"	5	[895]	1961	

RESISTANCE IN REACTIONS IN THE SOLID PHASE AND WITH NITROGEN (continued)

Reacting mixture	Temperature, °C	Principal reaction products	Duration of action, hr	Ref.	Year	Remarks
$LaB_6 + Mo_2C$ $ThB_6 + ThO_2$	1800—2100 2300	$LaB_6 + Mo_2C$ $ThB_6 + ThO_2$	2—5 —	[895] [1009]	1961 1961	
CeB_6, TiB_2, ZrB_2, ThB_6, NbB_2, TaB_2, Cr_2B, Cr_5B_3, CrB, (Cr_3B_2), Cr_3B_4, CrB_2, Mo_2B, Mo_2B_5, WB, W_2B_5, Fe_2B FeB, Co_2B, CoB, Ni_2B	2000—2200	Resistant to carbon	—	[51]	1952	[180]
CeB_4, Ti_2B, TiB, Ti_2B_5 (?), ZrB, ZrB_{12}, ThB_4, Nb_2B, NbB, Nb_3B_4, Nb_3B_2, Ta_3B_4, Ta_3B_2, Ta_3B_4, Cr_4B, Cr_2B_5 (?), W_2B_5	2000—2200	Nonresistant to carbon	—	[51]	1952	[180]
$2Mo_2B + C$ $W_2B + C$	~2000 ~2000	$Mo_2B + MoB + MoC$ $WB + WC$	— —	[180] [180]	1955 1955	

RESISTANCE IN REACTIONS IN THE SOLID PHASE AND WITH NITROGEN (continued)

Reacting mixture	Temperature, °C	Principal reaction products	Duration of action, hr	Ref.	Year	Remarks
$W_2B + ThO_2$	1800	$W_2B + ThO_2$	—	[1009]	1960	
$WB + ThO_2$	2100	$WB + ThO_2$	—	[1009]	1960	
$TiB_2(67\%) + B_4C(33\%)$	1600	TiB_2, B_4C	—	[51]	1952	
$TiB_2(67\%) + B_4C(33\%)$	2150	TiB_2, B_4C	—	[51]	1952	
$ZrB_2(67\%) + B_4C(33\%)$	2100	ZrB_2, B_4C	—	[51]	1952	
$VB_2(67\%) + B_4C(33\%)$	2000	$VB_2 + B_4C + C$	—	[51]	1952	[692]
$TaB(67\%) + B_4C(33\%)$	2000	$TaB + B_4C$	—	[51]	1952	
$TaB_2 + Nb$	1600	Reacts slightly*	5	[895]	1961	
$TaB_2 + Nb$	1800	Reacts*	2	[895]	1961	
$TaB_2 + Nb$	1800	"	5	[895]	1961	
$TaB_2 + Nb$	2000	Reacts strongly*	2	[895]	1961	
$TaB_2 + Nb$	2000	*	5	[895]	1961	
$TaB_2 + Nb$	2100	" "	2	[895]	1961	
$TaB_2 + Nb$	2100	" "	5	[895]	1961	
$TaB_2 + Ta$	1600	" "	5	[895]	1961	
$TaB_2 + Ta$	1800	Reacts slightly*	2	[895]	1961	
$TaB_2 + Ta$	1800	" "	5	[895]	1961	

311

RESISTANCE IN REACTIONS IN THE SOLID PHASE AND WITH NITROGEN (continued)

Reacting mixture	Temperature, °C	Principal reaction products	Duration of action, hr	Ref.	Year	Remarks
$TaB_2 + Ta$	2000	Reacts*	2	[895]	1961	
$TaB_2 + Ta$	2000	"	5	[895]	1961	
$TaB_2 + Mo$	1600	Reacts slightly*	5	[895]	1961	
$TaB_2 + Mo$	1800	"	2	[895]	1961	
$TaB_2 + Mo$	1800	Reacts*	5	[895]	1961	
$TaB_2 + Mo$	2000	"	2	[895]	1961	
$TaB_2 + Mo$	2000	"	5	[895]	1961	
$TaB_2 + Mo$	2100	Reacts strongly*	2	[895]	1961	
$TaB_2 + Mo$	2100	"	5	[895]	1961	
$TaB_2 + W$	1600	Reacts slightly*	5	[895]	1961	
$TaB_2 + W$	1800	"	2	[895]	1961	
$TaB_2 + W$	1800	Reacts*	5	[895]	1961	
$Cr_3B_2 + N_2$	800	$Cr_3B_2 + CrN$	—	[691]	1951	
$Cr_3B_2 + N_2$	900	$CrN + Cr_2N + BN$	—	[691]	1951	
$Cr_3B_2 + N_2$	1000	$CrN + Cr_2N + BN$	—	[691]	1951	
$Cr_3B_2 + N_2$	1180	$Cr_2N + BN$	—	[691]	1951	
$CrB + N_2$	550	CrB	—	[691]	1951	
$CrB + N_2$	800	CrB	—	[691]	1951	
$CrB + N_2$	900	$CrB + CrN$	—	[691]	1951	
$CrB + N_2$	1000	$CrB + CrN + BN$	—	[691]	1951	
$CrB + N_2$	1180	$Cr_2N + BN$	—	[691]	1951	
$Cr_3B_4 + N_2$	550	Cr_3B_4	—	[691]	1951	
$Cr_3B_4 + N_2$	900	Cr_3B_4	—	[691]	1951	
$Cr_3B_4 + N_2$	1050	$Cr_2N + Cr_3B_4$	—	[691]	1951	
$Cr_3B_4 + N_2$	1100	$Cr_2N + BN$	—	[691]	1951	

Reacting mixture	Temperature, °C	Principal reaction products	Duration of action, hr	Ref.	Year	Remarks
CrB_2	1000	CrB_2 + BN	—	[691]	1951	
CrB_2	1180	Cr_2N + BN	—	[691]	1951	
$W_2B + N_2$	700	W_2B	—	[691]	1951	
$W_2B + N_2$	800	W_2B	—	[691]	1951	
$W_2B + N_2$	850	W_2B + α-W + (γ + BN)	—	[691]	1951	
$W_2B + N_2$	900	α-W + W_2B + BN	—	[691]	1951	
$W_2B + N_2$	1000	α-W + W_2B + BN	—	[691]	1951	
$W_2B + N_2$	1100	α-W + BN	—	[691]	1951	
$WB + N_2$	700	WB	—	[691]	1951	
$WB + N_2$	750	WB	—	[691]	1951	
$WB + N_2$	800	WB + β + BN	—	[691]	1951	
$WB + N_2$	850	α-W + WB + (γ + BN)	—	[691]	1951	
$WB + N_2$	900	α-W + (WB) + BN	—	[691]	1951	
$WB + N_2$	1000	α-W + + BN	—	[691]	1951	
$WB + N_2$	1100	α-W + BN	—	[691]	1951	
$W_2B_5 + N_2$	700	WB_2	—	[691]	1951	
$W_2B_5 + N_2$	750	WB_2	—	[691]	1951	
$W_2B_5 + N_2$	850	α-W + WB_2 + (γ + BN)	—	[691]	1951	
$W_2B_5 + N_2$	900	α-W + (WB_2) + BN	—	[691]	1951	
$W_2B_5 + N_2$	1100	α-W + BN	—	[691]	1951	
$Fe_2B + N_2$	350	Fe_2B	—	[691]	1951	

313

Reacting mixture	Tempera-ture, °C	Principal reaction products	Duration of action, hr	Ref.	Year	Remarks
$Fe_2B + N_2$	400	ζ-phase + BN	—	[691]	1951	ζ-Fe_2N
$Fe_2B + N_2$	450	ε + BN	—	[691]	1951	γ-Fe_4N
$Fe_2B + N_2$	500	ε + γ' + BN	—	[691]	1951	
$Fe_2B + N_2$	550	γ' + ε + BN	—	[691]	1951	
$Fe_2B + N_2$	600	α-Fe + γ' + ε + BN	—	[691]	1951	
$Fe_2B + N_2$	700	α-Fe + BN	—	[691]	1951	
$Fe_2B + N_2$	770	α-Fe + BN	—	[691]	1951	
$FeB + N_2$	300	FeB	—	[691]	1951	
$FeB + N_2$	400	Fe_2B + ζ + BN	—	[691]	1951	
$FeB + N_2$	550	γ' + ε + BN	—	[691]	1951	
$FeB + N_2$	600	γ' + ε + α-Fe + BN	—	[691]	1951	
$FeB + N_2$	770	α-Fe + BN	—	[691]	1951	
$LaB_6 + MoSi_2$	1200—1500	$MoSi_2$ + X	—	[692]	1959	Vacuum
$NbB_2 + Ta$	1800—2000	NbB_2 + Ta	—	[692]	1959	
$TiB_2 + Mo$	1800—2000	$TiMoB_4$, $TiMo_2B_2$	—	[692]	1959	[1268]
$ZrB_2 + Mo$	1800—2000	ZrB_2 + Mo	—	[692]	1959	[1268]
$HfB_2 + Mo$	1600—2100	$HfMoB_2$ (?)	—	[1268]	1962	
$TiB_2 + Nb$	1600—2700	$TiNbB_2$	—	[692]	1959	
$TiB_2 + W$	1600—2400	$TiWB_2$	—	[692]	1959	
$ZrB_2 + W$	2100	ZrB_2 + W	72	[1268]	1962	[1268]
$HfB_2 + W$	2100	$HfWB_2$ (?)	—	[1268]	1962	
$Mo_2B_5 + Mo$	1000—1900	Mo_2B_5 + Mo	—	[692]	1959	
$Mo_2B_5 + MoSi_2 + Mo$	1800—1900	Mo_2B_5 + $MoSi_2$ + Mo	—	[692]	1959	Vacuum
UC + Be	950	Reacts on sintering at 15 kg/mm² pressure	12	[1014]	1959	

RESISTANCE IN REACTIONS IN THE SOLID PHASE AND WITH NITROGEN (continued)

Reacting mixture	Temperature, °C	Principal reaction products	Duration of action, hr	Ref.	Year	Remarks
UC + Zr	950	Reacts on sintering at 15 kg/mm² pressure	12	[1014]	1959	
UC + Si	1000	UC + USi₃	—	[1014]	1959	
UC + Ni	1000	UC + U₆Ni(and other phases of U—Ni system)		[1014]	1959	See Appendix
(TiC, ZrC, HfC, NbC, TaC) — C	—	Eutectic alloys		[996]	1961	
TiC + Nb	1600	TiC + Nb*	2	[895]	1961	
TiC + Nb	1800	Reacts slightly*	2	[895]	1961	
TiC + Nb	1600	TiC + Nb*	5	[895]	1961	
TiC + Nb	1800	Reacts slightly*	5	[895]	1961	
TiC + Nb	2000	Reacts*	5	[895]	1961	
TiC + Ta	1600—1800	TiC + Ta*	2—5	[895]	1961	
TiC + Ta	2000	Reacts*	5	[895]	1961	Vacuum
TiC + Mo	1600—2000	TiC + Mo*	2—5	[895]	1961	
TiC + W	1400—1800	TiC + W*	2—5	[895]	1961	
TiC + W	2000	Reacts slightly*	2	[895]	1961	
TiC + W	2000	"	5	[895]	1961	
TiC (56% + B₄C (44%)	1500—2150	TiB + B₄C	—	[51]	1951	
2 TiC + B₄C	1200	TiB₂ + C	—	[678]	1951	
2 TiC + B₄C	2000	TiB₂ + C + X	—	[678]	1951	[679]
ZrC + Mo	1000—2000	ZrC + Mo	2	[895]	1961	

315

RESISTANCE IN REACTIONS IN THE SOLID PHASE AND WITH NITROGEN (continued)

Reacting mixture	Temperature, °C	Principal reaction products	Duration of action, hr	Ref.	Year	Remarks
ZrC + Mo	2000	Reacts slightly*	5	[895]	1961	
ZrC + Mo	2200	Reacts*	2	[895]	1961	
ZrC + Mo	2200	Reacts slightly*	2	[895]	1961	
ZrC + W	1600—2000	ZrC + W*	2—5	[895]	1961	
ZrC (67%) + B$_4$C (33%)	1400—2150	ZrB$_2$ + B$_4$C + C	—	[51]	1952	
Cr$_7$C$_3$ + Ta	1800—2000	Cr$_7$C$_3$ + Ta*	—	[692]	1959	
Mo$_2$C + Ta	1400—1600	Mo$_2$C + Ta*	—	[692]	1959	
Mo$_2$C ± Ta	1700—1900	Reacts*	—	[692]	1959	
WC + Ta	1800—2000	WC + Ta*	—	[692]	1959	
HfC + Nb	1600	Reacts slightly*	2	[895]	1961	
HfC + Nb	1800	"	2	[895]	1961	
HfC + Nb	2000	Reacts*	2	[895]	1961	
HfC + Nb	2200	*	2	[895]	1961	
HfC + Ta	1400—1600	HfC + Ta	2—5	[895]	1961	
HfC + Ta	1800	Reacts slightly*	2	[895]	1961	
HfC + Ta	2000	Reacts*	2	[895]	1961	
HfC + Ta	2200	*	2	[895]	1961	
HfC + Mo	1000—1800	HfC + Mo*	2—5	[895]	1961	
HfC + Mo	1800	Reacts slightly*	5	[895]	1961	
HfC + Mo	2000	Reacts*	2	[895]	1961	
HfC + Mo	2200	"	2	[895]	1961	
HfC + W	1000—1800	HfC + W	2—5	[895]	1961	
HfC + W	2000	Reacts slightly*	2	[895]	1961	
HCf + W	2200	"	2	[895]	1961	

RESISTANCE IN REACTIONS IN THE SOLID PHASE AND WITH NITROGEN (continued)

Reacting mixture	Temperature, °C	Principal reaction products	Duration of action, hr	Ref.	Year	Remarks
VC (67%) + B₄C (33%)	1500—2100	$VB_2 + B_4C + C$ *	—	[51]	1952	
NbC + Nb	1600	NbC + Nb *	2	[895]	1961	
NbC + Nb	1800	Reacts*	2	[895]	1961	
NbC + Nb	2000	"	2	[895]	1961	
NbC + Nb	2200	"	2—5	[895]	1961	
NbC + Ta	1600	NbC + Ta	2	[895]	1961	
NbC + Ta	1800	Reacts slightly*	2	[895]	1961	
NbC + Ta	2000	Reacts*	2—5	[895]	1961	
NbC + Ta	2200	"	5	[895]	1961	
NbC + Mo	1000—1800	NbC + Mo	2	[895]	1961	
NbC + Mo	1800	Reacts slightly*	2	[895]	1961	
NbC + Mo	2000	"	2	[895]	1961	
NbC + Mo	2200	Reacts*	2	[895]	1961	
NbC + W	1600—2000	NbC + W		[895]	1961	
NbC + W	2200	Reacts slightly*		[895]	1961	
NbC (77%) + B₄C (23%)	1500—2100	$NbB_2 + B_4C + C$	—	[51]	1952	[996]
NbC + C	2900—3000	Eutectic	—	[884]	1961	
TaC + Nb	1600	Reacts slightly*	2	[895]	1961	
TaC + Nb	1800	"	2	[895]	1961	
TaC + Nb	2000	Reacts*	2	[895]	1961	
TaC + Nb	2200	"	2	[895]	1961	

RESISTANCE IN REACTIONS IN THE SOLID PHASE AND WITH NITROGEN (continued)

Reacting mixture	Temperature, °C	Principal reaction products	Duration of action, hr	Ref.	Year	Remarks
TaC + Mo	1000—1800	TaC + Mo*	2—5	[895]	1961	
TaC + Mo	2000	TaC + Mo*	2	[895]	1961	
TaC + Mo	2000	Reacts slightly*	5	[895]	1961	
TaC + W	1600—2000	TaC + W*	2—5	[895]	1961	
TaC + W	2000	Reacts slightly*	5	[895]	1961	
TaC + W	2200	Reacts*	2	[895]	1961	
TaC + C	3300	Eutectic	—	[884]	1961	[996]
TaC (67%) + B$_4$C (33%)	1500	TaB$_2$ + B$_4$C + C	—	[51]	1952	
Cr$_3$C$_2$ (67%) + B$_4$C (33%)	1700	CrB$_2$ + B$_4$C + C	—	[51]	1952	
Mo$_2$C (67%) + B$_4$C (33%)	1800	MoB$_2$ + B$_4$C + C	—	[51]	1952	
WC (77%) + B$_4$C (23%)	2100	W$_2$B$_5$ + B$_4$C + C	—	[51]	1952	
Fe$_3$C (71%) + B$_4$C (29%)	1500	FeB + B$_4$C + C	—	[51]	1952	
Cr$_3$C$_2$ + Cr$_2$O$_3$	1000—1100	Cr$_3$C$_2$ + Cr$_2$O$_3$	—	[179]	1961	
Cr$_3$C$_2$ + Cr$_2$O$_3$	1200—1700	Cr$_7$C$_3$ + Cr$_2$O$_3$ + C	—	[179]	1961	
Cr$_3$C$_2$ + Cr$_2$O$_3$	1400—1650	Cr + Cr$_2$O$_3$	—	[179]	1961	
ZrC + ZrO$_2$	1450	Start of reaction	—	[883]	1961	
ZrC + ThO$_2$	1700	” ” ”	—	[883]	1961	
ZrC + Nb	1600	Reacts slightly*	2	[895]	1961	[692]
ZrC + Nb	1800	”	2	[895]	1961	

RESISTANCE IN REACTIONS IN THE SOLID PHASE AND WITH NITROGEN (continued)

Reacting mixture	Temperature, °C	Principal reaction products	Duration of action, hr	Ref.	Year	Remarks
$ZrC + Nb$	2000	Reacts*	2	[895]	1961	
$ZrC + Nb$	2200	„ *	2	[895]	1961	
$ZrC + Ta$	1600	Does not react*	2	[895]	1961	
$ZrC + Ta$	1800	„ „ „ *	2	[895]	1961	
$ZrC + Ta$	2000	„ „ „ *	5	[895]	1961	
$ZrC + Ta$	2200	Reacts slightly*	2	[895]	1961	
$Cr_3C_2 + W$	1800—2000	$Cr_3C_2 + W$	—	[692]	1959	Vacuum
$Cr_7C_3 + Mo$	1800—2000	$Cr_7C_3 + Mo$	—	[692]	1959	
$Cr_7C_3 + W$	1800—2000	$Cr_7C_3 + W$	—	[692]	1959	
$Mo_2C + Mo$	1300—2000	$Mo_2C + Mo$	—	[692]	1959	
$Mo_2C + W$	1800—2000	$Mo_2C + Mo$	—	[692]	1959	
$Mo_2C + MoSi_2 + Mo$	1800—1900	$Mo_2C + MoSi_2 + Mo$	—	[692]	1959	Vacuum
$Mo_2C + Mo_2B_5 + Mo$	1800—1900	$Mo_2C + Mo_2B_5 + Mo$	—	[692]	1959	Vacuum
$W_2C + Mo$	1800—2000	$W_2C + Mo$	—	[692]	1959	Vacuum
$W_2C + B_2O_3$	800—1400	$W + B + CO$	—	[915]	1959	
$WC + Mo$	1800—2000	$WC + Mo$	—	[692]	1959	
$WC + 3W + B_4C$	2100—2200	$W_2B_5 + (WB)$	—	[677]	1959	
$WC + B_4C + C$	2100—2200	$W_2B_5 + (WB)$	—	[677]	1959	
$WC + B_2O_3$	1400	$W_2C + B + CO$	—	[915]	1959	
$Be_2C + N_2$	1250	$Be_3N_2 + C$	—	[467]	1952	
$TiC + 2B$	2400	TiB_2	—	[47]	1953	
$ZrC + 2B$	2600	ZrB_2	—	[47]	1953	
$VC + 2B$	2000	VB	—	[47]	1953	
$NbC + 2B$	2250	NbB_2	—	[47]	1953	
$TaC + 2B$	2400	TaB_2	—	[47]	1953	
$Cr_3C + 2BN$	1650	$Cr_2N + Cr$	—	[47]	1953	

RESISTANCE IN REACTIONS IN THE SOLID PHASE AND WITH NITROGEN (continued)

Reacting mixture	Temperature, °C	Principal reaction products	Duration of action, hr	Ref.	Year	Remarks
$Mo_2C + 2B$	2300	MoB	—	[47]	1953	
$WC + 2B$	2450	W_2B_5	—	[47]	1953	
$TiC + N_2$	1500—2500	TiC_xN_{1-x}	0.25—32	[1243]	1963	$\lg K_p = -5600/T + 2.78$
$ZrC + N_2$	1500—2500	ZrC_xN_{1-x}	0.25—32	[1243]	1963	$\lg K_p = -5290/T + 2.76$
$HfC + N_2$	1500—2500	HfC_xN_{1-x}	—	[1243]	1963	$\lg K_p = -3380/T + 2.00$
$UC + N_2$	{1100—1600	Does not react	—	[1275]	1962	
Disilicides of rare earth metals $(MeSi_2) + Al_2O_3$		Reacts	—	[545]	1959	
$TiSi_2 + 2B$	1600	TiB_2	—	[545]	1959	
$ZrSi_2 + 2B$	1450	ZrB_2	—	[47]	1953	
$CrSi_2 + 2B$	1650	CrB, some $CrSi_2$	—	[47]	1953	
$MoSi_2 + 2B$	1550	MoB, some $MoSi_2$	—	[47]	1953	
TiN + Nb	1750	Reacts slightly*	5	[895]	1961	
TiN + Nb	1800	Reacts*	2	[895]	1961	
TiN + Nb	2000	"	5	[895]	1961	
TiN + Nb	2000	Reacts strongly*	2	[895]	1961	
TiN + Nb	2100	"		[895]	1961	
TiN + Ta	2100	"	5	[895]	1961	
TiN + Ta	1600	TiN + Ta*	2	[895]	1961	
TiN + Ta	1800	TiN + Ta*	5	[895]	1961	
TiN + Ta	2000	Reacts*	2	[895]	1961	
TiN + Ta	2100	Reacts strongly*	5	[895]	1961	

320

RESISTANCE IN REACTIONS IN THE SOLID PHASE AND WITH NITROGEN (continued)

Reacting mixture	Temperature, °C	Principal reaction products	Duration of action, hr	Ref.	Year	Remarks
TiN + Mo	1600—1800	TiN + Mo*	2—5	[895]	1961	
TiN + Mo	2000	Reacts slightly*	2	[895]	1961	
TiN + Mo	2000	" *	5	[895]	1961	
TiN + Mo	2100	Reacts*	5	[895]	1961	
TiN + W	1600—1800	TiN + W*	2—5	[895]	1961	
TiN + W	2000	Reacts slightly*	2	[895]	1961	
TiN + W	2000	Reacts*	5	[895]	1961	
TiN + W	2100	*	2	[895]	1961	
TiN + W	2000	" Reacts slightly*	2	[895]	1961	
TiN + W	2000	Reacts*	5	[895]	1961	
TiN + W	2100	Reacts strongly*	2	[895]	1961	
ZrN + Ta	2000—2100	ZrN + Ta*	2—5	[895]	1961	
ZrN + Mo	1800—2100	ZrN + Mo*	2—5	[895]	1961	
ZrN + Mo	2100	Reacts slightly*	5	[895]	1961	
ZrN + W	2000—2100	ZrN + W	2—5	[895]	1961	
Ti + Si + N₂	1840	TiN	—	[200]	1956	
Ti + Si + N₂	1870	TiN + X	—	[200]	1956	
Zr + Si + N₂	1840	ZrN + ZrSi₂	—	[200]	1956	
Zr + Si₃N₄ + N₂	1340	X + Si	—	[200]	1956	
Ce + Si + N₂	1600	X	—	[200]	1956	
TaB₂ + W	1800	Reacts slightly*	5	[895]	1961	
TaB₂ + W	1800	"	2	[895]	1961	
TaB₂ + W	1800	Reacts*	5	[895]	1961	
TaB₂ + W	2000	"	2	[895]	1961	

RESISTANCE IN REACTIONS IN THE SOLID PHASE AND WITH NITROGEN (continued)

Reacting mixture	Temperature, °C	Principal reaction products	Duration of action, hr	Ref.	Year	Remarks
$TaB_2 + W$	2000	Reacts strongly*	5	[895]	1961	
$TaB_2 + W$	2100	" *	2	[895]	1961	
TaB (67%)+ B_4C (33%)	2000	$TaB_2 + B_4C$	—	[51]	1952	
CrB (67%)+ B_4C (33%)	1500—2000	$CrB + B_4C + C$	—	[51]	1952	
CrB (67%)+ B_4C (33%)	1700—2100	$CrB_2 + B_4C + C$	—	[51]	1952	
Mo_2B (67%)+ B_4C (33%)	1600	$MoB_2 + B_4C + C$	—	[51]	1952	
MoB (67%)+ B_4C (33%)	1800, 2000	$MoB_2 + B_4C + C$	—	[51]	1952	
Mo_2B_5 (67%)+ B_4C (33%)	1800, 2000	$MoB + B_4C + C$	—	[51]	1952	
$W_2B_5 + WC$	2350—2400	$WC + W_2B_5$	—	[677]	1959	
$Ce + 4B + C$	1800	$CeB_4 + (CeB_6)$	—	[180]	1955	
$Ce + 6B + 1/2C$	1800	$CeB_6 + CeB_4$	—	[180]	1955	
$Zr + 2B + C$	1800	$ZrB_2 + ZrC + X$	—	[180]	1955	$Ar, p = 0.5$ atm
$Nb + B + 1/2C$	1800	$NbB_2 + NbC$	—	[180]	1955	
$Nb + 2B + C$	1800	$NbB_2 + NbC$	—	[180]	1955	
$2Mo + B1/2C$	1800	$Mo_2B + \alpha\text{-}MoB$	—	[180]	1955	
$Mo + B + C$	1800	$\alpha\text{-}MoB + \beta\text{-}MoB$	—	[180]	1955	
$2Mo + 5B + C$	~1800	$MoB_2 + \alpha\text{-}MoB$	—	[180]	1955	
$Ti + B + BN + N_2$	~2000	$TiB_2 + TiN + (BN)$	—	[180]	1955	$N_2, p = 0.5$ atm

RESISTANCE IN REACTIONS IN THE SOLID PHASE AND WITH NITROGEN (continued)

Reacting mixture	Temperature, °C	Principal reaction products	Duration of action, hr	Ref.	Year	Remarks
$Ti + 3B + N_2$	~1550	$Ti + (TiB_2) + (BN)$	—	[180]	1955	
$Zr + 3B + N_2$	~1550	$ZrB_2 + (ZrN) + (BN)$	—	[180]	1955	
$Zr + 3B + 1/4BN + N_2$	~1550	$ZrB_2 + ZrN + (BN)$	—	[180]	1955	
$Zr + BN + B + N_2$	~2000	$ZrN + ZrB_2$	—	[180]	1955	
$W + 3B + 1/4BN + N_2$	~1550	$W_2B_5 + WB$	—	[180]	1955	N_2, p = 0.5 atm
$W + 3B + N_2$	~1550	$WB + W_2B + (W_2B_5)$	—	[180]	1955	
$Cr + 3B + N_2$	~1550	$Cr_2N + X$	—	[180]	1955	
$Cr + 3B + 1/4BN + N_2$	1500	$Cr_2 + CrB + BN$ (?)	2—3	[180]	1955	
$MgB_2 + N_2$	<900	Does not react	2—3	[1076]	1951	
$MgB_2 + N_2$	950—1000	Mg_3N_2	2—3	[1076]	1951	
$MgB_6 + N_2$	≦1350	Does not react	2—3	[1076]	1951	
$MgB_{12} + N_2$	≦1350	"	2—3	[1076]	1951	
$Mg_3B_2 + N_2$	900	Mg_3N_2		[1076]	1951	
$Nb + Si + N_2$	1840	$Nb_5Si_3N_7$	—	[200]	1956	
$(Ta, Nb) + Si + Si_3N_4 + N_2$	2080	$(TaNb)Si_2 + (Ta, Nb)_5Si_5N_7 + (Ta, Nb)Si_{0.6}$		[200]		
$Nb + Si + N_2$	1870	$NbSi_2 + Nb_5Si_3N_7$	—	[200]	1956	
$Ta + Si + N_2$	1330	$Ta_5Si_3N_2 + X$	—	[200]	1956	

323

RESISTANCE IN REACTIONS IN THE SOLID PHASE AND WITH NITROGEN (continued)

Reacting mixture	Temperature, °C	Principal reaction products	Duration of action, hr	Ref.	Year	Remarks
$Ta + Si + Si_3N_4 + N_2$	2080	$TaSi_2 + Ta_5Si_3N_7 + Ta_2N$	—	[200]	1956	
$Ta + Si + Si_3N_4 + N_2$	1380	$TaSi_2 + Ta_5Si_3N_7 + Ta_2N$	—	[200]	1956	
$Ta + Si + N_2$	1870	$X + TaSi_2$	—	[200]	1956	
$Ta_2N + Si_3N_4 + N_2$	1840	$TaSi_2 + Ta_5Si_3N_2$	—	[200]	1956	
$MoSi_2 + Mo$	1800—2000	Strong reaction	—	[692]	1959	
$MoSi_2 + Ta$	1900—2050	$Mo_3Si + TaSi$	—	[692]	1959	
$MoSi_2 + ZrO_2$	1700	Does not react	—	[895]	1961	
$MoSi_2 + ZrO_2$ **	1500	Reacts	—	[895]	1961	
$2B + Ti$	2400	TiB_2	—	[47]	1953	
$2B + Zr$	2600	ZrB_2	—	[47]	1953	
$2B + V$	2000	VB_2	—	[47]	1953	
$2B + Nb$	2250	NbB_2	—	[47]	1953	
$2B + Ta$	2400	TaB_2	—	[47]	1953	
$2B + Cr$	1650	$Cr_2B + C$	—	[47]	1953	
$2B + Mo$	2300	MoB	—	[47]	1953	
$2B + W$	2450	W_2B_5	—	[47]	1953	
$BeO + Nb$	1600—1700	$BeO + Nb$ *	0.5—1	[894]	1961	
$BeO + Nb$	1700	Reacts slightly*	1	[894]	1961	
$BeO + Nb$	1800	$BeO + Nb$ *	0.5	[894]	1961	
$BeO + Nb$	1800	Reacts	1	[894]	1961	
$BeO + Nb$	1900	"	0.5	[894]	1961	
$BeO + Nb$	1900	"	1	[894]	1961	

RESISTANCE IN REACTIONS IN THE SOLID PHASE AND WITH NITROGEN (continued)

Reacting mixture	Temperature, °C	Principal reaction products	Duration of action, hr	Ref.	Year	Remarks
BeO + Mo	1600	BeO + Mo*	0.5—1	[894]	1961	
BeO + Mo	1700	BeO + Mo*	0.5	[894]	1961	
BeO + Mo	1700	Reacts slightly*	1	[894]	1961	
BeO + Mo	1800	BeO + Mo*	0.5	[894]	1961	
BeO + Mo	1800	Reacts slightly*	1	[894]	1961	
BeO + W	1600—1700	BeO + W*	0.5—1	[894]	1961	
BeO + W	1800	Reacts*	1	[894]	1961	
BeO + W	1900	"	0.5	[894]	1961	
BeO + W	1900	"	1	[894]	1961	
MgO + N	1600—1900	MgO + Nb*	0.5—1	[894]	1961	
MgO + Nb	1900	Reacts slightly*	5	[894]	1961	
MgO + Nb	2000	MgO + Nb	0.5	[894]	1961	
MgO + Nb	2000	Reacts slightly*	1	[894]	1961	
MgO + Nb	2000	Reacts*	5	[894]	1961	
MgO + Mo	1600—2000	MgO + Mo*	0.5—5	[894]	1961	
MgO + W	1600—2000	MgO + W*	0.5—5	[894]	1961	
ZrO_2** + Nb	1600—2000	ZrO_2 + Nb*	0.5—5	[894]	1961	
ZrO_2** + Nb	2100	Reacts*	5	[894]	1961	
ZrO_2 + Mo	1600—2000	ZrO_2 + Mo*	0.5—5	[894]	1961	
ZrO_2 + W	1600	ZrO_2 + W*	0.5—5	[894]	1961	

RESISTANCE IN REACTIONS IN THE SOLID PHASE AND WITH NITROGEN (continued)

Reacting mixture	Tempera-ture, °C	Principal reaction products	Duration of action, hr	Ref.	Year	Remarks
$ZrO_2 + W$	1900	Reacts slightly*	1	[894]	1961	
$ZrO_2 + W$	1900	,,	5	[894]	1961	
$ZrO_2 + W$	2000	,,	0.5	[894]	1961	
$ZrO_2 + W$	2000	,,	1	[894]	1961	
$ZrO_2 + W$	2000	,,	5	[894]	1961	

Remarks: *denotes data for contact reaction (one or both of the reacting phases being in the compact state), as distinct from reaction in mixtures of powders or the reaction of gases on powders. **denotes stabilized ZrO_2. In the third column: X is a phase of unknown composition; phases secondary in content are given in parentheses.

Chapter VIII

EXAMPLES OF THE APPLICATION OF REFRACTORY COMPOUNDS

Phase	Fundamental properties	Fields of application	Ref.	Year
CaB_6	Refractoriness, low specific gravity, satisfactory high-temperature strength, low work factor in thermionic emission, high thermoelectric force	In the composition of light refractory alloys, for example, 10–35% CaB_6, 5–13% B, 60–80% B_4C (sp. gr. 2.48–2.49): transverse strength $= 30-37$ kg/mm² at 20°C; or 24–35% CaB_6, 60–70% ZrO_2, 6–7% C	[382] [774]	1958 1954
		In the composition of cathodes of electronic devices. For the production of electrodes of high-temperature thermocouples and devices for the conversion of heat energy into electrical energy	[25]	1951
SrB_6	Refractoriness, low specific gravity, low work factor in thermionic emission	In the composition of heat-resistant alloys, in the cathodes of electronic devices	[25]	1951
BaB_6	Good thermionic emission properties	Cathodes in electronics	[25]	1951
AlB_{12}	Refractoriness, semiconductor properties, capacity for absorbing neutrons	Semiconductor devices, nuclear engineering	[440]	1960
ScB_2	Low specific gravity, refractoriness	In the composition of light, heat-resistant alloys	[694]	1960
YB_6	Good thermionic emission properties	Cathodes in electronics	[25] [308]	1951 1959

EXAMPLES OF THE APPLICATION OF REFRACTORY COMPOUNDS (continued)

Phase	Fundamental properties	Fields of application	Ref.	Year
LaB_6	Good thermionic emission properties	Cathodes of ionic sources of current for cyclotrons and synchrophasotrons, magnetron rectifiers, electron beams for welding apparatus (for electron beam welding) and furnaces for the electron melting of metals and alloys	[25] [308] [284] [1062] [1336]	1951 1959 1960 1962 1963
		Cathodes for high-current microtrons (current density at $1600°C \sim 200$ A/cm^2)	[853]	1960
CeB_6	Thermionic emission properties, high electrical resistance	In the composition of cathodes, especially for increasing the electrical resistance of lanthanum boride cathodes	[25] [379]	1951 1959
SmB_6	Semiconductory properties, refractoriness	In semiconductor devices (for use at high temperatures)	[474]	1961
SmB_6, EuB_6	High neutron absorption cross section, oxidation-resistant	In nuclear power engineering	[285] [1091]	1959 1962
GdB_6	High neutron absorption, refractoriness	In nuclear power engineering	[285] [846]	1959 1960
TuB_6	Low electron work factor, γ-irradiator	In nuclear power engineering, electronics, devices for converting thermal energy into electrical energy	[474]	1961
ThB_4, ThB_6,	High-temperature strength, refractori-	In nuclear power engineering	[474]	1961

EXAMPLES OF THE APPLICATION OF REFRACTORY COMPOUNDS (continued)

Phase	Fundamental properties	Fields of application	Ref.	Year
UB_2, UB_4, UB_{12}	ness, nuclear properties			
TiB_2	High-temperature strength, refractoriness, resistance to scaling	In the composition of heat-resistant alloys, for example, TiB_2–CrB_2 (4:1 molar parts): sp.gr. 4.3–4.7; hardness 85 RA; bending strength (at 20°C) 35–40 kg/mm^2; stress-rupture transverse strength for 100 hr (at 1200°C) 20 kg/mm^2; modulus of elasticity 32,800 kg/mm^2 (20°C); shear modulus 10,000 kg/mm^2; coefficient of thermal expansion (20–1200°C) $8.5 \cdot 10^{-6}$; thermal conductivity 15.7 kcal/m·hr·deg; electrical resistance 32.8 $\mu\Omega$·cm; impact toughness 1.1–3.3 kg·m/cm^2	[424] [690]	1959 1960
TiB_2	High hardness and wear resistance	In the composition of cermets for metal cutting and rock drilling. Wear-resistant coatings	[455] [1079] [1080]	1960 1961 1962
		Grinding elements, wear-resistant linings, bearings, nozzles for sand-blasting apparatus	[1010]	1961
TiB_2	Refractoriness, resistance to the action of molten metals, linear dependence of electrical resistance on temperature	Electrodes for high-temperature thermocouples for measuring the temperature of molten metals and alloys	[487]	1959
		Sheaths for metal immersion thermocouples. Heating elements for high-temperature electrical resistance furnaces for use in neutral and reducing media and in vacuo	[489] [918]	1960 1961
		Lining of electrolyzers for the production of Al, pump components, spouts, runners in the production of zinc and other non-ferrous metals	[1010]	1961

EXAMPLES OF THE APPLICATION OF REFRACTORY COMPOUNDS (continued)

Phase	Fundamental properties	Fields of application	Ref.	Year
		Crucibles for precision melting. Pipes for conveying molten metals	[646]	1960
TiB$_2$	Satisfactory resistance to neutron irradiation	Cermets TiB$_2$–Ti for nuclear engineering	[924] [925]	1959 1959
ZrB$_2$	High-temperature strength, refractoriness, resistance to scaling	In the composition of heat-resistant alloys of the type of "Borolites." Fundamental properties of these alloys: Sp. gr. 5.2–5.4. Hardness 88–91 RA. Bending strength at 16°C, 48.0; at 1000°C, 46.0; at 1200°C, 40.0 kg/mm^2. Modulus of elasticity at 16°C, 22900; at 1000°C, 17400 kg/mm^2. Stress–rupture strength at 1000°C: 5 hr, 21.7; 10 hr, 18.2; 100 hr, 13.4; 1000 hr, 10.6 kg/mm^2. Coefficient of thermal expansion (25–1000°C) $3.2 \cdot 10^{-6}$. Thermal shock resistance, 200 heating–cooling cycles under the conditions 100–1300°C	[708] [710]	1955 1954
		In the composition of refractory Borolites with Mo, Cr bonds	[494]	1960
ZrB$_2$	Refractoriness, high resistance to the action of molten metals, alloys and slags, linear temperature dependence of electrical resistance over a wide temperature range	Electrode sheaths of high-temperature thermocouples for measuring the temperature of molten steel, cast iron, nonferrous and rare metals, and their alloys Sheaths for protection of metallic thermocouples	[487] [504] [874] [892] [893] [1018] [1160] [1165]	1959 1961 1961 1961 1961 1961 1963 1962

EXAMPLES OF THE APPLICATION OF REFRACTORY COMPOUNDS (continued)

Phase	Fundamental properties	Fields of application	Ref.	Year
		Furnace components and fittings in ferrous and nonferrous metal-lurgy	[489]	1960
		Heating elements for high-temperature electrical resistance furnaces	[62]	1961
		Crucibles for precision metallurgy. Boats for vacuum metallization by sputtering. Tubes for conveying molten metals	[122]	1961
ZrB_2	High electrical conductivity, resistance to the action of the electric arc	In the composition of electrical contacts (for example, in combination with silver) resistant to burning	[501]	1956
$\overline{H}fB_2$, VB_2, NbB_2, TaB_2	High high-temperature strength, refractoriness	In the composition of heat-resistant alloys, preparation of containers for liquid uranium and calcium	[440] [1276]	1960 1962
CrB, CrB_2	High high-temperature strength, refractoriness, resistance to scaling	In the composition of heat-resistant alloys of the type of Borolites. Fundamental properties: sp. gr. 6.77−7.31; hardness 77−88 RA; transverse strength (at 1000°C) 87.5−105 kg/mm^2; stress-rupture strength (at 1000°) for 2000−3000 hr, 10.6 kg/mm^2; impact toughness 6.6−8.8 $kg \cdot m/cm^2$; resistivity 27−54 $\mu\Omega \cdot cm$; thermal shock resistance, 200 heating-cooling cycles under the conditions 100−1300°C. Heat-resistant alloy 80% CrB + 20% bond	[710] [690]	1954 1960
CrB_2	High wear resistance	In the composition of wear-resistant hard-facing alloys, for example a hard-facing mixture of 50% Fe + 50% CrB, electrodes	[714] [529]	1949 1958

EXAMPLES OF THE APPLICATION OF REFRACTORY COMPOUNDS (continued)

Phase	Fundamental properties	Fields of application	Ref.	Year
		BKh-2 (80% CrB, 8% mica flour, 10% graphite, 2% potash) gives layers having a hardness of 78–79 RA. Mixture of KBKh with 60% ferrochrome, 30% Fe powder, 5% Cr_3C_2, and 5% CrB, increases the wear resistance of steel 10–12 times	[643]	1959
Mo_2B_5	Low vapor density, refractoriness, good alloying properties with Mo and W	For soldering W and Mo in radio engineering	[775]	1954
Mo_2B_5, W_2B_5	Resistance to the action of molten metals, thermal shock resistance, refractoriness, thermal conductivity	Crucibles and molds for precision metallurgy Heat-resistant alloys	[646] [440]	1960 1960
Borides of Mn, Fe, Co, Ni	Wear resistance, hardness	For the production of wear-resistant and corrosion-resistant coatings	[440]	1960
Borides of Fe	Wear resistance, hardness, oxidation resistance	In wear-resistant and corrosion-resistant coatings on steel components	[818] [1181] [1182]	1958 1963 1963
Borides	High chemical activity, crystal lattice strength	Catalysts for hydrogenation processes	[663]	1952

EXAMPLES OF THE APPLICATION OF REFRACTORY COMPOUNDS (continued)

Phase	Fundamental properties	Fields of application	Ref.	Year
All borides	Chemical resistance to the action of acids and their mixtures in both cold and hot states	Components of chemical apparatus	[530]	1959
Be_2C	Refractoriness, high-temperature strength, nuclear properties	In the composition of heat-resistant alloys, for example, 60% Be_2C + 40% BeO (compressive strength = 90 kg/mm²; tensile strength =15.9 kg/mm²; Poisson ratio 0.19; electrical resistance 10.9 $\Omega \cdot$ cm); in nuclear power engineering	[526]	1952
ThC, ThC_2	Low work function in thermionic emission, refractoriness	In the composition of electrodes in thermoelectronic devices for the conversion of heat energy into electrical energy	[562] [244]	1960 1961
UC, U_2C_3, UC_2	Refractoriness, nuclear properties	In nuclear power engineering (as uranium fuel and radiating constructional elements, in the composition of heat-liberating elements)	[251]	1955
TiC	Refractoriness, oxidation resistance, high-temperature strength	1. In the composition of heat-resistant alloys (cermets) for the production of gas-turbine blades, rotors, components of high-temperature test machines (clamps, rollers), for example, of WZ (Austria) alloys on the basis of TiC (35−75%) with a Ni−Co−Cr alloy bond: sp. gr. 6−6.95; Vickers hardness 600−1070 kg/mm²; transverse strength 120−190 kg/mm²; tensile strength 60−110 kg/mm²; impact toughness 0.4−0.97 kg·m/cm²; modulus of	[785] [1091]	1960 1962

EXAMPLES OF THE APPLICATION OF REFRACTORY COMPOUNDS (continued)

Phase	Fundamental properties	Fields of application	Ref.	Year
		elasticity 32,500–41,900 kg/mm² (all at a temperature of 20°C). Temperature dependence of transverse strength of the alloy WZ-12a with 50% TiC: 20°C, 141; 300°C, 140; 500°C, 139; 700°C, 112; 900°C, 70 kg/mm²; mean coefficient of linear expansion (20–1000°C) of the alloy WZ-12a (75% TiC), $9.9 \cdot 10^{-6}$; alloy WZ-12b (60% TiC), $9.2 \cdot 10^{-6}$; alloy WZ-12c (50% TiC), $10.6 \cdot 10^{-6}$; thermal shock resistance, 100 heating-cooling cycles under the conditions: 25 heating-cooling cycles 100–980°C, 25 heating and cooling cycles 100–1035°C; 25 heating and cooling cycles 100–1205°C; 25 heating and cooling cycles 100–1315°C.		
		Alloys of TiC (55–58%) + TiB_2 (17–18%) + Si (10%); tensile strength 37.45 kg/mm² at 20°C and 6.16 kg/mm² at 998°C.		
		Alloys 65% TiC, 15% solid solution of TiC–TaC–NbC and 20% Co: good thermal shock resistance, creep resistance up to 1050–1100°C	[778]	1955
		Alloys of 73% TiC + TiB_2 and 27% CoSi – tensile strength (short-time at 980°C) 19 kg/mm²	[819]	1954
		Alloys TiC (42.9–63.0%) + Cr_3C_2 (5.7–7.1%) + Ni (22.2–50.0%) + Co (7.4–25.9%) + Cr (7.4–11.1%) for temperatures below 1000°C	[707]	1953
		Cermets on the basis of TiC and other carbides of refractory metals, as well as SiC, B_4C, their alloys with oxides for the protective coating of rocket elements, including jets of powder rocket engines and nose parts of rockets	[499]	1961

EXAMPLES OF THE APPLICATION OF REFRACTORY COMPOUNDS (continued)

Phase	Fundamental properties	Fields of application	Ref.	Year
		Alloys of TiC with steel bond (or with a bond of carbon steel with an addition of 2.9% chromium and molybdenum; in this case, the composition of the final alloy is 26% Ti, 6.5% C, 1.8% Cr, 1.8% Mo; the rest iron) The alloy is prepared by sintering followed by annealing in a neutral medium and quenching at 955°C. Hardness of the alloy after annealing 38–43, after quenching 68–71 Rockwell; density 6.58 g/cm^3; modulus of elasticity 31,000 kg/mm^2; compressive strength 252 kg/mm^2; coefficient of thermal expansion $9.8 \cdot 10^{-6}$ (20–700°C); good impact toughness and good forging, rolling, pressing and bending properties.	[978]	1960
		2. In the composition of wear and corrosion-resistant coatings on cast iron and steel	[805]	1953
		3. In the composition of friction disks for aircraft construction (high permissible service temperatures of up to 1000°C; thermal conductivity, frictional properties 50% higher than in the case of ordinary brake disks; total service life of the disks 5 times that of ordinary disks)	[806]	1953
TiC	High wear resistance and hardness	In the composition of cermets for steel cutting (alloys T15K6, T30K4, etc., where the figure following T is the TiC content, and the figure following K is the cobalt bond content, remainder tungsten carbide)	[2] [544]	1957 1960

EXAMPLES OF THE APPLICATION OF REFRACTORY COMPOUNDS (continued)

Phase	Fundamental properties	Fields of application	Ref.	Year
TiC	High electrical conductivity and refractoriness, low rate of evaporation	Arc lamp electrodes. Electrodes of TiC with stabilizing coating of silicon nitride or boron nitride for underwater electro-oxygen steel cutting. Specific consumption of such electrodes 0.09—0.11 mm/m cut (compared with 0.70 m/m cut for metal electrodes of type EPR-1)	[808] [885]	1910 1962
TiC	Resistance to the action of reducing gases, linear temperature dependence of electrical resistance, high strength	Electrode sheaths of thermocouples for measurement of temperatures up to 2500°C in furnaces with reducing and inert media and in vacuo, sheaths of metal thermocouples Components and fittings of metallurgical furnaces	[487] [816]	1959 1960
TiC	Resistance to the action of molten metals	Crucibles in nuclear power engineering for the production of heat-exchangers, for example of the alloy 80% TiC + 5% WC, or TaC and 15% Co, resistant to molten sodium (900°C, 188 hr) and bismuth (1000°C, more than 180 hr)	[546] [807] [1185]	1954 1930 1963
ZrC	High-temperature strength, high resistance to oxidation, refractoriness	In the composition of heat-resistant alloys	[1] [809]	1957 1952
ZrC	Low neutron absorption cross section (purified	In nuclear power engineering	[547]	1959

EXAMPLES OF THE APPLICATION OF REFRACTORY COMPOUNDS (continued)

Phase	Fundamental properties	Fields of application	Ref.	Year
	from HfC), refractoriness, high-temperature strength			
ZrC	Thermal shock resistance, satisfactory thermionic emission properties	In the composition of cathodes of the alloy UC–ZrC having a high work factor for thermoelectric devices for the direct conversion of thermal energy into electrical energy	[562] [1062] [1130] [1259]	1960 1962 1962 1962
ZrC	Resistance to the action of molten metals	Crucibles, boats, tubes	[97]	1952
HfC	Exceptionally high temperature	In the composition of special refractories (for example, for the lining of crucibles for melting refractory metals)	[776]	1958
ZrC, NbC	High thermionic emission properties	Thermal emission transducer	[1161]	1962
VC, NbC, TaC	High hardness and wear resistance	As alloying additions to cermets on a WC and TiC basis (increase in the life of cutting tools by 10–20%)	[2] [784]	1957 1960
NbC, TaC	Resistance to the action of molten metals and metal vapor, satisfactory strength at high	1. For the production of heating elements, evaporating apparatus for Al (TaC, service life 4–7 hr; NbC, service life 1 hr at 1500°C)	[587]	1960
		2. Lining of crucibles (TaC) for melting refractory metals (Ti, etc.)	[776]	1958

EXAMPLES OF THE APPLICATION OF REFRACTORY COMPOUNDS (continued)

Phase	Fundamental properties	Fields of application	Ref.	Year
	temperatures, low vapor tension, good radiation properties	3. Heating elements for high-temperature electrical resistance furnaces	[1]	1957
		4. Coatings of TaC on metal W and Re bases for special incandescent elements of electric lamps	[814] [815]	1924 1930
Cr_3C_2	High wear resistance and hardness	In the composition of hard-facing alloys, for example KhR-19 electrodes with a coating of 80% Cr_3C_2, 10% CrB, 10% graphite	[1] [657]	1957 1961
Cr_3C_2	High chemical resistance and resistance to oxidation	In the composition of heat-resistant and oxidation-resistant alloys, for example with a Ni bond, especially the alloys $TiC-Cr_3C_2-Ni$ and $Cr_3C_2-WC-Ni$ (83 : 2 : 15); hardness 88 RA, coefficient of thermal expansion $6.4 \cdot 10^{-6}$, tensile strength 65 kg/mm²	[657]	1961
		For the production of filters in chemical industry and electrodes for electrochemical processes. High-temperature solders for electronics, for example for fixing cathodes of LaB_6 to cores of Mo, W, Ta (composition of solder Cr_3C_2 + 1% CaF_2 or NaF)	[810]	1952
Mo_2C		Solders for high-temperature soldering in electronics (for example soldering ThO_2 to metals)	[692]	1959
		In the composition of heat-resistant and hard alloys	[1, 2]	1957
Mo_2C	Catalytic properties	Catalyst for the dehydrogenation of alcohols, cyclohexane, etc.	[890]	1961
WC	High hardness and wear resistance	In the composition of cermets of VK type for machining cast iron, bronze, brass, marble, porcelain, plastics (VK2, VK3, VK6; the number following K is the Co bond content, the remainder is WC)	[544]	1960

EXAMPLES OF THE APPLICATION OF REFRACTORY COMPOUNDS (continued)

Phase	Fundamental properties	Fields of application	Ref.	Year
WC		Reinforcing elements of crowns for rock drilling, diamond substi-tutes for truing grinding wheels (alloys WC + W$_2$C, likar* or relit*), facing work (especially facing drill crowns)	[2] [544] [717]	1957 1960 1951
Be$_3$N$_2$	Catalytic properties	Catalyst for the dehydrogenation of alcohols, cyclohexanone, etc.	[890]	1961
	Refractoriness	Special refractories	[97]	1952
AlN	Refractoriness, thermal shock resistance, re-sistance to Al melts, low coefficient of thermal expansion, satisfactory thermal conductivity	Refractories (especially for melting and synthesis of semicon-ductor alloys)	[792] [1184]	1959 1962
UN	Refractoriness, chem-ical resistance to molten metals, nu-clear properties	Special refractories (crucibles, boats), in nuclear power engineer-ing	[97]	1952
TiN	High-temperature re-sistance, refractori-ness, hardness, wear	1. In the composition of special refractories	[804]	1952
		2. In the composition of refractory alloys, for example MgO+TiN	[97] [2]	1952 1957

*Transliteration of the Russian. These are apparently trade names and the correct English spellings may differ from those given here.

339

EXAMPLES OF THE APPLICATION OF REFRACTORY COMPOUNDS (continued)

Phase	Fundamental properties	Fields of application	Ref.	Year
	resistance, resistance to the action of molten metals	(high thermal shock resistance, strength at 1090°C 30% higher than at room temperature)	[793]	1953
		3. In the composition of grinding wheels	[794]	1938
		4. In the composition of coatings on titanium and graphite components	[795]	1956
			[1203]	1961
		5. Dusting of molds for the production of clean castings	[1]	1957
		6. Nitriding of titanium cores for electrical measuring instruments	[1022]	1961
TiN, ZrN	High electrical conductivity, non-arcing	1. Conducting elements of thorium cathodes	[796]	1949
		2. Rectifier igniters (25% TiN + 75% BeO)	[796]	1949
		3. In the composition of high-ohmic resistances (TiN + Cr_2N)	[799]	1955
NbN	High electrical conductivity, ability to become superconducting at 15°K, oxidation resistance	1. Detector devices	[747]	1946
			[798]	1949
		2. Bolometers	[800]	1946
			[801]	1949
		3. Electron tubes for image transmission	[802]	1950
			[803]	
Nitrides of Cr, Fe	High hardness and wear resistance	In the composition of wear-resistant coatings on steel	[282]	1960
$BaSi_2$, $LaSi_2$	Semiconductors, oxidation resistance	Semiconductor devices	[299]	1960
			[314]	1960

EXAMPLES OF THE APPLICATION OF REFRACTORY COMPOUNDS (continued)

Phase	Fundamental properties	Fields of application	Ref.	Year
CeSi$_2$			[315]	1960
DySi$_2$	Oxidation resistance, neutron absorber	In nuclear power engineering (atomic reactor control rods)	[285]	1959
TiSi$_2$, Ti$_5$Si$_3$	High-temperature strength, oxidation resistance	In the composition of heat-resistant alloys Ti$_5$Si$_3$–SiC, TiSi$_2$–SiC	[717] [773]	1959 1959
V$_3$Si	Superconductivity ($T_c = 17°K$)	Technical physics and automation	[117]	1959
Cr$_3$Si	High-temperature strength, oxidation resistance	In the composition of heat-resistant alloys, for example 50% Cr$_3$Si + 50% Cr$_2$, tensile strength (at 980°C) = 91 kg/mm^2, high resistance to oxidation	[778] [787]	1955 1955
MoSi$_2$	High reactability in contact with refractory metals and silicon, refractoriness, resistance to scaling	High-temperature soldering by means of MoSi$_2$ in electronics, for example cathodes of LaB$_6$ with cores of Mo, Ta, W	[692]	1959
		For the same purposes, it is possible to use a mixture of powders of Mo (93–99.5%) and Si (0.5–7%)	[713]	1955
		Oxidation-resistant alloys for gas-turbine components, combustion chambers of jet engines, guided missiles, nozzles of sand-blasting apparatus, components of metallurgical furnaces, hot-pressing and drawing dies, soldering devices, for example 75% MoSi$_2$ + 25% Al$_2$O$_3$, good thermal shock resistance; 90% MoSi$_2$ + 10% Co, 90% MoSi$_2$ + 6% Co, tensile strength = 53 kg/mm^2 at 980°C, loss in weight on oxidation 2 mg/cm^2 (after 100 hr at 1095°C)	[144] [778]	1953 1955

EXAMPLES OF THE APPLICATION OF REFRACTORY COMPOUNDS (continued)

Phase	Fundamental properties	Fields of application	Ref.	Year
MoSi₂, WSi₂	High resistance to oxidation and to the action of other chemical reagents, stability in different gaseous media, linear temperature dependence of electrical resistance, high thermo-emf, resistance to molten chlorides	Electrode sheaths of high-temperature thermocouples for measurement of temperatures in air up to 1700–1800°C, temperatures of molten salts and so forth (for example, thermocouples $MoSi_2$–B_4C, $MoSi_2$–borided graphite, $MoSi_2$–WSi_2)	[487] [504] [1165]	1959 1960 1962
		Electrode sheaths of thermocouples for the measurement of temperatures of reducing media up to 1850°C and oxidizing media up to 1700°C	[715]	1960
		Heating elements for high-temperature electric resistance furnaces for service in air at temperatures of up to 1650–1700°C	[789–791] [811]	1957– 1958 1960 1953
		Protective nonscaling coatings on molybdenum heating elements and other molybdenum components	[811] [919] [117]	1953 1961 1959
		In the composition of high-temperature heating elements of complex composition, for example 50–90% Mo, 15–50% Si, 1.50% Al		
MoSi₂	High work factor in thermionic emission	Anti-emission (grid) coatings in electronics	[117]	1959
MoSi₂	Catalytic properties	Catalyst in the dehydrogenation of alcohols, cyclohexanone, etc.	[890]	1961
MnSi, MnSi₂	Semiconductors	Semiconductor devices	[314] [315] [268]	1960 1961 1961
		Electrodes of thermogenerators with an efficiency of 5–13%	[1137]	1961

EXAMPLES OF THE APPLICATION OF REFRACTORY COMPOUNDS (continued)

Phase	Fundamental properties	Fields of application	Ref.	Year
ReSi$_2$	Semiconductor properties. Resistance to scaling up to 1600–1700°C	Semiconductor devices	[314] [315] [347]	1960 1961
Iron silicides	Oxidation resistance, chemical stability	In the composition of corrosion-resistant and oxidation-resistant coatings on steel components	[117] [818]	1959
Cu$_3$P	Formation of fluid alloys with copper, having good wetting properties	Soldering of brass components (instead of silver solder), reducing the cost of solder to a quarter or a third	[245]	1960
Fe$_2$P	Catalytic properties	Catalysis in chemical industry	[245]	1960
Phases of the Ni–P system	High hardness and wear resistance	Production of hard and wear-resistant coatings on steel (microhardness of coatings up to 950 kg/mm^2)	[245]	1960
Fe$_2$P, Co$_2$P, (Fe, Co)$_2$P	Magnetic properties	In making permanent magnets	[1279]	1963

EXAMPLES OF THE APPLICATION OF REFRACTORY COMPOUNDS (continued)

Phase	Fundamental properties	Fields of application	Ref.	Year
SiP	Semiconductor properties	Semiconductor devices	[245]	1960
CeS, Ce$_2$S$_3$, LaS, La$_2$S$_3$, ThS, Th$_2$S$_3$, Th$_4$S$_7$, etc.	Refractoriness, low vapor tension, resistance to the action of molten metals	Production of refractories for the melting of refractory metals (Ce, U, Th, Ti, Zr, etc.)	[304]	1961
Ce$_2$S$_3$, La$_2$S$_3$, Th$_2$S$_3$, etc.	Semiconductor properties	Semiconductor devices and thermogenerators	[304] [1114]	1961 1961
B$_4$C in powder form	High abrasive properties, hardness	1. Grinding and polishing of hard materials (industrial jewels, minerals, alloys, glass, ceramics, quartz); productivity 50–70% of that of diamond	[440]	1960
		2. Sharpening and honing of hard alloy tool bits	[718]	1941
		3. Lapping work	[725]	1951
		4. Metallographic work	[726]	1949
B$_4$C in	High hardness, abrasive	1. Grinding and cutting wheels	[440]	1960

EXAMPLES OF THE APPLICATION OF REFRACTORY COMPOUNDS (continued)

Phase	Fundamental properties	Fields of application	Ref.	Year
the form of sintered components and alloys with other carbides and metals	properties, wear resistance, high-temperature strength, oxidation resistance, high neutron absorption cross section	2. Heat- and oxidation-resistant alloys, for example, the alloy 64% B_4C + 34% Fe: tensile strength at 20°C =24.5; 870°C, 22.8; 1093°C, 22.4; 1316°C, 17.6 kg/mm²; high oxidation resistance (loss in weight $5.8 \cdot 10^{-5}$ g/cm²·hr at 870°C and $91.2 \cdot 10^{-5}$ g/cm²·hr at 1093°C; sp.gr. 3.2–3.29 g/cm³)	[383] [440]	1952 1960
		3. Tool bits for boring drills and tools for machining hard materials	[440]	1960
		4. Nozzle tips for welding in protective gases, practically no reaction with metal splashes	[440]	1960
		5. Tools for truing grinding wheels, high truing accuracy; performance 5 to 10 times that of other diamond substitutes (of hard alloys, carborundum, alumina)		
		6. Sand-blast nozzles, life 300 times that of cast-iron nozzles	[440]	1960
		7. Gases, life 100–200 times that of steel gages	[440]	1960
		8. Filters for the textile and chemical industries, thread guides in the viscose industry	[440]	1960
		9. Dies for drawing rods of abrasive materials, welding electrodes, etc.	[440]	1960
		10. Chemical ware	[440]	1960
		11. Neutron absorbers, cermets for nuclear engineering (for example B_4C + Al_2O_3)	[738]	1951
B_4C	High electrical resistance, chemical stability, semiconductor	Boron carbide nonwire resistance of film type produced by diffusion treatment of anthracite with boron (nominal resistance values from 1 Ω to 300 kΩ and dissipating power from 0.05 to	[926] [711]	1958 1959

EXAMPLES OF THE APPLICATION OF REFRACTORY COMPOUNDS (continued)

Phase	Fundamental properties	Fields of application	Ref.	Year
	properties, non-arcing properties	300 W, specific dissipating power ~ 0.4 W/cm²; temperature coefficient $\leq 5.10^{-4}$; in low-ohmic resistances, temperature coefficient one order less)	[740]	1937
		Semiconductor igniters for ignitrons	[712]	1959
		Nonlinear high-ohmic resistors (alloys and chemical compounds of B_4C and SiC, so-called borosilicocarbides $B_xSi_yC_z$)	[788]	1960
		In the composition of contacts on a silver basis (reduced rate of burning through arcing)	[501]	1956
SiC	High electrical resistance, resistance to thermal shock, semiconductor properties, resistance to the action of chemical reagents	Electrical engineering components:		
		1. Waveguide absorbers (in waveguides for power transmission above 1000 Mc) for the separation of part or the complete absorption of the incoming power	[271]	1956
		2. Miniature, nonlinear inertia-free resistors (the resistance of which depends sharply on the direction of the electric field)	[1063]	1961
		3. Igniters for ignitrons (rectifiers with a mercury cathode, in which the cathode spot on the surface of the mercury is a source of free electrons and is produced periodically on the passage of current pulses through the semiconductor igniter (in this case SiC)	[1073]	1940
		4. Dischargers		
		5. Temperature compensators		
SiC	High chemical resistance to acids, alkalis, molten metals and metallic vapors	Components of pumps for pumping cold and hot acid solutions; production of refrigerators, scrubbers used for hot corrosive gases, tips of nozzles for spraying chemically active liquids; agitators resistant to corrosion as well as the abrasive action of	[530]	1959
			[786]	1957
			[1000]	1961

EXAMPLES OF THE APPLICATION OF REFRACTORY COMPOUNDS (continued)

Phase	Fundamental properties	Fields of application	Ref.	Year
SiC	High hardness, abrasive properties	solid components of suspensions or pulps; diffusers, in particular resistant to the action of phosphoric acid; collectors, cyclones subjected to intense abrasive action of powders or dusts		
SiC		Grinding, polishing; abrasive components of different shapes and purposes	[786]	1957
SiC	Reducing ability	Deoxidation of steels, production of ferroalloys	[786]	1957
SiC	Refractoriness, thermal shock resistance, resistance to scaling	Heat-resistant alloys, for example, $SiC + C$, $SiC + Si + C$ (tensile strength $1-9$ kg/mm^2, transverse strength $4.2-5.3$ kg/mm^2), $SiC + Co$, $SiC + B$, and others.	[812] [714] [788]	1960 1949 1960
α-BN	Refractoriness, high electrical resistance and semiconductor properties, lubricating properties	1. Thermal insulation of high-frequency vacuum induction furnace	[440]	1950
		2. Refractory coatings for molds and crucibles, refractory holders in automatic welding, crucibles for precision metallurgy, refractory pouring spouts for mixers and converters	[742] [769] [779] [272]	1958 1950 1948 1955
		3. High-temperature lubricant	[1041]	1961
		4. Light-load plain bearings operating under conditions of strong corrosive action by acid solutions	[1042]	1962
		5. In the composition of dielectrics		
		6. In the composition of high-temperature semiconductor materials		

EXAMPLES OF THE APPLICATION OF REFRACTORY COMPOUNDS (continued)

Phase	Fundamental properties	Fields of application	Ref.	Year
β-BN (borazon)	Refractoriness, high hardness, high-temperature strength, semiconductor properties	7. Refractories for cryolite–alumina baths in the electrolysis of aluminum	[44	1960
		1. As diamond substitute	[780]	1958
		2. In the composition of heat-resistant alloys	[1090]	1962
		3. High-temperature semiconductors		
Si_3N_4	Refractoriness, high thermal shock resistance, good oxidation resistance up to 1200–1300°C	In the composition of heat-resistant alloys, for example SiC + SiC_3N, B_4C + Si_3N_4, SiC + Si_3N_4 + Fe	[117] [782] [783]	1959 1950 1954
Si_3N_4	High chemical resistance to the action of acids, alkali solutions, molten metals, salts, and slags	Refractory components, parts of systems for pumping molten metals and salts, pumps for liquid metals, linings of baths for the production of aluminum by electrolysis from cryolite–alumina melts. High refractory properties are possessed by components made from SiC with a bond of Si_3N_4 (porosity 15–19%), compressive strength (at 20°C) 20.6 kg/mm², tensile strength (1000°C) 1.9–1.96 kg/mm², high thermal shock resistance and resistance to the action of molten metals	[781] [817] [1158]	1960 1959 1960
		Sheaths of Si_3N and alloys of Si_3N_4 with SiC for the protection of thermocouples for measuring the temperatures of molten fluoride baths in aluminum electrolyzers (life 100 hr at 940–970°C)	[888]	1961

EXAMPLES OF THE APPLICATION OF REFRACTORY COMPOUNDS (continued)

Phase	Fundamental properties	Fields of application	Ref.	Year
Si_3N_4	High electrical resistance, semiconductor properties	In the composition of bulk resistors, high-temperature thermistors, thermistors of different types	[813]	1960
B_6Si, B_4Si, B_3Si and other boron–silicon alloys	Refractoriness, resistance to scaling	In the composition of heat-resistant alloys	[440] [913]	1960 1960
BP	Semiconductor properties, resistance to scaling	Semiconductor devices	[103]	1960
Pyrographite	High resistance to oxidation, strength, and resistance to corrosion	In rocket construction, for the production of metal-melting crucibles, ceramics, as semiconductor	[979]	1960

349

APPENDIX

Yttrium—boron system [1111]

Aluminum—boron system [1108, 1154, 1155]

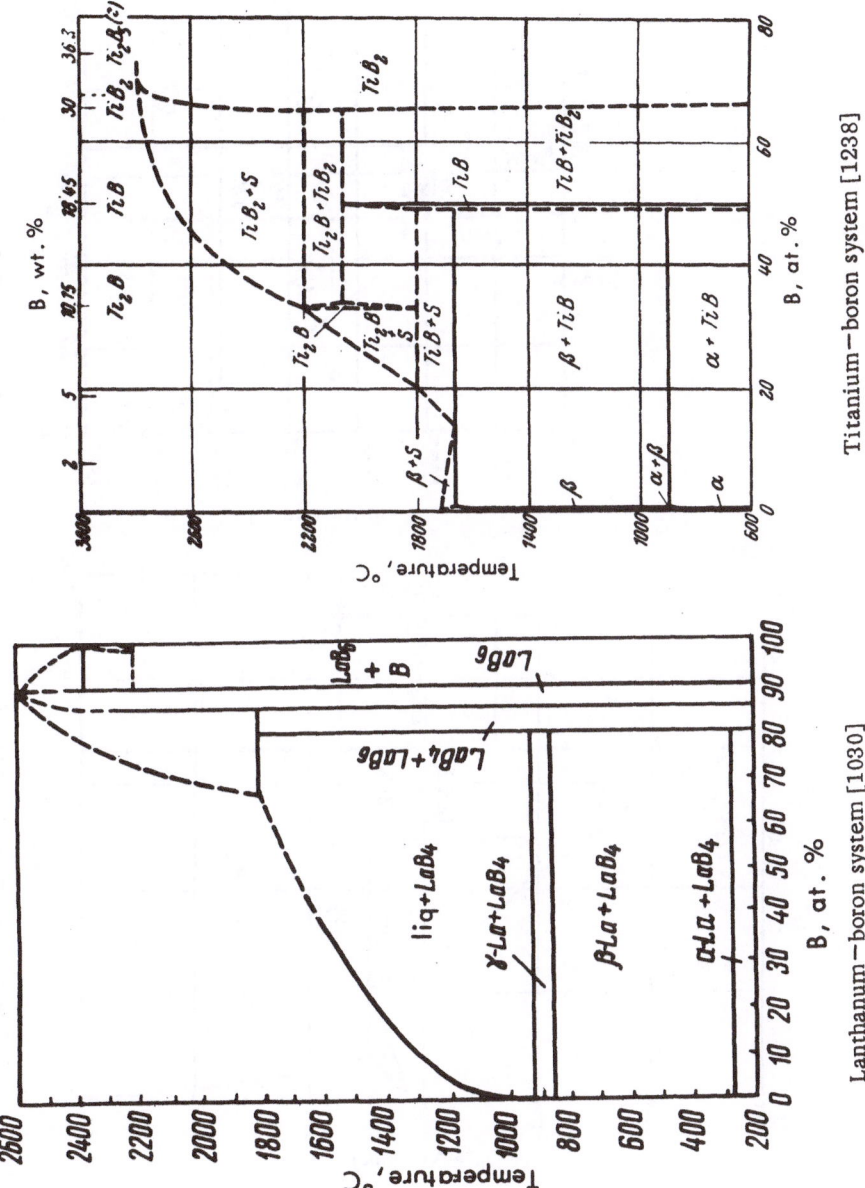

Titanium—boron system [1238]

Lanthanum—boron system [1030]

353

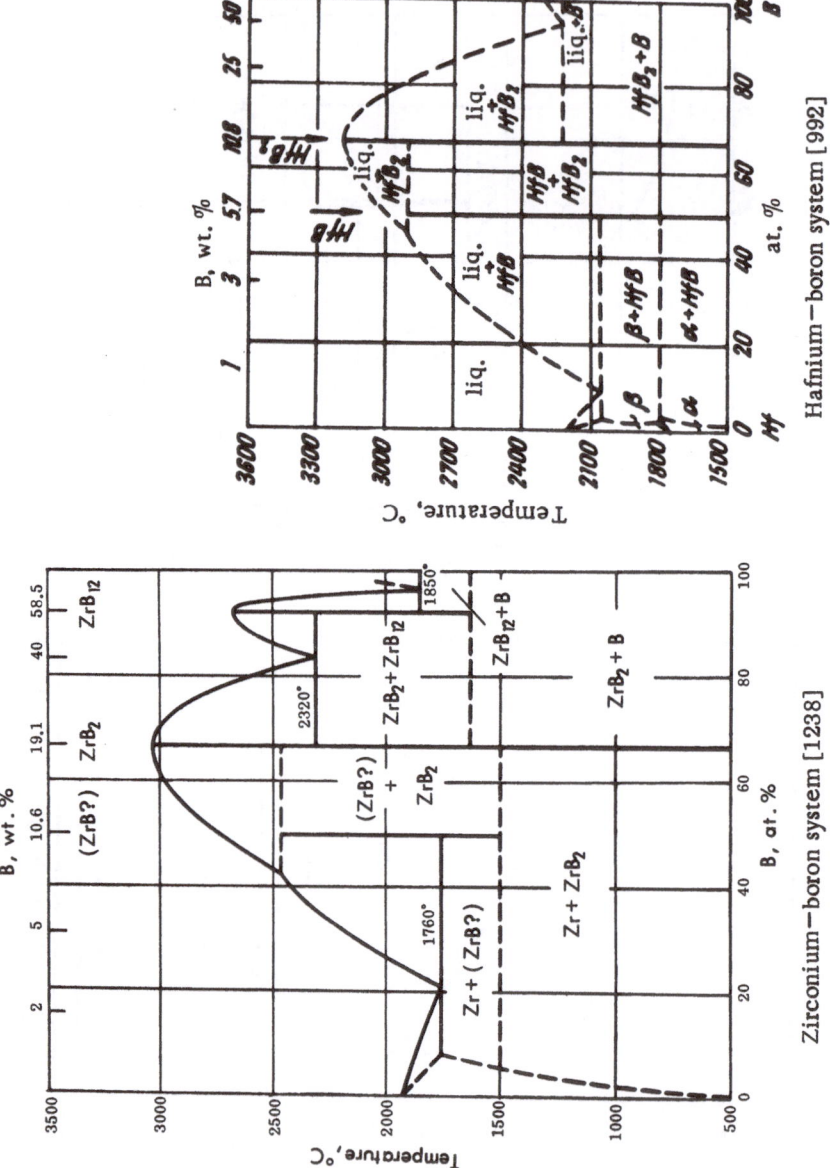

Hafnium–boron system [992]

Zirconium–boron system [1238]

354

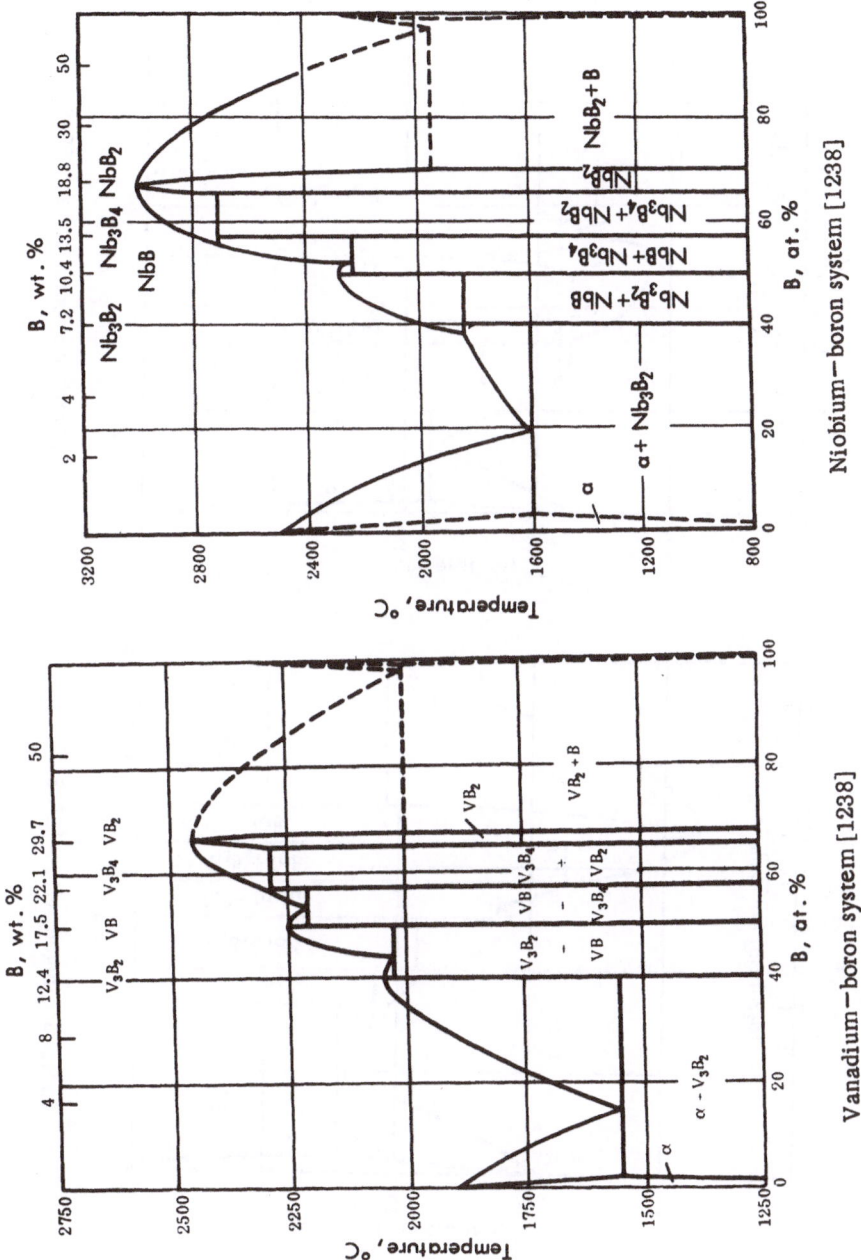

Niobium–boron system [1238]

Vanadium–boron system [1238]

355

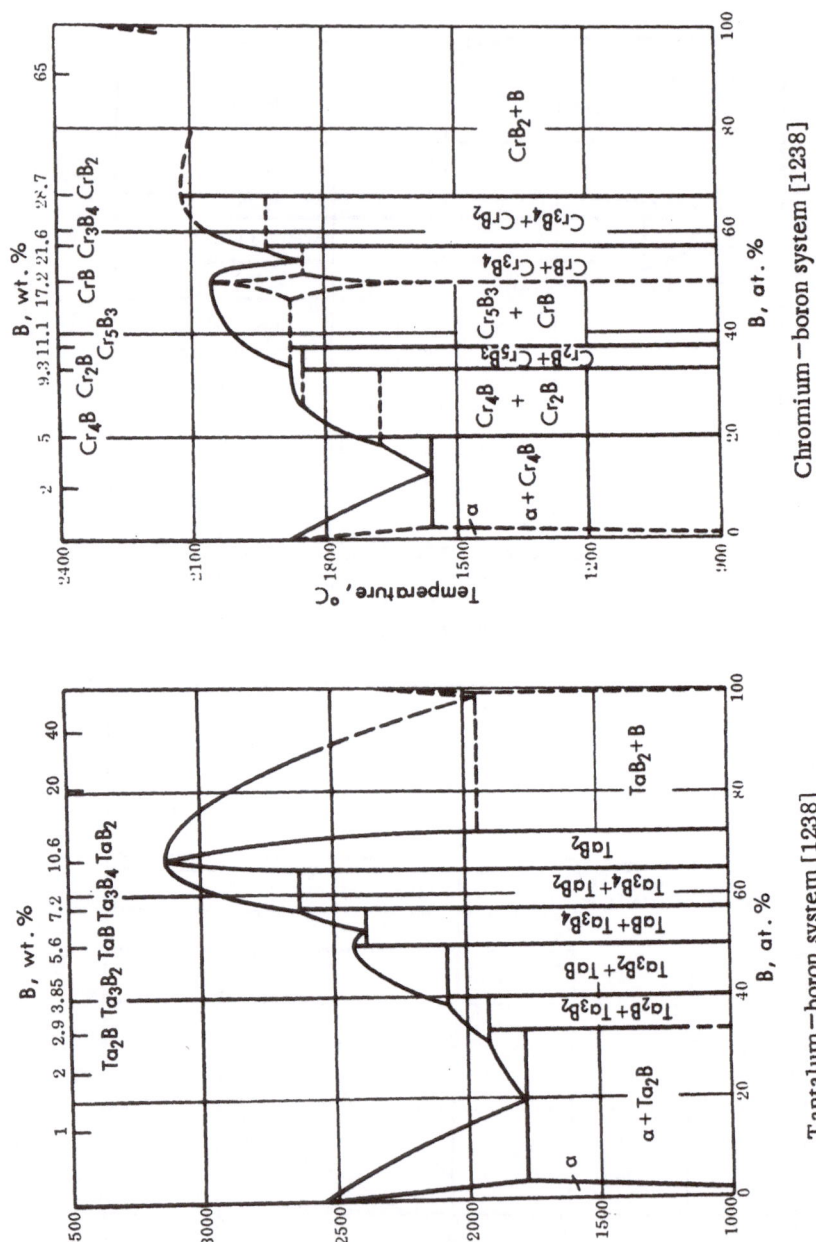

Chromium−boron system [1238]

Tantalum−boron system [1238]

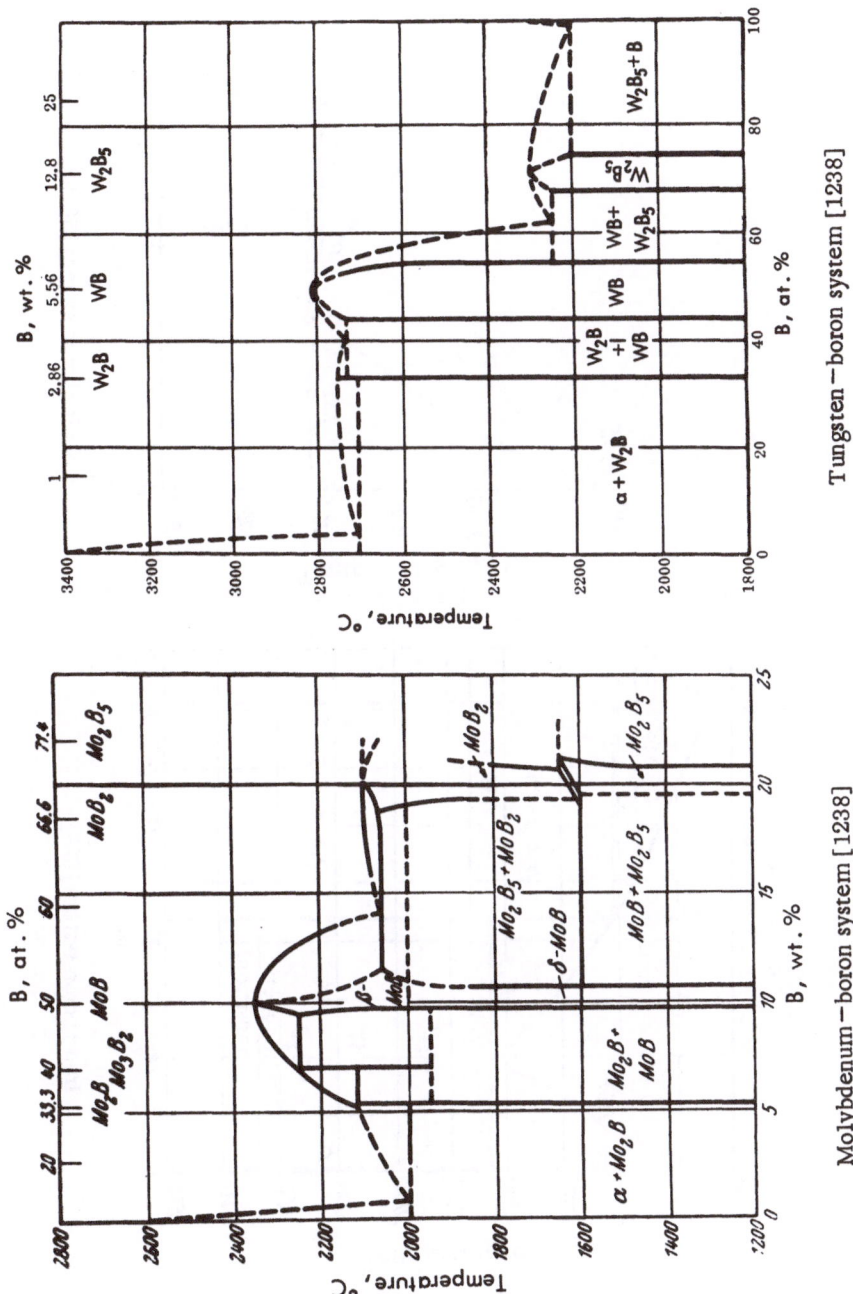

Tungsten—boron system [1238]

Molybdenum—boron system [1238]

357

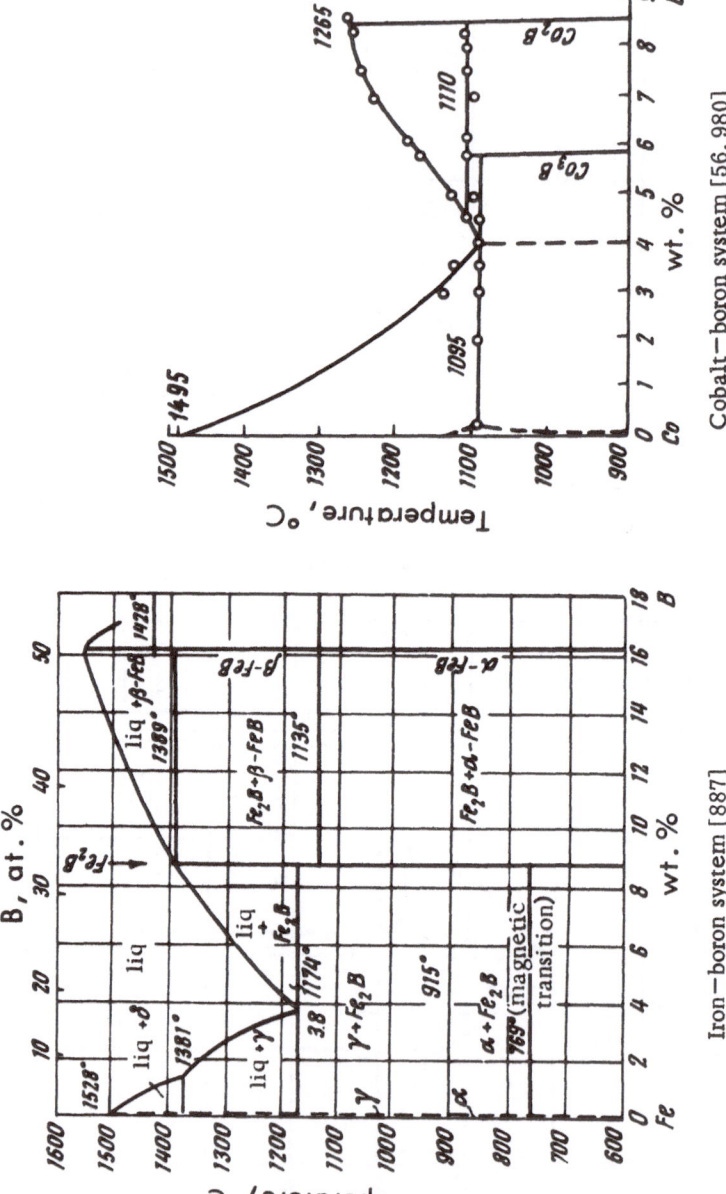

Cobalt–boron system [56, 980]

Iron–boron system [887]

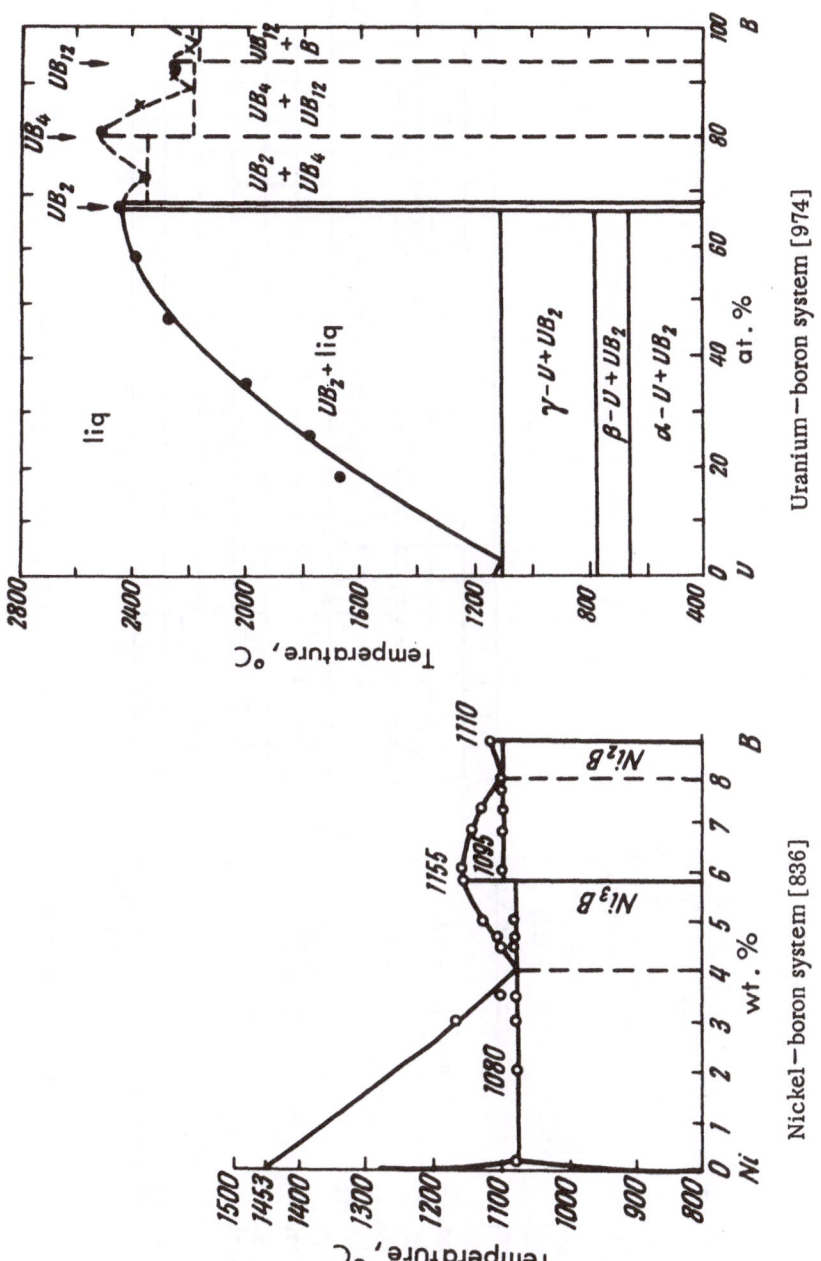

Uranium–boron system [974]

Nickel–boron system [836]

359

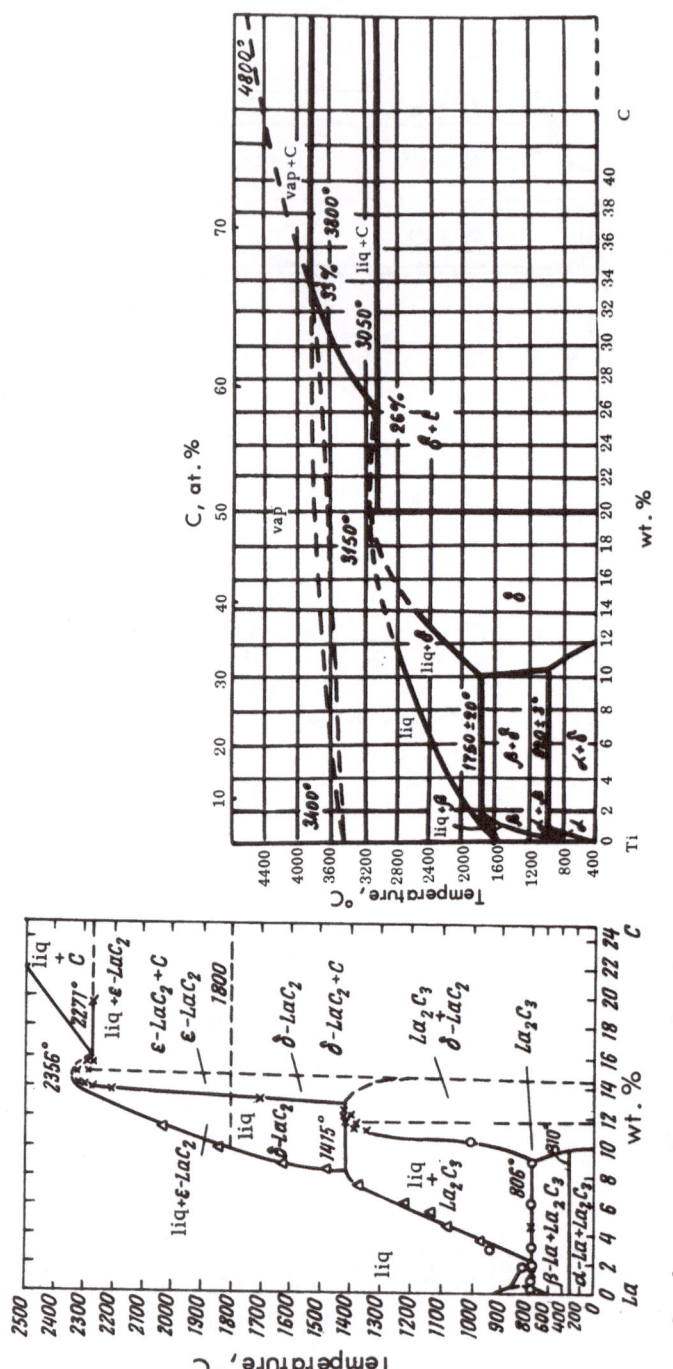

Titanium–carbon system [981, 1137]

Lanthanum–carbon system [570]

360

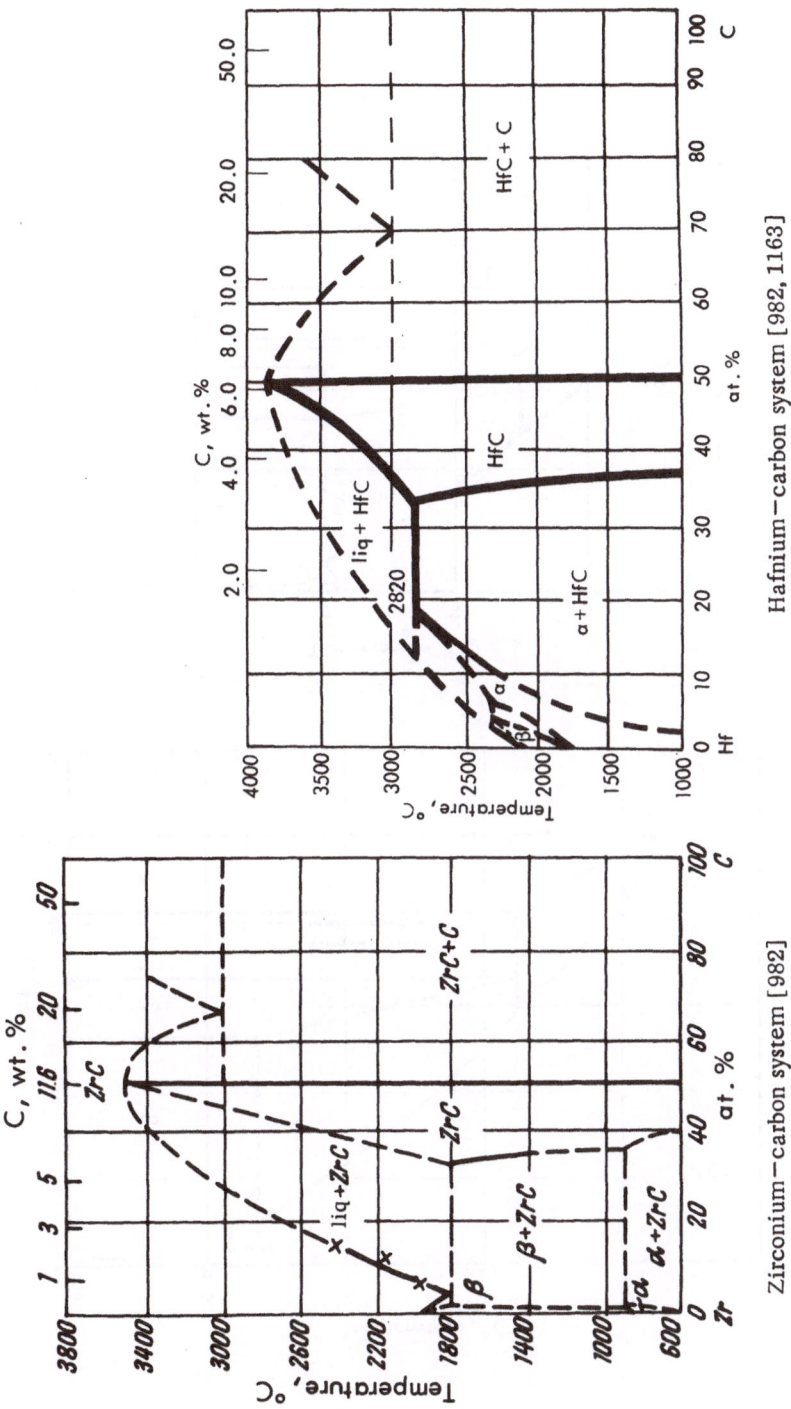

Hafnium–carbon system [982, 1163]

Zirconium–carbon system [982]

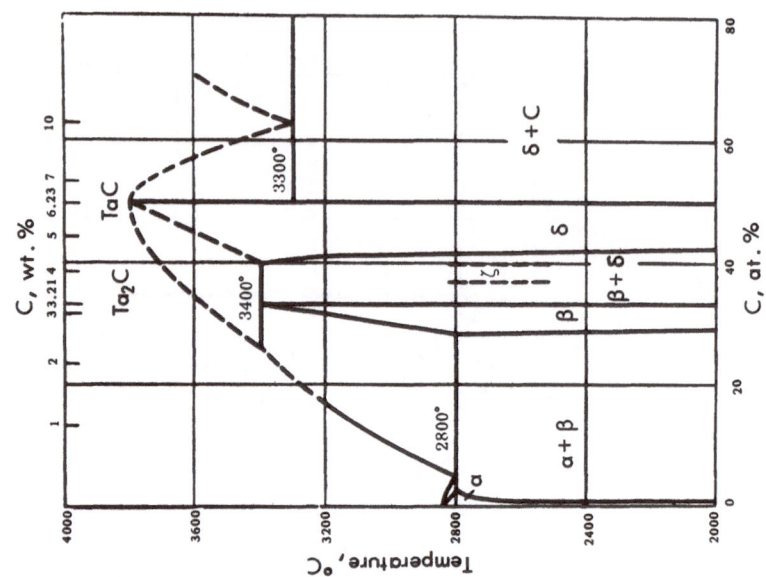

Tantalum—carbon system [1238]

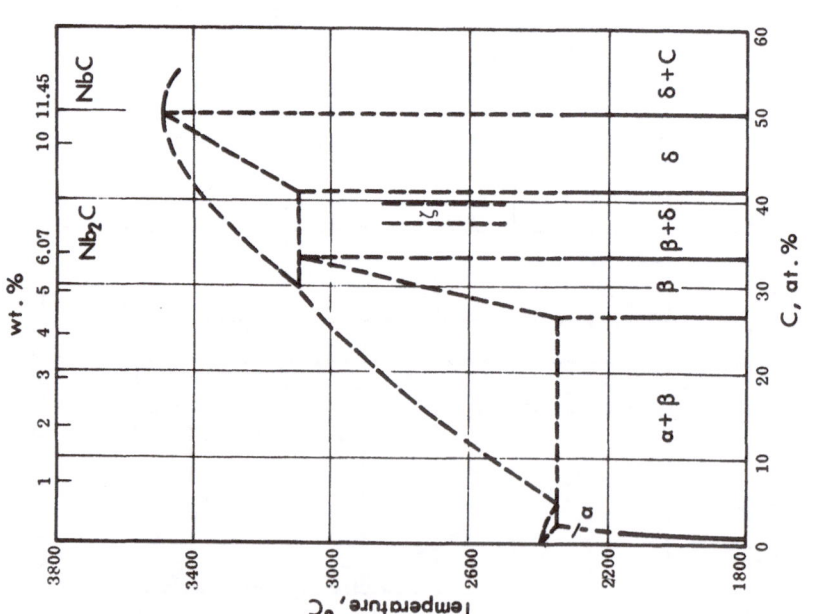

Niobium—carbon system [1238]; see [1083]

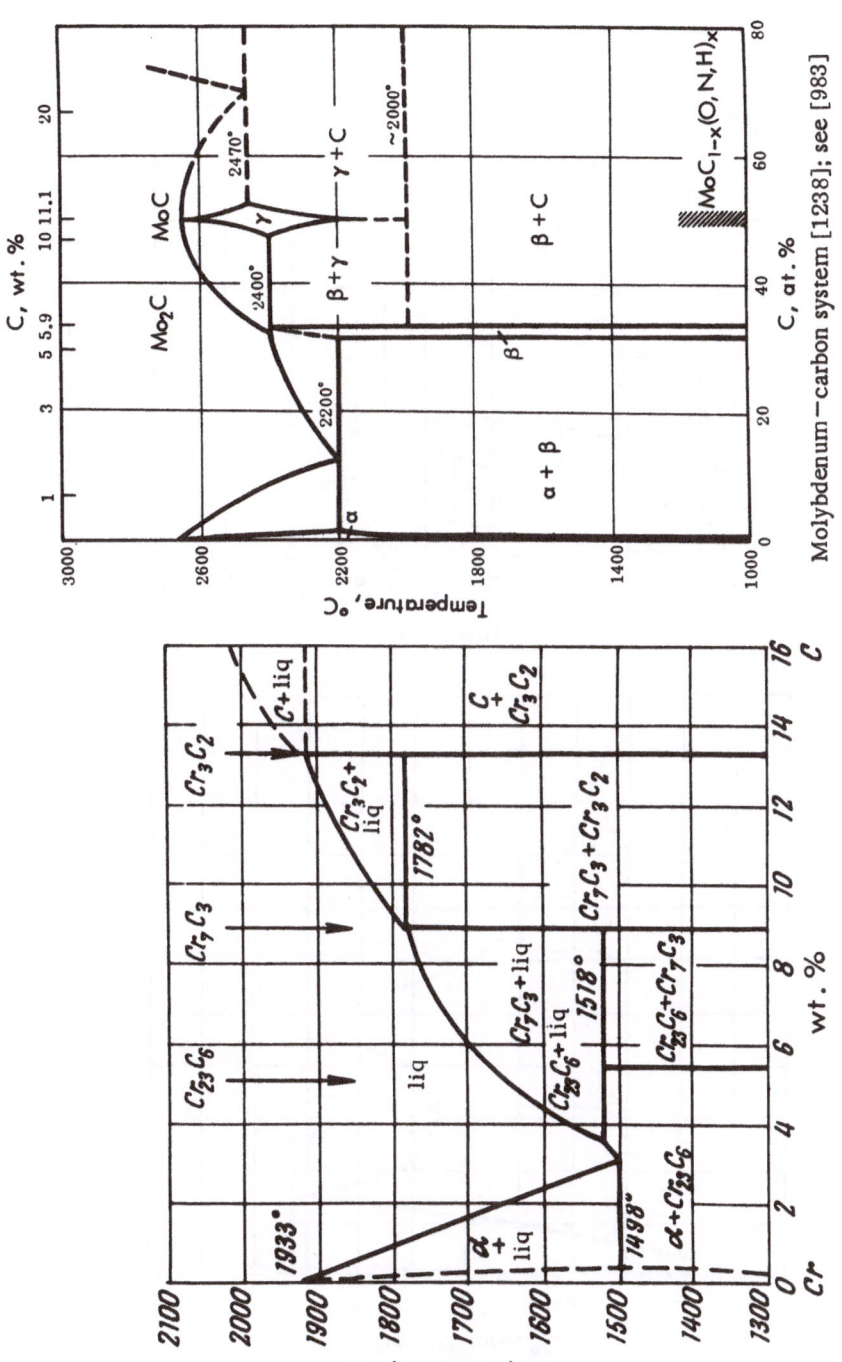

Molybdenum-carbon system [1238]; see [983]

Chromium-carbon system [887]

363

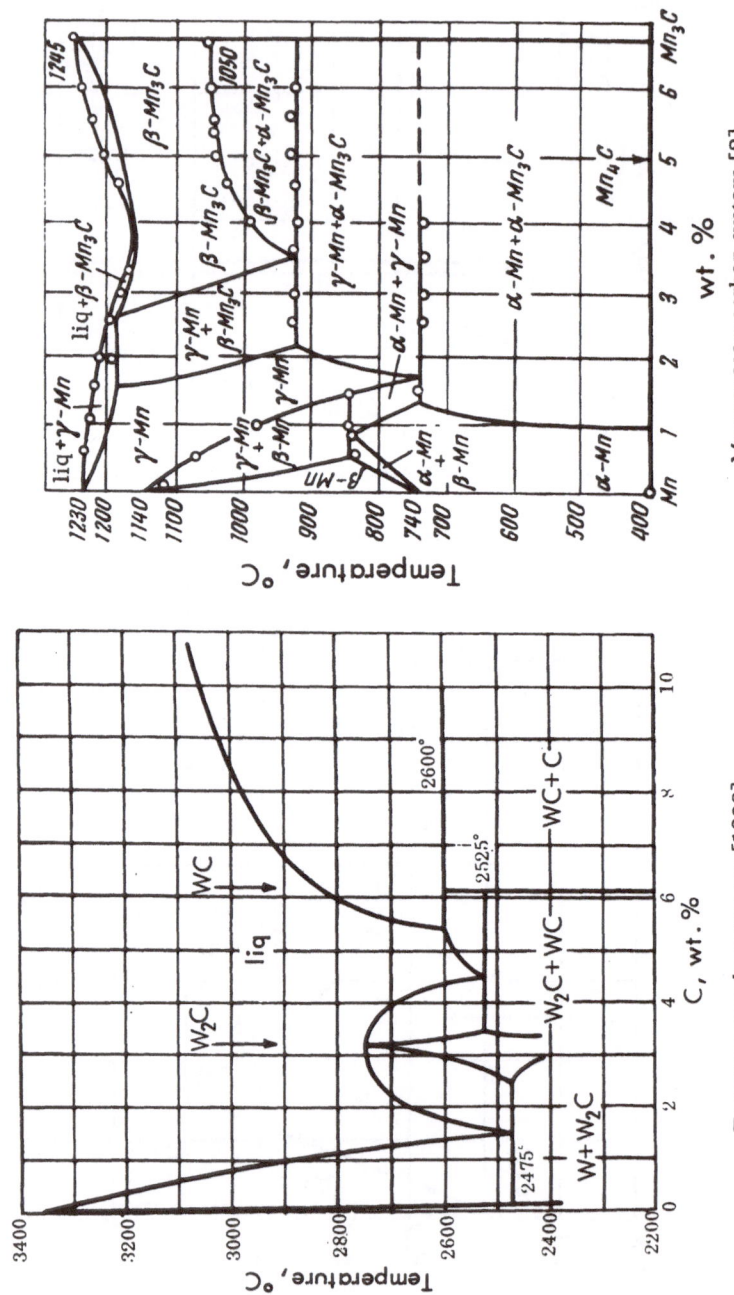

Manganese−carbon system [2]

Tungsten−carbon system [1238]

364

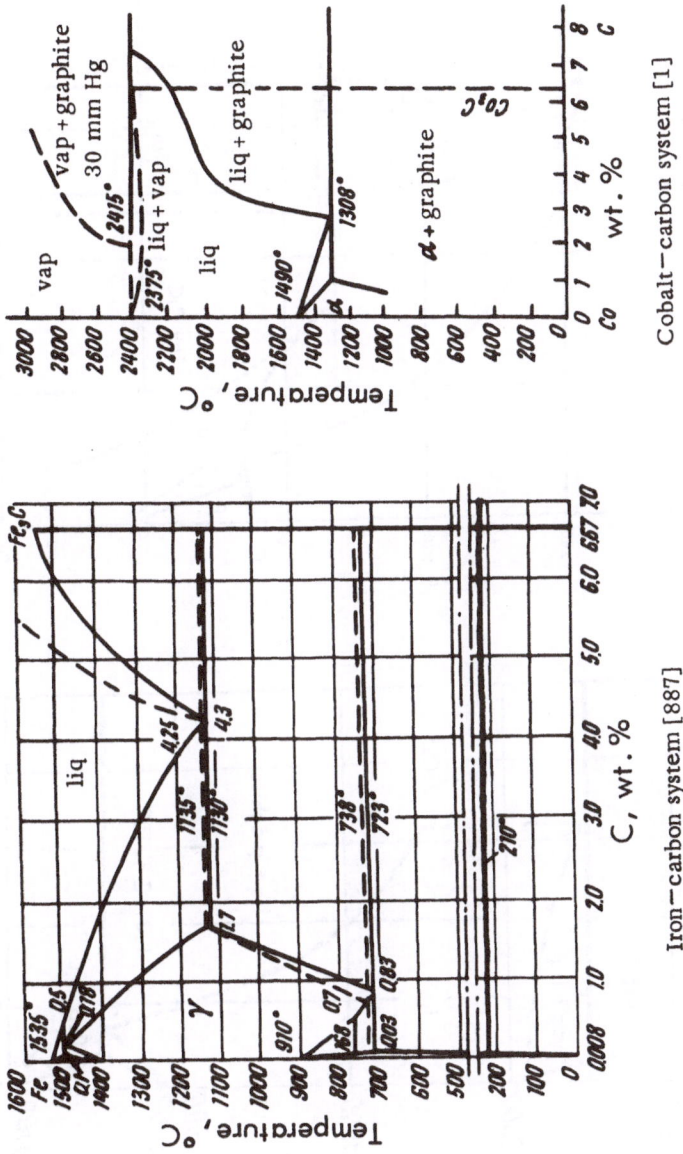

Cobalt—carbon system [1]

Iron—carbon system [887]

365

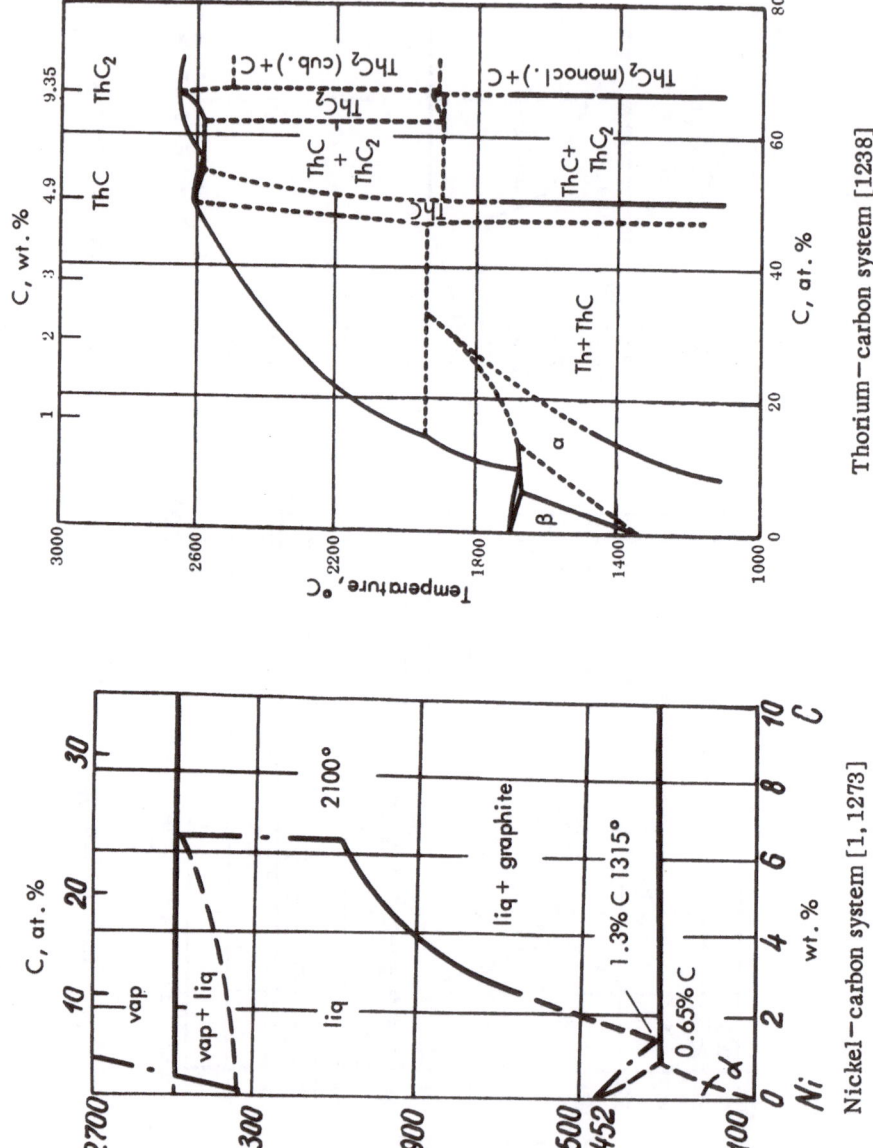

Thorium–carbon system [1238]

Nickel–carbon system [1, 1273]

366

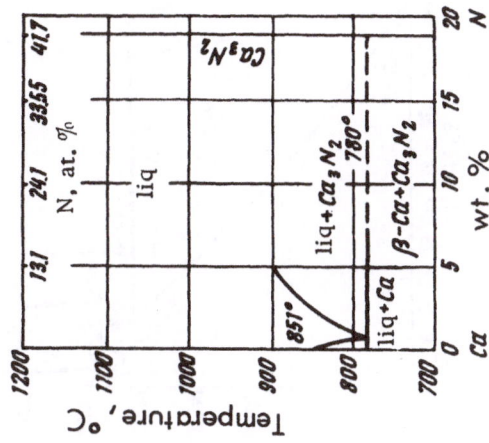

Calcium−nitrogen system [697]

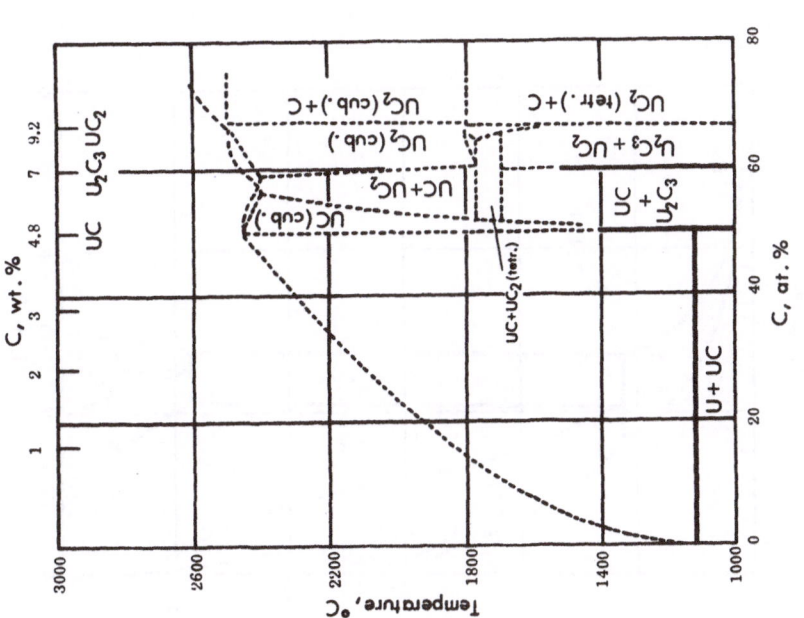

Uranium−carbon system [1238]

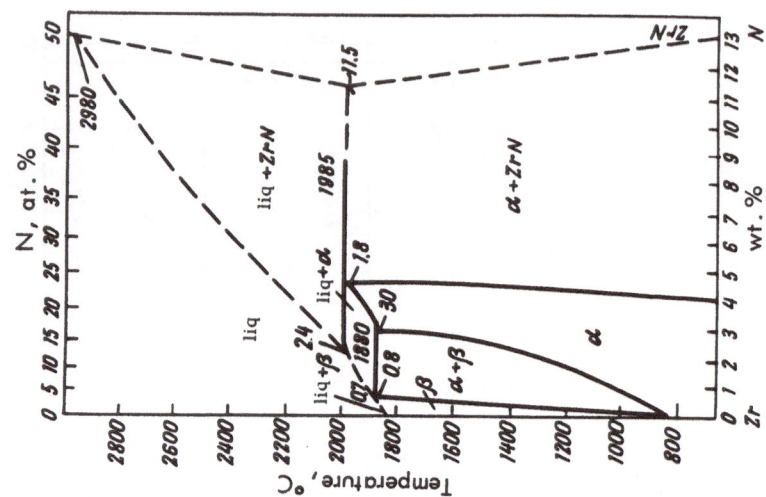

Zirconium−nitrogen system [592]

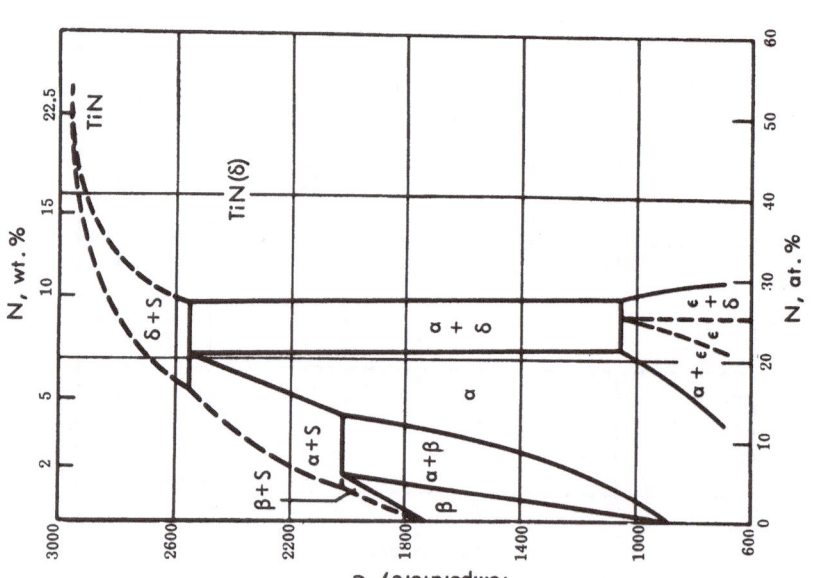

Titanium−nitrogen system [1238]; see [87]

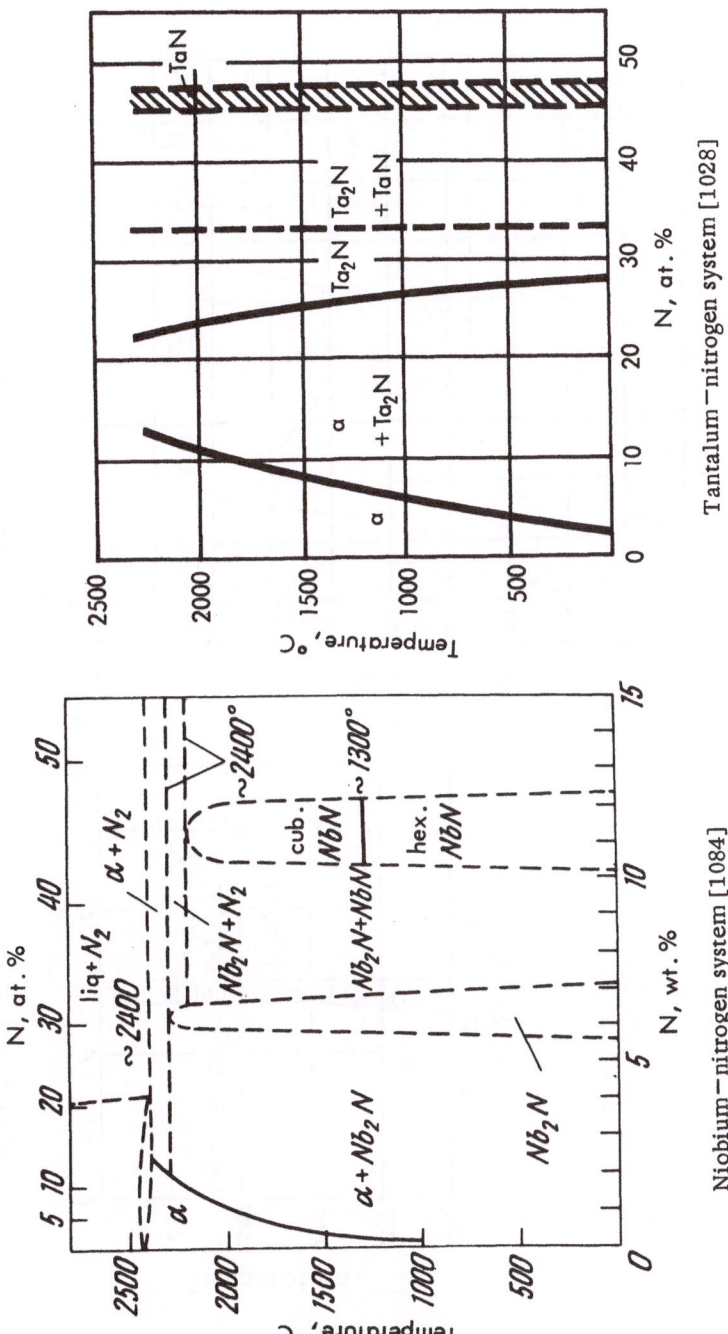

Tantalum—nitrogen system [1028]

Niobium—nitrogen system [1084]

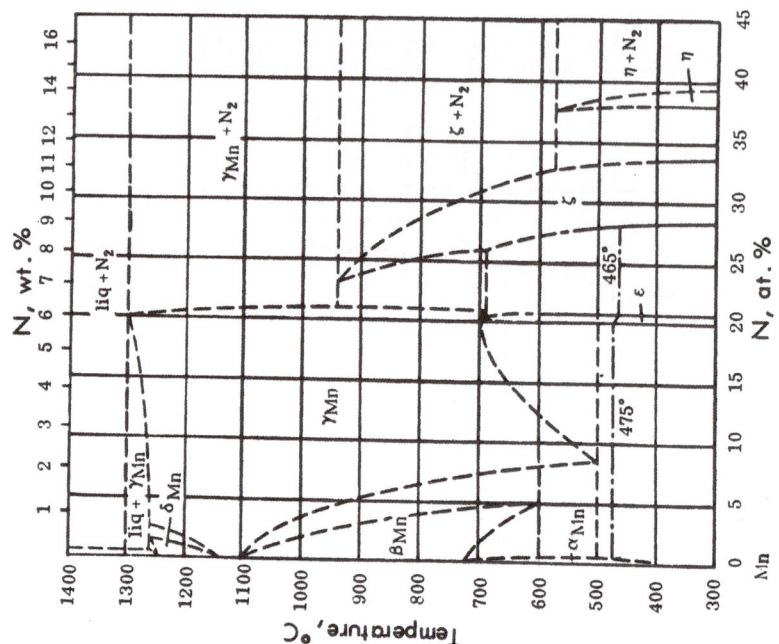

Manganese−nitrogen system [1333]; see [697]

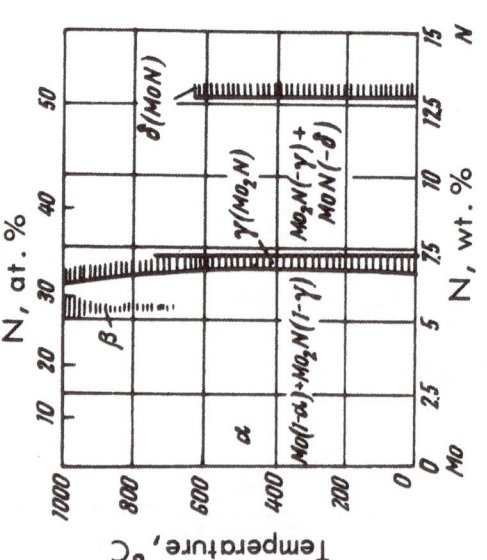

Molybdenum−nitrogen system [94]

370

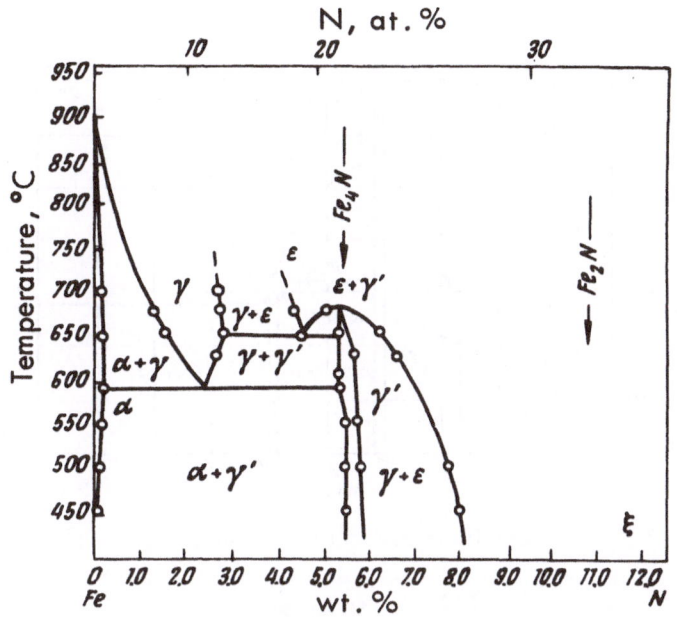

Iron—nitrogen system [697]

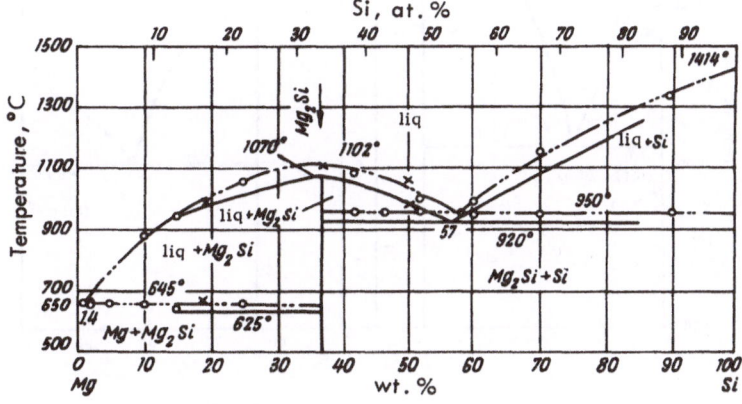

Magnesium—silicon system [566, 568]

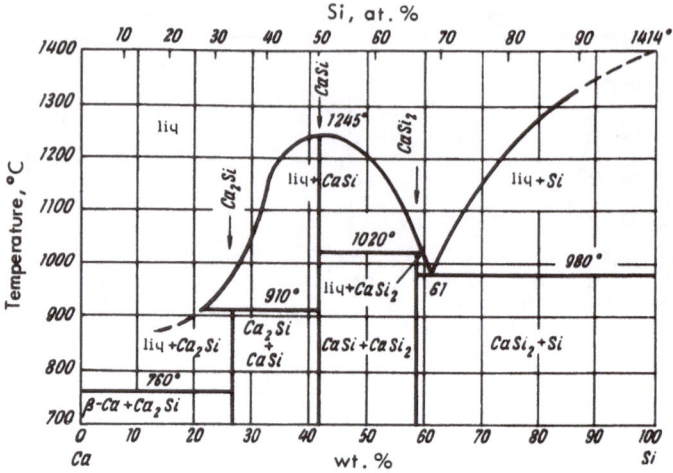

Calcium—silicon system [566]

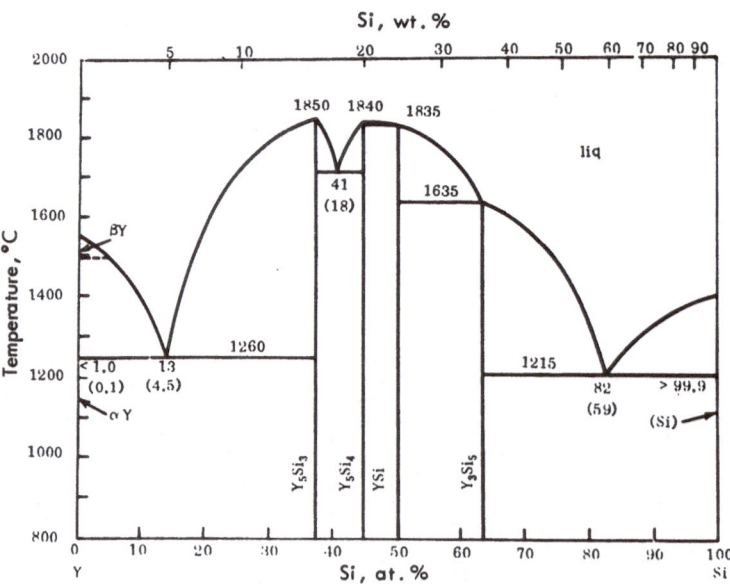

Yttrium—silicon system [1111]

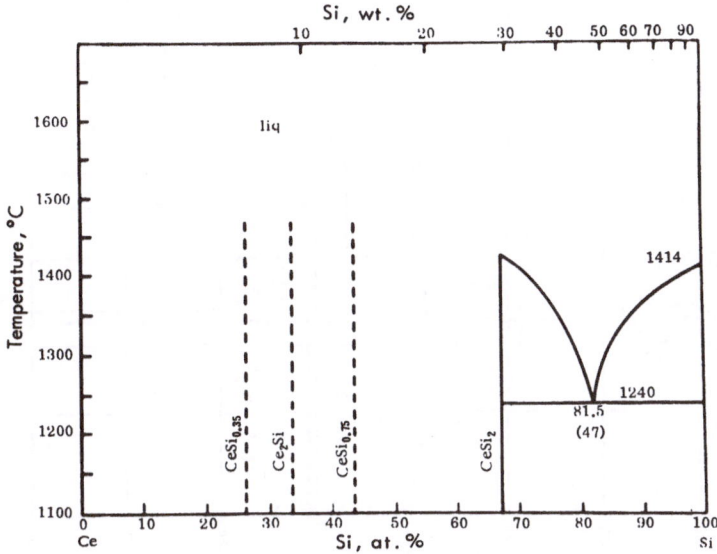

Cerium—silicon system [1111]

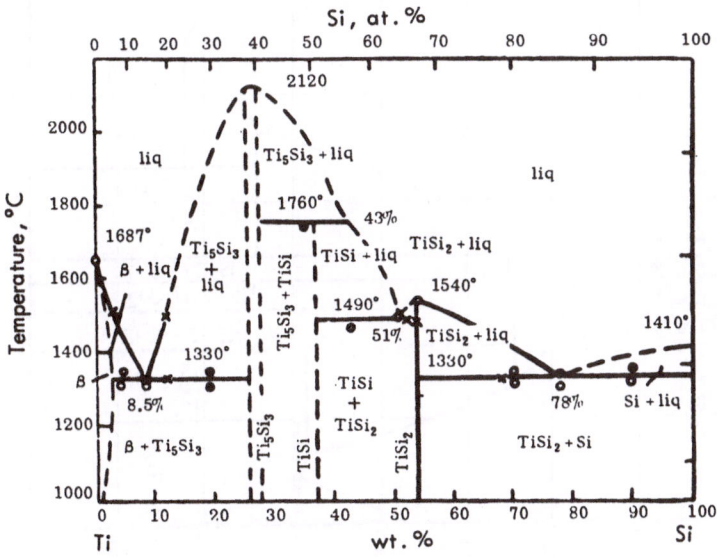

Titanium—silicon system [241]

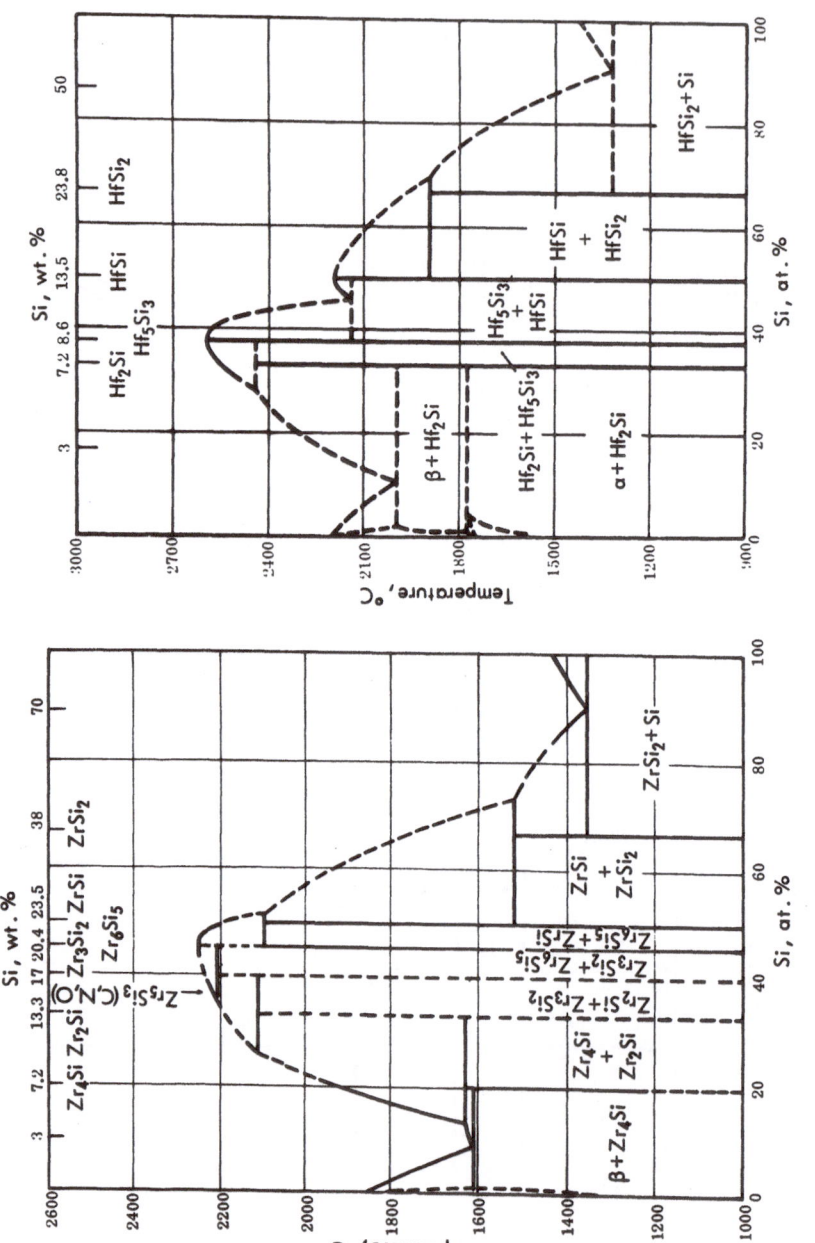

Hafnium−silicon system [1238]

Zirconium−silicon system [1238]; see [986]

374

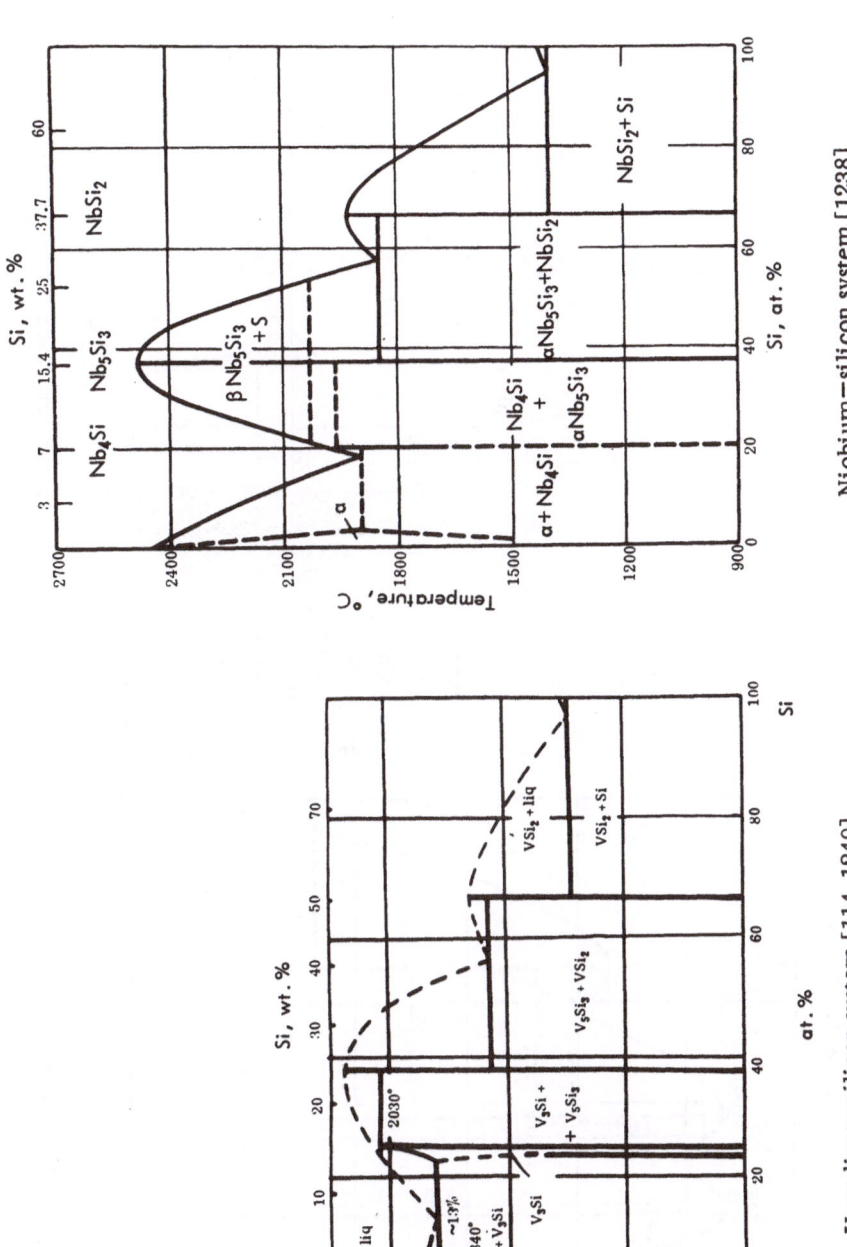

Niobium—silicon system [1238]

Vanadium—silicon system [114, 1249]

375

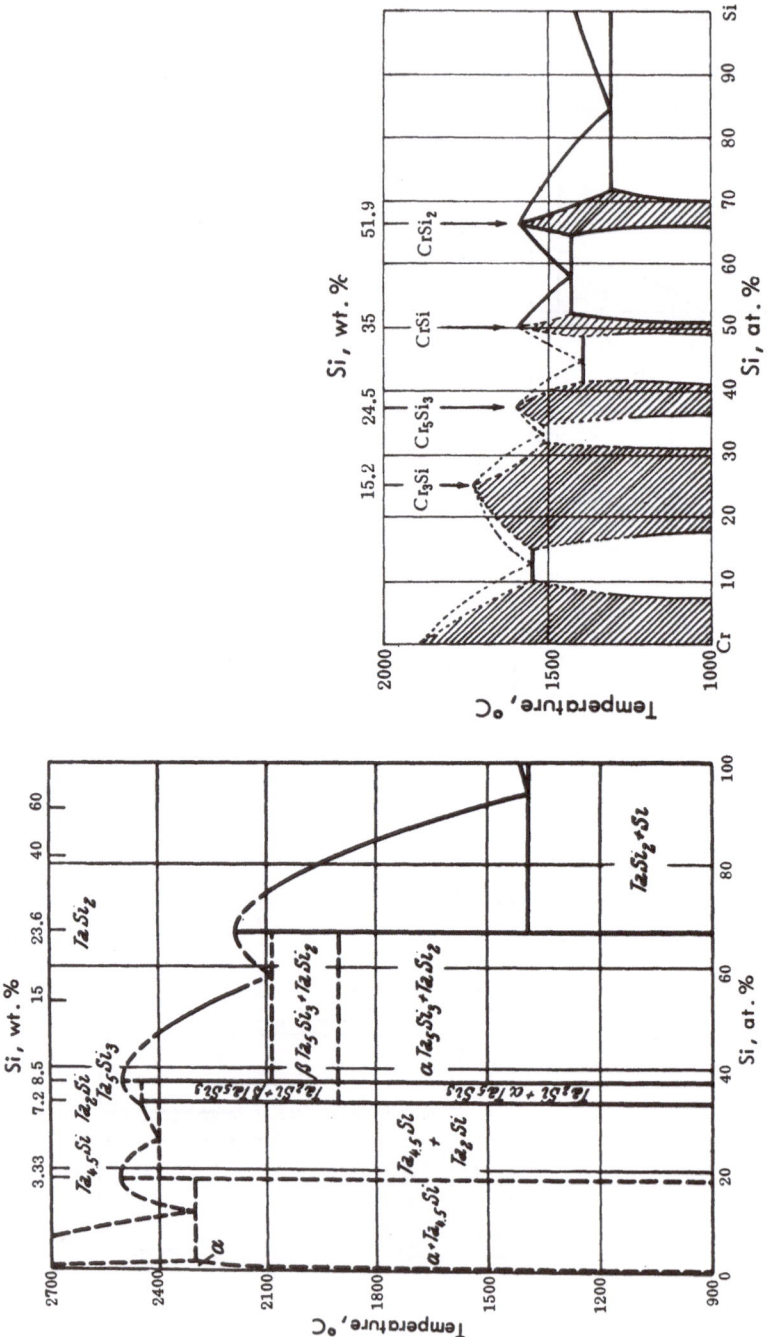

Chromium–silicon system [1238]; see [978, 1329]

Tantalum–silicon system [1238]; see [419]

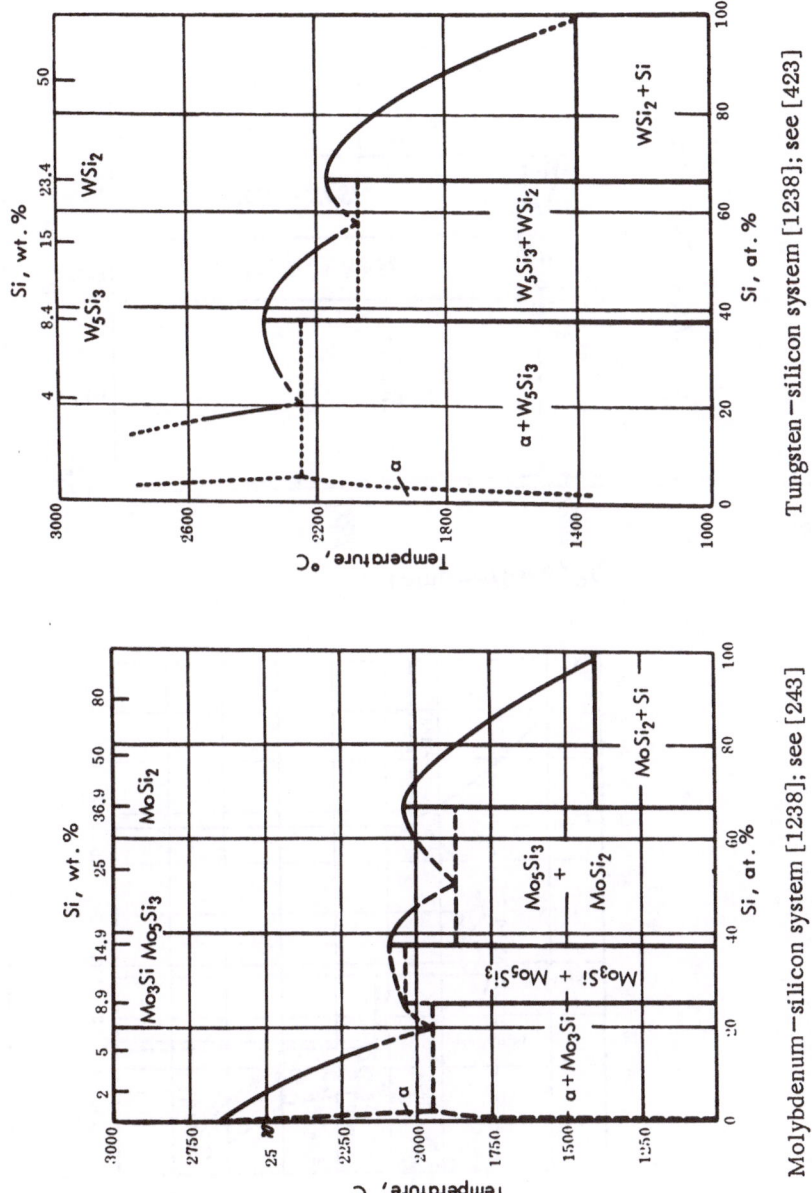

Tungsten–silicon system [1238]; see [423]

Molybdenum–silicon system [1238]; see [243]

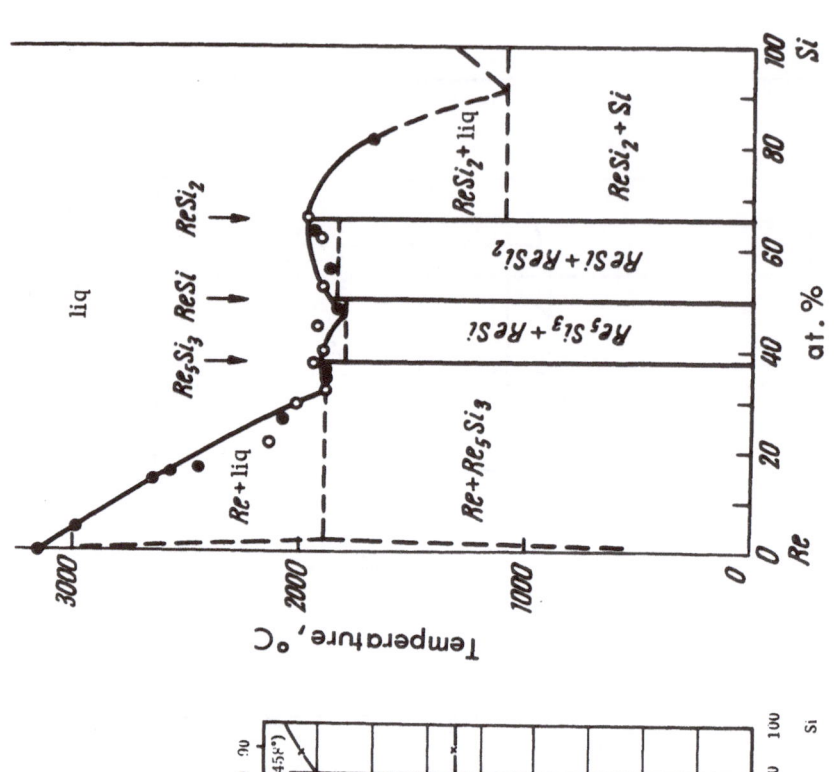

Rhenium–silicon system [127]

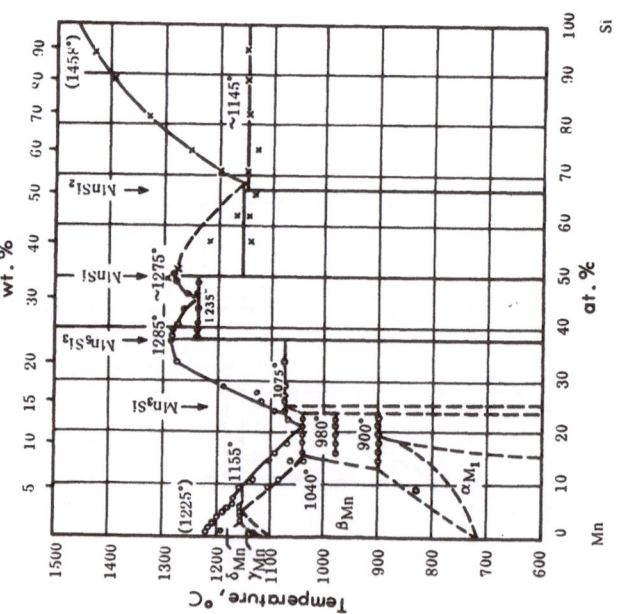

Manganese–silicon system [1333]

378

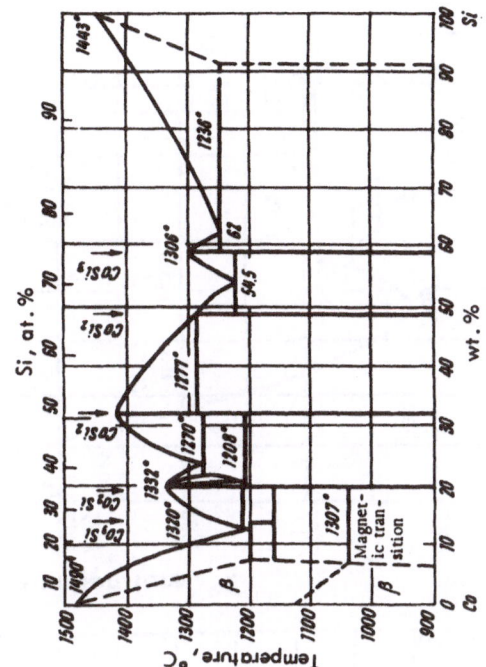

Cobalt−silicon system [887]

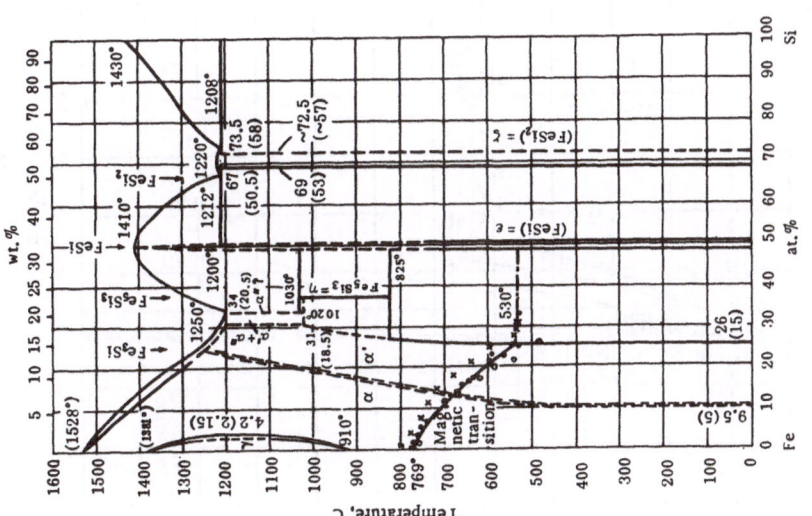

Iron−silicon system [1333]

379

Uranium—silicon system [117]

Nickel—silicon system [1333]

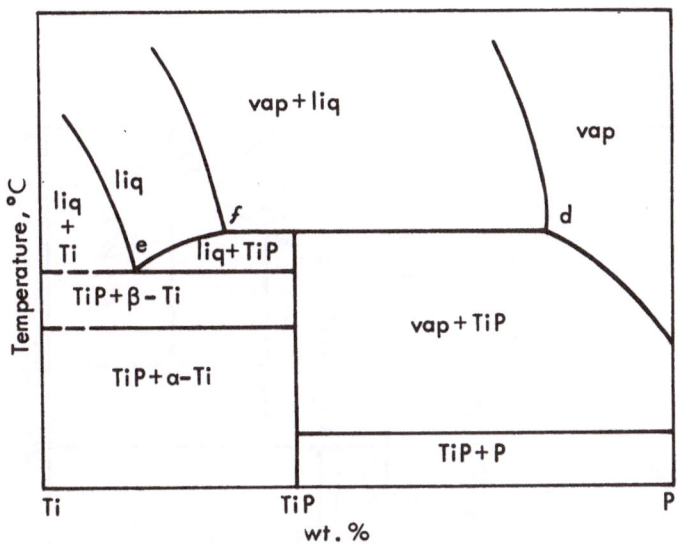

Titanium—phosphorus system (hypothetical diagram) [948]

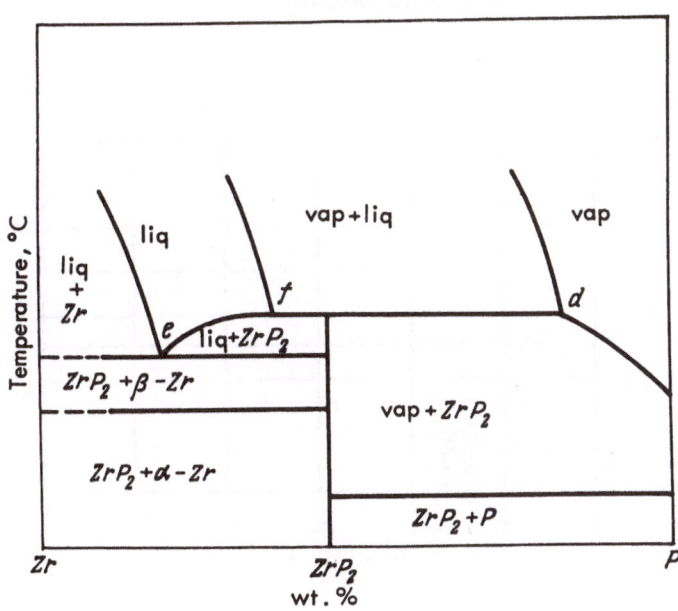

Zirconium—phosphorus system (hypothetical diagram)
[949]

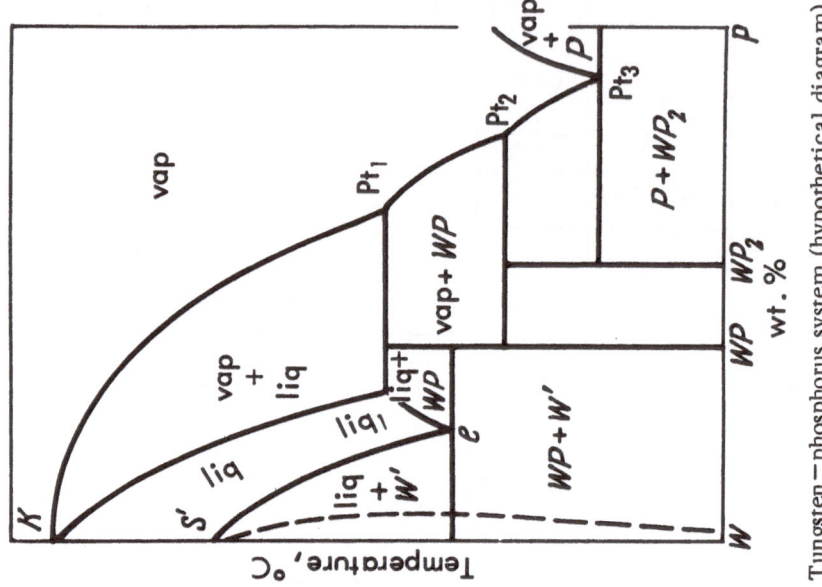

Tungsten−phosphorus system (hypothetical diagram)
[950]

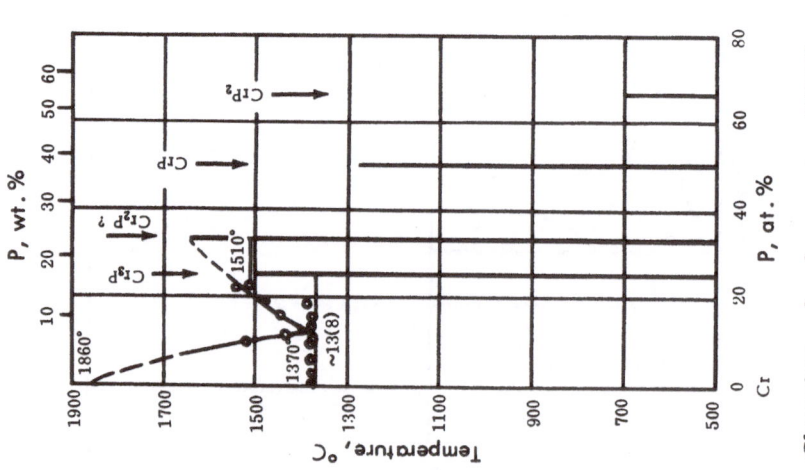

Chromium−phosphorus system [1333]

382

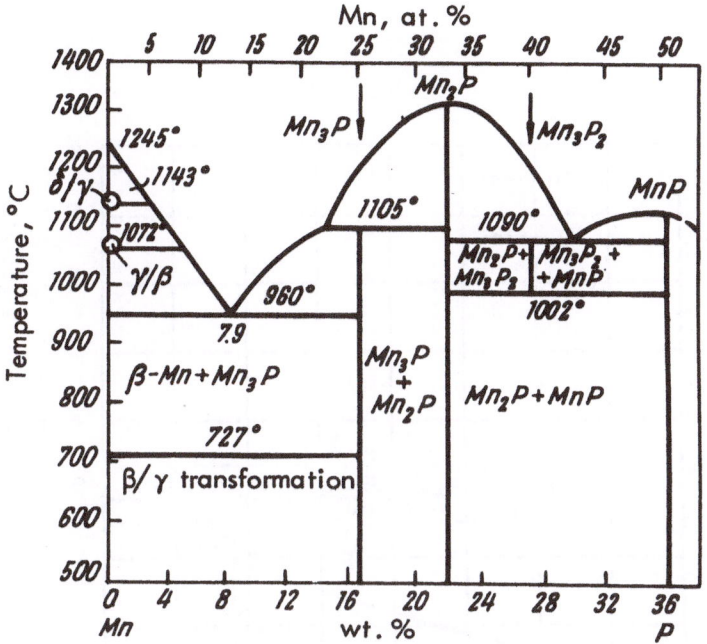

Manganese−phosphorus system [951]

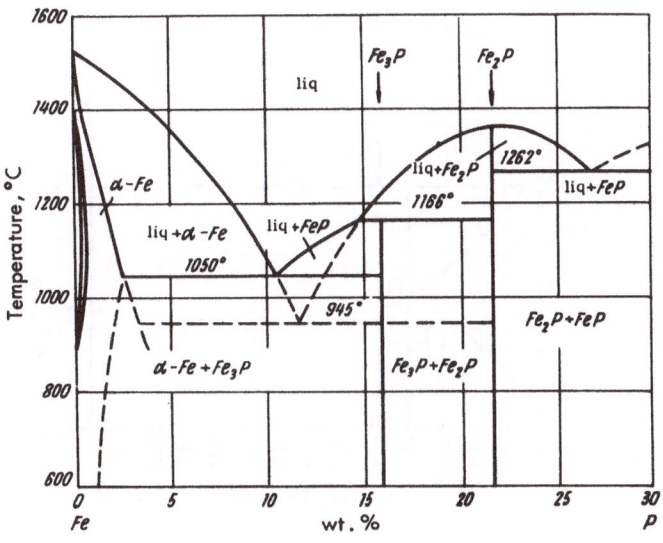

Iron−phosphorus system [952]

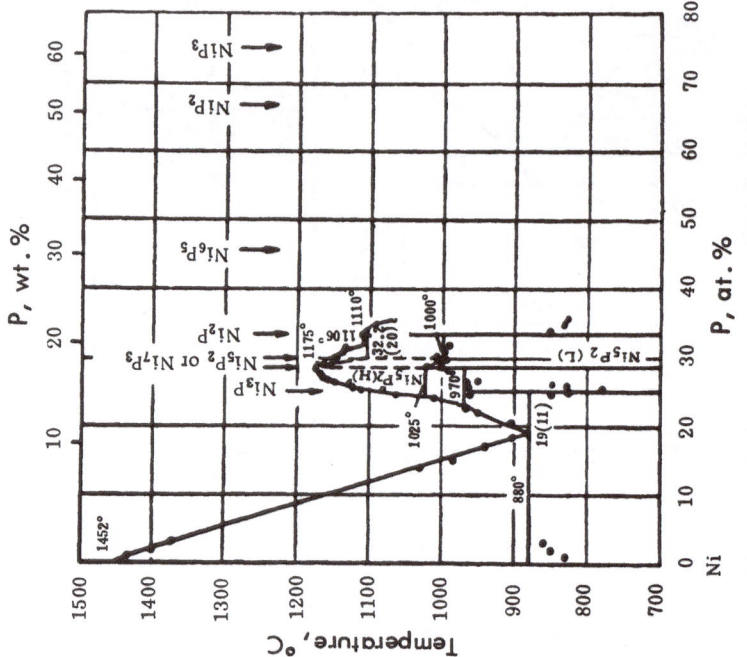

Nickel−phosphorus system [1333]; see [954]

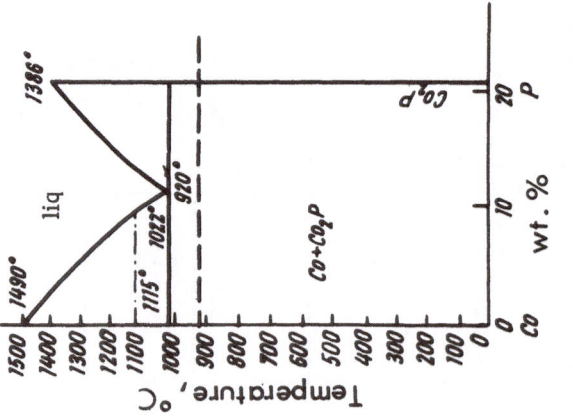

Cobalt−phosphorus system [953]

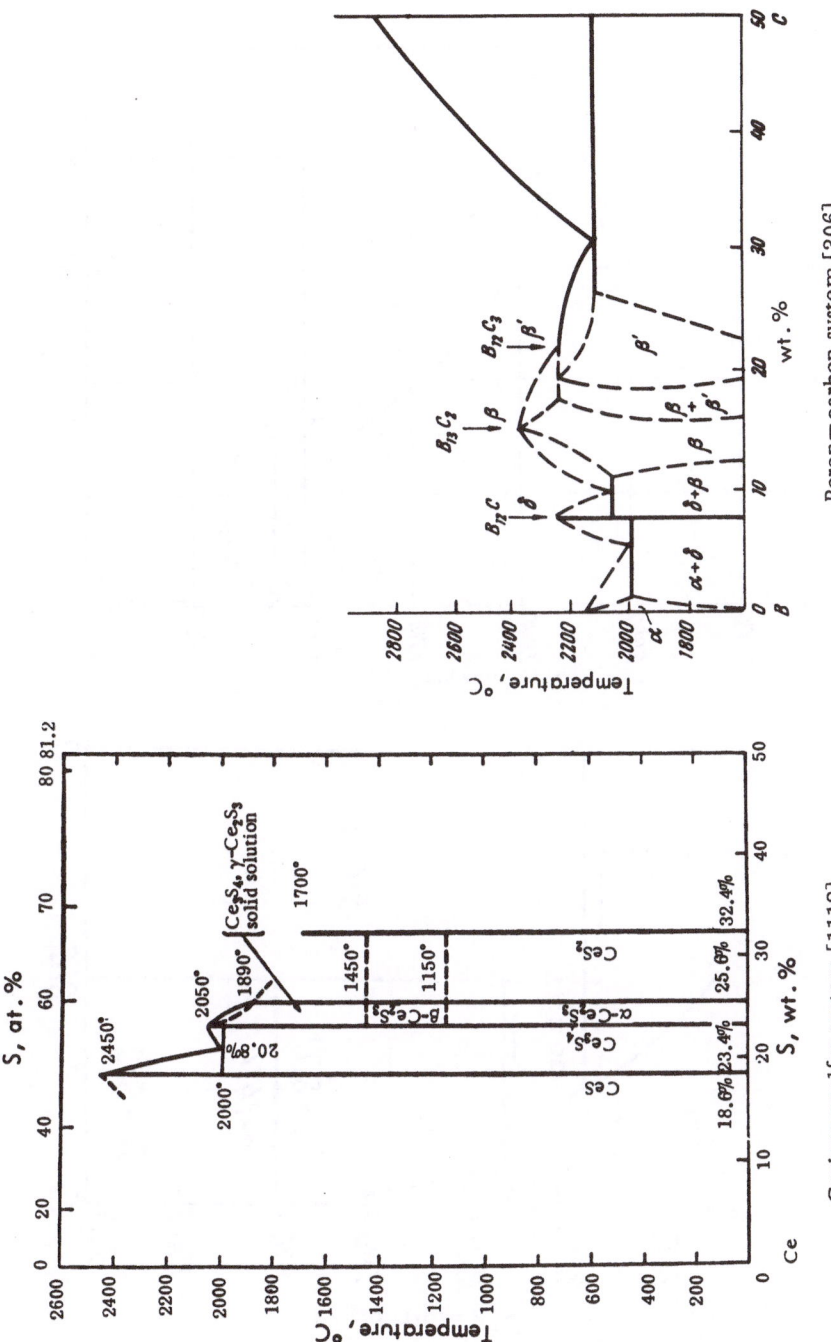

Cerium–sulfur system [1112]

Boron–carbon system [306]

385

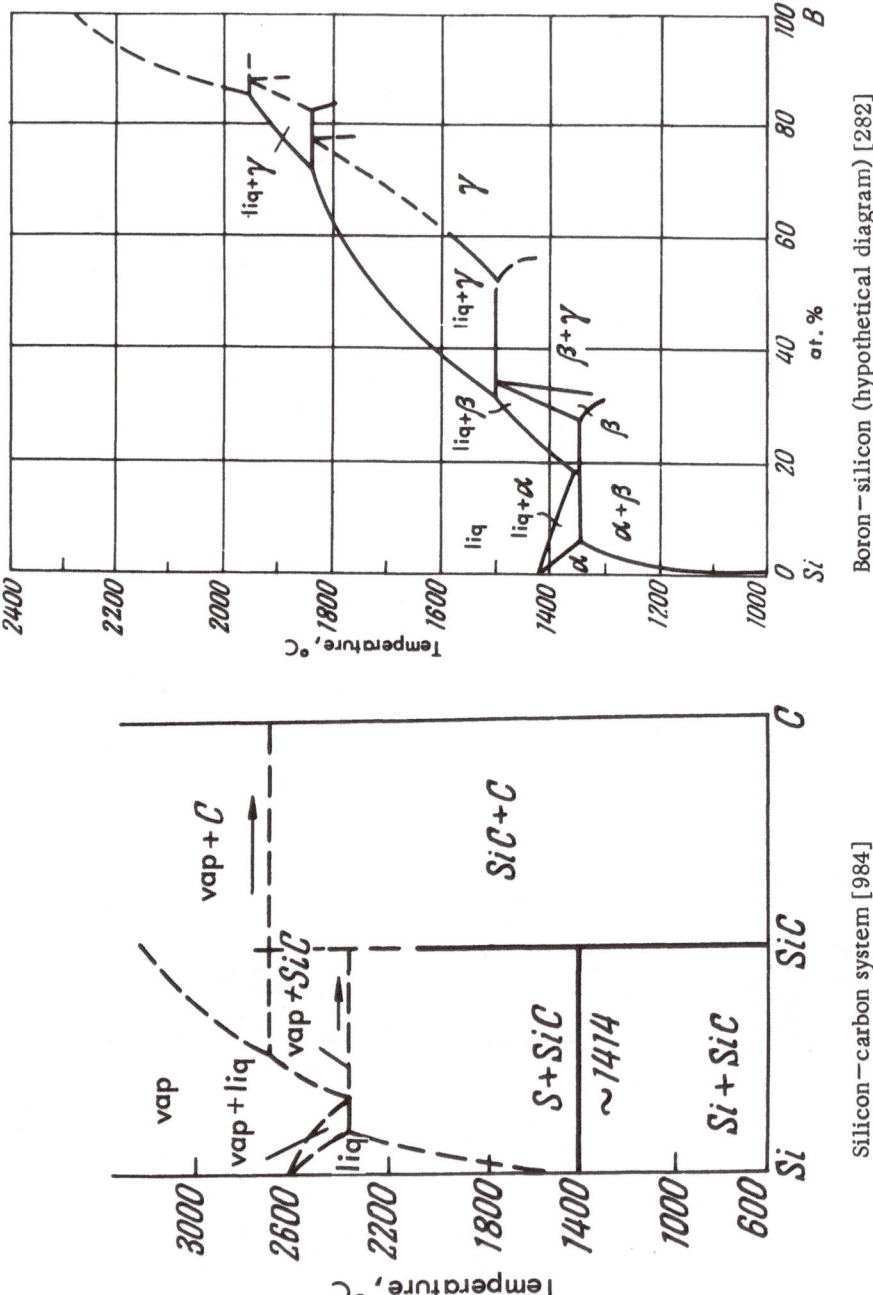

Boron—silicon (hypothetical diagram) [282]

Silicon—carbon system [984]

LITERATURE CITED

1. G. V. Samsonov, Ya. S. Umanskii. Hard Compounds of Refractory Metals, Metallurgizdat, Moscow, 1957.
2. R. Kieffer, P. Schwarzkopf. Hard Alloys [Russian translation], Metallurgizdat, Moscow, 1957.
3. G. V. Samsonov, A. E. Grodshtein. Zhur. Fiz. Khim. 30:379, 1956.
4. M. Stackelberg, P. Neuman. Z. physik. Chem. 19:314, 1952.
5. G. S. Zhdanov, N. N. Zhuravlev, A. E. Stepanova, M. M. Umanskii. Kristallografiya 2:289, 1957.
6. N. N. Zhuravlev, A. A. Stepanova. Kristallografiya 3:82, 1958.
7. G. A. Kudintseva, V. S. Neshpor, Yu. B. Paderno, G. V. Samsonov, B. M. Tsarev. High-Temperature Cermets, Izd. AN UkrSSR, Kiev, 1961.
8. I. Binder. Powder Met. Bull. 7:74, 1956.
9. B. Post, D. Moskowitz, F. Glaser. J. Chem. Soc. 78:1800, 1956.
10. G. A. Kudintseva, M. D. Polyakova, G. V. Samsonov, B. M. Tsarev. Fiz. Metall. i Metalloved. 6:272, 1958.
11. R. Kiessling. Acta Chem. Scand. 4:209, 1950.
12. G. Allard. Bull. soc. chim. France 50:79, 1932.
13. N. N. Tvorogov. Zhur. Neorg. Khim. 4:1961, 1959.
14. E. Felten, I. Binder, B. Post. J. Am. Chem. Soc. 80:3479, 1958.
15. P. Blum, F. Bertaut. Acta Cryst. 7:81, 1954.
16. V. S. Neshpor, G. V. Samsonov. Zhur. Fiz. Khim. 32:1328, 1958.
17. A. Zalkin, D. Templeton. J. Chem. Phys. 18:391, 1950.
18. G. V. Samsonov, N. N. Zhuravlev, Yu. B. Paderno, V. R. Melik-Adamyan. Kristallografiya 4:538, 1959.
19. V. P. Dzeganovskii, G. V. Samsonov, I. A. Samsonov, I. A. Semashko. Doklady Akad. Nauk 119:505, 1958.
20. A. A. Stepanova, N. N. Zhuravlev. Kristallografiya 3:94, 1958.
21. Yu. B. Paderno, T. I. Serebryakova, G. V. Samsonov. Kristallografiya 4:542, 1959.
22. Yu. B. Paderno, T. I. Serebryakova, G. V. Samsonov. Doklady Akad. Nauk SSSR 125:317, 1959.
23. V. S. Neshpor, G. V. Samsonov. Dopovidi Akad. Nauk UkrRSR, No. 5, p. 478, 1957.
24. Yu. B. Paderno, G. V. Samsonov. Zhur. Strukt. Khim. 2:213, 1961.
25. J. Lafferty. J. Appl. Phys. 22, 299, 1951.
26. H. Allard. Compt. rend. 189:108, 1929.
27. G. V. Samsonov, O. I. Zorina. Zhur. Neorg. Khim. 1:2260, 1956.
28. L. Brewer, D. Sawyer, D. Templeton, C. Dauben. J. Am. Ceram. Soc. 34:173, 1951.
29. P. Bertaut, P. Blume. Compt. rend. 229:666, 1949.
30. B. Post, F. Glaser. J. Chem. Phys. 20:1050, 1952.
31. P. Ehrlich. Z. anorg. Chem. 259:1, 1949.
32. F. Glaser, B. Post. J. Metals 5:1119, 1953.
33. J. Norton, H. Blumenthal, S. Sindeband. Powder Met. Bull. 4:157, 1949.
34. V. A. Épel'baum, M. A. Gurevich. Zhur. Fiz. Khim. 32:2275, 1958.

35. B. Post, F. Glaser. J. Metals 4:631, 1952.
36. F. Glaser, D. Moskowitz, B. Post. J. Metals, Sect. I, 5(1):1119, 1953.
37. H. Nowotny, A. Wittmann. Monatsh. Chem. 89:220, 1958.
38. H. Blumenthal. J. Am. Chem. Soc. 74:2942, 1952.
39. D. Moskowitz. J. Metals. Sect. II, 8:1325, 1956.
40. J. Norton, H. Blumenthal, S. Sindeband. J. Metals (Trans. AIME) 185:749, 1949.
41. L. Andersson, R. Kiessling. Acta Chem. Scand. 4:160, 1950.
42. R. Kiessling. Acta Chem. Scand. 3:603, 1949.
43. F. Bertaut, P. Blume. Compt. rend. 236:1055, 1953.
44. R. Kiessling. Acta Chem. Scand. 3:595, 1949.
45. S. Sindeband. J. Metals (Trans. AIME) 185:198, 1949.
46. V. A. Épel'baum, N. G. Sevost'yanov, M. A. Gurevich, B. F. Ormont, G. S. Zhdanov. Zhur. Neorg. Khim. 3:2545, 1958.
47. P. Schwarzkopf, F. Glaser. Z. Metallk. 44:353, 1953.
48. R. Kiessling. Acta Chem. Scand. 1:893, 1947.
49. R. Steinitz, J. Binder, D. Moskowitz. J. Metals 4:148, 1952.
50. F. Bertaut, P. Blume. Acta Cryst. 4:72, 1951.
51. F. Glaser. J. Metals 4:391, 1952.
52. B. Kiessling. Acta Chem. Scand. 4:168, 1950.
53. V. S. Neshpor, Yu. B. Paderno, G. V. Samsonov. Doklady Akad. Nauk SSSR 118:515, 1958.
54. S. Rundquist. Nature 181:259, 1958.
55. S. Rundquist. Acta Chem. Scand. 12:658, 1958.
56. P. T. Kolomytsev. Doklady Akad. Nauk 124:1247, 1959.
57. B. F. Ormont. Structures of Inorganic Substances, GITTL, 1950.
58. G. B. Bokii. Introduction to Crystallography, Izd. MGU, Moscow, 1954.
59. T. Bjürstrom. Arkiv Kemi, Mineral. Geol. 11A:1, 1933.
60. P. Blume. J. phys. rad. 13:430, 1952.
61. E. Ryshkewitsh. Treatise of Refractory Oxides, Academic Press, N.Y., 1960.
62. G. V. Samsonov, P. S. Kislyi, A. D. Panasyuk. Izmeritelnaya Tekh. No. 10, p. 32, 1961.
63. G. V. Samsonov, N. S. Rozinova. Izvest. Sektora Fiz.-Khim. Anal. Akad. Nauk SSSR 27:243, 1953.
64. P. Cotter, I. Kohn. J. Am. Ceram. Soc. 37:415, 1954.
65. N. Schönberg. Acta Chem. Scand. 8:624, 1954.
66. G. Brauer, H. Renner, H. Wernet. Z. anorg. Chem. 277:249, 1954.
67. V. I. Smirnova, B. F. Ormont. Doklady Akad. Nauk SSSR 100:127, 1955.
68. A. Westgren. Nature 132:480, 1933.
69. A. Westgren. Jernkontorets Ann. 119:231, 1935.
70. K. Hellborn, A. Westgren. Svensk. Kem. Tidskr. 7:141, 1933.
71. K. Kuo, G. Hägg. Nature 170:245, 1952.
72. N. F. Lashko. Doklady Akad. Nauk SSSR 81:605, 1951.
73. E. Ohman. Jernkontorets Ann. 128:13, 1944.
74. M. Isole. Sci. Repts. Research Insts. Tohoku Univ. Ser. A, 3:468, 1951.
75. G. Hägg. Z. Krist. Vol. 89, 1934.
76. H. Goldschmidt. J. Iron Steel Inst. 160:345, 1948.
77. G. Bredig, E. Bergkampf. Z. physik. Chem., Badenstein Anniversary Volumes, 1931, p. 177.
78. B. Jacobson, A. Westgren. Z. physik. Chem. B20:362, 1933.
79. W. Mayer. Metallwirtschaft 17:413, 1938.
80. H. Wilhelm, P. Chiotti. Trans. Am. Soc. Met. 42:1295, 1950.
81. E. Hunt, R. Rundle. J. Am. Chem. Soc. 73:4777, 1951.
82. R. Rundle, N. Baenziger, A. Wilson, R. McDonald. J. Am. Chem. Soc. 70:99, 1948.

83. W. Mallett, A. Gerds, D. Vaughan. J. Electrochem. Soc. 92:505, 1952.
84. U. Esch, A. Schneider. Z. anorg. Chem. 257:254, 1948.
85. W. Zachariasen. Acta Cryst. 2:388, 1949.
86. W. Zachariasen. Acta Cryst. 5:17, 1956.
87. A. Polty, H. Margolin, J. Nilsen. Trans. Am. Soc. Met. 46:312, 1954.
88. P. Duwez, F. Odell. J. Electrochem. Soc. 97:299, 1950.
89. K. Becker, F. Ebert. Z. Physik 31:368, 1925.
90. H. Hahn. Z. anorg. Chem. 258:58, 1949.
91. N. Schönberg. Acta Chem. Scand. 8:202, 1954.
92. N. Schönberg. Acta Chem. Scand. 8:199, 1954.
93. S. Eriksson. Jernkontorets Ann. 118:530, 1934.
94. G. Hägg. Z. physik. Chem. B7:339, 1930.
95. N. Schönberg. Acta Chem. Scand. 8:204, 1954.
96. G. Hägg. Z. physik. Chem. B6:346, 1929.
97. P. Chiotti. J. Am. Ceram. Soc. 35:123, 1952.
98. P. Eckerlin, E. Wölfel. Z. anorg. Chem. 280:3215, 1955.
99. G. Brauer, H. Hägg. Z. anorg. Chem. 267:198, 1952.
100. F. Bertaut, P. Blume. Acta Cryst. 3:319, 1950.
101. W. Zachariasen. Acta Cryst. 2:94, 1949.
102. E. Jackobsen, B. Treemen, A. Tarp, A. A. Searcy. J. Am. Chem. Soc. 78:4850, 1956.
103. B. Stone, D. Hill. Phys. Rev. Letters 4:282, 1960.
104. A. Jandelli, R. Ferro. Ann. Chim. (Rome) 42:598, 1952.
105. O. Runnals, R. Boucher. Acta Cryst. 8:592, 1955.
106. P. Pietrokowsky, P. Duwez. J. Metals 3:1777, 1951.
107. N. V. Ageev, V. P. Samsonov. Doklady Akad. Nauk SSSR 112:681, 1957.
108. H. Schachner, H. Nowotny, H. Kudielka. Monatsh. Chem. 85:1140, 1954.
109. H. Seyferth. Z. Krist. 67, 295, 1928.
110. S. Naray-Szabo. Z. Krist. A97:223, 1937.
111. B. Post, F. Glaser, D. Moskowitz. J. Chem. Phys. 22:1264, 1954.
112. P. Cotter, J. Kohn, B. Potter. J. Am. Ceram. Soc. 39:11, 1956.
113. H. Wallbaum. Z. Metallk. 31:363, 1939.
114. R. Kieffer, F. Benešovsky, H. Schmid. Z. Metallk. 47:247, 1956.
115. E. Parthé, H. Nowotny, H. Schmid. Monatsh. Chem. 86:413, 1955.
116. H. Wallbaum. Z. Metallk. 33:378, 1941.
117. G. V. Samsonov. Silicides and Their Application in Technology, Izd. AN UkrSSR, Kiev, 1959.
118. E. Parthé, B. Lux, H. Nowotny. Monatsh. Chem. 86:857, 1955.
119. C. Dauben, D. Templeton, C. Myers. J. Phys. Chem. 60:443, 1956.
120. B. Boren. Arkiv Kemi, Mineral. Geol. 11A:2, 1933.
121. D. Templeton, C. Dauben. Acta Cryst. 3:261, 1950.
122. G. V. Samsonov. Technical Digest 1:18, 1962.
123. N. V. Martynenko, L. N. Éfimenko, D. N. Solonikin. Fiz. Metal. i Metalloved. 8:8, 1959.
124. B. Aronsson. Acta Chem. Scand. 9:1107, 1955.
125. R. Vogel, H. Bedarff. Arch. Eisenhuttenw. 7:423, 1933-34.
126. R. McNees, A. Searcy. J. Am. Chem. Soc. 77:5290, 1955.
127. A. Knapton. Hochschmelzende Metalle, 3, Plansee-Seminar, De Re Metallica, Wien, 1959, p. 412.
128. J. Buddery. Thesis, Imperial College, London, 1931.
129. N. Schönberg. Acta Chem. Scand. 8:1460, 1954.
130. N. Schönberg. Acta Chem. Scand. 8:226, 1954.
131. B. Aronsson. Acta Chem. Scand. 2:549, 1948.
132. H. Nowotny, E. Hehglein. Monatsh. Chem. 79:385, 1948.
133. O. Arstad, H. Nowotny. Z. phys. Chem. B38:356, 1957.

134. I. Berak. T. Neumann. Z. Metallk. 41:19, 1950.
135. G. Hägg. Z. Krist. 68:470, 1928.
136. S. Rundquist, F. Jellineck. Acta Chem. Scand. 13:425, 1959.
137. S. Hendricks, P. Kosting. Z. Krist. 74:522, 1930.
138. S. Rundquist, F. Jellineck. Acta Chem. Scand. 13:551, 1959.
139. B. Aronsson. Acta Chem. Scand. 9:137, 1955.
140. A. F. Tereshchenko. Investigation of the High-Temperature Characteristics of the Strength and Plasticity of Heat-Resistant Materials on Heating of Specimens by the Electrical Resistance Method, author's abstract of dissertation, Kiev, 1961.
141. P. V. Gel'd, V. D. Lyubimov. Izvest. Akad. Nauk SSSR, Otdel. Tekh. Nauk, Metallurgiya i Topliva, No. 6, p. 120, 1961.
142. E. Strotzer, W. Biltz, K. Meisel. Z. anorg. Chem. 238:69, 1938.
143. T. Strotzer, W. Biltz, K. Meisel. Z. anorg. Chem. 239:216, 1938.
144. A. Gorum. Acta Cryst. 10:143, 1957.
145. I. Flahaut, M. Guittard. Compt. rend. 242:1318, 1956.
146. I. Flahaut, M. Guittard. Compt. rend. 243:1210, 1956.
147. I. Flahaut, M. Guittard. Compt. rend. 241:1775, 1955.
148. A. Jandelli. Gazz. Chim. Ital. 85:881, 1955.
149. M. Picon, N. Patrie. Compt. rend. 242:1521, 1956.
150. N. P. Zvereva. Doklady Akad. Nauk SSSR 113:333, 1957.
151. W. Zachariasen. Acta Cryst. 2:57, 1949.
152. W. Zachariasen. Acta Cryst. 2:60, 1949.
153. M. Picon, M. Patrie. Compt. rend. 242:516, 1956.
154. H. Eick. J. Am. Chem. Soc. 80:43, 1958.
155. W. Zachariasen. Acta Cryst. 2:291, 1949.
156. E. Eastman, J. Brewer, L. Bromley, P. Gilles, N. Longfreen. J. Am. Chem. Soc. 72:2248, 1950.
157. I. Flahaut, M. Guittard. Compt. rend. 243:1419, 1956.
158. E. Eastman, L. Brewer, L. Bromley, P. Gilles, N. Longfreen. J. Am. Chem. Soc. 73:3896, 1951.
159. M. Picon, I. Flahaut. Compt. rend. 243:2074, 1956.
160. I. Flahaut, M. Guittard, J. Loriers, M. Patrie. Compt. rend. 245:2191, 1957.
161. L. Domange, I. Flahaut, M. Guittard, J. Loriers. Compt. rend. 247:1614, 1958.
162. W. Zachariasen. Acta Cryst. 2:189, 1949.
163. W. Zachariasen, A. Plettinger. Report ANL, 4565, 1950.
164. M. Picon, I. Flahaut. Compt. rend. 240:784, 1955.
165. M. Picon, I. Flahaut. Compt. rend. 237:1160, 1955.
166. M. Picon, I. Flahaut. Compt. rend. 240:2150, 1955.
167. E. Parthé. Acta Cryst. 12:559, 1959.
168. G. S. Zhdanov, G. A. Meerson, N. N. Zhuravlev, G. V. Samsonov. Zhur. Fiz. Khim. 28:1076, 1954.
169. R. Paese. Acta Cryst. 5:356, 1952.
170. R. Wentorff. J. Chem. 26:956, 1957.
171. A. Neuhaus, H. Meyer. Angew. Chem. 69:556, 1957.
172. S. Ruddleston, P. Popper. Acta Cryst. 11:465, 1958.
173. G. V. Samsonov, V. P. Latysheva. Doklady Akad. Nauk SSSR 105:499, 1955.
174. R. Adamsky. Acta Cryst. 11:744, 1958.
175. H. Nowotny, F. Pigger, R. Kieffer, F. Benesovsky. Monatsh. Chem. 89:611, 1958.
176. Lien Ching-K'uei. Heats of Formation and Heat Capacities in the Chromium–Silicon System, author's abstract of dissertation, IMET AN SSSR, Moscow, 1959.
177. L. Brewer, et al. Chemistry and Metallurgy of Miscellaneous Materials, Thermodynamics and Physical Properties of Miscellaneous Materials, edited by L. L. Quill, McGraw-Hill Book Co., N.Y., 1950.

178. O. Whittemore. J. Can. Ceram. Soc. 28:43, 1959.
179. T. Ya. Kosolapova. Production and Properties of Chromium Carbides, author's abstract of dissertation, MITKhT, Moscow, Kiev, 1961.
180. L. Brewer, H. Haraldsen. J. Eletrochem. Soc. 102:399, 1955.
181. G. V. Samsonov. Doklady Akad. Nauk SSSR 83:689, 1953.
182. V. A. Épel'baum, M. P. Starostina. In coll. Boron, Transactions of a Conference on the Chemistry of Boron and Its Compounds, GKhI, Moscow, 1958, p. 102.
183. G. V. Samsonov. Zhur. Fiz. Khim. 30:2057, 1956.
184. G. V. Samsonov. Zhur. Priklad. Khim. 28:1018, 1955.
185. O. Kubaschewski, E. Evans. Thermochemistry in Metallurgy [Russian translation], IL, Moscow, 1954.
186. G. Humphrey. J. Am. Chem. Soc. 73:2261, 1951.
187. F. Richardson. J. Iron Steel Inst. 175:33, 1953.
188. H. Kelley, P. Boerickl, G. Moore, E. Huffmann, W. Bengert. Bull. U.S. Bur. Mines, No. 622, 1944.
189. A. Mah, R. Boyle. J. Am. Chem. Soc. 77:6512, 1955.
190. V. I. Smirnova, B. F. Ormont. Doklady Akad. Nauk SSSR 100:127, 1955.
191. V.I. Smirnova, B. F. Ormont. Doklady Akad. Nauk SSSR 96:1017, 1954.
192. A. A. Baikov. Sbornik Trudov Akad. Nauk SSSR 11:70, 1948.
193. W. Kroll. Z. anorg. Chem. 234:42, 1937.
194. A. Mah, N. Gellert. J. Am. Chem. Soc. 78:3261, 1956.
195. O. Kubaschewski, M. Vills. Z. Elektrochem. 53:32, 1949.
196. L. Wöhler, F. Schuff. Z. anorg. Chem. 209:33, 1932.
197. D. Robins, J. Jenkins. Acta Met. 3:598, 1955.
198. Yu. M. Golutvin. Zhur. Fiz. Khim. 30:2251, 1956.
199. A. Searcy. J. Am. Ceram. Soc. 40:431, 1957.
200. L. Brewer, O. Krikorian. J. Electrochem. Soc. 103:38, 1956.
201. S. A. Shchukarev, M. P. Morozova, Li Miao-hsiu. Zhur. Obshchei Khim. 29:3142, 1959.
202. S. A. Shchukarev, G. Grossman, M. P. Morozova. Zhur. Obshchei Khim. 29:655, 1959.
203. S. A. Shchukarev, M. P. Morozova, Li Miao-hsiu. Zhur. Obschei Khim. 29:2465, 1959.
204. B. F. Ormont. Zhur. Neorg. Khim. 4:2176, 1959.
205. F. Ephraim. Anorganic Chemistry, Nordeman Publishing Co., N.Y., 1939.
206. L. Ya. Markovskii, D. L. Orshanskii, V. P. Pryanishnikov. Chemical Electrothermics, Goskhimizdat, Moscow, 1952.
207. D. Smith, A. Dworkin, D. Sasmor, E. Van Artsdalen. J. Chem. Phys. 72:2654, 1955.
208. A. Dworkin, D. Sasmor, E. Van Artsdalen. J. Chem. Phys. 22:837, 1954.
209. K. S. Evstrop'ev, K. A. Toropov. Chemistry of Silicon and Physical Chemistry of Silicides, 1950, p. 223.
210. W. De Sorbo. J. Am. Chem. Soc. 75:1825, 1953.
211. E. King, A. Christensen. J. Phys. Chem. 62:499, 1958.
212. K. Kelley. J. Am. Chem. Soc. 63:1137, 1941.
213. Yu. M. Golutvin. Zhur. Fiz. Khim. 33:1798, 1959.
214. G. V. Samsonov. Ukrain. Khim. Zhur. 23:287, 1957.
215. G. A. Kudintseva, B. M. Tsarev. Radiotekh. i Elektron. 3:428, 1958.
216. T. I. Serebryakova, Yu. B. Paderno, G. V. Samsonov. Optika i Spektroskopiya 8:410, 1960.
217. L. Andrieux. Ann. chim. 12:422, 1929.
218. L. Andrieux, A. Barbetti. Compt. rend. 194:1573, 1932.
219. Voprosy Radiolokatsionnoi Tekhniki 6:81, 1951.
220. G. V. Samsonov, Yu. B. Paderno, V. S. Fomenko. Problems of Powder Metallurgy and the Strength of Materials, Izd. AN UkrSSR, Kiev, No. 8, 1960, p. 66.

221. A. Accary, P. Blum. Nuclear Power 5:122, 1960.
222. H. Nowotny, F. Benešovsky, R. Kieffer. Z. Metallk. 50:258, 1959.
223. H. Nowotny, F. Benešovsky, R. Kieffer. Z. Metallk. 50:417, 1959.
224. R. Kieffer, F. Benešovsky. Powder Metallurgy, No. 1/2, 145, 1958.
225. V. S. Neshpor, P. S. Kislyi. Ogneupory 23:231, 1959.
226. P. Gilles, B. Polock. J. Metals 5:1537, 1953.
227. P. T. Kolomytsev, N. V. Moskaleva, S. A. Strekopytov. Collection of Scientific Papers on Hard Alloys, Izd. VVIA im. N. E. Zhukovskogo, No. 734, 1958, p. 40.
228. H. Giebelhausen. Z. anorg. Chem. 91:251, 1915.
229. R. Kieffer, W. Kölbl. Powder Met. Bull. 4:4, 1949.
230. Welding Engr. No. 4, 1958.
231. B. A. Generozov. Zavodskaya Laboratoriya 13:314, 1947.
232. E. Friederich, L. Sittig. Z. anorg. Chem. 144:169, 1925.
233. C. Agte, K. Moers. Z. anorg. Chem. 198:233, 1931.
234. F. Ellinger. Trans. Am. Soc. Metals 31:89, 1943.
235. D. Bloom, N. Grant. J. Metals 188:41, 1950.
236. C. Agte, H. Alterthum. Z. tech. Physik II(148):216, 1930.
237. W. Sykes. Trans. Am. Soc. Steel Treating 18:968, 1930.
238. W. Sidgwick. Chemical Elements and Their Compounds, Oxford, 1950.
239. T. Vasilos, W. Kingery. J. Am. Ceram. Soc. 37, 1954.
240. V. Paranjpe, M. Cohen, M. Bever, C. Floe. J. Metals 188:261, 1950.
241. M. Hansen, H. Kessler, D. McPherson. Am. Soc. Metals, Preprint, No. 4, 1951.
242. A. Knapton. Nature 175:730, 1955.
243. R. Kieffer, E. Cerwenka. Z. Metallk. 43:101, 1952.
244. N. D. Morgulis, Yu. P. Korchevoi. Atomnaya Energ. 10:49, 1961.
245. G. V. Samsonov, L. L. Vereikina. Phosphides, Izd. AN UkrSSR, Kiev, 1961.
246. M. Picon, J. Gonge. Compt. rend. 193:585, 1931.
247. W. Klemm, K. Meisel, H. Vogel. Z. anorg. Chem. 190:123, 1930.
248. E. Strotzer, M. Zumbusch. Z. anorg. Chem. 247:415, 1941.
249. G. V. Samsonov, N. M. Popova. Zhur. Obshchei Khim. 27:3, 1957.
250. E. Eastman, L. Brewer, L. Gromley, P. Gilles, N. Longfreen. J. Am. Chem. Soc. 72:4019, 1950.
251. K. Siborg, D. Katz. Actinides [Russian translation], IL, Moscow, 1955.
252. M. S. Vendrikh. In coll. Brief Communications on Scientific Research Work, 1952-1957, M. I. Kalinin Nonferrous Metals Institute, Metallurgizdat, Moscow, 1959, p. 165.
253. A. N. Krestovnikov, M. S. Vendrikh. Pure Metals and Semiconductors, Metallurgizdat, Moscow, 1959, p. 165.
254. B. Naylor. J. Am. Chem. Soc. 68:370, 1946.
255. E. King. J. Am. Chem. Soc. 71:316, 1949.
256. Gmelins Handbuch der anorg. Chemie, System No. 14, Silizium, Berlin, 1954.
257. Bor, Gmelins Handbuch der anorg. Chemie, Ergänzungsband, Verlag Chemie, Weinheim, 10, 1954.
258. A. A. Stepanova, M. M. Umanskii. In coll. Boron, Transactions of a Conference on the Chemistry of Boron and Its Compounds, GKhI, Moscow, 1958, p. 102.
259. G. Beckmann, R. Kiessling. Nature 178:1341, 1956.
260. V. S. Neshpor, G. V. Samsonov. Fiz. Metal. i Metalloved. 4:181, 1957.
261. A. M. Belikov, Ya. S. Umanskii. Kristallografiya 4:684, 1959.
262. A. M. Belikov, Ya. S. Umanskii. Nauch. Doklady Vysshei Shkoly 1:192, 1958; see also author's abstract of dissertation of Belikov's X-Ray Determination of the Constants of the Quasi-Electric Force of Thermal Vibrations and the Coefficients of Thermal Expansion of Refractory Metal Phases, Moscow, 1958.

263. I. Gangler. J. Am. Ceram. Soc. 33:367, 1950.
264. K. Becker. Refractory Compounds and Their Use in Technology [Russian translation], ONTI, 1934.
265. P. Alexander. Metals & Alloys 9:179, 1938.
266. G. V. Samsonov, E. V. Petrash. Metalloved. i Obrabotka Metal. 1:19, 1955.
267. Materials & Methods 43:131, 1956; Metallurgia, No. 318, 175, 1956.
268. G. V. Samsonov, G. V. Lashkarev. Doklady Akad. Nauk UkrSSR, No. 9, p. 36, 1961.
269. R. Ridgway. Trans. Am. Electrochem. Soc. 56:117, 1934.
270. J. Collins, R. Gerby. J. Metals 7:612, 1955.
271. N. P. Bogoroditskii, V. V. Pasynkov, G. F. Kholuyanov, D. A. Yas'kov. Izvest. Akad. Nauk SSSR, Ser. Fiz. 20:1571, 1956.
272. K. Taylor. Ind. Eng. Chem. 47:2506, 1955.
273. G. A. Kudintseva, B. M. Tsarev, V. A. Épel'baum. In coll. Boron, Transactions of a Conference on the Chemistry of Boron and Its Compounds, Moscow, 1958, p. 112.
274. N. N. Sirota, S. N. Chizhevskaya. In coll. Physics and Physico-Chemical Analysis, Metallurgizdat, Moscow, 1957, p. 175.
275. N. N. Zhuravlev. Kristallografiya 1:666, 1956.
276. A. N. Krestovnikov, M. S. Vendrikh. Izvest. Vyssh. Ucheb. Zaved. Tsvetnaya Metallurgiya, No. 1, p. 73, 1958.
277. G. A. Meerson, G. V. Samsonov. Izvest. Sektora Fiz.-Khim. Anal. Akad. Nauk SSSR 22:92, 1953.
278. P. Pascal. Traité de chimie minérale, Vol. IV, 1933. p. 608.
279. C. Enfberg, E. Zehma. J. Am. Ceram. Soc. 42:38, 1959.
280. M. Kh. Karapet'yants. Chemical Thermodynamics, Goskhimizdat, Moscow, 1953.
281. Yu. B. Paderno, G. V. Samsonov. Doklady Akad. Nauk SSSR 137:646, 1961.
282. G. V. Samsonov, K. I. Portnoi. Alloys Based on Refractory Compounds, Moscow, 1961.
283. Yu. B. Paderno, G. V. Samsonov. Doklady Akad. Nauk UkrSSR, No. 11, p. 1215, 1959.
284. Yu. B. Paderno, G. V. Samsonov. Elektronika, Seriya 1, No. 7, p. 123, 1960.
285. J. Binder, R. Steinitz. Planseeber. Pulvermet. 7:18, 1959.
286. G. V. Samsonov. Zhur. Tekh. Fiz. 26:716. 1956.
287. S. N. L'vov, V. F. Nemchenko, G. V. Samsonov. Doklady Akad. Nauk SSSR 135:577, 1960.
288. S. Sindeband, P. Schwarzkopf. Powder Met. Bull. 5:42, 1950.
289. F. Glaser. Powder Met. Bull. 6:51, 1951.
290. Yu. B. Paderno, T. I. Serebryakova, G. V. Samsonov. Tsvetnye Metal. No. 11, p. 48, 1959.
291. K. Moers. Z. anorg. Chem. 198:243, 1931.
292. F. Glaser, W. Ivanick. J. Metals 4:387, 1952.
293. F. Glaser, D. Moskowitz. Powder Met. Bull. 6:178, 1953.
294. G. V. Samsonov, V. N. Paderno. Zhur. Priklad. Khim. 34:693, 1961.
295. M. Andrews. J. Am. Chem. Soc. 54:1845, 1932.
296. S. N. L'vov, V. F. Nemchenko, T. Ya. Kosolapova, G. V. Samsonov. Fiz. Metal. i Metalloved. 11:143, 1961.
297. C. Agte, K. Moers. Z. anorg. Chem. 198:233, 1931.
298. P. Clausing. Z. anorg. Chem. 208:401, 1932.
299. V. S. Neshpor, G. V. Samsonov. Fiz. Tverd. Tela 2:2101, 1960.
300. D. Robins. Phil. Mag. 3:313, 1958.
301. E. Galistl. Diss. techn. Hochschule Granz., 1951.
302. L. N. Guseva, B. I. Ovechkin. Doklady Akad. Nauk SSSR 112:681, 1957.
303. E. N. Nikitin. Zhur. Tekh. Fiz. 28:26, 1958.
304. G. V. Samsonov, S. V. Radzikovskaya. Uspekhi Khim. 30:60, 1961.

305. F. McTaggart. Australian J. Chem. 11:471, 1958.
306. N. N. Zhuravlev, G. N. Makarenko, G. V. Samsonov, V. S. Sinel'nikova, G. G. Tsebulya. Izvest. Akad. Nauk SSSR, Otdel. Tekh. Nauk, Ser. Metallurgiya i Toplivo, No. 1, p. 133, 1960.
307. G. Fetterley. J. Electrochem. Soc. 24:746, 1957.
308. G. V. Samsonov. Tsvetnye Metal. 32:58, 1959.
309. G. V. Samsonov, G. G. Tsebulya. Ukrain. Fiz. Zhur. No. 5, p. 35, 1960.
310. N. N. Sirota, G. V. Samsonov, N. S. Strel'nikova. In coll. Physics and Physicochemical Analysis, Metallurgizdat, Moscow, 1957.
311. N. V. Kolomoets, V. S. Neshpor, G. V. Samsonov, S. A. Semenkovich. Zhur. Tekh. Fiz. 28:2382, 1958.
312. H. Juretschke, R. Steinitz. Phys. Rev. Solids 4:118, 1958.
313. Ya. S. Umanskii. Carbides of Hard Alloys, Metallurgizdat, Moscow, 1947.
314. V. S. Neshpor, G. V. Samsonov. Doklady Akad. Nauk SSSR 133:817, 1960.
315. V. S. Neshpor, G. V. Samsonov. Doklady Akad. Nauk SSSR 134:1337, 1960.
316. V. S. Neshpor, V. F. Nemchenko, S. N. L'vov, G. V. Samsonov. Ukrain. Fiz. Zhur. 5:839, 1960.
317. G. V. Samsonov, V. S. Sinel'nikova. In coll. High-Temperature Cermet Materials, Izd. AN UkrSSR, Kiev, 1962.
318. W. Meissner, H. Franz. H. Westerhoff. Z. Physik 75:521, 1953.
319. W. Meissner, H. Franz. Z. Physik 65:30, 1930.
320. G. V. Samsonov, V. S. Neshpor. Technology of Nonferrous Metals (collected papers of the Mintsvetmetzolota), No. 29, Metallurgizdat, Moscow, 1958, p. 361.
321. N. Korak (see R. Kieffer, P. Schwarzkopf). Hard Alloys [Russian translation], Metallurgizdat, Moscow, 1957.
322. W. Ziegler, R. Young. Oxford Conference of Low Temperature Physics, 1951.
323. J. Hulm, B. Matthias. Phys. Rev. 82:273, 1954.
324. W. Meissner, H. Franz. Z. Physik 63:39, 1930.
325. W. Meissner. Ergeb. exakt. Naturw. 11, 1935.
326. Ya. G. Dorfman, I. K. Kikoin. Physics of Metals, GTTI, 1934, p. 405.
327. B. Matthias, E. Corenzwit, C. Miller. Phys. Rev. 93:1415, 1954.
328. W. Meissner, H. Franz. Z. Physik 63:558, 1930.
329. W. Meissner. Z. Ges. Kälte-Ind. Beitr. 39:104, 1932.
330. W. Meissner, H. Franz, H. Westerhoff. Ann. Physik 17:593, 1933.
331. I. Millennesu, J. Allen, J. Milhelm. Trans. Roy. Soc. Can. 25:13, 1954.
332. G. Aschermann, E. Friederich, E. Justi, I. Kramer. Z. Physik 42:349, 1941.
333. F. Horn, W. Ziegler. J. Am. Electrochem. Soc. 69:2762, 1947.
334. D. Cook, A. Zemansky, H. Boorse. Phys. Rev. 79:7021, 1950.
335. G. Hardy, J. Hulm. Phys. Rev. 89:884, 1953.
336. V. R. Golik, B. G. Lazarev, V. I. Khotkevich. Zhur. Eksptl. i Teoret. Fiz. 19:202, 1949.
337. R. Decker, D. Stebbins. J. Appl. Phys. 26:1004, 1955.
338. G. V. Samsonov, V. S. Neshpor, G. A. Kudintseva. Radiotekh. i Elektron. 2:631, 1957.
339. D. Goldwater, R. Haddod. J. Appl. Phys. 22:70, 1951.
340. R. Ridd, et al. J. Appl. Phys. 30:1575, 1959.
341. T. Hanley. J. Appl. Phys. 21:1193, 1950.
342. P. Schwarzkopf, S. Sindeband. Vortrag Electrochem. Soc. Cleveland, 1950.
343. R. Kieffer, F. Benešovsky. Metall 6:243, 1952.
344. E. Reed. Met. Abstr. 18:102, 1950/51.
345. R. Long. Met. Abstr. 1954.
346. Nuclear Reactors, Vol. III. Reports of the U.S. Atomic Energy Commission [Russian translation], Moscow, 1956.
347. G. W. C. Kaye, T. H. Laby. Tables of Physical and Chemical Constants and Mathematical Functions [Russian translation], IL, Moscow, 1949.

348. K. Becker. Z. Metallk. 20:487, 1928.
349. K. Becker, H. Ewest. Z. tech. Phys. II(148):216, 1930.
350. G. V. Samsonov, Yu. B. Paderno, S. U. Kreingol'd. Zhur. Priklad. Khim. 34:10, 1961.
351. R. Tate. U.S. Atomic Energy Commission Publ., 1948 (EPA-531); Met. Abstr. 16:102, 1950/51.
352. O. Kubaschewsky, E. Evans. Metallurgical Thermochemistry, London, 1951.
353. G. V. Samsonov, T. I. Serebryakova, A. S. Bolgar. Zhur. Neorg. Khim. 6:2243, 1961.
354. V. S. Neshpor, G. V. Samsonov. Zhur. Neorg. Khim. 33:993, 1960.
355. F. Richardson, Trans. Faraday Soc. 4:244, 1948.
356. H. Maxwell, A. Hayse. J. Am. Chem. Soc. 48:584, 1926.
357. K. Kelly. Bull. U.S. Bur. Mines, No. 406, 1957.
358. R Benoit, P. Blum. Compt. rend. 234:2428, 1952.
359. E. Klemm, W. Schuth, M. Stackelberg. Z. physik. Chem. B19:321, 1932.
360. R. Benoit. J. chim. phys. 52:119, 1955.
361. B. Matthias, J. Hulm. Phys. Rev. 87:799, 1952.
362. Ya. G. Dorfman. Magnetic Properties and the Structure of Matter, GITTL, Moscow, 1955.
363. W. Klemm, W. Schütz. Z. anorg. Chem. 201:24, 1931.
364. F. Trombe. Compt. rend. 219:162, 1944.
365. C. Kriessman. Phys. Rev. 94:837, 1954.
366. G. Foëx. Helv. Phys. Acta 26:199, 1954.
367. G. Foëx. J. phys. radium 8:37, 1948.
368. G. Foëx. Constantes sélectionnes, Diamagnétisme et paramagnétisme, Paris, 1957.
369. L. Domange, I. Flahaut, M. Guittard. Compt. rend. 249:697, 1959.
370. W. Klemm, H. Senft. Z. anorg. Chem. 241:259, 1939.
371. M. Picon, I. Flahaut. Compt. rend. 241:655, 1955.
372. G. V. Samsonov, V. S. Neshpor. Problems of Powder Metallurgy and the Strength of Materials, No. 5, Izd. AN UkrSSR, Kiev, 1958, p. 3.
373. C. Goetzel. Treatise on Powder Metallurgy, Vol. 2, N.Y., 1950, p. 80.
374. A. Lomas. Mach. Lloyd, Europ. Edit. 25:103, 1953.
375. W. Koster, W. Rauscher. Z. Metallk. 39:111, 1948.
376. I. Wachtmann, D. Lam. J. Am. Ceram. Soc. 42:254, 1959.
377. G. V. Samsonov. Production and Application of Articles of Refractory Compounds, Izd. Inst. Informatsii GNTK RSFSR, No. M-60 (99) 5, 1960.
378. L. L. Vereikina, V. N. Rudenko, G. V. Samsonov. Zavodskaya Laboratoriya 26:620, 1960.
379. G. V. Samsonov, V. S. Neshpor, T. I. Serebryakova. Inzhen. Fiz. Zhur. 2:118, 1959.
380. Gracham. Mech. Design 26:159, 1954.
381. C. Lowe. U.S. Patent 2647061, 1953.
382. J. Everhart. Materials & Methods 40:90, 1954.
383. H. Hamijan, W. Lidman. J. Am. Ceram. Soc. 35:44, 1952.
384. I. S. Brokhin, V. F. Funke. Ogneupory, No. 12, p. 562, 1957.
385. A. Bobrowsky. Trans. Am. Soc. Mech. Engrs. 11:621, 1949.
386. I. S. Brokhin, I. S. Zolotarev, A. I. Baranov. Tsvetnye Metal. No. 9, p. 61, 1958.
387. G. Humphrey. J. Am. Chem. Soc. 75:2806, 1953.
388. P. F. Gel'd, F. G. Kusenko. Izvest. Akad. Nauk SSSR, Otdel. Tekh. Nauk, Ser. Metallurgiya i Toplivo, No. 2, p. 79, 1960.
389. W. Ziegler, R. Young. Phys. Rev. 90:115, 1953.
390. P. Bridgman. Proc. Am. Acad. Arts. Sci. 66:255, 1952.
391. A. Lieberman, W. Grandall. J. Am. Ceram. Soc. 35:304, 1952.
392. C. Cline. J. Am. Electrochem. Soc. 106:52, 1959.

393. G. A. Meerson, G. V. Samsonov, R. B. Kotel'nikov, M. S. Boinova, I. P. Evteeva, S. D. Krasnenkova. Zhur. Neorg. Khim. 3:898, 1958.
394. L. Andrieux. Rev. met. 45:49, 1948.
395. E. Friederich. Z. physik. Chem. 18, No. 12, 1926.
396. L. Andrieux. Diss. Univ. Paris, 1929.
397. G. Weiss. Ann. chim. 1:446, 1956.
398. A. Kh. Breger. Zhur. Fiz. Khim. 10:593, 1959.
399. R. Kieffer, F. Kölbl. International Powder Metallurgy Conference, Graz, 1948, No. 28.
400. V. N. Eremenko. Titanium Carbide and Heat-Resistant Alloys Based on It, Izd. AN UkrSSR, Kiev, 1954.
401. A. Williams. Metal Treatment 18:445, 1951.
402. E. Engle, J. Wulff. Powder Metallurgy, Cleveland, 436, 1942.
403. R. Kieffer, F. Benešovsky, E. Honak. Z. anorg. Chem. 268:191, 1951.
404. G. V. Samsonov. Doklady Akad. Nauk SSSR 86:319, 1952.
405. G. V. Samsonov, L. Ya. Markovskii. Uspekhi Khim. 25:190, 1956.
406. G. V. Samsonov. Doklady Akad. Nauk 113:1299, 1957.
407. A. E. Koval'skii, L. A. Kanova. Zavodskaya Laboratoriya 16:1362, 1950.
408. A. E. Koval'skii, L. A. Petrova. In coll. Microhardness, Izd. AN SSSR 170, 1951.
409. R. Kieffer, F. Kölbl. Powder Met. Bull. 4:4, 1949.
410. C. Curtis, L. Doncy. J. Am. Ceram. Soc. 37:458, 1954.
411. J. Hinnüber. Z. Ver. deut. Ingr. 92:111, 1950.
412. L. Foster, et al. J. Am. Ceram. Soc. 33:27, 1950.
413. J. Hinnüber, O. Rüdiger. Arch. Eisenhuttenw. 24:267, 1953.
414. R. Kieffer, F. Benešovsky, H. Schroth. Z. Metallk. 44:437, 1953.
415. V. F. Funke, G. V. Samsonov. Zhur. Obshchii Khim. 28:267, 1957.
416. A. Searcy, R. McNess. J. Am. Chem. Soc. 75:1578, 1953.
417. K. P. Davydov, P. V. Gel'd. Fiz. Metal. i Metalloved. 2:192, 1959.
418. H. Nowotny, E. Laube, R. Kieffer, F. Benesovsky. Monatsh. Chem. 89:701, 1958.
419. R. Kieffer, F. Benesovsky, H. Nowotny, H. Schachner. Z. Metallk. 44:242, 1953.
420. I. Hamaker, G. Sheline. Report, CN-654, May 1943.
421. G. Brauer, A. Mitius. Z. anorg. Chem. 249:325, 1942.
422. C. Whitsett. Iowa State Coll. J. Sci. 31:541, 1957.
423. R. Kieffer, F. Benešovsky, E. Galistl. Z. Metallk. 43:284, 1952.
424. K. I. Portnoi, G. V. Samsonov. Boride Alloys, Izd. GNKT RSFSR, Moscow, 1959.
425. F. Glaser. J. Metals 1:475, 1949.
426. F. Bertaut, P. Blume. Compt. rend. 231:626, 1950.
427. K. Toman. Acta Cryst. 4:462, 1951.
428. R. Kieffer, F. Benešovsky, R. Maschenschaek. Z. Metallk. 45:493, 1954.
429. M. V. Kamentsev. Artificial Abrasive Materials, Mashgiz, Vol. II, 1950, p. 81.
430. G. V. Samsonov, A. S. Bolgar, T. S. Verkhoglyadova. Izvest. Akad. Nauk SSSR, Otdel. Tekh. Nauk, Metallurgiya i Toplivo, No. 1, p. 145, 1961.
431. B. Polock. J. Phys. Chem. 63:587, 1959.
432. M. Hoch. J. Appl. Phys. 29:1588, 1958.
433. W. Chupke, J. Berkowitz, C. Giese, M. Inghram. J. Phys. Chem. 62:5, 1958.
434. K. Kelley. Bull. U.S. Bur. Mines, No. 407, Washington, 1937.
435. I. E. Campbell (ed.). High-Temperature Technology [Russian translation], IL, Moscow, 1959.
436. C. Krans, C. Hurd. J. Am. Chem. Soc. 45:2559, 1923.
437. C. Dashman. Scientific Principles of Vacuum Technique [Russian translation], IL, Moscow, 1950.
438. C. Myers, A. Searcy. J. Am. Chem. Soc. 79:526, 1957.

439. R. Slade, G. Higson. J. Am. Chem. Soc. 115:215, 1919.
440. G. V. Samsonov, L. Ya. Markovskii, A. F. Zhigach, M. G. Balyashko. Boron, Its Compounds and Alloys, Izd. AN UkrSSR, Kiev, 1960.
441. W. Hincke, L. Brantley. J. Am. Chem. Soc. 52:48, 1930.
442. A. S. Berezhnoi. Silicon and Its Binary Systems, Izd. AN UkrSSR, Kiev, 1959.
443. R. B. Kotel'nikov. Investigation of Some Properties of Alloys of the Systems TiB_2-ZrB_2, TiB_2-CrB_2, $TiB_2-W_2B_5$, $ZrB-CrB_2$, author's abstract of dissertation, Mintsvetmetzoloto, Moscow, 1955.
444. V. P. Elyutin, Yu. A. Pavlov, B. E. Levin. Ferroalloys, Metallurgizdat, Moscow, 1951.
445. G. V. Samsonov, V. P. Latysheva. Fiz. Metal. i Metalloved. 2:309, 1956.
446. P. Busby, N. Warga, C. Wells. J. Metals, Sect. I, 5:1463, 1953.
447. S. G. Kreimer, L. D. Efros, E. A. Voronkova. Zhur. Tekh. Khim. 22:858, 1952.
448. W. Thomas, G. Leak. Phil. Mag. 45:986, 1954.
449. I. Stamley. J. Metals 185:752, 1949.
450. F. Seitz. The Physics of Metals [Russian translation], GITTL, Moscow, 1947, p. 205.
451. P. L. Gruzin, Yu. A. Polikarpov, M. A. Shumilov. Zavodskaya Laboratoriya 21:417, 1955.
452. R. Wasilewski, C. Kehl. J. Inst. Metals 3:94, 1954.
453. E. Gulbransen, A. Andrew. J. Metals 1:515, 1949; 1:741, 1948.
454. E. Gulbransen, A. Andrew. J. Metals 1:586, 1950.
455. V. F. Funke, S. I. Yudkovskii, G. V. Samsonov. In collected papers of VNIITS, Hard Alloys, Metallurgizdat, Moscow, 1960.
456. A. Dravnieks. J. Am. Chem. Soc. 72:3568, 1950.
457. G. V. Samsonov. In coll. Physics and Physicochemical Analysis, Metallurgizdat, Moscow, 1957, p. 192.
458. G. V. Samsonov, L. A. Solonnikova. Fiz. Metal. i Metalloved. 5:565, 1957.
459. G. V. Samsonov, M. S. Koval'chenko, T. S. Verkhoglyadova, Inzhen. Fiz. Zhur. 2:62, 1959.
460. W. Batz, H. Mead, C. Birchenall. J. Metals 4:170, 1952.
461. G. V. Samsonov, M. S. Koval'chenko, T. S. Verkhoglyadova. Zhur. Neorg. Khim. 4:2759, 1959.
462. P. S. Kislyi, G. V. Samsonov. Fiz. Tverd. Tela 2:692, 1960.
463. C. Fuller, I. Ditzenberger. J. Appl. Phys. 25:1439, 1950.
464. G. V. Samsonov, V. M. Sleptsov. Doklady Akad. Nauk UkrSSR, No. 10, p. 1116, 1959.
465. E. Rudy, F. Benešovsky. Planseeber. Pulvermetal. 8:72, 1960.
466. Gmelins Handbuch anorg. Chem., Titan, System No. 41, 1951.
467. M. P. Slavinskii. Physico-Chemical Properties of the Elements, Metallurgizdat, Moscow, 1952.
468. S. M. Nikolaeva, Ya. S. Umanskii. Izvest. Akad. Nauk SSSR, Otdel. Tekh. Nauk, Ser. Fiz. 20:631, 1956.
469. V. S. Neshpor, G. V. Samsonov. Fiz. Metal. i Metalloved. 4:181, 1957.
470. G. V. Samsonov, V. S. Neshpor. Inzhen. Fiz. Zhur. 1:30, 1958.
471. I. N. Frantsevich. Problems of Powder Metallurgy and the Strength of Materials, Izd. AN UkrSSR, Kiev, No. 3, 1956, p. 14.
472. G. V. Samsonov, V. S. Neshpor, L. M. Khrenova. Fiz. Metal. i Metalloved. 4:622, 1959.
473. V. S. Neshpor. Fiz. Metal. i Metalloved. 7:559, 1959.
474. G. V. Samsonov, Yu. B. Paderno. Borides of the Rare-Earth Metals, Izd. AN UkrSSR, Kiev, 1961.
475. L. Pauling, S. Weinbaum. Z. Krist. 87:181, 1934.
476. F. Laves. Z. physik. Chem. B22:114, 1933.
477. H. Eick, P. Gilles. J. Am. Chem. Soc. 81:5030, 1959.
478. A. Binet du Jassoneix. Compt. rend. 141:191, 1905.

479. A. Zalkin, D. Templeton. Acta Cryst. 6:269, 1953.
480. L. Andrieux, P. Blume. Compt. rend. 229:210, 1949.
481. W. Köster, W. Mühlfinger. Z. Metallk. 30:348, 1938.
482. A. Westgren, G. Fragmen. Z. anorg. Chem. 156:27, 1926.
483. L. Lander, L. Germer. Am. Inst. Mining Met. Engrs., Tech. Pub. No. 2259, 1947.
484. P. Duwez, F. Odell. J. Am. Electrochem. Soc. 97:299, 1950.
485. G. Brauer, K. Zapp. Z. anorg. Chem. 277:1291, 1954.
486. D. Templeton, C. Dauben. Acta Cryst. 3:261, 1950.
487. P. S. Kislyi, G. V. Samsonov. Izvest. Akad. Nauk SSSR, Otdel. Tekh. Nauk, Seriya Metallurgiya i Toplivo, No. 6, p. 133, 1959.
488. E. M. Savitskii. Influence of Temperature on the Mechanical Properties of Metals and Alloys, Izd. AN SSSR, 1957.
489. G. V. Samsonov, P. S. Kislyi, L. V. Grudinina. Mashinostroenie, Izd. GNTK UkrSSR, Kiev, No. 6, p. 65, 1960.
490. E. Zintl, E. Husemann. Z. physik. Chem. 21:143, 1933.
491. A. Jaboin. Compt. rend. 129:763, 1900.
492. F. Faller, W. Biltz. Z. anorg. Chem. 248:209, 1941.
493. H. Heinerth, W. Biltz. Z. anorg. Chem. 198:168, 1931.
494. M. S. Koval'chenko, G. V. Samsonov, G. A. Yasinskaya. Izvest. Akad. Nauk SSSR, Otdel. Khim. Nauk, Ser. Metallurgiya i Toplivo, No. 2, p. 115, 1960.
495. H. Haraldsen. Z. anorg. Chem. 221:997, 1935.
496. W. Biltz, M. Heimbrecht. Z. anorg. Chem. 241:349, 1939.
497. W. Klemm, K. Meisel, H. Vogel. Z. anorg. Chem. 190:123, 1930.
498. M. Picon, I. Cogne. Compt. rend. 193:595, 1951.
499. R. Kieffer, F. Benešovsky. Werkstoffe für Raketentriebwerke, Akademische Verlags-Gesellschaft, Athenaion, Konstanz, 1961.
500. S. V. Vonsovskii. Zhur. Tekh. Fiz. 18:131, 1948.
501. A. Keil. Z. Metallk. 47:243, 1956.
502. J. Flahaut, J. Guittard, I. Lories, M. Patrie. Chimie hautes températures, Paris, CNRS, 1959, p. 51.
503. S. V. Vonsovskii. Doklady Akad. Nauk SSSR 26:564, 1940.
504. G. V. Samsonov, P. S. Kislyi, A. D. Panasyuk, A. G. Strel'chenko, I. G. Khavrunyak, G. N. Serikova. Ogneupory, No. 2, p. 72, 1961.
505. O. Glemser, K. Beltz, P. Naumann. Z. anorg. Chem. 291:51, 1957.
506. V. N. Eremenko, Yu. V. Naidich. Wetting of the Surfaces of Refractory Compounds by the Rare Metals, Vid. AN UkrRSR, Kiev, 1958.
507. D. Livey, P. Murray. Warmfeste und korrosionsbeständige Sinterwerkstoffe, 2, Plansee-Seminar, Reutte, Tirol, 1956.
508. A. I. Belyaev, E. A. Zhemchuzhina. Surface Phenomena in Metallic Processes, Metallurgizdat, Moscow, 1952.
509. M. Humenik, W. Kingery. J. Am. Ceram. Soc. 37:18, 1954.
510. J. Gurland, L. Norton. J. Metals 4:1051, 1952.
511. J. Baxter, A. Roberts. Powder Metallurgy Symposium, London, Iron and Steel Inst. 1954, p. 63.
512. M. Stackelberg, R. Schnorrenberg. Z. physik. Chem. B27:37, 1934.
513. M. Stackelberg. Z. physik. Chem. B9:437, 1930.
514. M. Bredig. J. Phys. Chem. 46:801, 1942.
515. Gmelins Handbuch der anorg. Chem., System No. 30, Verlag Chemie, Berlin, 1932, p. 300.
516. M. Mallett, E. Durbin, M. Udy, D. Vaughan. Center U.S. Atomic Energy, Comm. Publ., BM (MWM), 5, 1953.
517. R. Teitel. Structure Reports 12:22, 1949.
518. C. Dauben. University of California Relation Laboratory, Private Comm., U.S. Atomic Energy Comm. Publ., USRL-2888, 1955, p. 30.
519. I. Clarke, K. Jack. Chem. Ind. No. 46, 1004, 1951.

520. R. Juza, H. Ruff. Naturwissenschaften 38:331, 1951.
521. K. Becker. Z. Physik 34:185, 1933.
522. P. McKenna. Ind. Eng. Chem. 28:767, 1936.
523. L. Doney. Ceramic Age 63:March 21, 1954.
524. C. E. Curtis, L. Doney, I. Johnson. J. Am. Ceram. Soc. 37:458, 1954.
525. T. Gibas. Spieki ceramicznei cermetale, Wydawnictwa techniczna, Warszawa, 1961.
526. I. Coobs, W. Koshuba. J. Electrochem. Soc. 99:115, 1952.
527. I. Staritzky. Analyt. Chem. 28:915, 1956.
528. W. Schupka, et al. J. Phys. Chem. 62:611, 1958.
529. I. I. Iskol'dskii, S. L. Cherkashina. Zhur. Priklad. Khim. 31:25, 1958.
530. S. Ya. Plotkin, G. V. Samsonov. Khim. Mashinostroenie, No. 4, p. 37, 1959.
531. P. Chiotti. Iowa State Coll. J. Sci. 26:185, 1952.
532. R. Powers, M. Doyle. Acta Met. 6:643, 1958.
533. F. Morgan. J. Appl. Phys. 22:108, 1951.
534. Technical Progress Review, Reactor Core Materials 1:18, 1958.
535. G. Hardy, J. Hulm. Phys. Rev. 93:1004, 1954.
536. Yu. A. Priselkov, Yu. A. Sapozhnikov, A. V. Tseplyaeva. Izvest. Akad. Nauk SSSR, Metallurgiya i Toplivo, No. 1, p. 134, 1960.
537. H. Obermüller. Métall 9:38, 1955.
538. E. Parthé. Acta Cryst. 12:559, 1959.
539. L. Ya. Markovskii, N. V. Vekshina, R. A. Shtrikhman. Ogneupory, No. 1, p. 42, 1957.
540. L. Ya. Markovskii, Yu. D. Kondrashov, G. V. Kaputovskaya. Zhur. Obshchii Khim. 25:1045, 1955.
541. L. Ya. Markovskii, Yu. D. Kondrashov, G. V. Kaputovskaya. Doklady Akad. Nauk 100:1095, 1955.
542. L. Ya. Markovskii, Yu. D. Kondrashov, G. V. Kaputovskaya. Zhur. Obschii Khim. 25:432, 1955.
543. J. H. Westbrook (ed.). Mechanical Properties of Intermetallic Compounds, J. Wiley & Sons, N.Y., 1960.
544. V. S. Rakovskii, G. V. Samsonov, I. I. Ol'khov. Production of Hard Alloys, Metallurgizdat, Moscow, 1960.
545. J. Perri, I. Binder, B. Post. J. Phys. Chem. 63:616, 1959.
546. E. Reed. J. Am. Ceram. Soc. 37:146, 1954.
547. G. A. Meerson, Yu. V. Gagarinskii (eds.). Metallurgy of Zirconium [Russian translation], IL, Moscow, 1959.
548. K. Narite, K. Mori. Bull. Chem. Soc. Japan 32:417, 1959.
549. A. Brown, J. Morreys. Nature 183:673, 1959.
550. E. Wedekind. Berichte 46:1198, 1923.
551. S. Rundquist. Acta Chem. Scand. 13:1193, 1959.
552. I. Redmond, E. Smith. J. Inst. Metals 987, 1947.
553. P. Norton. Refractories, Third ed., McGraw-Hill, N.Y., 1950.
554. H. Wenzel, W. Roeger, L. Barbrow, F. Coldwell. J. Res. Natl. Bur. Stand. 6:325, 1931.
555. P. Ehrlich, H. Hein. Electrochem. 57:70, 1953.
556. R. Edwards, G. Malloy. J. Phys. Chem. 62:45, 1958.
557. R. Juza, W. Sachse. Z. anorg. Chem. 251:201, 1943.
558. W. Hahn, A. Konrad. Z. anorg. Chem. 264:181, 1951.
559. R. Swift, D. White. J. Am. Chem. Soc. 79:3641, 1957.
560. B. Welker, C. Ewing, R. Miller. J. Phys. Chem. 61:1682, 1957.
561. R. Fruchart, A. Michel. Compt. rend. 245:171, 1957.
562. N. D. Morgulis. Uspekhi Fiz. Nauk 70:679, 1960.
563. G. S. Markevich, Yu. D. Kondrashev, L. Ya. Markovskii. Zhur. Neorg. Khim. 5:1783, 1960.
564. E. Rudy, H. Nowotny, F. Benešovsky, R. Kieffer, A. Neckel. Monatsh. Chem. 91:1761, 1960.

565. R. Kiessling. Acta Chem. Scand. 3:90, 1949.
566. L. Wöhler, O. Schliphake. Z. anorg. Chem. 11:1951, 1926.
567. E. Laube, H. Nowotny. Monatsh. Chem. 89:312, 1958.
568. R. Vogel. Z. anorg. Chem. 61:46, 1909.
569. J. Drain, A. Michel. Bull. soc. chim. France, No. 7-8, 517, 1951.
570. F. Spedding, K. Gschneider, A. Daane. Trans. Metallurg. Soc., AIME 215:192, 1959.
571. H. Nowotny, et al. Monatsh. Chem. 85:255, 1954.
572. A. E. Koval'skii, S. V. Semenovskaya. Kristallografiya 4:923, 1959.
573. L. Y. Markovskii, G. S. Markevich. Zhur. Neorg. Khim. 33:1667, 1960.
574. E. Felten. J. Am. Chem. Soc. 72:5977, 1956.
575. S. Naray-Szabo. Z. Krist. 94:367, 1936.
576. F. Halla, H. Weil. Z. Krist. 101:435, 1939.
577. J. Cohn, J. Kartz, A. Giardini. Z. Krist. 111:53, 1948.
578. G. V. Samsonov, N. N. Zhuravlev. Fiz. Metal. i Metalloved. 1:564, 1956.
579. W. Zachariasen, A. Plettinger. Report ANL-4545, 1950.
580. R. Juza, A. Rabenau. Z. anorg. Chem. 285:212, 1956.
581. M. Müller, H. Knott. Acta Cryst. 11:751, 1958.
582. P. Sthapitanonda, J. Margrave. J. Phys. Chem. 60:1628, 1956.
583. I. I. Zhukov. Izvest. Sekt. Fiz.-Khim. Anal. Akad Nauk SSSR 3(1), 1926.
584. B. Neumann, C. Kröger, H. Kunz. Z. anorg. Chem. 207:133, 1932.
585. D. Mithal. Ind. Eng. Chem. 41:2027, 1949.
586. Ya. S. Umanskii. Zhur. Fiz. Khim. 14:334, 1940.
587. S. V. Glebov. Ogneupory, No. 7, p. 336, 1960.
588. E. Schröder. Z. Naturforsch. 12a:247, 1957.
589. V. I. Khitrova, Z. G. Pinsker. Kristallografiya 2:545, 1958.
590. E. Gebhardt, H. Seghezzi, W. Durrschmabel. Z. Metallk. 49(11), 1958.
591. R. Rowerst, M. Doyle. Acta Metallurgica 4:233, 1956.
592. R. Domagala, D. McPherson, M. Hansen. J. Metals 8:88, 1956.
593. V. I. Khitrova, Z. G. Pinsker. Kristallografiya 4:545, 1959.
594. C. Lautz, E. Schröder. Z. Naturforsch. 11a:517, 1956.
595. R. Juza, H. Ruff. Z. Electrochem. 61:810, 1957.
596. S. M. Ariya, M. S. Erofeeva, G. P. Molchanov. Zhur. Obshchii Khim. 27:1740, 1957.
597. S. M. Ariya, E. A. Prokof'ev, I. I. Matveeva. Zhur. Obshchii Khim. 25:634, 1955.
598. K. Kelley. Bull. U.S. Bur. Mines, No. 407, 1957.
599. P. Ehrlich, H. Hein. Z. Elektrochem. 57:710, 1953.
600. H. Hahn, A. Konrad. Z. anorg. Chem. 264:174, 1951.
601. V. S. Mosgovoi, A. M. Samarin. Izvest. Akad. Nauk SSSR, Otdel. Tekh. Nauk, No. 10, p. 1929, 1950.
602. Z. G. Pinsker, S. V. Kaverin. Doklady Akad. Nauk SSSR 96:529, 1954.
603. Z. G. Pinsker, S. V. Kaverin. Doklady Akad. Nauk SSSR 95:797, 1954.
604. R. Juze, W. Sachse. Z. anorg. Chem. 253:95, 1945.
605. E. Klemm, G. Winkelmann. Z. anorg. Chem. 288:87, 1956.
606. H. Eick, N. Baenziger, L. Eyring. J. Am. Chem. Soc. 78:2987, 1956.
607. C. Kempter, N. Krikorian, J. McGuire. J. Phys. Chem. 61:1237, 1957.
608. F. Endter. Z. anorg. Chem. 257:127, 1948.
609. M. Mallett, A. Gerds. Bull. Memorial Inst. Columbus, Ohio, 1954.
610. A. Mah. J. Am. Chem. Soc. 80:2954, 1958.
611. T. Renner. Z. anorg. Chem. 298:22, 1959.
612. C. Neugebauer, J. Margrave. Z. anorg. Chem. 290:82, 1957.
613. L. Cronin. Am. Ceram. Soc. Bull. 30:234, 1951.
614. A. Gerds, M. Mallett. J. Electrochem. Soc. 101:175, 1954.
615. R. Elson, S. Fried, R. Sellers. Report, ANI-4545, 1950.

616. B. Abraham, N. Davidson, E. Westrum. Transuranium Elements, Part II [Russian translation], IL, 1949, p. 945.
617. W. Zachariasen. Ibid., p. 1448.
618. Belker. Acta Cryst. 11:30, 1958.
619. A. Mah. J. Amer. Chem. Soc. 80, 3872, 1958.
620. M. Mallett, I. Belle, B. Cleland. J. Electrochem. Soc. 101:1, 1955.
621. M. Mallett, E. Baroody, H. Nelson, C. Papp. J. Electrochem. Soc. 100:105, 1953.
622. D. Vaughan. J. Metals 8:78, 1956.
623. A. Jandelli. Z. anorg. Chem. 288:81, 1956.
624. Z. G. Pinsker, S. V. Kaverin. Kristallografiya 2:386, 1957.
625. D. Hardie, K. Jack. Nature 180:322, 1957.
626. R. Juza, H. Puff, F. Wagenknecht. Z. Elektrochem. 61:804, 1957.
627. G. Brauer, S. Leser. Z. Metallk. 50, 8, 1959.
628. O. Kubaschevsky. J. Inst. Metals 1:405, 1952.
629. F. Lihl, P. Jenitschek. Z. Metallk. 44:414, 1953.
630. M. Jamazaki. J. Chem. Phys. 27:746, 1957.
631. S. S. Shalyt. In coll. Semiconductors in Science and Technology, Vol. 1, 1957, p. 82.
632. C. Goodman. In coll. New Semiconductor Materials [Russian translation], IL, 1958, p. 35.
633. A. F. Ioffe. Physics of Semiconductors, Izd. AN SSSR, Moscow-Leningrad, 1957.
634. J. Lagrenaudie. J. Chem. Phys. et Phys. Chem. Biol. 54:222, 1954.
635. B. F. Ormont. Zhur. Neorg. Khim. 4:2176, 1959.
636. W. Shaw, D. Hudson, G. Danielson. Phys. Rev. 107:419, 1957.
637. J. Lagrenaudie. J. phys. radium 50:352, 1953.
638. E. Brame, J. Margrave, V. Meloche. J. Inorg. & Nuclear Chem. 5:48, 1957.
639. E. I. Akumov. Zhur. Priklad. Khim. 21:227, 1948.
640. P. Pietrokovsky. Acta Cryst. 7:435, 1954.
641. S. Isserow. Angew. Chem. 70:136, 1958.
642. M. Arvin, S. Tipsord. J. Phys. and Chem. Solids 9:336, 1959.
643. I. I. Iskol'dskii, S. L. Cherchinskaya. In collection: Hard Alloys, Metallurgizdat, Moscow, 1959, p. 116.
644. K. D. Modylevskaya, G. V. Samsonov. Ukrain. Khim. Zhur. 25:55, 1959.
645. L. Ya. Markovskii, G. V. Kaputovskaya. Zhur. Priklad. Khim. 33:569, 1960.
646. G. V. Samsonov, G. A. Yasinskaya, T'ai-Shou-wei. Ogneupory, No. 1, p. 35, 1960.
647. M. Diesen, G. Hüttig. Planseeber. Pulvermetal. 4:10, 1956.
648. E. Eastman, et al. J. Am. Ceram. Soc. 34:128, 1951.
649. E. Fitzer, J. Schwab. Metall 9:1062, 1955.
650. R. Kieffer, F. Benešovsky, H. Schmid. Z. Metallk. 47:247, 1956.
651. G. S. Markovich, L. Ya. Markovskii. Zhur. Priklad. Khim. 33:1008, 1960.
652. G. A. Meerson, G. V. Samsonov, R. B. Kotel'nikova, M. S. Voinova, I. P. Evteeva, S. D. Krasnenkova. Collected Scientific Papers of the Mintsvetmetzoloto, No. 29, Technology of Nonferrous Metals, Metallurgizdat, Moscow, 1958, p. 323.
653. V. S. Neshpor, G. V. Samsonov. Ibid., p. 349.
654. K. I. Portnoi, G. V. Samsonov, K. I. Frolova. Zhur. Priklad. Khim. 33:577, 1960.
655. G. V. Samsonov, V. S. Neshpor, V. A. Ermakova. Zhur. Neorg. Khim. 3:868, 1958.
656. K. I. Portnoi, G. V. Samsonov, K. I. Frolova. Izvest. Akad. Nauk SSSR, Otdel. Tekh. Nauk, Ser. Metallurgiya i Toplivo, No. 2, p. 117, 1959.
657. V. V. Grigor'eva, V. N. Klimenko. Alloys Based on Chromium Carbide, Izd. AN UkrSSR, Kiev, 1961.

658. W. Kinna, O. Rüdiger. Arch. Eisenhüttenw. 24:535, 1953.
659. I. Hinnüber, O. Rüdiger. Arch. Eisenhüttenw. 24:257, 1953.
660. T. Ya. Kosolapova, G. V. Samsonov. Zhur. Fiz. Khim. 35:363, 1961.
661. B. F. Ormont. Connection Between Chemical and Mechanical Strength (Hardness) of Very Hard Brittle Substances (Carbides, Nitrides, etc.). Paper read at the Permanent Colloquium on Hard Phases of Variable Composition, FKhI im. Karpova, No. 3, Moscow, 1956.
662. A. Newkirk. J. Am. Chem. Soc. 77:4521, 1955.
663. N. Joseph. Ind. Eng. Chem. 44:1006, 1952.
664. H. Montgomery. U. S. Patent 2496671, 1950; J. Res. Natl. Bur. Standards 23:39, 1939.
665. M. V. Smirnov, Yu. N. Krasnov. Zhur. Neorg. Khim. 5:1241, 1960.
666. M. V. Smirnov, B. F. Ormont. Doklady Akad. Nauk SSSR 96:557, 1954.
667. V. P. Kopylova. Resistance of Metal-like Carbides to the Action of Acids and Alkalis, Informpis'mo IMSS AN UkrSSR, No. 140, Izd. UkrSSR, Kiev, 1958.
668. V. P. Kopylova. Zhur. Priklad. Khim. 34:1936, 1961.
669. T. Ya. Kosolapova. Resistance of the Carbides Cr_3C_2 and Cr_7C_3 to the Action of Acids and Alkalis, Informpis'mo IMSS AN UKrSSR, No. 113, Izd. AN UkrSSR, Kiev, 1958.
670. G. T. Kabanik. Chemical Resistance of Refractory Nitrides in Acids and Alkalis, Informpis'mo IMSS AN UkrSSR, No. 125, Izd. AN UkrSSR, Kiev, 1958.
671. T. Ya. Kosolapova. Chemical Resistance of Silicides of Refractory Metals in Acids and Alkalis, Informpis'mo IMSS AN UkrSSR, No. 158, Izd. AN UkrSSR, Kiev, 1958.
672. T. N. Nazarchuk. Zhur. Neorg. Khim. 4:2665, 1959.
673. A. I. Miklashevskii. Carborundum, Its Chemical Analysis and Properties, GONTI, Moscow-Leningrad, 1938.
674. K. Taylor, C. Lenie. J. Electrochem. Soc. 107:308, 1960.
675. S. Amberg. Monatsh. Chem. 91:412, 1960.
676. B. Mathias, W. Zachariasen. Phys. Chem. Solids 7:98, 1958.
677. G. V. Samsonov. Problems of Powder Metallurgy and Strength of Materials, No. 7, Izd. AN UkrSSR, Kiev, 1959, p. 72.
678. S. Nelson, T. Willmore, R. Womelsdorf. J. Electrochem. Soc. 98:465, 1951.
679. H. Greenhouse, O. Accountis, H. Sisler. J. Am. Chem. Soc. 73:5086, 1951.
680. R. Kieffer, H. Schmid, F. Benešovsky. Warmfeste und korrosionbeständige Sinterwerkstoffe, II, Plansee-Seminar, 1956, p. 154.
681. L. Ya. Markovskii, N. V. Vekshina. Zhur. Neorg. Khim. 2:1692, 1957.
682. M. E. Levina, S. K. Safonova. Vestnik MGU, Ser. Mat., Astron., Fiz., i Khim., No. 2, p. 161, 1956.
683. M. E. Levina. Vestnik MGU, Ser. Mat., Astron., Fiz., i Khim., No. 1, p. 245, 1956.
684. M. Zumbusch, W. Biltz. Z. anorg. Chem. 249:1, 1942.
685. F. Faller, W. Biltz. Z. anorg. Chem. 248:209, 1941.
686. M. Zumbusch. Z. anorg. Chem. 245:402, 1941.
687. I. Sheft, S. Fried. J. Am. Chem. Soc. 75:1236, 1953.
688. M. Picon, J. Flahaut. Compt. rend. 241:655, 1955.
689. J. Flahaut, M. Guittard, M. Patrie. Bull. soc. chim. France, No. 7:990, 1958.
690. D. Westbrook. Problemy Sovremennoi Metallurgii, No. 4, p. 111, 1960.
691. R. Kiessling, J. Liu. J. Metals 3:639, 1951.
692. Yu. B. Paderno, G. V. Samsonov, L. M. Khrenova. Elektronika, No. 4, p. 165, 1959.
693. V. I. Alekseev, L. A. Shvartsman. Doklady Akad. Nauk SSSR 133:1331, 1960.
694. G. V. Samsonov. Doklady Akad. Nauk SSSR 133:1344, 1960.
695. S. I. Alyamovskii, P. V. Gel'd, G. P. Shveikin. Collected papers of the S. M. Kirov Ural Polytechnic Institute, No. 92, 1959, p. 125.
696. P. Emmet, S. Hendricke, S. Braunauer. J. Am. Chem. Soc. 52:1456, 1930.

697. A. E. Vol. Structure and Properties of Binary Metal Systems, Vol. I, Fiz-matgiz, Moscow, 1959.
698. K. Kelley, U.S. Department of the Interior, Bureau of Mines Bull. No. 476, Washington, 1949.
699. G. F. Kosolapova. Metallurg. No. 11, p. 79, 1936.
700. G. Wiener, J. Berger. J. Metals, Sect. 2, 7:360, 1955.
701. R. Bridelle. Ann. Chem. p. 824, Sept.-Oct. 1955.
702. B. Neumann, C. Kroger, H. Haebler. Z. anorg. Chem. 218:379, 1934.
703. A. Sieverts, G. Zapf. Z. anorg. Chem. 229:161, 1936.
704. M. Stackelberg, E. Schnorrenberg, R. Paulus, K. Spiesse. Z. physik. Chem. 175:1936.
705. J. Neely, O. Teeter, J. Trice. J. Am. Ceram. Soc. 33:363, 1950.
706. W. Nodge, R. Evans, A. Haskins. J. Metals 7:824, 1955.
707. W. Havekotte. Metal Prog. 64:67, 1953.
708. F. Pfaffinger. Planseeber. Pulvermetal. 3:17, 1955.
709. J. Harwood. Metal Powder Assoc. No. 4, 36, 1952.
710. G. Ault, G. Deutsch. J. Metals, Sect. I, 6:1214, 1954.
711. H. Henninger. Nachrtech. No. 11, 514, 1959.
712. V. V. Aleksandrov, V. I. Pruzhinina, A. I. Rekov, T. S. Tarakanova, E. A. Teplov. Fiz. Tverd. Tela 1:1587, 1959.
713. L. Cronin. U.S. Patent 2725287, 1955.
714. H. Greenwood. Engineer, No. 4862, 349, 1949.
715. O. Kröckel. Silikat Tech. 11:108, 1960.
716. G. Kohn, D. Eckert. Anal. Chem. 32:296, 1960.
717. Nondiamond Truing of Grinding Wheels, Collection, Mashgiz, Moscow, 1951.
718. Ya. B. Mindlin. Grinding and Sharpening of Tools with Hard Alloys, Oboron-giz, 1941.
719. B. McDonald, W. Stuart. Acta Cryst. 13:447, 1960.
720. B. Decker, I. Kasper. Acta Cryst. 7:77, 1954.
721. G. A. Meerson, G. V. Samsonov, R. B. Kotel'nikova, N. Ya. Tseitina. Sbornik Trudov Mintsvetmetzoloto 25:209, 1955.
722. A. Pruch. Acta Cryst. 4:66, 1951.
723. B. Aronsson, E. Stenberg, I. Aselius. Acta Chem. Scand. 14:733, 1960.
724. G. V. Samsonov, G. G. Tsebulya. Ukrain. Fiz. Zhur. 5:35, 1960.
725. I. L. Zagyanskii, M. I. Imbritskii. Rabochii-énergetik, No. 4, p. 8, 1951.
726. I. N. Chaporova. Zavodskaya Laboratoriya 15:799, 1949.
727. W. Mallett, A. Gerds, H. Nelson. J. Electrochem. Soc. 99:197, 1952.
728. A. Osawa, M. Oya. Sci. Repts. Research Insts. Tohuku Univ. 19:25, 1930.
729. W. Rostoker, A. Jamamoto. Trans. ASM 46:1136, 1954.
730. K. Hellborn, A. Westgren. Svensk. Kem. Tidskr. 45:141, 1933.
731. K. Kuo, L. Persson. J. Iron. Steel Inst. 178:39, 1954.
732. L. Hofer, E. Cohn, W. Reebles. J. Am. Chem. Soc. 71:189, 1949.
733. H. Lipson, W. Petsch. J. Iron Steel Inst. 142:95, 1940.
734. L. Hofer, Z. Cohn, W. Reebles. J. Phys. & Colloid Chem. 54:1161, 1950.
735. R. Bernier. Ann. chim. 6:104, 1951.
736. H. Hartmann, U. Fröhlich. Z. anorg. Chem. 218:190, 1934.
737. G. Foëx, et al. U.S. Atomic Energy Comm. Publ., 1952, ISC-224; J. Inst. Metals 1952/53, No. 4, Met. Abstr. 256.
738. C. Cheer, J. Tittman. Rev. Sci. Instr. 22:837, 1951.
739. G. Brauer. Z. Elektrochem. 46:397, 1940.
740. U.S. Patent 2095769, 1937.
741. G. Brauer, J. Jander, H. Rogener. Z. Physik 134:432, 1953.
742. V. M. Sleptsov, G. V. Samsonov. Problems of Powder Metallurgy and the Strength of Materials, No. 5, 1958, p. 65.
743. O. Zwicker. Z. Mettallk. 42:274, 1951.
744. G. Wiener, G. Berger. J. Metals 7:360, 1955.

745. A. A. Burdesé. Metallurgia italiana 47:357, 1955.
746. V. Paranjpe, M. Cohler, M. Bever, C. Floe. J. Metals 188:261, 1950.
747. N. Terso. Naturwiss. 45:620, 1958.
748. K. Jack. Acta Cryst. 3:392, 1950.
749. I. Perri, E. Banks, B. Post. J. Phys. Chem. 63:2073, 1959.
750. H. Schachner, H. Nowotny, R. Maschenschalk. Monatsh. Chem. 84:677, 1953.
751. H. Schachner, H. Nowotny, R. Maschenschalk. Monatsh. Chem. 85:1, 1954.
752. C. Lundin, D. McPherson, M. Hansen. Amer. Soc. Metals, Preprint No. 41, 1952.
753. F. Laves, H. Wallbaum. Z. Krist. A101:78, 1939.
754. H. Schachner, E. Cerwenka, H. Nowotny. Monatsh. Chem. 85:245, 1954.
755. E. Parthé, H. Schachner, H. Nowotny. Monatsh. Chem. 86:183, 1955.
756. B. Post, E. Philips, W. Herz. Powder Met. Bull. 1:149, 1956.
757. W. Zachariasen. Z. physik. Chem. 128:39, 1927.
758. E. Fitzer. Z. Metallk. 44:462, 1953.
759. K. Toman. Acta Cryst. 5:329, 1952.
760. S. Rundquist, F. Jellinek. Acta Chem. Scand. 13:425, 1959.
761. W. Klemm, K. Meisel, H. Vogel. Z. anorg. Chem. 190:123, 1930.
762. A. Kh. Breger, G. S. Zhdanov. Doklady Akad. Nauk 28:629, 1940.
763. P. Popper, T. Ingles. Nature 179:1075, 1957.
764. V. Matkovich. Acta Cryst. 13:679, 1960.
765. B. Vassiliu, F. Wilde. Nature 179:435, 1957.
766. W. Forgang, B. Decker. Trans. Met. Soc. AIME 212:343, 1958.
767. G. Perry, S. La Place, B. Post. Acta Cryst. 11:310, 1958.
768. E. Hellner. Z. anorg. Chem. 261:226, 1950.
769. L. Foster, et al. J. Am. Ceram. Soc. 33:27, 1950.
770. V. I. Zhelankin, V. S. Kutsev, B. F. Ormont. Zhur. Fiz. Tekh. 33:1988, 1959.
771. H. Wilhelm, P. Chiotti, A. Show, A. Daane. J. Am. Chem. Soc. 2:318, 1949.
772. Z. G. Pinsker, S. V. Kaverin. Kristallografiya 1:66, 1956.
773. V. I. Kudryavtsev, G. V. Safronov. In coll. Transactions of the Seminar on Oxidation-Resistant Materials, No. 5, Izd. AN UkrSSR, Kiev, 1960, p. 52.
774. British Patent 711444, 1954.
775. P. Schwarzkopf, F. Glaser. Iron Age 173:138, 1954.
776. Financial Times, Jan. 1, 1958.
777. Engineering 187(4845):91, 1959.
778. R. Long. Metal Prog. 68:123, 1955.
779. G. Summer. Mech. Eng. Jan. 1948.
780. G. V. Samsonov. Vestnik Akad. Nauk UkrSSR, No. 5, 66, 1958.
781. I. S. Kainarskii, E. V. Dekhtyareva, V. A. Kuchtenko. Ogneupory, No. 4, p. 175, 1960.
782. C. Finlay. U.S. Patent 2529333, 1950.
783. A. Abbey. British Patent 716836, 1954.
784. V. I. Tret'yakov, I. P. Karabasov, A. B. Platov. In coll. Hard Alloys, II, Metallurgizdat, Moscow, 1960, p. 79.
785. A. I. Avgustinik. Zhur. VKhO im. Mendeleeva 5:156, 1960.
786. G. Butter. J. Electrochem. Soc. 104:641, 1957.
787. W. Arbiter, Wright Air Development Center, Rep. 1953 (WADC-TR-53-190), p. 85.
788. K. I. Portnoi, G. V. Samsonov, L. A. Solonnikova. Zhur. Neorg. Khim. 5:2032, 1960.
789. R. Kieffer, F. Benešovsky..Planseeber. Pulvermetal. 5:56, 1957.
790. E. Fitzer, O. Rubisch, F. Selka. Elektrowärme, No. 6, 1958.
791. H. Moissan, H. Hoffmann. Ber. chem. Ges. 34:3324, 1904.
792. G. Long, L. Foster. J. Am. Ceram. Soc. 42:1, 1959.
793. Chem. Eng. 60:186, 1953.
794. V. P. Remin. Vestnik Metalloprom. 18:57, 1958.

795. K. Bungardt, R. Runding. Z. Metallk. 47:577, 1956.
796. O. Bus, R. Vandergrift, T. Hanley. J. Appl. Phys. 20:295, 1949.
797. Chem. Eng. News 24:3361, 1946.
798. Chem. Abstracts 43:4957, 1949.
799. E. Olson, E. Layer, A. Middleton. J. Electrochem. Soc. 102:73, 1955.
800. H. Milton. Chem. Rev. 39:419, 1946.
801. N. Tuson. J. Appl. Phys. 20:59, 1949.
802. I. D. Konozenko. Uspekhi Fiz. Nauk 56:283, 1955.
803. Chem. Abstracts 44:10524, 1950.
804. L. Foster. ASM Rev. Metal Lit. 9:737, 1952.
805. A. Münster, W. Ruppert. Z. Elektrochem. 57:564, 1953.
806. H. Hamijan, W. Lidman. J. Metals 5:696, 1953.
807. O. Myer. Ber. deut. chem. Ges. 11:333, 1930.
808. German Patents 231231 and 234446, 1910.
809. W. Lidman, H. Hamijan. J. Am. Ceram. Soc. 35:236, 1952.
810. I. Kennedy. Materials & Methods 36:166, 1952.
811. E. Fitzer. Pulvermetallurg, I Plansee-Seminar De re metallica, Wien, 1953, p. 244.
812. V. V. Grigor'eva, V. N. Eremenko. Problems of Powder Metallurgy and the Strength of Materials, VIII, Izd. AN UkrSSR, Kiev, 1960, p. 38.
813. G. V. Samsonov. Zhur. Vsesoyuznee Khimicheskoe Obshchestvo im. Mendeleeva, No. 6, p. 515, 1960.
814. German Patent 437165, 1924.
815. German Patent 536749, 1930.
816. V. F. Bochkov. Ogneupory, No. 1, p. 39, 1960.
817. R. Brown, C. Landback. Am. Ceram. Soc. Bull. 352, 1959.
818. N. S. Gorbunov. Diffusion Coatings on Iron and Steel, Izd. AN SSSR, Moscow, 1958.
819. H. Greenhouse, R. Stoops, T. Shevlin. J. Am. Ceram. Soc. 37:203, 1954.
820. G. V. Samsonov, G. N. Makarenko, T. Ya. Kosolapova. Zhur. Priklad. Khim. 34:1444, 1961.
821. S. S. Kabalkina, L. F. Vereshchagin. Doklady Akad. Nauk SSSR 134:330, 1960.
822. E. S. Sarkisov. Zhur. Fiz. Khim. 28:627, 1954.
823. O. I. Shulishova. In coll. High-Temperature Cermet Materials, Izd. AN UkrSSR, Kiev, 1962.
824. H. Bittner, H. Goretzki. Monatsh. Chem. 91:616, 1960.
825. E. Kauer, A. Rabenau. Z. Naturforsch. 12a:942, 1957.
826. P. Vaughan, Braout. Abstract of paper presented at meeting of the Am. Cryst. Assoc., June 27, 1955; C. Dauben, J. Electrochem. Soc. 104:521, 1957.
827. C. Lundin, D. McPherson, H. Hansen. Trans. Am. Soc. Metals 45:901, 1953.
828. A. Weil. Nature 152:413, 1943.
829. E. Parthé. Powder Met. Bull. 8:23, 1957.
830. E. Parthé, H. Nowotny, H. Schmid. Monatsh. Chem. 86:385, 1955.
831. V. F. Funke, A. P. Shurshakov, S. P. Yudkovskii, V. I. Shulepov, Yu. N. Yurkevich. Fiz. Metal. i Metalloved. 10:207, 1960.
832. R. Kissling. In coll. Heat-Resistant and Corrosion-Resistant Cermet Materials, Oborongiz, 1959, p. 194.
833. C. Kline, P. Sands. Nature 185:456, 1960.
834. D. Robins, I. Jenkins. In coll. Heat-Resistant and Corrosion-Resistant Cermet Materials, Oborongiz, 1959, p. 195.
835. E. Gebhardt, H. Seghezzi, W. Durschnabel. Powder Met. Bull. 8:94, 1959.
836. P. T. Kolomytsev. Izvest. Akad. Nauk SSSR, Otdel Khim. Nauk, Ser. Metallurgiya i Toplivo, No. 3, p. 83, 1960.
837. F. Spedding, K. Gschneider, A. Daane. J. Am. Chem. Soc. 80:4499, 1958.
838. M. Stackelberg. Z. Electrochem. 37:542, 1931.
839. E. Rauh, R. Thorn. J. Chem. Phys. 31:1481, 1959.

840. R. Vickery, R. Sedlaček. J. Chem. Soc. No. 2, 503, 1959.
841. N. Z. Miryasov, A. P. Parsanov. Vestnik MGU, Ser. Mat., Mekh., Astron., Fiz., No. 1, p. 43, 1959.
842. W. Borchert, M. Röder. Z. anorg. Chem. 302:253, 1959.
843. R. Vickery, R. Sedlaček, A. Ruben, J. Chem. Soc. No. 2, 498, 1959.
844. M. Atoji, K. Gschneider, A. Daane, R. Rundle, F. Spedding. J. Am. Chem. Soc. 80:1804, 1958.
845. Warf. Palineck. Status Report, July 20, 1955; US Office of Ordnance Research, Project 683; Contract DA-04-495-Ord., 1955/6; cited in [843].
846. I. Binder. J. Am. Ceram. Soc. 43:287, 1960.
847. J. Lagrenaudie. J. Chem. Phys. et Phys. Chem. Biol. 50:352, 1953.
848. Yu. M. Golutvin, T. M. Kozlovskaya. Zhur. Fiz. Khim. 34:2350, 1960.
849. B. Aronsson, M. Bäckman, S. Rundquist. Acta Chem. Scand. 14:1001, 1960.
850. E. Parthé. Acta Cryst. 13:868, 1960.
851. V. I. Khitrova, Z. G. Pinsker. Kristallografiya 5:711, 1960.
852. A. Seybolt. Trans. Am. Soc. Metals 52:971, 1960.
853. S. P. Kapitsa, V. P. Bykov, V. N. Melekhin. Zhur. Eksptl. i Teoret. Fiz. 39:997, 1960.
854. E. Tyrkdohan, P. Bills, V. Tippett. J. Appl. Phys. 8:296, 1958.
855. G. L. Gal'chenko, A. N. Kornilov, S. M. Skuratov. Zhur. Neorg. Khim. 5:2651, 1960.
856. G. A. Meerson, R. B. Kotel'nikov, S. N. Bashlykov. Atomnaya Energ. 9:387, 1960.
857. E. Pokorny. Mines et Metallurgie, No. 3529, 359, 1959.
858. A. N. Krestovnikov, M. S. Vendrikh. Izvest. Vyssh. Ucheb. Zaved., Ser. Tsvetnaya Metallurgiya, No. 3, p. 13, 1960.
859. A. N. Krestovnikov, M. S. Vendrikh. Izvest. Vyssh. Ucheb. Zaved., Ser. Tsvetnaya Metallurgiya, No. 2, p. 54, 1959.
860. V. N. Igishev. Electrical Conductivity and Nature of Ferrosilicon Alloys at High Temperatures, author's abstract of dissertation, Sib. Met. Inst. im. S. Ordzhonikidze, 1960.
861. V. S. Aleksashin, V. S. Mikheev. Zhur. Neorg. Khim. 5:2216, 1960.
862. L. Kleinman, J. Philips. Phys. Rev. 117:460, 1960.
863. G. S. Markevich. Investigation of the Beryllium—Boron System, author's abstract of dissertation, Chemical Faculty, Leningrad State University, Leningrad, 1961.
864. A. Searcy, A. Tharp. J. Phys. Chem. 64:1939, 1960.
865. D. Sands, C. Cline, A. Zalkin, L. Holnog. Acta Cryst. 14:309, 1961.
866. G. S. Markevich, L. Ya. Markovskii. Sbornik Trudov GIPKh, Leningrad, No. 45, p. 139, 1960.
867. F. G. Kusenko, P. V. Gel'd. Izvest. Sib. Otdel. Akad. Nauk SSSR, No. 2, p. 46, 1960.
868. R. Oriani, W. Murphy. J. Am. Chem. Soc. 76:343, 1954.
869. M. Nadler, C. Kempter. J. Phys. Chem. 64:1471, 1960.
870. E. Storms, N. Krikorian. J. Phys. Chem. 64:1471, 1960.
871. K. P. Yatsimirskii. Zhur. Neorg. Khim. 6:518, 1961.
872. B. Coles, D. Griffits. Proc. Phys. Soc. 77:213, 1961.
873. C. Quillard, J. Wyart. Rev. Metallurgia 45:271, 1948.
874. A. I. Rudnaya, V. G. Tishchenko. Automation and Instrument Construction, Nauch.-Tekhn. Sb. Inst. Avtomatiki Gosplana UkrSSR, Kiev, No. 1, p. 83, 1961.
875. H. Auer-Welsbach, H. Nowotny. Monatsh. Chem. 92:198, 1961.
876. V. A. Trigubenko, B. M. Tsarev. Radiotekh. i Elektron. 6(11):1900, 1961.
877. V. M. Alekseev, L. A. Shvartsman. Fiz. Metal. i Metalloved. 11:545, 1961.
878. H. Becher. Z. anorg. Chem. 308:13, 1961.
879. G. V. Samsonov, T. Ya. Kosolapova, G. N. Makarenko. Zhur. Neorg. Khim. 7:975, 1962.

880. G. V. Samsonov, O. I. Shulishova. Izvest. Akad. Nauk SSSR, Otdel. Tekh. Nauk, No. 3, p. 53, 1962.
881. L. L. Vereikina, G. V. Samsonov. Ukrain. Khim. Zhur. 35:17, 1962.
882. A. Michel. Bull. soc. chim. France, No. 1, 143, 1961.
883. A. I. Avgustinik, V. M. Troyanov. Papers read at the Scientific and Technical Conference of the Lomonosov Leningrad Technological Institute, GNTIKhL, Leningrad, 1961, p. 183.
884. M. S. Koval'chenko, G. V. Samsonov. Doklady Akad. Nauk UkrSSR, No. 11, p. 76, 1961.
885. V. V. Pen'kovskii, G. V. Samsonov. Avtomat. Svarka, No. 2, p. 39, 1962.
886. T. S. Verkhoglyadova, S. N. L'vov, V. F. Nemchenko, G. V. Samsonov. Fiz. Metal. i Metalloved. 12:622, 1961.
887. M. Hansen. Structure of Binary Alloys [Russian translation], Metallurgizdat, Moscow, 1941.
888. V. I. Lakh, V. Ya. Prokhorenko, L. S. Terebukh, P. S. Kislyi, A. D. Panasyuk, G. V. Samsonov. Tsvetnye Metal. No. 8, p. 38, 1961.
889. G. V. Samsonov, V. S. Sinel'nikova. Ukrain. Fiz. Zhur. 6:105, 1961.
890. G. A. Gaziev, O. V. Krylov, S. Z. Roginskii, G. V. Samsonov, E. A. Fokina, M. I. Yanovskaya. Doklady Akad. Nauk 140(4):863, 1961.
891. N. N. Zhuravlev, A. A. Stepanova, Yu. B. Paderno, G. V. Samsonov. Kristallografiya 6:791, 1961.
892. V. S. Kocho, A. D. Panasyuk, G. V. Samsonov, A. G. Strel'chenko, I. G. Khavrunyak. Stal', No. 4, p. 317, 1962.
893. V. V. Fesenko. Poroshkovaya Metallurgiya 1:85, 1961.
894. G. V. Samsonov, G. A. Yasinskaya, E. A. Shiller, L. V. Strashinskaya. Ogneupory, No. 7, p. 335, 1961.
895. G. V. Samsonov, G. A. Yasinskaya, E. A. Shiller, L. V. Strashinskaya. Izvest. Akad. Nauk SSSR, Otdel. Tekh. Nauk, No. 6, 1962.
896. Analysis of Refractory Compounds, Group of Authors, Metallurgizdat, Moscow, 1962.
897. H. Katz. J. Appl. Phys. 24:597, 1953.
898. V. S. Fomenko. Radiotekh. i Elektron. 6:1406, 1961.
899. G. A. Kudintseva, V. S. Neshpor, G. V. Samsonov, B. M. Tsarev, Yu. B. Paderno. In coll. High-Temperature Cermet Materials, ed. G. V. Samsonov, Izd. AN UkrSSR, Kiev, 1962.
900. J. M. Lafferty. Phys. Rev. 79:1012, 1950.
901. R. Dekker, D. Stebbins. J. Appl. Phys. 26:1004, 1955.
902. G. A. Kudintseva. Elektron. 4:193, 1960.
903. G. V. Samsonov, V. S. Neshpor, G. A. Kudintseva. Radiotekh. i Elektron. 2:631, 1957.
904. G. A. Kudintseva, B. M. Tsarev, V. A. Epel'baum. In coll. Boron, Transactions of a Conference on the Chemistry of Boron and Its Compounds, Goskhimizdat, Moscow, 1958, p. 106.
905. D. L. Goldwater, R. Haddad. J. Appl. Phys. 22:70, 1951.
906. G. A. Haas, J. T. Jensen. J. Appl. Phys. 31:1231, 1960.
907. G. V. Samsonov, T. S. Verkhoglyadova. Zhur. Neorg. Khim. 5:1231, 1961.
908. G. V. Samsonov, T. S. Verkhoglyadova. Zhur. Strukt. Khim. 2(5):617, 1961.
909. W. Obrowski. J. Inst. Metals 5:65, 1960.
910. R. Fries, C. Kempter. Anal. Chem. 32:1998, 1960.
911. S. La Place, B. Post. Planseeber. Pulvermetal. 9:109, 1961.
912. H. Rizzo, L. Bidwell. J. Am. Ceram. Soc. 43:550, 1960.
913. H. Rizzo, B. Weber, M. Schwarz. J. Am. Ceram. Soc. 43:5, 1960.
914. Kh. I. Gol'dberg. Electrical Properties of Alloys of Iron and Silicon Containing Leboite, author's abstract of dissertation, Novokuznetsk, 1960.
915. A. Erb. Ann. chim. 14:713, 1959.
916. V. S. Neshpor, G. V. Samsonov. Fiz. Metal. i Metalloved. 11:683, 1961.

917. V. S. Neshpor. Investigation of the Production Conditions and Some Physical Properties of Silicides of the Transition Metals, author's abstract of dissertation, Kiev Polytech. Inst., Kiev, 1961.
918. Industrial Heating 28:137, 1961.
919. K. Sedlatschek, H. Stadler. Planseeber. Pulvermetal. 9:39, 1961.
920. H. Nowotny, E. Laube. Planseeber. Pulvermetal. 9:54, 1961.
921. E. Hilpert, M. Ornstein. Ber. chem. Ges. 43:1672, 1913.
922. D. Moskowitz, M. Humenik. Planseeber. Pulvermetal. 9:60, 1961.
923. B. Beck. Planseeber. Pulvermetal. 9:96, 1961.
924. L. Prus, E. Byron, F. von Plinsky, S. Porembka. Nuclear Sci. and Eng. 6:167, 1959.
925. E. Byron, F. von Plinsky, S. Porembka. Nuclear Sci. and Eng. 6:361, 1959.
926. D. Dunning. Nuclear Sci. and Eng. 4:419, 1958.
927. H. Becher. Z. anorg. Chem. 306:266, 1960.
928. O. Kubaschewski, E. Evans. Metallurgical Thermochemistry, London, 1958.
929. T. S. Verkhoglyadova. Investigation of the Production Conditions and Some Physico-Chemical Properties of Nitrides of the Transition Metals, author's abstract of dissertation, Kiev Polytech. Inst., Kiev, 1962.
930. P. S. Kislyi, S. N. L'vov, V. F. Nemchenko, G. V. Samsonov. Izvest. Akad. Nauk SSSR, Otdel. Tekh. Nauk, Ser. Metallurgiya i Toplivo, No. 6, 1962.
931. S. N. L'vov, V. F. Nemchenko, G. V. Samsonov. Poroshkovaya Metallurgiya 1(6):68, 1961.
932. G. V. Samsonov, G. A. Yasinskaya, E. A. Shiller. Metallography and Production Technology of Cermets, Refractory Metals and Compounds Based on Them, Metallurgizdat, 1962, p. 156.
933. V. S. Fomenko, Yu. B. Paderno, G. V. Samsonov. Ogneupory, No. 1, p. 40, 1962.
934. Tichter, Spengel. Z. anorg. Chem. 82:195, 1913.
935. B. Aronsson. Arkiv Kemi 16:379, 1960.
936. B. Aronsson. Modern Materials, Advances in Research and Applications 2:143, 1960.
937. M. A. Gurevich, B. F. Ormont. Zhur. Anal. Khim. 11:177, 1956.
938. B. Aronsson. Acta Chem. Scand. 14:1414, 1960.
939. E. I. Gladyshevskii. Doklady Akad. Nauk UkrSSR, No. 3, p. 294, 1959.
940. L. N. Kugai, T. N. Nazarchuk. Zhur. Anal. Khim. 16:203, 1961.
941. M. Billey. Ann. chim. 4:795, 1959.
942. I. G. Shafran, M. V. Pavlova. Zavodskaya Laboratoriya 7:1241, 1938.
943. J. W. Mellor. A Comprehensive Treatise of Inorganic and Theoretical Chemistry 5:882, 1925.
944. Electronics 32:124, 1959.
945. Selected Values of Chemical Thermodynamic Properties, National Bureau of Standards Circular 500, 1952.
946. S. Fujishiro, N. Gokcen. J. Phys. Chem. 65:161, 1961.
947. J. Kempbelle, G. Powell, D. Novicki, B. Conser. J. Electrochem. Soc. 96:318, 1949.
948. R. Vogel, B. Giepen. Arch. Eisenhüttenw. 30:565, 1955.
949. R. Vogel, R. Dobbener. Arch Eisenhüttenw. 29:129, 1958.
950. R. Schneider, R. Vogel. Arch Eisenhüttenw. 26:483, 1955.
951. J. Berak, T. Neumann. Z. Metallk. 41:19, 1950.
952. R. Vogel, H. Gontermann. Arch. Eisenhüttenw. 3:369, 1922.
953. S. F. Schemtschushny, J. Shepelev. Z. anorg. Chem. 64:245, 1909.
954. N. Konstantinov. Z. anorg. Chem. 60:405, 1908.
955. G. Brauer. J. Less Common Metals 2:131, 1960.
956. V. A. Epel'baum, N. G. Sevost'yanov, M. A. Gurevich, G. S. Zhdanov. Zhur. Strukt. Khim. 1:20, 1960.
957. H. Moissan, P. Williams. Compt. rend. 125:629, 1897.

958. E. Wedekind. Berichte 46:1198, 1913.
959. A. I. Miklashevskii. Zavodskaya Laboratoriya 7:168, 1938.
960. O. Mühlhausen. Z. anorg. Chem. 5:105, 1894.
961. G. Remi. Silicon and Its Compounds [Russian translation], ONTI, Moscow, 1938, p. 3.
962. G. V. Kukolev. Chemistry of Silicon and Physical Chemistry of Silicates, 1951, p. 196.
963. K. D. Modylevskaya, M. D. Lyutaya, T. N. Nazarchuk. Zavodskaya Laboratoriya 27(11):1345, 1961.
964. H. Blumental. Anal. Chem. 23:192, 1951.
965. E. Friederich, L. Sittig. Z. anorg. Chem. 143:708, 1924.
966. E. Friederich, L. Sittig. Z. anorg. Chem. 143:308, 1925.
967. P. Pascal. Traité de chimie minérale 5:607, 1932.
968. O. Ruff, R. Wemsch. Z. anorg. Chem. 85:292, 1914.
969. S. V. Illarionov. Methods of Production, Physical Properties and Electronic Structure of Refractory Metals, Their Compounds and Alloys, Theses of Conference Papers, Izd. AN UkrSSR, Kiev, 1961.
970. Yu. B. Paderno. Investigation of the Production Conditions of Hexaborides of the Rare-Earth Metals and Study of Their Physical Properties, author's abstract of dissertation, Akad. Nauk UkrSSR, Otdel. Tekh. Nauk, Kiev, 1962.
971. L. D. Dudkin, E. S. Kuznetsova. Investigation of the Conditions of Production, Physical Properties and Electronic Structure of Refractory Metals, Their Compounds and Alloys, Theses of Conference Papers, Izd. AN UkrSSR, Kiev, 1961.
972. H. Martens, L. Jaffe. J. Appl. Phys. 31:1122, 1960.
973. B. Howlett. J. Inst. Metals 88:91, 1959-1960.
974. B. Howlett. J. Inst. Metals 88:467, 1959-1960.
975. M. Hoch, D. P. Dingledy, H. Johnson. J. Am. Chem. Soc. 77:304, 1954.
976. S. Wise, J. L. Margrave, R. L. Altman. J. Phys. Chem. 64:915, 1960.
977. M. Stackelberg, R. Paulus. Z. physik. Chem. 22B:305, 1933.
978. Technische Rundschau, No. 38, 45, 1960.
979. Technische Rundschau, No. 34, 2, 1960.
980. P. T. Kolomytsev. Doklady Akad. Nauk SSSR 130:767, 1960.
981. I. Cadoff, J. Nielsen. J. Metals 5:248, 1953.
982. F. Benešovsky, E. Rudy. Planseeber. Pulvermetal. 8:66, 1960.
983. W. Sykes, K. Horn, C. Tucker. Trans. Am. Inst. Mining, Met. Engrs. 117:1935.
984. I. S. Brokhin, V. F. Funke. Zhur. Neorg. Khim. 3:847, 1958.
985. R. Vogel. Z. anorg. Chem. 61:46, 1909.
986. R. Kieffer, F. Benešovsky, R. Maschenschelk. Z. Metallk. 45:493, 1954.
987. R. Kieffer, F. Benešovsky, H. Schroth. Z. Metallk. 44:457, 1953.
988. E. Vaughan. Trans. Faraday Soc. 55:2025, 1959.
989. G. V. Samsonov, T. S. Verkhoglyadova. Doklady Akad. Nauk UkrSSR, No. 12, 1961.
990. L. Corlis, N. Eliot, J. Hastings. Phys. Rev. 117:929, 1961.
991. G. V. Samsonov, T. S. Verkhoglyadova. Doklady Akad. Nauk SSSR 138:342, 1961.
992. H. Nowotny, H. Braun, F. Benešovsky. Radex Rundschau, No. 6, 367, 1960.
993. M. Wilkinson, H. Schild, J. Cable, E. Wollan, W. Kocher. J. Appl. Phys. 31:3585, 1960.
994. H. Nowotny. Chemical and Thermodynamic Properties at High Temperatures, XVIII International Congress on Pure and Applied Chemistry, Montreal, Canada, August 6-12, 1961.
995. Yu. B. Paderno. Atomnaya Energ. 10:396, 1961.
996. K. I. Portnoi, Yu. V. Levinskii, V. I. Fadeeva. Izvest. Akad. Nauk SSSR, Otdel. Tekh. Nauk, Ser. Metallurgiya i Toplivo, No. 2, p. 147, 1961.
997. G. V. Samsonov, T. S. Verkhoglyadova. Doklady Akad. Nauk 142(3):612, 1962.

998. G. F. Silina, Yu. I. Zarembo, L. E. Bertina. Beryllium, Chemical Technology and Metallurgy, Atomizdat, Moscow, 1960, p. 55.

999. J. Drowart, G. de Maria. Proceedings of Conference on Silicon Carbide, Boston, Massachusetts, Pergamon Press, New York, 1960, p. 16.

1000. R. Dial, G. Mangsen. Corrosion 17:107, 1961.

1001. F. G. Kusenko, P. V. Gel'd. In coll. Physico-Chemical Principles of Steel Production, Izd. AN SSSR, Moscow, 1961, p. 41.

1002. Yu. M. Getman, P. V. Gel'd. In coll. Physico-Chemical Principles of Steel Production, Izd. AN SSSR, Moscow, 1961, p. 52.

1003. A. B. Lyashchenko, P. I. Mel'nichuk, I. N. Frantsevich. Poroshkovaya Metallurgiya 1(5):10, 1961.

1004. S. Grisaffe. J. Am. Ceram. Soc. 43:494, 1960.

1005. M. Bredig. J. Am. Ceram. Soc. 43:493, 1960.

1006. M. Heller, G. Danielson. Abstr. International Conference on Semiconductor Physics, Prague, 1960, p. 125.

1007. A. Chretien, G. Hellgorsky. Compt. rend. 252:742, 1961.

1008. S. Nakamura, N. Azuna, S. Arai. Nagoya Kogyo Gijutsu Shikensho Hokoku 9:59, 1960.

1009. D. Pitman, D. Das. J. Elektrochem. Soc. 107:763, 1960.

1010. Materials in Design Engineering 53:12, 1961.

1011. G. Drowart, G. De Maria. Proceedings of Conference on Silicon Carbide, Boston, Massachusetts, Pergamon Press, New York, 1960.

1012. J. Flahaut, L. Domange, M. Guittard, J. Loriers. Bull. soc. chim. France, No. 1, 102, 1961.

1013. M. Budig. Z. anorg. Chem. 310:338, 1961.

1014. A. Betcher, G. Schneider. Transactions of the Second International Conference on the Peaceful Uses of Atomic Energy, Geneva, 1958, Nuclear Fuel and Nuclear Materials [Russian translation], 1959, p. 269.

1015. A. N. Minkevich. Metalloved. i Termicheskaya Obrabotka Metal. No. 8, p. 9, 1961.

1016. S. Rundquist. Acta Chem. Scand. 15:451, 1961.

1017. S. Rundquist. Acta Chem. Scand. 15:342, 1961.

1018. V. S. Kocho, A. G. Strel'chenko, I. G. Khavrunyak, V. N. Korotkevich, E. P. Dryapik, E. A. Ploshchenko. Byull. TsIIN ChM, No. 10(414), 36, 1961.

1019. V. S. Neshpor. Kristallografiya 6:466, 1961.

1020. Chem. Process 7:20, 1961.

1021. E. Colton. J. Inorg. & Nuclear Chem. 17:108, 1961.

1022. A. D. Nachinkov. Nitriding of Titanium and Its Alloys at Low Partial Pressure of the Nitrogen, author's abstract of dissertation, Leningrad, 1961.

1023. A. V. Smirnov, A. D. Nachinkov. Metalloved. i Termicheskaya Obrabotka Metal. No. 3, p. 22, 1960.

1024. A. V. Smirnov, A. D. Nachinkov. Metalloved. i Termicheskaya Obrabotka Metal. No. 7, p. 42, 1960.

1025. A. Brown. Acta Cryst. 14:860, 1961.

1026. J. Wasilewskii, G. Kehl. Metallurgia, Vol. 50, 1954.

1027. E. M. Savitskii. Mechanical Properties of Intermetallic Compounds, edited by J. H. Westbrook, J. Wiley & Sons, N.Y., 1960, p. 106.

1028. E. Gebhardt, H. Seghezzi, E. Fromm. Z. Metallk. 52:464, 1961.

1029. C. Brukl, H. Nowotny, O. Schob, F. Benešovsky. Monatsh. Chem. 92:781, 1961.

1030. R. Johnson, A. Daane. J. Phys. Chem. 65:909, 1961.

1031. B. Polock. J. Phys. Chem. 65:731, 1961.

1032. E. Storms, N. Krikozian, C. Kempter. Anal. Chem. 32:1722, 1960.

1033. V. Matkovich. J. Am. Chem. Soc. 83:1804, 1961.

1034. E. Sellier. Metal Powder Report 15:81, 1961.

1035. N. Parr. Research 13:261, 1960.

1036. K. N. Davydov, F. A. Sidorenko, P. V. Gel'd. Fiz. Metal. i Metalloved. 12:424, 1961.
1037. V. A. Korshunov, P. V. Gel'd. Izvest. Vysshikh Uchebnykh Zavedenii, Fizika, No. 4, p. 146, 1961.
1038. N. V. Vekshina, L. Ya. Markovskii. Zhur. Priklad. Khim. 34:2171, 1961.
1039. V. I. Alekseev, L. Ya. Shvartsman. Doklady Akad. Nauk SSSR 141:346, 1961.
1040. C. McCabe, R. Hudson. J. Metals, No. 1a, 1957.
1041. D. Belforti, S. Blum, B. Bavarnik. Nature 190(4779):907, 1961.
1042. O. Guetert, R. Mozzi. Nature 193(4815):570, 1962.
1043. B. Roberts. Superconducting Materials and Some of Their Properties. Pub. by Research Information Section, The Knolls, Schenectady, N.Y., 1961.
1044. Yu. M. Golutvin, T. M. Kozlovskaya. Zhur. Fiz. Khim. 36:362, 1962.
1045. V. I. Khitrova, Z. G. Pinsker. Kristallografiya 6:882, 1961.
1046. E. N. Nikitin, V. G. Bazanov, V. I. Tarasov. Fiz. Tverd. Tela 3:1645, 1961.
1047. J. Japan. Inst. Metals, 25:289, 1961, from Abstract Journal "Metallurgiya," No. 6G, 248, 1962.
1048. J. Graham, F. McTaggart. Australian J. Chem. 13:67, 1960.
1049. V. A. Korshunov. Electrical Conductivity and Thermoelectromotive Force of Solid and Liquid Manganese—Silicon Alloys, author's abstract of dissertation, Agrophysical Institute, Leningrad, 1962.
1050. W. Williams. J. Phys. Chem. 65:2213, 1961.
1051. R. McDonald, D. Stull. J. Phys. Chem. 65:1918, 1961.
1052. B. Magnusson, C. Brosset. Acta Chem. Scand. 16:449, 1962.
1053. I. V. Fedoseev, O. G. Nemkova. Zhur. Neorg. Khim. 7:980, 1962.
1054. S. Rundquist. Acta Chem. Scand. 16:287, 1962.
1055. F. G. Kusenko, P. V. Gel'd. Izvest. VUZov, Tsvetnaya Metallurgiya, No. 2, p. 43, 1961.
1056. V. A. Korshunov, P. V. Gel'd. Fiz. Metal. i Metalloved. 11:945, 1961.
1057. R. P. Krentsis, P. V. Gel'd. Fiz. Metal. i Metalloved. 13:319, 1962.
1058. S. I. Alyamovskii, P. V. Gel'd, I. I. Matveenko. Zhur. Strukt. Khim. 2:445, 1961.
1059. V. I. Zhelankin, V. S. Kutsev, B. F. Ormont. Zhur. Fiz. Khim. 35:2608, 1961.
1060. M. Mah, E. King, W. Weller, A. Christensen. Bureau of Mines, Report No. 5716, U.S. Department of the Interior, 1961.
1061. S. La Placa, B. Post. Acta Cryst. 15:97, 1962.
1062. B. I. Mikhailovskii. Ukrain. Fiz. Zhur. 7:75, 1962.
1063. L. G. Babak, S. A. Bochek, S. M. Genkina, S. O. Dobrolezh, V. A. Zhidkov, R. Z. Smushkevich. Ukrain. Fiz. Zhur. 6:541, 1961.
1064. P. Jourgenssen, M. Wadswarth, O. Cutler. J. Am. Ceram. Soc. 44:258, 1961.
1065. T. Whalen, M. Humenic. Trans. Metal. Soc. AIME 18:952, 1960.
1066. L. N. Kugai, T. N. Nazarchuk. Zhur. Anal. Khim. 16:205, 1961.
1067. V. N. Igishev. In coll. Physical Properties of Alloys, Trudy Ural'skogo Politekhnicheskogo Inst. im. S. M. Kirova, No. 114, Sverdlovsk, 1961, p. 67.
1068. F. A. Sidorenko, L. B. Dubrovskaya. Ibid., p. 107.
1069. S. I. Alyamovskii, P. V. Gel'd, I. I. Matveenko. Ibid., p. 149.
1070. V. A. Korshunov, P. V. Gel'd. Ibid., p. 164.
1071. C. Fragmen. J. Iron Steel Inst. 114:397, 1926.
1072. M. Picon, L. Domange, J. Flahaut, M. Guittard, M. Patric. Bull. soc. chim. France 2:223, 1960.
1073. M. Bonnet. Rev. mét. 37:16, 1940.
1074. V. I. Tumanov, V. F. Funke, L. I. Belen'kaya. Collection of Materials on the Metallography and Production Technology of Cermets, Refractory Metals and Compounds Based on Them, Part I, TsIIN TsM, Moscow, 1962, p. 167.
1075. L. Ya. Markovskii, E. T. Bezruk. Zhur. Priklad. Khim. 35:491, 1962.
1076. L. Ya. Markovskii, G. V. Kapustovskaya. Zhur. Priklad. Khim. 35:723, 1962.

1077. R. Dolloff. Research Study to Determine the Phase Equilibrium Relations of Selected Metal Carbides at High Temperatures, Research Laboratory National Carbon Company, Wadd Technical Report, Parme, Ohio, 1960, p. 60.
1078. G. V. Samsonov, G. N. Makarenko, T. Ya. Kosolapova. Doklady Akad. Nauk SSSR 144:1062, 1962.
1079. V. F. Funke, S. I. Yudkovskii, G. V. Samsonov. Zhur. Priklad. Khim. 34:1031, 1961.
1080. V. F. Funke, S. I. Yudkovskii, G. V. Samsonov. Hard Alloys, Transactions of VNIITS, No. 4, Metallurgizdat, 1962, p. 92.
1081. I. E. Campbell (ed.). High-Temperature Technology [Russian translation], IL, Moscow, 1959.
1082. G. V. Samsonov, K. I. Portnoi. Alloys on the Basis of Refractory Compounds, Oborongiz, Moscow, 1961.
1083. R. Elliott. Trans. Am. Soc. Met. 53:13, 1961.
1084. R. Elliott, S. Komjathy. In col. Columbium Symposium, Met. Soc. Conf., Vol. 10, Interscience Publishing Co., N.Y., 1961, p. 367.
1085. A. S. Zaimovskii, L. A. Chudnovskaya. Magnetic Materials, Gosénergoizdat, Moscow, 1957.
1086. V. N. Eremenko, V. V. Fesenko. In coll. Surface Phenomena in Metals and Alloys and Their Part in Powder Metallurgy Processes, Izd. AN UkrSSR, Kiev, 1961, p. 178.
1087. O. Hönigschmid. Karbide und Silizide, Halle/Saale, 1914.
1088. Gmelins Handbuch der anorg. Chemie, No. 59, 1220, 1932.
1089. Gmelins Handbuch der anorg. Chemie, No. 58, 251, 1932.
1090. R. Wentorff. J. Chem. Phys. 36:1990, 1962.
1091. V. A. Kirillin and A. E. Sheindlin (eds.). High-Temperature Research [Russian translation], IL, Moscow, 1962.
1092. J. Smith, D. Bayley. Acta Cryst. 10:341, 1957.
1093. C. Curtis, L. Doney, J. Johnson. J. Am. Ceram. Soc. 37:458, 1954.
1094. V. Dufek, Technické Zprávy VUPM, No. 44, 1962.
1095. D. Brown, J. Stobo. Pulvermetallurgie in der Atomkerntechnik, 4, Planseeseminar, Springer-Verlag, 1962, p. 279.
1096. D. Keller, J. Fackelmann, E. Speidel, S. Paprocki. Ibid., p. 304.
1097. K. Matterson, H. Jones, N. Moore. Ibid., p. 329.
1098. A. Ogard, W. Pritchard, D. Douglass, J. Leary. Ibid., p. 365.
1099. U. Essen, W. Klemm. Z. anorg. Chem. 317:25, 1962.
1100. J. Daon. Compt. rend. 250:3635, 1960.
1101. F. Gaume-Mahn. Bull. soc. chim. France, No. 11-12, 1862, 1956.
1102. E. Huber, E. Head, C. Holley. J. Phys. Chem. 61:497, 1957.
1103. E. Huber, C. Matthews, C. Holley. J. Am. Chem. Soc. 77:6493, 1955.
1104. E. Huber, C. Holley. J. Am. Chem. Soc. 77:1444, 1955.
1105. E. Huber, E. Head, C. Holley. J. Phys. Chem. 60:1457, 1956.
1106. E. Huber, E. Head, C. Holley. J. Phys. Chem. 60:1582, 1956.
1107. V. K. Kharchenko, L. I. Struk. Poroshkovaya Metallurgiya, No. 2, p. 87, 1962.
1108. V. T. Serebryanskii. Author's abstract of dissertation, FKhI im. L. Ya. Karpova, 1962.
1109. M. Przybylska, A. Reddoch, G. Ritter. J. Am. Chem. Soc. 85:407, 1963.
1110. C. Lundin. Rare Earth Metal Phase Diagrams, Chicago, 1959.
1111. F. Spedding, A. Daane. The Rare Earths, J. Wiley, New York, London, 1961.
1112. K. Schneider. Rare Earth Alloys, Princeton, Toronto, New York, London, 1961.
1113. S. N. L'vov, V. F. Nemchenko, Yu. B. Paderno. Doklady Akad. Nauk SSSR 142:1371, 1963.
1114. S. Kurnik, M. Cutler, R. Fitzpatrick, J. Leavy. "High-Temperature Semiconductors for Thermoelectric Conversion," in Rare Earth Research, ed. E. V. Kleber, Macmillan, New York, 1961.

1115. R. Johnson, A. Daane. J. Chem. Phys. 38:425, 1963.
1116. E. I. Gladishevskii. Dopovidi Akad. Nauk UkrSSR, No. 7, p. 886, 1963.
1117. H. Holleck, F. Benešovsky, E. Laube, H. Nowotny. Monatsh. Chem. 93:1075, 1962.
1118. J. Norreys, M. Wheeler. Trans. Brit. Ceram. Soc. 62:183, 1963.
1119. K. Andrews. Acta Cryst. 16:68, 1963.
1120. B. Sharma. Acta Cryst. 16:322, 1963.
1121. F. Galasso, I. Pyle. Acta Cryst. 16:228, 1963.
1122. J. Arbuckel, E. Parthé. Acta Cryst. 15:1205, 1962.
1123. K. Ashbee, W. Eeles. Acta Cryst. 15:1312, 1962.
1124. E. Gillam. Acta Cryst. 15:1183, 1962.
1125. R. Chang. Acta Cryst. 14:1097, 1961.
1126. W. B. Pearson. Handbook of Lattice Spacings and Structures of Metals and Alloys, Pergamon Press, London, 1958.
1127. J. Leciejewicz. Acta Cryst. 14:200, 1961.
1128. Er. Parthé, Ed. Parthé. Acta Cryst. 16:71, 1963.
1129. A. Iandelli. Rare Earth Research, ed. E. V. Kleber, Macmillan, New York, 1961, p. 135.
1130. Atomnaya Energ. 13:291, 1962.
1131. V. S. Neshpor, M. I. Reznichenko. Ogneupory, No. 3, p. 134, 1963.
1132. V. I. Marchenko, G. V. Samsonov. Fiz. Metal. i Metalloved. 15:631, 1963.
1133. I. Cadoff, E. Miller (eds.). Thermoelectric Materials and Devices, New York, 1959-60.
1134. R. P. Krentsis. Specific Heat, Enthalpy, and Entropy of Iron Silicides and Some Steels, author's abstract of dissertation, S. M. Kirova Ural Polytechnic Institute, Sverdlovsk, 1962.
1135. P. V. Gel'd, R. P. Krentsis. Fiz. Metal. i Metalloved. 15:63, 1963.
1136. V. A. Korshunov, P. V. Gel'd. Izvest. Vyssh. Ucheb. Zaved., Ser. Fiz., No. 4, p. 146, 1961.
1137. N. Nishimma, H. Kimura. J. Japan. Inst. Metals 20:528, 1956.
1138. V. A. Korshunov, P. V. Gel'd. Izvest. Vyssh. Ucheb. Zaved., Ser. Fiz., No. 6, p. 29, 1960.
1139. W. Trzebiatowski, R. Troć, J. Leciejewicz. Bull. acad. polon. sci., sér. chim., 10:395, 1962.
1140. L. Ya. Markovskii. Poroshkovaya Metallurgiya, No. 4, p. 39, 1962.
1141. L. B. Dubrovskaya, P. V. Gel'd. Zhur. Neorg. Khim. 7:145, 1962.
1142. S. I. Alyamovskii, P. V. Gel'd, I. I. Matveenko. Zhur. Neorg. Khim. 7:836, 1962.
1143. F. A. Sidorenko, P. V. Gel'd, P. C. Rempel'. Izvest. Vyssh. Ucheb. Zaved., Ser. Chern. Metallurgiya, No. 4, p. 102, 1962.
1144. R. P. Krentsis, P. V. Gel'd. Izvest. Vyssh. Ucheb. Zaved., Ser. Chern. Metally, No. 11, p. 12, 1962.
1145. P. V. Gel'd, V. A. Lipatova, F. A. Sidorenko, T. S. Shubina. Fiz. Metal. i Metalloved. 14:298, 1962.
1146. A. M. Belikov, A. A. Savinskaya. Fiz. Metal. i Metalloved. 14:299, 1962.
1147. V. D. Oreshkin. Trudy Khim. Met. Inst. Akad. Nauk SSSR, No. 14, p. 17, 1960.
1148. S. A. Shukarev, M. P. Morozova, G. V. Pron'. Zhur. Obshchei Khim. 32:2069, 1962.
1149. V. I. Tumanov, V. F. Funke, L. I. Belen'kaya. Zhur. Fiz. Khim. 36:1574, 1962.
1150. N. N. Zhuravlev, A. A. Stepanova. Atomnaya Energ. 13:183, 1962.
1151. R. Raylor. J. Am. Ceram. Soc. 45:353, 1962.
1152. W. Hofmann, W. Jäniche. Z. Metallk. 28:1, 1936.
1153. F. Lihl, P. Enitschek. Z. Metallk. 44:414, 1953.
1154. V. T. Serebryanskii, V. A. Épel'baum, G. S. Zhdanov. Doklady Akad. Nauk SSSR 141:884, 1961.
1155. V. T. Serebryanskii, V. A. Épel'baum. Zhur. Strukt. Khim. No. 2, p. 748, 1961.

1156. J. Kohn. Boron (Synthesis, Structure, and Properties), Proceedings of the Conference on Boron, J. Kohn, W. Nye, G. Gaulé (eds.), Plenum Press, New York, 1960, p. 75.

1157. S. Fujishiro. J. Japan. Soc. Powder Metallurgy 8:73, 1961.

1158. Engineering, No. 4906, p. 572, 1960.

1159. V. S. Kocho, V. N. Korotkevich, K. R. Vlasov. Automation of Industrial Processes in Machine Construction, Gostekhizdat UkrSSR, Kiev, 1963, p. 47.

1160. W. Eubank, L. Pruitt, H. Thurnauer. Boron (Synthesis, Structure, and Properties), Proceedings of the Conference on Boron, J. Kohn, W. Nye, G. Gaulé (eds.), Plenum Press, New York, 1960, p. 116.

1161. B. V. Bondarenko, S. V. Ermakov. Radiotekh. i Elektron. 7:2099, 1962.

1162. S. V. Ermakov, B. M. Tsarev. Radiotekh. i Elektron. 7:2102, 1962.

1163. R. G. Avarbé, A. I. Avgustinik, Yu. N. Vil'k, Yu. D. Kondrashov, S. S. Nikol'-skii, Yu. A. Omel'chenko, S. S. Ordan'yan. Zhur. Priklad. Khim. 35:1976, 1962.

1164. Têng Fêng-Hsiang. Investigation of Some Hard Alloys of the System Boron—Silicon—Carbon, author's abstract of dissertation, Moscow Institute of Steels and Alloys, Moscow, 1963.

1165. P. S. Kislyi. Poroshkovaya Metallurgiya, No. 4, p. 50, 1962.

1166. S. N. L'vov, N. F. Nemchenko, P. S. Kislyi, T. S. Verkhoglyadova, T. Ya. Kosolapova. Poroshkovaya Metallurgiya, No. 4, p. 20, 1962.

1167. J. Leitnaker, M. Bowman. J. Electrochem. Soc. 109:441, 1962.

1168. J. Leitnaker, M. Bowman. J. Chem. Phys. 36:350, 1962.

1169. J. Leitnaker, W. Witteman. J. Chem. Phys. 36:1445, 1962.

1170. P. Schissel, W. Williams. Bull. Am. Phys. Soc. 2(4):139, 1959.

1171. M. L. Baskin, V. I. Tret'yakov, I. N. Chaporova. Fiz. Metal. i Metalloved. 14:422, 1962.

1172. R. Rundle, N. Baenziger, A. Wilson, R. McDonald. J. Am. Chem. Soc. 70:1718, 1948.

1173. L. Litz, A. Garrett, F. Croxton. J. Am. Chem. Soc. 70:1718, 1948.

1174. M. Malett, A. Gerds, H. Nelson. J. Electrochem. Soc. 99:197, 1952.

1175. N. V. Ageev, Yu. M. Golutvin, V. P. Samsonov. Zhur. Neorg. Khim. 4:1864, 1959.

1176. M. P. Morozova, M. K. Khripun, S. M. Ariya. Zhur. Obshchei Khim. 32:2072, 1962.

1177. R. Kipley. J. Less Common Metals 4:496, 1963.

1178. G. V. Samsonov. Zhur. Strukt. Khim. 4:395, 1963.

1179. G. V. Samsonov, S. N. Endrzheevskaya. Zhur. Obshchei Khim. 33:2803, 1963.

1180. M. Kh. Karapet'yants, M. L. Karapet'yants. Tables of Some Thermodynamic Properties of Various Materials, Trudy Moskov. Khim. Tekh. Inst. im. D. I. Mendeleeva, No. 39, Moscow, 1961.

1181. I. N. Frantsevich, R. F. Voitovich, V. A. Lavrenko. High-Temperature Oxidation of Metals and Alloys, Gostekhizdat UkrSSR, Kiev, 1963.

1182. G. V. Samsonov, A. P. Épik. Coatings of Refractory Compounds, Metallurg-izdat, Moscow, 1963.

1183. H. Becher. Z. anorg. Chem. 321:217, 1963.

1184. G. Long, L. Foster. J. Electrochem. Soc. 109:1176, 1962.

1185. V. P. Samsonov, I. A. Shcherbakov. Metallurgy and Chemistry of Titanium, No. 9, Akad. Nauk SSSR, Moscow, 1963, p. 274.

1186. I. Wolf. Z. anorg. Chem. 87:120, 1914.

1187. W. Kleber, H. D. Witzke. Naturwissenschaften 50:372, 1963.

1188. H. Ott. Z. Physik 22:21, 1924.

1189. A. Addamiano. J. Electrochem. Soc. 108:72, 1961.

1190. T. Inglese, P. Popper. The Preparation and Properties of Boron Nitride, Heywood and Co. Ltd., London, 1961.

1191. A. S. Dobrolezh, S. M. Zubkova, V. A. Kravets, V. Z. Smushkevich, K. B. Tolpygo, I. N. Frantsevich. Silicon Carbide, Gostekhizdat UkrSSR, Kiev, 1963.
1192. L. Ramsdell, J. Kohn. Acta Cryst. 5:215, 1952.
1193. D. Lundquist. Acta Chem. Scand. 2:177, 1948.
1194. L. Ramsdell. Am. Mineralogist 32:64, 1947.
1195. L. Ramsdell, J. Kohn. Acta Cryst. 4:111, 1951.
1196. E. B. Gasilova. Doklady Akad. Nauk SSSR 101:671, 1955.
1197. R. Mitchell. J. Chem. Phys. 22:1977, 1954.
1198. R. Mitchell, N. Bakarat, E. Sharly. Z. Krist. 111:63, 1958.
1199. R. Adamsky, K. Merz. J. Chem. Soc. 81:250, 1959.
1200. S. I. Alyamovskii, G. P. Shveikin, P. V. Gel'd. Zhur. Neorg. Khim. 8:2000, 1963.
1201. H. Becher, A. Schäfer. Z. anorg. Chem. 318:304, 1962.
1202. Metal Ind. 97:87, 1960.
1203. Metal Ind. 80:143, 1961.
1204. I. V. Fedoseev, O. G. Nemkova. Zhur. Neorg. Khim. 7:980, 1962.
1205. B. Holmberg. Acta Chem. Scand. 16:1253, 1962.
1206. W. Olson, R. Mulford. J. Phys. Chem. 67:295, 1963.
1207. F. Lihl, P. Ettmayer, A. Kutzelnigg. Z. Metallk. 53:715, 1963.
1208. V. I. Khitrova. Kristallografiya 8:39, 1963.
1209. M. Przybylska, A. Reddoch, G. Ritter. J. Am. Chem. Soc. 85:407, 1963.
1210. V. V. Zubenko, S. S. Kvitko, M. M. Umanskii. Kristallografiya 4:244, 1959.
1211. A. Giorgi, E. Szklarz, E. Storms, A. Bowman, B. Matthias. Phys. Rev. 125:837, 1962.
1212. G. Lautz, D. Schneider. Z. Naturforsch. 172A:54, 1962.
1213. B. Matthias, E. Wood, E. Corenzwit, V. Bala. J. Phys. Chem. Solids 1:188, 1956.
1214. B. Matthias, E. Corenzwit, W. Zachariasen. Phys. Rev. 112:89, 1958.
1215. B. Chandrasekhar, I. Hulm. J. Phys. Chem. Solids 4:259, 1958.
1216. V. S. Neshpor, G. V. Samsonov. Zhur. Priklad. Khim. 33:993, 1960.
1217. V. I. Marchenko, G. V. Samsonov. Ukrain. Fiz. Zhur. No. 1, 110, 1963.
1218. V. I. Marchenko, G. V. Samsonov. Zhur. Neorg. Khim. 8(9), 1963.
1219. V. I. Marchenko, G. V. Samsonov. Poroshkovaya Metallurgiya, No. 2, p. 60, 1963.
1220. V. I. Marchenko, G. V. Samsonov. Dopovidi Akad. Nauk UkrRSR, No. 4, p. 463, 1963.
1221. V. I. Marchenko, G. V. Samsonov. Fiz. Metal. i Metalloved. 15:631, 1963.
1222. V. I. Marchenko, G. V. Samsonov, V. S. Fomenko. Radiotekh. i Elektron. 8:1076, 1963.
1223. V. I. Marchenko, G. V. Samsonov, V. S. Fomenko. Zhur. Tekh. Fiz. No. 10, 1963.
1224. V. I. Marchenko. Physical Properties of Sulfides of Rare-Earth Metals of the Cerium Group, author's abstract of dissertation, Institute of Cermets and Special Alloys, Akad. Nauk UkrSSR, Kiev, 1963.
1225. M. Picon, M. Patrie. Compt. rend. 243:1769, 1956.
1226. A. A. Men'kov, L. N. Komissarova, Yu. P. Simanov, Vikt. I. Spitsin. Doklady Akad. Nauk SSSR 141:364, 1961.
1227. J. Flahaut, M. Guittard, M. Patrie. Bull. soc. chim. France, p. 1917, 1959.
1228. M. Picon, L. Domange, J. Flahaut, M. Guittard. Bull. soc. chim. France, p. 221, 1960.
1229. M. Picon, J. Flahaut, M. Guittard, M. Patrie. 16th Internat. Congr. Pure et Appl. Chemie (Paris, 1957); Inorg. Chemie, Butterworth Sci. Publ., London, 1958 (cited by [1112]).
1230. L. Litz. In coll. High-Temperature Technology, Standard Research Institute, Menlo Park, California, 1959 (cited by [1112]).

1231. V. S. Neshpor, G. V. Samsonov. In coll. High-Temperature Cermet Materials, Akad. Nauk UkrSSR, Kiev, 1962, p. 113.
1232. F. A. Sidorenko, P. V. Gel'd, L. B. Dubrovskaya. In coll. High-Temperature Cermet Materials, Akad. Nauk UkrSSR, Kiev, 1962, p. 124.
1233. V. N. Igishev, P. V. Gel'd. In coll. High-Temperature Cermet Materials, Akad. Nauk UkrSSR, Kiev, 1962, p. 133.
1234. A. I. Gol'dberg. V. A. Lipatova, P. V. Gel'd. In coll. High-Temperature Cermet Materials, Akad. Nauk UkrSSR, Kiev, 1962, p. 140.
1235. G. V. Samsonov, V. S. Neshpor, Yu. B. Paderno. In coll. Rare-Earth Elements, Akad. Nauk SSSR, Moscow, 1963, p. 22.
1236. G. V. Samsonov. Uspekhi Khim. 31:1478, 1962.
1237. P. Elliott, C. Kempter. J. Phys. Chem. 62:630, 1958.
1238. R. Kieffer, F. Benešovsky. Hartstoffe, Wien, Springer, 1963.
1239. W. Jeitschko, H. Nowotny. Monatsh. Chem. 93:1197, 1962.
1240. V. S. Neshpor, G. V. Samsonov. Izvest. Akad. Nauk SSSR, Otdel, Tekh. Nauk, Ser. Metallurgiya i Gornoe Delo, No. 1, p. 147, 1963.
1241. V. S. Neshpor, I. G. Barantseva. Inzh.-Fiz. Zhur. 6:109, 1963.
1242. V. N. Paderno. Izvest. Akad. Nauk SSSR, Otdel. Tekh. Nauk, Ser. Metallurgiya i Toplivo, No. 6, p. 176, 1962.
1243. K. I. Portnoi, Yu. V. Levinskii. In coll. Investigation of Alloys of Nonferrous Metals, No. 9, Akad. Nauk SSSR, Moscow, 1963, p. 279.
1244. A. N. Zelikman, N. N. Gorovits. Zhur. Priklad. Khim. 23:689, 1950.
1245. V. S. Neshpor, V. L. Yupko. Poroshkovaya Metallurgiya, No. 2, p. 55, 1963.
1246. N. G. Vannerberg. Acta Chem. Scand. 16:1212, 1962.
1247. M. Atoji, C. Medrud. J. Chem. Phys. 31:332, 1959.
1248. N. G. Vannerberg. Acta Chem. Scand. 15:769, 1961.
1249. Yu. V. Efimov. Zhur. Neorg. Khim. 8:1522, 1963.
1250. A. F. Bessonov, V. G. Vlasov. Fiz. Metal. i Metalloved. 14:478, 1962.
1251. E. Parthé, V. Sadagopan. Acta Cryst. 16:202, 1963.
1252. L. N. Butorina, Z. G. Pinsker. Kristallografiya 5:560, 1960.
1253. P. Schaffer. J. Am. Ceram. Soc. 46:77, 1963.
1254. M. Coupland. Proc. Phys. Soc. 73:577, 1959.
1255. Shun-ichi Mackawa. J. Phys. Soc. Japan 17:1592, 1962.
1256. P. V. Gel'd, V. D. Lyubimov. Poroshkovaya Metallurgiya, No. 4, p. 76, 1963.
1257. C. Wert. J. Appl. Phys. 21:1196, 1950.
1258. P. Powers, M. Doyle. J. Metals 9:1285, Sec. 2, 1957.
1259. V. A. Grodko, V. S. Zolotarevskii, B. N. Markar'yan, I. M. Rubanovich. Poroshkovaya Metallurgiya, No. 4, p. 79, 1963.
1260. T. Hiroaki, I. Kikon. Bull. Chem. Soc. Japan 35:1425, 1962.
1261. B. S. Kul'varskaya, V. A. Grodko, B. N. Markar'yan, I. M. Rubanovich. Radiotekh. i Elektron. 8:675, 1963.
1262. B. N. Mikhailovskii, R. I. Marchenko. Radiotekh. i Elektron. 8:680, 1963.
1263. N. G. Vannerberg. Acta Chem. Scand. 15:769, 1961.
1264. G. Pilström. Acta Chem. Scand. 15:893, 1961.
1265. An Hsi-Yung. Investigation of Equilibrium Phase Diagrams of Alloys of Molybdenum with Chromium and Silicon, author's abstract of dissertation, Moscow Institute of Steel, Moscow, 1962.
1266. P. M. Arzhanyi, R. M. Volkova, D. A. Prokoshkin. Trudy Inst. Metallurgiya im. A. A. Baikova 11:78, 1962.
1267. J. Piazza, M. Sinnott. J. Chem. Eng. Data 7:451, 1962.
1268. J. Leitnaker, N. Krikozian, M. Krupka. J. Electrochem. Soc. 109:66, 1962.
1269. E. Storms, R. McNeal. J. Phys. Chem. 66:1401, 1962.
1270. E. Storms, N. Krikozian. J. Chem. Phys. 64:1471, 1960.
1271. G. Röcktäschel, A. Weiss. Z. anorg. Chem. 316:231, 1962.
1272. L. Wöhler, W. Schuff. Z. anorg. Chem. 209:33, 1932.

1273. A. A. Vertman, A. M. Samarin, E. S. Filippov. Doklady Akad. Nauk SSSR 148:342, 1963.
1274. A. Searcy, L. Finnie. J. Am. Ceram. Soc. 45:268, 1962.
1275. K. Peakall, J. Antill. J. Less Common Metals 4:426, 1962.
1276. I. Jenkins, N. Keen. J. Less Common Metals 4:387, 1962.
1277. S. Rundquist, T. Lundstrem. Acta Chem. Scand. 17:37, 1963.
1278. C. Kempter, N. Krikozian. J. Less Common Metals 4:244, 1962.
1279. K. Vos, W. Velga, M. Steeg, N. Zijlstra. Z. angew. Phys. 15:256, 1963.
1280. I. Senateur, R. Fruchart, A. Michev. Compt. rend. 255:1615, 1962.
1281. G. A. Meerson, R. B. Kotelnikov, S. N. Bashlykov. Reactor Sci. and Tech. 16:485, 1962.
1282. F. Bundy, R. Wentorf. J. Chem. Phys. 38:1144, 1963.
1283. S. La Placa, I. Binder, B. Post. J. Inorg. & Nuclear Chem. 18:113, 1961.
1284. R. LaBotz, M. Donald. J. Electrochem. Soc. 110:121, 1963.
1285. C. Hilsum, A. C. Rose-Innes. Semiconducting III-V Compounds, Pergamon Press, Oxford, London, New York, Paris, 1961.
1286. M. Hoch, J. Vardi. J. Am. Ceram. Soc. 46:245, 1963.
1287. R. Taylor. J. Am. Ceram. Soc. 44:525, 1961.
1288. Yu. M. Golutvin, T. M. Kozlovskaya, É. G. Maslennikova. Zhur. Fiz. Khim. 37:1362, 1963.
1289. V. S. Neshpor, V. L. Yupko. Zhur. Priklad. Khim. 36:1139, 1963.
1290. V. E. Ivanov, A. I. Somov, V. G. Yarovoi. Zhur. Priklad. Khim. 35:1960, 1962.
1291. V. V. Fesenko, A. S. Bolgar. Poroshkovaya Metallurgiya, No. 1, p. 17, 1963.
1292. P. Schissel, O. Trulson. J. Phys. Chem. 66:1492, 1962.
1293. C. Alcock, P. Grieveson. Thermodynamics of Nuclear Materials, Vienna, 1962, p. 563.
1294. A. S. Mikulinskii, F. S. Maron. Zhur. Priklad. Khim. 33:835, 1960.
1295. D. Meschi, A. Searcy. J. Phys. Chem. 63:1175, 1959.
1296. S. Davis, D. Anthrop, A. Searcy. J. Chem. Phys. 34:659, 1963.
1297. O. Ruff, M. Konschak. Z. Elektrochem. 32:517, 1933.
1298. O. Ruff, P. Grieger. Z. anorg. Chem. 211:145, 1933.
1299. J. Drowart, G. DeMaria, M. Inghram. J. Chem. Phys. 29:1015, 1958.
1300. J. Drowart, G. DeMaria. Proceedings of the Conference on Silicon Carbide, Oxford, London, New York, Paris, 1950.
1301. S. Fujishiro. Trans. Japan. Inst. Metals 1:125, 1960.
1302. G. Kibler, T. Lyon, G. Vidal, M. Linevsky. AD-245233, September 30, 1960, p. 11 (cited by [1303]).
1303. R. Fries. J. Chem. Phys. 37:320, 1962.
1304. M. Hoch, P. Blackburn, D. Dingledy, H. Johnston. J. Phys. Chem. 59:97, 1955.
1305. G. Kibler, T. Lyon, G. Vidale, M. Linevsky. WADD-TR-60-646, July, 1960, p. 23 (cited by [1303]).
1306. S. Fujishiro, N. Gocken. Trans. Met. Soc. AIME 221:275, 1961.
1307. G. Kibler, T. Lyon, G. Vidale, M. Linevsky. NP-9791, December 31, 1960, p. 15 (cited by [1303]).
1308. D. Jackson, G. Barton, O. Krikozian, R. Newbury. Thermodynamics of Nuclear Materials, Vienna, 1962, p. 530.
1309. H. Lonsdale, J. Gvanes. Thermodynamics of Nuclear Materials, Vienna, 1962, p. 601.
1310. V. E. Ivanov, A. A. Kruglykh, S. V. Pavlov, P. P. Kovtun, V. M. Antonenko. Thermodynamics of Nuclear Materials, Vienna, 1962, p. 735.
1311. H. Eick, E. Rank. Thermodynamics of Nuclear Materials, Vienna, 1962, p. 549.
1312. C. Alcock, P. Grieverson. Thermodynamics of Nuclear Materials, Vienna, 1962, p. 563.

1314.* R. Mulford, J. Ford, I. Hoffmann. Thermodynamics of Nuclear Materials, Vienna, 1962, p. 517.

1315. P. Panenka, L. Elliot. Trans. Met. Soc. AIME 215:781, 1959.

1316. A. Seybold, R. Griani. J. Metals, 8:556, Sec. II, 1956.

1317. C. Alcock, P. Grieverson. J. Inst. Metals 90:304, 1962.

1318. I. Binder, S. La Placa, B. Post. Boron (Synthesis, Structure, and Properties), Proceedings of the Conference on Boron, J. Kohn, W. Nye, G. Gaulé (eds.), Plenum Press, New York, 1960, p. 86.

1319. E. Greiner. Boron (Synthesis, Structure, and Properties), Proceedings of the Conference on Boron, J. Kohn, W. Nye, G. Gaulé (eds.), Plenum Press, New York, 1960, p. 105.

1320. F. Williams. Am. Chem. Soc., Abstracts of Papers, Boston Meeting, April 7, 1959 (cited by [1319]).

1321. A. Giardini, J. Kohn, L. Toman, D. Eckert. Boron (Synthesis, Structure, and Properties), Proceedings of the Conference on Boron, J. Kohn, W. Nye, G. Gaulé (eds.), Plenum Press, New York, 1960, p. 140.

1322. L. Ramsdell, R. Mitchell. Am. Mineralogist 38(1, 2), 1953.

1323. O. Guentezt. J. Chem. Phys. 37:884, 1962.

1324. E. Dancy, L. Everett, C. McCable. Trans. Met. Soc. AIME 224(6), 1962.

1325. V. I. Tumanov, V. F. Funke, L. I. Belen'kaya. Poroshkovaya Metallurgiya, No. 5, p. 43, 1963.

1326. A. P. Épik. In coll. Surface Phenomena in Fusion and Processes of Powder Metallurgy, Izd. Akad. Nauk UkrSSR, Kiev, 1963.

1327. A. P. Épik. Poroshkovaya Metallurgiya, No. 5, p. 21, 1963.

1328. P. Duhart. Ann. Chim. 7(13):339, 1963.

1329. H. Goldschmidt, I. Brand. J. Less Common Metals 3:34, 1961.

1330. R. Shulman, B. Wyluda. J. Phys. Chem. Solids 23:166, 1962.

1331. V. Ya. Leonidov. Heat of Formation of Higher Oxides and Carbides of Niobium and Tantalum, author's abstract of dissertation, Khimfak MGU, Moscow, 1963.

1332. A. Giorgi, E. Szklarz, E. Storms, A. Bowman. Phys. Rev. 129:1524, 1963.

1333. M. Khansen, K. Anderko. Structures of Binary Alloys, Metallurgizdat, Moscow, 1962.

1334. G. V. Samsonov, V. S. Fomenko, Yu. B. Paderno. Ukrain. Fiz. Zhur. No. 6, 700, 1963.

1335. S. I. Alyamovskii. Structural Features of Some Special Oxides, Carbides, and Oxycarbides of Vanadium and Niobium, author's abstract of dissertation, S. M. Kirova Ural Polytechnic Institute, Sverdlovsk, 1963.

1336. Yu. B. Paderno, L. I. Pomanyuk, V. S. Fomenko. Ukrain. Fiz. Zhur. No. 6, p. 707, 1963.

1337. I. I. Kovenskii. Ukrain. Fiz. Zhur. No. 7, p. 797, 1963.

*Reference [1313] missing in Russian original.